T0202374

MULTIVARIABLE CALCULUS

Multivariable Calculus

ROLLAND TRAPP

OXFORD
UNIVERSITY PRESS

OXFORD
UNIVERSITY PRESS

Great Clarendon Street, Oxford, OX2 6DP,
United Kingdom

Oxford University Press is a department of the University of Oxford.
It furthers the University's objective of excellence in research, scholarship,
and education by publishing worldwide. Oxford is a registered trade mark of
Oxford University Press in the UK and in certain other countries

© Rolland Trapp 2019

The moral rights of the author have been asserted

First Edition published in 2019

Impression: 1

All rights reserved. No part of this publication may be reproduced, stored in
a retrieval system, or transmitted, in any form or by any means, without the
prior permission in writing of Oxford University Press, or as expressly permitted
by law, by licence or under terms agreed with the appropriate reprographics
rights organization. Enquiries concerning reproduction outside the scope of the
above should be sent to the Rights Department, Oxford University Press, at the
address above

You must not circulate this work in any other form
and you must impose this same condition on any acquirer

Published in the United States of America by Oxford University Press
198 Madison Avenue, New York, NY 10016, United States of America

British Library Cataloguing in Publication Data
Data available

Library of Congress Control Number: 2019947623

ISBN 978–0–19–883517–2 (hbk.)
ISBN 978–0–19–883518–9 (pbk.)

Printed and bound by
CPI Group (UK) Ltd, Croydon, CR0 4YY

Links to third party websites are provided by Oxford in good faith and
for information only. Oxford disclaims any responsibility for the materials
contained in any third party website referenced in this work.

Preface

This book is a basic introduction to multivariable calculus, intended for a one-semester course. For the student, my hope is that it sparks an appreciation and (dare I say?) enthusiasm for both this subject and further investigations in mathematics. For the instructor, my hope is that this book serves as a resource to augment the irreplaceable contribution your own energy and expertise makes to your students' education. For all who use it, I hope you engage with and benefit from this text.

The transition to three dimensions is a significant hurdle for most students of multivariable calculus. Students spend years learning about two dimensions, from plotting points and connecting the dots in middle school to using derivatives to discuss concavity in calculus. Upon entering a course in multivariable calculus students have had substantial exposure to, and experience with, all things planar. As a result, students have developed a fair intuition for thinking in two dimensions, an intuition and comfort level that is usually underdeveloped in dimension three. Several features of this text are intended to facilitate the transition to three dimensions.

One of these features is that the first chapter is an attempt to familiarize students with three dimensions. Curves and surfaces are studied from a variety of perspectives, and students should come away with the ability to relate analytic formulas to the geometric objects they represent. A wide range of descriptions for curves and surfaces are presented, not all of which need emphasis, and a judicious choice of techniques to focus on will streamline this chapter. Since the graph of a function is a special case of a parametric surface, formulas for graphs (e.g. normal vectors, flux integrals, etc.) are presented as special cases of those for parametric surfaces later in the text.

A second feature that helps develop three-dimensional acumen is the use of Math Apps regularly throughout the text. Math Apps are interactive graphics, developed using Maple software, that highlight geometric aspects of the topic under discussion. Readers must have Maple Player (a free download), or a full version of Maple, to take advantage of the Math Apps. Math Apps can be downloaded from the companion website, please see below for further instructions.

In addition to the geometric appeal of multivariable calculus, the subject has many applications to other fields of study. This text uses applications to motivate and illustrate mathematical techniques. Streamlines from an ideal fluid flow, for example, are used to illustrate level curves, while the integral of a vector field on a surface is motivated by the flux of a flow. Bézier curves, used in computer imaging, are introduced as an application of vector algebra, and the physical notion of work as an application of the dot product. Applications, while not the focus of this text, are sprinkled throughout to demonstrate the utility of the subject. Students understanding the material here should be well-poised to pursue interests in related fields.

Students: It has been my goal to make this a text you can *read* and learn from, and I hope that I have accomplished that to a certain extent. My recommendation is to grab a coffee and allow yourself the luxury of ruminating over the topics in this (and any other) text. There are a lot of formulas in this text, which I've tried to motivate to varying degrees. Do not content yourselves with memorization, rather ponder the concept behind the formalism. Some of the ideas might require two cups of coffee, and that's Ok.

Faculty: I've tried to produce a valuable resource to supplement your course. The topics are standard, and hopefully the text is flexible enough for you to put your own spin on it, and to take advantage of the plethora of other resources now available to educators.

Maple Player: A word about viewing Math Apps is in order before we get started. Maple Player is a free application that allows you to use the Math Apps you encounter in this text. There are downloadable versions of Maple Player for Windows, Macintosh, and Linux which can be found on the Maplesoft website (a quick search for Maple Player will find it for you). After downloading the application, go to where it is downloaded and double-click to start the installation wizard. Following the prompts should get you up and running.

Math Apps: The Math Apps accompanying this text must be downloaded from the companion website:

https://global.oup.com/booksites/content/9780198835172/

Once downloaded, eBook users should make sure the text and the Math App files are in the same folder. Simply clicking the figure hyperlink in the text will open the corresponding Math App in Maple Player. As they encounter a Math App in the text, hardcopy users will have to open the corresponding Math App file manually on their computer.

I apologize in advance for the mistakes that undoubtedly lurk in the following pages (hopefully they don't "glare"), and thank you in advance for forwarding corrections to me, and for your patience.

Acknowledgments

I am very grateful to many people who have helped me while writing this book. Professionally I'd like to thank Jeremy Aiken, Corey Dunn, Giovanna Llosent, Jeff Meyer, Lynn Scow, and Wenxiang Wang. I've benefitted from discussions with these folks, and from their willingness to pilot various versions of these notes as they developed. I also thank the many students who were willing (forced?) to follow some circuitous routes as I formalized these ideas. Their patience, questions, and input are greatly appreciated. I am thankful to Katherine Ward and Dan Taber at Oxford University Press for their patience and guidance, and for their efforts on my behalf.

On a personal level, thanks goes to Rich and Adele Kehoe, Harold and Karen Sprague, Kevin and Patti Hogan, Claudia Bouslough, and Chuck and Cindy Peterson—a small group of friends from church. Their prayers and encouragement were invaluable during the final year of preparing this manuscript—as was the fact that they never asked "Aren't you done yet?"!

Thanks goes to my kids, Ben, Jake and Ellen. They *did* ask "Aren't you done yet?", but I could take it from them. Thanks to Ellen for encouragement and prayer, to Jake for offering to market my book in his sphere of influence, and to Ben, who gave me the idea of using Math Apps during a conversation we had driving back from school.

My wife, Becky, deserves my greatest thanks. Bec enriches every aspect of my life, and is my best friend. She supported me when I started this project, encouraged me when it didn't progress as I thought it should, and sacrificed a great deal to allow me to finish. I'm looking forward to getting reacquainted! Thanks for your help, Bec. This book is dedicated to you.

Contents

1 Introduction to Three Dimensions

In this chapter we begin our foray into multivariable calculus by getting comfortable with three dimensions. Section one introduces coordinate systems for describing points in three space. We see that Cartesian and polar coordinates in the plane extend naturally to Cartesian and cylindrical coordinates for space. A third coordinate system, spherical coordinates, is also introduced, rounding out those systems most used in this text. Sections 1.2, 1.3 and 1.5 study three different methods of describing surfaces in $\mathbb{R}^3$. Section 1.2 focuses on surfaces arising from graphs of functions of two variables, using level curves to aid understanding. Level curves turn out to be the equivalent of a topographic map for the graph of $z = f(x, y)$. Solution sets of equations in three variables also give rise to surfaces in $\mathbb{R}^3$, and these are considered in Section 1.3. Parametric curves are the topic of Section 1.4, and parametric surfaces are that of Section 1.5. Each method of defining surfaces yields a different insight into the geometry of $\mathbb{R}^3$, and will be used regularly when discussing differentiation and integration. The chapter ends with a section on describing regions in space using systems of inequalities. This skill will be useful when determining limits of integration for multiple integrals.

1.1 Describing Points in 3-Space

Before describing coordinate systems in three dimensions we recall what we know about points in $\mathbb{R}^2$, the Cartesian plane. The common coordinate systems in $\mathbb{R}^2$ will extend naturally to coordinates for $\mathbb{R}^3$.

Coordinates for $\mathbb{R}^2$ In two dimensions there are two familiar methods for describing points. The most common coordinate system is Cartesian coordinates in which a point is described by how far horizontally and vertically it is from the origin. To walk to the point (x, y) from the origin $(0, 0)$ simply walk x units horizontally, then y units vertically.

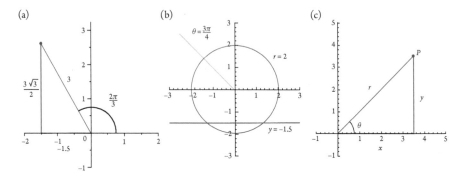

Figure 1.1.1 Review of planar coordinates

Polar coordinates also describe points in the plane. Rather than using horizontal and vertical distances, polar coordinates tell you how far to walk and in what direction. To walk from $(0,0)$ to the point with polar coordinates (r,θ), simply walk r units at an angle θ with the positive x-axis. Thus in polar coordinates, r is the distance to the origin and θ is the angle with the positive x-axis.

There is some ambiguity when using polar coordinates to describe points in the plane. Typically one chooses $r \geq 0$, but we also make the convention that if $r < 0$, go $|r|$ units in the opposite direction from θ. Thus $\left(3, \frac{2\pi}{3}\right)$ and $\left(-3, \frac{5\pi}{3}\right)$ represent the same point in polar coordinates, namely, the point with Cartesian coordinates $\left(-\frac{3}{2}, \frac{3\sqrt{3}}{2}\right)$. You'll also remember that polar coordinates with the same r and angles that differ by a multiple of 2π represent the same point. In general this ambiguity will not cause confusion.

Constant Coordinate Curves: Equations for curves in the plane can be given in either coordinate system, and we recall some particularly simple ones here. In particular, consider the curves obtained by fixing just one of the coordinates. Solution sets to the Cartesian equations $x = c$ and $y = d$ are vertical and horizontal lines, respectively (here c and d are constants). Moreover, since r denotes the distance to the origin, the polar equation $r = c$ denotes a circle centered at $(0,0)$ of radius c. Finally, the polar equation $\theta = d$ defines a ray emanating from the origin and making an angle of d with the positive x-axis (we assume $r \geq 0$). Since we get these curves by fixing one coordinate and letting the other vary, we call them constant coordinate curves.

Finally, the triangle pictured in Figure 1.1.1(c) verifies the relationships between Cartesian and polar coordinates. Trigonometry leads to the familiar change of coordinate formulas:

Change of Coordinates		
	Polar to Cartesian	Cartesian to Polar
	$x = r\cos\theta$	$r = \sqrt{x^2 + y^2}$
	$y = r\sin\theta$	$\theta = \tan^{-1}\left(\frac{y}{x}\right)$

The conversion $\theta = \tan^{-1}(y/x)$ is valid as long as $x > 0$, since the range of the arctangent is $-\pi/2 < \theta < \pi/2$. If $x < 0$, you must use $\theta = \tan^{-1}(y/x) + \pi$.

Coordinates for $\mathbb{R}^3$ To describe points in three dimensions, it stands to reason that a third coordinate is needed. The most direct way is to add a third coordinate to the two-dimensional coordinate systems just discussed. The result will be Cartesian and cylindrical coordinates for $\mathbb{R}^3$. We will describe a third coordinate system for $\mathbb{R}^3$, called spherical coordinates, which is useful as well.

Cartesian Coordinates

To obtain Cartesian coordinates for $\mathbb{R}^3$, start with an xy-plane and add a third axis through the origin which is perpendicular to both the x- and y-axes. Call the third axis the z-axis. Typically we think of the xy-plane as lying horizontally in space, and the z-axis as being the vertical direction. Cartesian coordinates (x, y, z) of the point P in $\mathbb{R}^3$ mean the same as they did in two dimensions, with the z-coordinate giving the height of P above or below the xy-plane.

Constant Coordinate Surfaces: In three dimensions, the Cartesian equation $z = c$ represents all points a fixed height c from the xy-plane. Thus $z = c$ is an equation for a plane parallel to the xy-plane but c units from it. This is analogous to the two-dimensional situation, where the equation $y = c$ describes a line parallel to, and c units from, the x-axis.

There are three coordinate planes in $\mathbb{R}^3$, the xy-, xz-, and yz-planes, which slice space into octants, pictured in Figure 1.1.2(a). Can you think of equations for them (Hint: the equation for the y-axis in $\mathbb{R}^2$ is $x = 0$)? It should be clear that the three-dimensional Cartesian equation $x = c$ describes a plane parallel to, and c units from, the yz-plane. See Figure 1.1.2(b) for the planes obtained by fixing a single Cartesian coordinate. Notice that

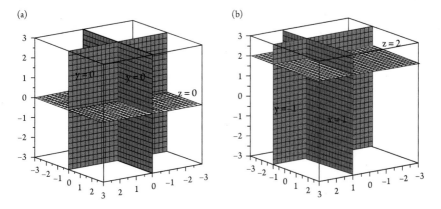

Figure 1.1.2 Cartesian Coordinates

the solution set to the equation $x = c$ depends on what dimension you're in. In $\mathbb{R}^2$ it is a line while in $\mathbb{R}^3$ it's a plane. The context of the situation will dictate which interpretation to use.

Example 1.1.1. *Describing a plane in* $\mathbb{R}^3$

The phrase "the horizontal plane two units above the xy-plane" describes the horizontal plane in Figure 1.1.2 in English. Geometrically this surface is a plane, and analytically it can be described as (the solution set of) the equation $z = 2$. We now know that equations like $z = c$ describe planes! ▲

Example 1.1.2. *Describing a line in* $\mathbb{R}^3$

The simplest lines to describe in $\mathbb{R}^3$ are the coordinate axes. The x-axis can be described as the set of all points whose y- and z-coordinates are both zero, so the set of all points of the form $(x, 0, 0)$. Similarly the y- and z-axes are all points of the form $(0, y, 0)$ and $(0, 0, z)$, respectively. We now consider lines parallel to the coordinate axes.

In $\mathbb{R}^3$, fixing one coordinate gives a plane. Fixing two coordinates, however, will give a line. For example, the solution set of the Cartesian system of equations $x = 1, y = -1$ is the set of all points $(1, -1, z)$ where z is a variable. Thus it is a line parallel to the z-axis, and is the intersection of the planes $x = 1$ and $y = -1$ pictured in Figure 1.1.2(b). It is interesting that a single Cartesian equation yields a surface in $\mathbb{R}^3$ (e.g. $y = -1$ is a plane), while a system of two Cartesian equations yields a curve (e.g. $x = 1, y = -1$ describes a line). We call $\mathbf{C}(t) = (1, -1, t)$, $-\infty < t < \infty$ parametric equations for the line. ▲

Example 1.1.3. *Describing solids in space—Cartesian coordinates*

We describe the portion of $\mathbb{R}^3$ defined by the system of inequalities

$$0 \le x \le 2$$
$$0 \le y \le 3$$
$$0 \le z \le 5$$

The restrictions on x indicate the solid is between the planes $x = 0$ and $x = 2$. Similarly it is between the xz-plane and the plane $y = 3$, as well as between $z = 0$ and $z = 5$. Thus it is a rectangular box (see Figure 1.1.3). ▲

We will usually think of the yz-plane as the plane of the paper, with the x-axis pointing out of the paper toward you. This is helpful to keep in mind when viewing static pictures, but software allows more flexibility.

We also mention that the distance formula for $\mathbb{R}^3$ is a natural generalization of the two-dimensional one. Let (x_1, y_1, z_1) and (x_2, y_2, z_2) be points in space, then the distance d between them is given by

$$d = \sqrt{(x_1 - x_2)^2 + (y_1 - y_2)^2 + (z_1 - z_2)^2}.$$

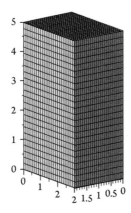

Figure 1.1.3 A rectangular box

Example 1.1.4. *Equations for spheres*

This distance formula gives rise to Cartesian equations for spheres in $\mathbb{R}^3$. Indeed, a sphere is all points a fixed distance from a given point. The equation for a sphere radius r and centered at (a, b, c) is

$$(x-a)^2 + (y-b)^2 + (z-c)^2 = r^2,$$

where we squared the distance formula to simplify the equation. Note the similarity between equations for spheres in three dimensions and those of circles in two. ▲

Cylindrical Coordinates

Cylindrical coordinates for $\mathbb{R}^3$ are obtained by adding the z-coordinate to polar coordinates for the xy-plane in space. To get from the origin to the point (r, θ, z), first walk in the xy-plane r units at an angle of θ with the x-axis. Then jump z units vertically. Note that, while r represented distance to the origin in polar coordinates, it represents distance to the z-axis in cylindrical coordinates. The right triangle pictured in Figure 1.1.4(a) indicates that the distance from (r, θ, z) to the origin in $\mathbb{R}^3$ is given by $\rho = \sqrt{r^2 + z^2}$.

Constant Coordinate Surfaces: As in Cartesian coordinates, let's analyze what we get by fixing one cylindrical coordinate. Since Cartesian and cylindrical have the same z-coordinate, fixing z yields a horizontal plane. Letting r be constant describes the set of all points a fixed distance from the z-axis. To get a feel for what this is, recall that in polar coordinates fixing r gave a circle. In three dimensions, that circle can be translated up and down the z-axis without changing the distance to the z-axis (i.e. without changing r). Thus fixing r yields a cylinder in $\mathbb{R}^3$ whose axis of symmetry is the z-axis. Hence the name "Cylindrical" coordinates. Finally, if you fix θ in polar coordinates, you get a ray emanating from the origin. As in the case of the cylinder, translate this up and

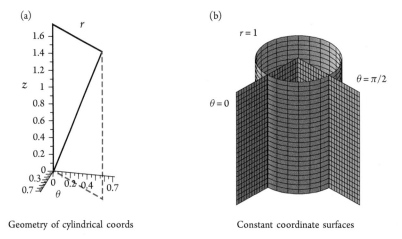

Geometry of cylindrical coords   Constant coordinate surfaces

Figure 1.1.4 Cylindrical coordinates

down the z-axis to find what you get in space. The result is a half-plane that makes an angle of θ with the positive x-axis, and whose boundary is the z-axis.

Math App 1.1.1. *Cylindrical constant coordinate surfaces*

Throughout this text Math Apps will be used to enhance geometric understanding. Please refer to Page vi of the preface for instructions on downloading Math Apps. Print users then open the app manually while eBook users click the figure hyperlink below.

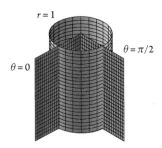

Since both Cartesian and Cylindrical coordinates for $\mathbb{R}^3$ extend coordinate systems for $\mathbb{R}^2$, converting between them is the same as between Cartesian and polar. There is the obvious addition that the z-coordinates are the same. Thus the Cartesian coordinates for the cylindrical coordinates (r, θ, z) are

$$(x, y, z) = (r\cos\theta, r\sin\theta, z). \tag{1.1.1}$$

Example 1.1.5. *Constant Coordinate Surfaces*

We again emphasize the English, geometric, and analytic descriptions of a surface. The infinite cylinder with radius one and z-axis as core is pictured in Figure 1.1.4(b). The given cylinder is the solution set of the cylindrical equation $r = 1$.

Cylinders with z-axis as core are constant coordinate surfaces when using cylindrical coordinates, as are half-planes with z-axis as boundary. The half-plane that contains the positive y-axis is given by the equation $\theta = \pi/2$. ▲

Example 1.1.6. *Curves of intersection in cylindrical coordinates*

In Cartesian coordinates we found the solution set to the system $x = 1$, $y = -1$ was the line of intersection of the corresponding planes. Interesting curves also arise from intersecting constant coordinate surfaces in cylindrical coordinates.

Figure 1.1.4(b) indicates that the system of cylindrical equations $r = 1$, $\theta = \pi/2$ describes the intersection of the cylinder $r = 1$ and the half-plane $\theta = \pi/2$. The result is a vertical line. ▲

Remark: While discussing cylindrical coordinates, we should mention that there is nothing special about using the z-axis as the third coordinate. We could have just as easily used polar coordinates in the yz-plane, and the x-axis as the third coordinate. Then r would be the distance to the x-axis, θ the angle with the positive y-axis and x would just be x. Unless otherwise stated, however, cylindrical coordinates will mean (r,θ,z).

We finish our initiation into cylindrical coordinates by looking at the solution set of a system of inequalities. The idea of describing regions in space using a system of inequalities will be useful when setting up limits of integration in triple integrals.

Example 1.1.7. *Describing solids in space—cylindrical coordinates*

We describe the portion of $\mathbb{R}^3$ defined by the system of inequalities

$$0 \leq r \leq 2$$
$$0 \leq \theta \leq \frac{\pi}{2}$$
$$0 \leq z \leq 5$$

The restrictions on r, which is the distance to the z-axis, describe an infinite solid cylinder with core along the z-axis and radius 2. The restrictions on z cut it down to a solid cylinder radius 2 and height 5, with base in the xy-plane. Finally, the restriction on θ reduces it to that portion which is in the first octant (See Figure 1.1.5). ▲

Spherical Coordinates

In cylindrical coordinates, two coordinates describe distances and one describes a direction. We now introduce spherical coordinates, in which two describe directions and only

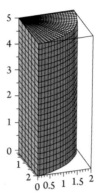

Figure 1.1.5 The solid wedge

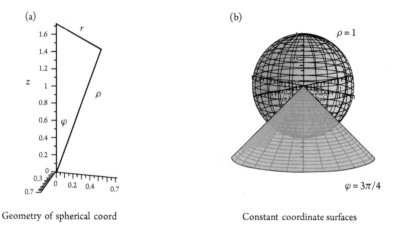

Geometry of spherical coord   Constant coordinate surfaces

Figure 1.1.6 Spherical coordinates

one is a distance. Spherical coordinates are denoted (ρ,θ,ϕ), where ρ is the distance to the origin, θ is our old friend from polar and cylindrical coordinates, and ϕ is the angle with the positive z-axis. See Figure 1.1.6.

Constant Coordinate Surfaces: Fixing θ just gives us a half-plane with boundary on the z-axis and making an angle of θ with the positive x-axis, as before. Fixing ρ focuses on all points a fixed distance from the origin; take a guess at what shape that might be. The set of all points in $\mathbb{R}^3$ satisfying $\phi = c$ is the set of all points making a fixed angle with the positive z-axis. This is actually a cone with vertex at the origin and making an angle of c with the positive z-axis.

Math App 1.1.2. *Spherical constant coordinate surfaces*

Click the following hyperlink, or print users open manually, to view and manipulate a Math App illustrating constant coordinate surfaces $\phi = c$.

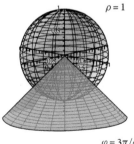

$\varphi = 3\pi/4$

Now consider curves of intersection of constant coordinate surfaces. Fixing both θ and ϕ results in a ray through the origin. Indeed, fixing θ results in a half-plane, while fixing ϕ yields a cone. Fixing both is equivalent to taking the intersection of the half-plane and cone (convince yourself I'm not lying), which is a ray in the half-plane that makes the given angle with the z-axis.

Example 1.1.8. *Spherical equation from geometric description*

Find a spherical equation for the cone with vertex at the origin and that makes an angle of $\pi/3$ with the positive z-axis.

Since the angle with the positive z-axis is the coordinate ϕ in spherical coordinates, the spherical equation is $\phi = \pi/3$. ▲

Example 1.1.9. *Curve of intersection between constant coordinate surfaces*

Describe, as carefully as possible, the curve of intersection of the surfaces $\phi = 3\pi/4$ and $\rho = 1$.

The surfaces are pictured in Figure 1.1.6(b), and the intersection of the cone and sphere will be a circle. Since the cone is pointing straight down, it will be a horizontal circle. Further, using the trigonometry of the triangle corresponding to that of Figure 1.1.6(a) we can determine the radius and height below the xy-plane. In the triangle we have hypotenuse 1, since $\rho = 1$, and an angle of $3\pi/4$ with the positive z-axis. Trigonometry implies that $r = \sqrt{2}/2$ and $z = -\sqrt{2}/2$.

In summary, the intersection of $\phi = 3\pi/4$ and $\rho = 1$ is a horizontal circle at height $z = -\sqrt{2}/2$ with radius $r = \sqrt{2}/2$. ▲

Example 1.1.10. *Describing solids in space—spherical coordinates*

We describe the portion of $\mathbb{R}^3$ defined by the system of inequalities

$$2 \leq \rho \leq 3$$
$$0 \leq \phi \leq \pi/2$$
$$0 \leq \theta \leq 3\pi/2$$

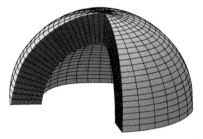

Figure 1.1.7 A solid shell

The restrictions on ρ indicate that the solid is between spheres of radius 2 and 3 centered at the origin. The region of $\mathbb{R}^3$ described by $0 \leq \phi \leq \pi/2$ is the top half of space, while the restriction $0 \leq \theta \leq 3\pi/2$ lets you got three-quarters of the way around the z-axis. Thus it is the solid pictured in Figure 1.1.7. ▲

Coordinate Conversion: Converting between spherical and other coordinates is easily achieved using trigonometry and the right triangle illustrated in Figure 1.1.6(a). Notice that $r = \rho \sin \phi$ and $z = \rho \cos \phi$, giving the conversion from spherical to cylindrical. A simple substitution then gives $(x, y, z) = (\rho \sin \phi \cos \theta, \rho \sin \phi \sin \theta, \rho \cos \phi)$. We summarize converting between the different coordinate systems in the following table.

Converting between coordinate systems

	Cartesian		Cylindrical		Spherical
x	$=$	$r \cos \theta$	$=$	$\rho \sin \phi \cos \theta$
y	$=$	$r \sin \theta$	$=$	$\rho \sin \phi \sin \theta$
z	$=$	z	$=$	$\rho \cos \phi$

Other useful conversions

$$r^2 = x^2 + y^2, \quad \rho^2 = x^2 + y^2 + z^2, \quad r = \rho \sin \phi$$

It will frequently be helpful to translate between coordinate systems, and these conversions facilitate that translation.

Example 1.1.11. *Verifying a conversion analytically*

By direct substitution, and simplification, we verify $x^2 + y^2 + z^2 = \rho^2$.

$$x^2 + y^2 + z^2 = (\rho \sin \phi \cos \theta)^2 + (\rho \sin \phi \sin \theta)^2 + (\rho \cos \phi)^2$$
$$= \rho^2 \sin^2 \phi \cos^2 \theta + \rho^2 \sin^2 \phi \sin^2 \theta + \rho^2 \cos^2 \phi$$

$$= \rho^2 \sin^2 \phi \left(\cos^2 \theta + \sin^2 \theta\right) + \rho^2 \cos^2 \phi = \rho^2 \sin^2 \phi + \rho^2 \cos^2 \phi$$
$$= \rho^2. \ \blacktriangle$$

Example 1.1.12. *Equations in different coordinate systems*

The conversions in the above table allow one to translate an equation from one coordinate system to another. For example, the Cartesian equation for a sphere centered at the origin radius 2 is $x^2 + y^2 + z^2 = 4$. Using the appropriate above conversion, we replace $x^2 + y^2 + z^2$ with ρ^2 to get the spherical equation $\rho^2 = 4$, which simplifies to $\rho = 2$.

Similarly, we can derive cylindrical and spherical equations for the Cartesian equation $x = 5$. Replacing x with $r\cos\theta$ we obtain the cylindrical equation $r\cos\theta = 5$ for the plane $x = 5$. Using the appropriate spherical substitution we find the spherical equation is $\rho \sin\phi \cos\phi = 5$. $\blacktriangle$

Example 1.1.13. *Equations for surface described geometrically*

Find an equation in each coordinate system for the plane parallel to the xz-plane, and 7 units to the right of it.

Since such planes have Cartesian equations $y = c$ for some constant c, we have $y = 7$ as the Cartesian equation of the plane. Substitutions yield:

Cartesian	$y = 7$
Cylindrical	$r\sin\theta = 7$
Spherical	$\rho \sin\phi \sin\theta = 7$ $\blacktriangle$

In this section we've introduced coordinate systems in three dimensions. We now summarize the important points.

Things to know/Skills to have

- The interpretation of each coordinate in Cartesian, cylindrical and spherical coordinates (e.g. z is the height above the xy-plane, while r is the distance to the z-axis).
- Converting between different coordinate systems, and why they work using trig.
- Be able to sketch and describe in English constant coordinate surfaces in each coordinate system.
- Be able to describe in English and sketch the intersection of two constant coordinate surfaces.
- Be able to give equations in each coordinate system for constant coordinate surfaces.
- Be able to use a system of inequalities to describe solids in $\mathbb{R}^3$.

Exercises

1. Sketch the following constant coordinate surfaces; include the coordinate planes in your sketch.
   (a) $z = -2$.
   (b) $x = 4$.
   (c) $y = -5$.

2. Sketch the following constant coordinate surfaces; include the coordinate axes in your sketch.
   (a) $z = -2$.
   (b) $r = 4$.
   (c) $\theta = -3\pi/4$.

3. Sketch the following constant coordinate surfaces; include the coordinate axes in your sketch.
   (a) $\rho = 2$.
   (b) $\phi = \pi/3$.
   (c) $\theta = -3\pi/4$.

4. Sketch the constant coordinate surfaces $x = -1$, $x = 0$, and $x = 3$ on the same set of axes.

5. Sketch the constant coordinate surfaces $r = 1$, $r = 3$, and $r = 5$ on the same set of axes.

6. Sketch the constant coordinate surfaces $\theta = 0$, $\theta = \pi/4$, and $\theta = 3\pi/4$ on the same set of axes.

7. Sketch the constant coordinate surfaces $\phi = \pi/6$, $\phi = \pi/2$, and $\phi = 3\pi/4$ on the same set of axes.

8. Sketch the constant coordinate surfaces $\rho = 1$, $\rho = 2$, and $\rho = 5$ on the same set of axes.

9. Give a one-sentence English description of each of the following constant coordinate surfaces:
   (a) $z = 5$.
   (b) $r = 4$.
   (c) $\rho = 2$.
   (d) $\phi = \pi/3$.
   (e) $\theta = -3\pi/4$.

10. Find Cartesian equations for the following surfaces.
    (a) $\rho = 2$.
    (b) $\phi = \pi/2$.
    (c) $r = 3$.

11. Find cylindrical equations for the following surfaces.
    (a) $x^2 + y^2 = 9$.
    (b) $z = 3$.
    (c) $\rho \sin \phi = 6$.

12. Find spherical equations for the following surfaces.
    (a) $r^2 + z^2 = 4$.
    (b) $z = 3$.
    (c) $x^2 + y^2 + z^2 = 9$.

13. Find a Cartesian equation for the set S of all points 6 units above the xy-plane. Now find cylindrical and spherical equations for S.

14. Find a Cartesian equation for the set S of all points 3 units to the left of the xz-plane. Now find cylindrical and spherical equations for S.

15. Find a cylindrical equation for the set of all points in $\mathbb{R}^3$ that are 4 units from the z-axis.

16. Find a spherical equation for the set S of all points 5 units from the origin. Now find Cartesian and cylindrical equations for S.

17. Find a Cartesian equation for the sphere centered at $(0, 0, 1)$ with radius 1. Now find a cylindrical equation for it.

18. Describe, as carefully as you can, the intersection of the constant coordinate surfaces given below. Include what geometric shape it is (e.g. a line, ray, circle, etc.), and how it sits in $\mathbb{R}^3$ (e.g. horizontally, parallel to the y-axis, etc.).
    (a) $y = 3, z = -2$.
    (b) $z = 5, r = 2$.
    (c) $z = -2, \theta = \frac{3\pi}{4}$.
    (d) $r = 5, \theta = -\frac{\pi}{3}$.
    (e) $\rho = 3, \phi = \frac{\pi}{4}$.
    (f) $\rho = 3, \phi = \frac{\pi}{2}$.
    (g) $\rho = 5, \theta = \frac{4\pi}{3}$.
    (h) $\phi = \frac{\pi}{6}, \theta = \frac{\pi}{6}$.

19. Use the Pythagorean theorem to prove the distance formula in $\mathbb{R}^3$.

20. Justify in English the conversion $z = \rho \cos \phi$. (Hint: use Figure 1.1.6)

21. Justify in English the conversion $r = \rho \sin \phi$. (Hint: use Figure 1.1.6)

22. Justify, analytically and in English, the conversion $r^2 = x^2 + y^2$.

23. Justify, in English, the conversion $\rho^2 = x^2 + y^2 + z^2$.

24. Sketch the solid determined by the system of inequalities:

$$0 \le x \le 3; \quad 0 \le y \le 5; \quad 0 \le z \le 1.$$

25. Sketch the solid determined by the system of inequalities:

$$0 \leq x \leq 4; \quad -2 \leq y \leq 0; \quad 0 \leq z \leq 3.$$

26. Sketch the solid determined by the system of inequalities:

$$0 \leq r \leq 4; \quad 0 \leq \theta \leq \pi; \quad 0 \leq z \leq 3.$$

27. Sketch the solid determined by the system of inequalities:

$$2 \leq r \leq 3; \quad 0 \leq \theta \leq \pi/2; \quad 0 \leq z \leq 4.$$

28. Sketch the solid determined by the system of inequalities:

$$0 \leq \rho \leq 2; \quad 0 \leq \theta \leq \pi/2; \quad 0 \leq \phi \leq \pi.$$

29. Sketch the solid determined by the system of inequalities:

$$1 \leq \rho \leq 2; \quad 0 \leq \theta \leq \pi; \quad 0 \leq \phi \leq \pi/2.$$

30. A cylindrical can has radius 5 and height 2. A coordinate system is introduced so that the center of mass of the can is at the origin, and its axis is the z-axis. What system of inequalities on cylindrical coordinates describes the region of space occupied by the can?

31. A spherical shell is centered at the origin. Its inner radius is 2 and it is half a unit thick. What system of inequalities in spherical coordinates describes the region of space occupied by the shell?

32. Define a different set of cylindrical coordinates, where r is the distance to the x-axis and θ is the angle made with the positive y-axis. What are the change-of-coordinate functions from this system to Cartesian coordinates?

33. Let T be rotation of space counterclockwise around the z-axis through an angle of $\frac{\pi}{2}$, and let $(\rho, \theta, \phi) = \left(2, -\frac{\pi}{3}, \frac{\pi}{4}\right)$ be the spherical coordinates of the point P. Find the spherical coordinates of the rotated point $T(P)$.

1.2 Surfaces from Graphs

In single-variable calculus, considerable effort is spent on studying curves defined as graphs of functions $y = f(x)$. The derivative $f'(x)$ is the instantaneous rate of change of f, and can be used to determine when the graph of f is increasing or decreasing, the concavity of f, and extreme values of f. The integral of f can represent area or distance traveled, and can be used to find physical quantities like arclength and centers of mass. Indeed, much of single-variable calculus is concerned with analyzing properties of functions $f(x)$ and their graphs. In multivariable calculus we will be concerned with functions of several variables, and we begin our study in this section with analyzing graphs of functions of two variables.

Recall that the graph of a function, say

$$f(x) = \sinh x = \frac{e^x - e^{-x}}{2},$$

is the set of all points of the form $(x, f(x))$ as pictured in Figure 1.2.1.

In this context the domain of the function is the horizontal axis—a subset of the plane. The range consists of y-values, which are on the vertical axis. Combining the domain and range in the same space is so familiar, we rarely think about it. We take our cue from the two-dimensional case, and make the following definition:

Definition 1.2.1. *The graph of a function $f(x, y)$ is the set of all points in $\mathbb{R}^3$ of the form $(x, y, f(x, y))$.*

To sketch the graph of $z = f(x, y)$, consider the domain to be the xy-plane in $\mathbb{R}^3$ and the function value $f(x, y)$ determines the height above it. For example, if $f(x, y) = x^2 - y^2$, the point $(2, 1, f(2, 1)) = (2, 1, 3)$ is on the graph of f. We illustrate the graphs of two functions below, then describe the level curve technique for understanding the graph of a function of two variables.

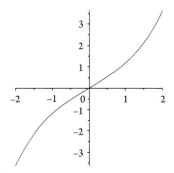

Figure 1.2.1 The hyperbolic sine function

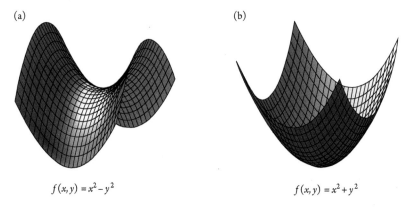

(a)

(b)

$$f(x,y) = x^2 - y^2$$

$$f(x,y) = x^2 + y^2$$

Figure 1.2.2 The graphs of two functions

Example 1.2.1. *The graphs of two functions*

We give, without justification, the graphs of two functions. These surfaces represent all points in $\mathbb{R}^3$ of the form $(x,y,f(x,y))$ for the respective functions. ▲

Our current task is to develop a method for determining the pictures of Figure 1.2.2 from the formulas. This method uses level curves, which we will want to understand analytically, geometrically, and conversationally. We motivate level curves by considering topographic maps.

A topographic map is one that also illustrates the topography of a region using what are called contour lines. Contour lines are constant altitude curves, or curves that connect points of the same altitude. Figure 1.2.3 is a topographic map of the area around Longs Peak in Rocky Mountain National Park, Colorado. The contour lines, together with shading and colors, help give us an idea of what the terrain is like. For example, consecutive darker curves represent an altitude gain of 250 feet. The closer they are together, the steeper the terrain in that area. The label on a curve tells you its altitude, and we can use them to determine which peak is higher, Longs Peak in the lower center of the region, or Mount Meeker, south east of Longs. One could also use contour lines to determine the direction a hiker at the middle of the map should walk to get downhill the fastest. A lot of information is encoded in a topographic map.

We wish to use similar techniques to understand graphs of functions $z = f(x,y)$. The "mathematical mountain" in Figure 1.2.4(a), for example, is represented by the topographic map in Figure 1.2.4(b). Notice that the contour lines are labeled with their corresponding altitudes. If we had a big can of paint and a lot of time we could paint the curves on the mountain that correspond to the contour lines on the map, as in Figure 1.2.4(c). The contour lines of the map are not actually on the mountain, they live in two dimensions. However, looking only at the topographic map, the contour lines do give us an idea of what the mountain looks like.

To construct a topographic map from a formula, note that the constant altitude curves *on the mountain* in Figure 1.2.4(c) can be thought of as the curve of intersection of the mountain and a horizontal plane. For example, the highest curve is 630 feet above sea

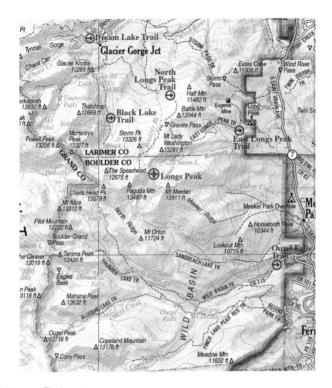

Figure 1.2.3 Longs Peak area

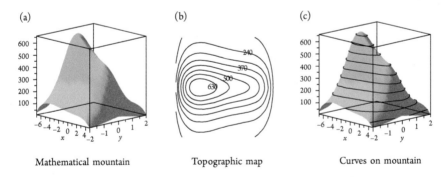

(a)   (b)   (c)

Mathematical mountain   Topographic map   Curves on mountain

Figure 1.2.4 Topographic maps motivate level curves

level, and can be thought of as the intersection of the mountain with the plane $z = 630$. Thus constant altitude curves on the mountain $z = f(x, y)$ at height c can be thought of as the solution set in $\mathbb{R}^3$ of the system of equations

$$\begin{cases} z = f(x, y) \\ z = c. \end{cases}$$

(1.2.1)

The contour lines of Figure 1.2.4(b) are obtained by dropping the curves on the mountain down into the xy-plane.

This "dropping" is more precisely called *projecting* into the xy-plane, and is accomplished analytically by eliminating the z-coordinate. To eliminate the z-coordinates from the system of equations 1.2.1, merely turn them into a single equation by substituting c for z, obtaining $c = f(x,y)$. Thus we have accomplished our goal. We have determined how to find equations for contour lines from a formula for the function f! Let's look at a concrete example before going further.

Example 1.2.2. *A contour line, or level curve*

Let $f(x,y) = x^2 + y^2$, and find a contour line at height 4 for the graph of $z = x^2 + y^2$.

As described above, the curve on the "mountain" $z = x^2 + y^2$ at height 4 is the solution set to the system of equations

$$\begin{cases} z = x^2 + y^2 \\ z = 4 \end{cases}$$

To find the contour line, merely substitute 4 for z in the first equation yielding

$$x^2 + y^2 = 4.$$

The curve in the xy-plane defined by this equation is the desired contour line, and we will call it the *level curve* of the function $z = x^2 + y^2$ at level 4. It is obtained by projecting into the xy-plane the curve of intersection of the surface $z = x^2 + y^2$ and the plane $z = 4$. Figure 1.2.5 illustrates what's going on geometrically. ▲

After this concrete example, we are ready to make a general definition. We remark that, although the term "contour line" is used in the context of topographic maps, we will transition to using the term *level curve*.

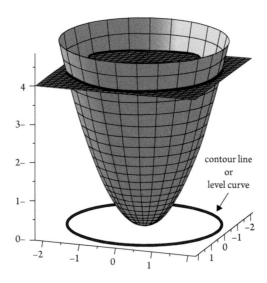

Figure 1.2.5 Geometric understanding of the $\mathbb{R}^3$ problem

Definition 1.2.2. *The* level curve *of* $f(x,y)$ *at level* c *is the curve in the xy-plane given by the equation* $c = f(x,y)$.

We found that the level curve of $f(x,y) = x^2 + y^2$ at level $c = 4$ is the circle $x^2 + y^2 = 4$ (see Figure 1.2.5). The equation for the level curve is found by setting $f(x,y)$ equal to the given level. Equivalently, we substitute the level c for z in the equation $z = f(x,y)$. Let's investigate this further with some examples.

Example 1.2.3. *Level curves of* $f(x,y) = x^2 + y^2$

Using the above strategy we find that the level curve at level $c = 0$ is the solution set of

$$x^2 + y^2 = 0,$$

which is a single point—the origin. Thus a level curve need not be a curve at all, but can be a single point. In fact, the situation can be more extreme. Level curves can be intersecting lines, multiple curves, or even fail to exist. For example, the level curve of $f(x,y) = x^2 + y^2$ at level $c = -1$ does not exist since it is the solution set of the equation

$$x^2 + y^2 = -1.$$

If one is considering several level curves simultaneously, it is sometimes convenient to summarize the equations in tabular form. For example, we see the following equations corresponding to different levels:

Level	Level Curve
$c = -1$	$x^2 + y^2 = -1$
$c = 0$	$x^2 + y^2 = 0$
$c = 0.5$	$x^2 + y^2 = 0.5$
$c = 1$	$x^2 + y^2 = 1$
$c = 2$	$x^2 + y^2 = 2$
$c = 4$	$x^2 + y^2 = 4$

Analytically, then, it is easy to find equations for level curves—just let the function equal the level. To develop a greater geometric understanding of the graph of $z = f(x,y)$, we plot the level curves in the xy-plane. When plotting level curves in the plane, it is customary to label each curve with its corresponding level, as in a topographic map (see Figure 1.2.6). It is also common to consider several level curves at the same time. Doing this allows us to analyze the surface $z = f(x,y)$. For example, the level curves in Figure 1.2.6 are concentric circles centered at the point at level zero. Take a moment to compare the surface in Figure 1.2.5 with its corresponding level curves in Figure 1.2.6.▲

Math App 1.2.1. *Visualizing level curves*

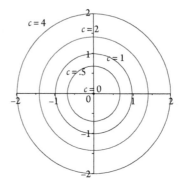

Figure 1.2.6 Level curves of $f(x,y) = x^2 + y^2$

Click the hyperlink below, or print users open manually, to visualize the process of slicing the surface $z = f(x,y)$ and graphing the level curve side by side. Use the slider to change the level, and see how the level curves change.

Surface with Plane

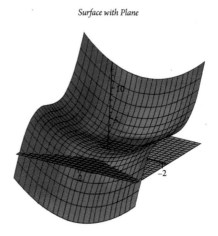

Example 1.2.4. *Level curves of* $f(x,y) = x^2 - y^2$

Let's now analyze the surface in Figure 1.2.2(a). The level curve of $f(x,y) = x^2 - y^2$ at level $c = 1$ is the hyperbola $x^2 - y^2 = 1$. It has asymptotes $y = \pm x$, vertices $(\pm 1, 0)$, and opens sideways. The level curves of $f(x,y) = x^2 - y^2$ at levels $c = -1, 0, 1, 2$ are the curves in the xy-plane given by the equation $f(x,y) = c$. They are

Level	Level Curve
$c = -1$	$x^2 - y^2 = -1$
$c = 0$	$x^2 - y^2 = 0$
$c = 1$	$x^2 - y^2 = 1$
$c = 2$	$x^2 - y^2 = 2$

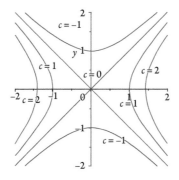

Figure 1.2.7 Level curves of $f(x,y) = x^2 - y^2$

Most of these curves are hyperbolas in the plane, with one exception. At level $c = 0$, the level curve is given by $x^2 - y^2 = 0$. The left-hand side factors to $(x+y)(x-y) = 0$, which holds when either $y = x$ or $y = -x$. Thus the level curve at level zero is really a pair of intersecting lines. Plotting the level curves in the xy-plane yields Figure 1.2.7. ▲

Math App 1.2.2. *Level curves of quadratic functions*

The previous two examples have both involved quadratic functions. Click the hyperlink below, or print users open manually, to explore level curves of the function $f(x,y) = Ax^2 + Bxy + Cy^2$ for different values of A, B, C. It turns out that the discriminant $4AC - B^2$ is a magic number! See if you can tell what the level curves look like when $4AC - B^2$ is positive, negative, and zero.

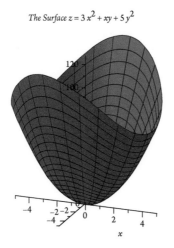

The Surface $z = 3x^2 + xy + 5y^2$

You now have two examples of level curves under your belt, and it's time to see how to use them to visualize the surface $z = f(x,y)$. Recall that Figures 1.2.6 and 1.2.7 are topographic maps of their corresponding surfaces. To use these maps to visualize your surface, you just need to add the third dimension. We now illustrate how level curves can tell us when a surface is above or below the xy-plane.

Example 1.2.5. *Interpreting level curves of $f(x,y) = x^2 - y^2$.*

In Figure 1.2.7 the level curves at level 0 cut the plane into four pieces: the left, right, top, and bottom. On each piece the function $f(x,y) = x^2 - y^2$ is either always positive or always negative, telling us if the graph of f is above or below that portion of its domain. The levels are positive on the left and right pieces, indicating the graph is above the xy-plane over those regions. Similarly, the graph of f is below the xy-plane in the top and bottom regions. ▲

The surface of Figure 1.2.2(a) is called a saddle (it kinda looks like one, doesn't it: the sides go down but the front and back go up), and the origin is a saddle point of the surface. Now if a circus wanted to design a saddle for a monkey who rode horses in the show, they'd have to consider its tail. A *monkey saddle* would have to go down in the back so its tail could relax and the monkey would be comfortable. The surface given by $f(x,y) = x(y-x)(x+y)$ is a monkey saddle. What are the level curves at level $c = 0$ for the monkey saddle? These curves divide the xy-plane into regions on which the graph of f is always above or always below the xy-plane. Over which regions is it above? below?

We now include an example where a little algebra is helpful before sketching level curves.

Example 1.2.6. *Level curves of $f(x,y) = x^2 - 2x + y^2 + 4y + 1$*

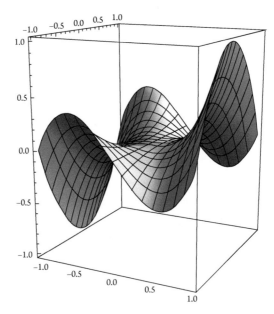

Figure 1.2.8 A monkey saddle

Sketch the level curves of $f(x,y) = x^2 - 2x + y^2 + 4y + 1$ at levels $c = -1, 0, 1$. First, we complete the square on x and y to see that $f(x,y) = (x-1)^2 + (y+2)^2 - 4$. Setting the function equal to each level gives the equation for the level curve in the xy-plane. We have

Level	Level Curve	Simplified Equation
$c = -1$	$-1 = (x-1)^2 + (y+2)^2 - 4$	$3 = (x-1)^2 + (y+2)^2$
$c = 0$	$0 = (x-1)^2 + (y+2)^2 - 4$	$4 = (x-1)^2 + (y+2)^2$
$c = 1$	$1 = (x-1)^2 + (y+2)^2 - 4$	$5 = (x-1)^2 + (y+2)^2$

Thus the level curves are all circles centered at the point $(1,-2)$ with radii $\sqrt{3}, 2, \sqrt{5}$, respectively. See Figure 1.2.9(a) for a sketch of the level curves. It turns out that this surface is again a paraboloid, seen in Figure 1.2.9(b), but be forewarned that not every surface with circular level curves is a paraboloid! ▲

Example 1.2.7. *Level curves and optimization*

In the previous example we sketched some level curves of $f(x,y) = x^2 - 2x + y^2 + 4y + 1$. In this example we show how level curves can be used to solve constrained optimization problems. In a constrained optimization problem there is an *objective function*, which needs to be maximized or minimized. The "constrained" part of the problem means we want to optimize our objective function, but subject to some *constraint equation*. A general method for doing so, called the method of Lagrange multipliers, will be outlined in Section 3.8. For now we content ourselves with an example.

We address the following question:

An ant crawls on the surface $z = f(x,y) = x^2 - 2x + y^2 + 4y + 1$ directly "above" the unit circle in the xy-plane. What is the maximum altitude that the ant achieves? This is a constrained optimization problem, where the objective function is $f(x,y)$, and we want to maximize it on the unit circle. The constraint equation is therefore $x^2 + y^2 = 1$ (see Figure 1.2.9(a) for the level curves together with the constraint equation).

This question can be completely rephrased in terms of level curves. The ant is one unit up when it is above the intersection of the unit circle and the level curve $c = 1$. More generally, the level of a level curve through a point on the unit circle tells you the height of the ant there. Woah! Stop and ruminate on that sentence—it's worth understanding.

To find its maximum height, then, one has to find the highest level of any level curve intersecting the unit circle. Figure 1.2.9(b) shows the actual path of the ant on the surface "above" the constraint curve.

In this case, level curves are concentric circles centered at $(1,-2)$ and the levels are increasing as the radius increases. If a level curve $\mathcal{C}$ intersects the unit circle, it either does so in two points or one. If $\mathcal{C}$ intersects the unit circle in two points, then there are level curves with slightly larger radius that still intersect the unit circle. Therefore the largest level curve intersecting the unit circle will be the large one tangent to it. This occurs when the radius of the large circle centered at $(1,-2)$ goes through the center of the unit

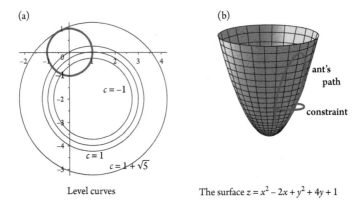

(a) Level curves

(b) The surface $z = x^2 - 2x + y^2 + 4y + 1$

ant's path

constraint

$c = -1$

$c = 1$

$c = 1 + \sqrt{5}$

Figure 1.2.9 A shifted paraboloid

circle. Since the distance from $(1, -2)$ to the origin is $\sqrt{5}$ units and the distance from the origin to the unit circle is 1, we have that the radius of the level curve giving the maximum altitude of the ant is $1 + \sqrt{5}$.

We now have to find the level associated with this radius. The equation of the level curve at level c is $c + 4 = (x - 1)^2 + (y + 2)^2$. In general, then, the relationship between the radius r and level c is given by $r^2 = c + 4$. Substituting $1 + \sqrt{5}$ for r gives $c = 2 + 2\sqrt{5}$. Thus the maximum height achieved by the ant is $2 + 2\sqrt{5}$.

We pause to point out the critical observation in solving this problem. We wanted to maximize the function $f(x, y) = x^2 - 2x + y^2 + 4y + 1$, subject to the constraint that $x^2 + y^2 = 1$. The key observation was that $f(x, y)$ would be maximized at a point where the constraint curve (the unit circle) was tangent to the level curve. More generally, the maximum (or minimum) of a function subject to a constraint will occur at a point where the constraint curve and level curves are tangent to each other. This will be expanded on in Section 3.8. ▲

Level curves and critical points: One can look at a topographic map and tell where the mountain peaks are. Similarly, level curves can tell us where relative maxima and minima of our function are, as well as saddle points. More precisely, level curves near extreme values look like concentric ellipses (for nice functions $f(x, y)$), while near a saddle they look more like hyperbolas. We illustrate with two examples.

Example 1.2.8. *Extreme values from level curves*

Some level curves of the function $f(x, y) = e^{-x^2}(y^3 - y)$ are given in Figure 1.2.10, and are labeled with their corresponding heights. The points P and Q in this example are degenerate level curves. Since the level curves are elliptical in shape around the points P and Q, those points are relative extreme points for the surface. The levels are decreasing toward P, so P is a relative minimum of the surface. Similarly, the levels are increasing toward Q so it is a relative maximum. ▲

In the previous example we used the fact that if a level curve is a point (like P or Q above), and the curves around them are elliptical in shape, the points correspond to

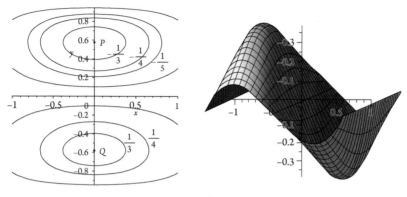

Figure 1.2.10 Some level curves of $f(x,y) = e^{-x^2}(y^3 - y)$

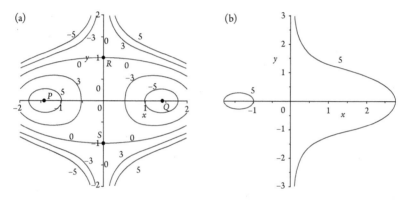

Figure 1.2.11 Some level curves of $f(x,y) = x^3 + 6xy^2 - 6x$

relative extrema of $f(x,y)$ (some additional assumptions on the function f are necessary, but we ignore the details for now). We've also seen that near a saddle point the level curves look like hyperbolas (see Figure 1.2.7). We use these facts to interpret the following level curve diagram.

Example 1.2.9. *Interpreting level curves, again*

The level curves in Figure 1.2.11 are for the function $f(x,y) = x^3 + 6xy^2 - 6x$. Notice that they are not connected. For example, in Figure 1.2.11(a) the level curves at level 5 appear in the upper and lower right, then close to the point P on the left. There are actually only two components of the level curve $5 = x^3 + 6xy^2 - 6x$, as in Figure 1.2.11(b), the display just wasn't wide enough to show it. In any case, this illustrates that level curves can have several pieces.

Reasoning as in the previous example and discussion, we wish to classify the points P, Q, R, S as relative maxima, minima, or saddles. Since levels are increasing toward P, $f(x,y)$ attains a relative maximum there. Similarly, $f(x,y)$ has a relative minimum at Q. The level curve at level 0 intersects itself at R and S, and look like hyperbolas nearby, so R and S are saddles of f. ▲

Fluid Flow: We introduce two-dimensional fluid flow here, since streamlines (the paths of particles in the flow) are level curves of the *stream function* corresponding to the flow. After defining some terminology from the field of fluid mechanics, we discuss a model for ideal fluid flow around a corner (for a more thorough discussion see [9]).

Viscosity is a property of the fluid itself, rather than the flow, which basically describes how thick it is. Pouring caramel on a bowl of ice cream or honey in a cup of tea illustrates that these are high-viscosity fluids. Water and air are low viscosity fluids, and pour quite readily. A second property of fluids is that of *incompressibility*. Air and helium are compressible in that they can be forced into smaller spaces at the expense of increasing pressure. Conversely, compressed gasses can expand to fill larger spaces, reducing pressure. You witness this every time you inflate a balloon from a tank of helium. Water, on the other hand, is incompressible. Viscosity and compressibility are properties that help describe and distinguish fluids.

In order to understand a fluid flow one must also consider properties of the flow itself, not just of the fluid. A flow can cause particles to both move along with it, as well as spin while moving. An example illustrates what I mean. The surface of a pond can be thought of as a two-dimensional fluid flow. In the fall, if you drop a leaf into the pond, the leaf will begin to drift along with the flow. If the leaf starts spinning as well, the flow is *rotational*. An *irrotational* flow is one in which all leaves drift along, but none spin.

It turns out that models that are both incompressible and irrotational closely imitate the flow of low viscosity fluids. We will consider such models, which are called *ideal fluid flows*, as they are excellent applications of much of the mathematics discussed in this book. (By the way, the notions of incompressible and irrotational will be made quite precise when we discuss divergence and curl in Section 5.5.)

Example 1.2.10. *Ideal flow around a right-angle*

With a bit of intuition in hand, we begin a more formal discussion of fluid flow. An ideal fluid flow is described using two functions: the velocity potential and stream functions. We focus on the two-dimensional ideal flow around a corner. The velocity potential and stream functions for this situation are

$$\varphi(x,y) = x^2 - y^2, \text{ and}$$
$$\psi(x,y) = 2xy,$$

respectively. The level curves of the stream function ψ are called *streamlines*, and are the paths particles travel along in the flow. Thus the flow is along hyperbolas with equations $2xy = c$, which we sketch in the first quadrant in Figure 1.2.12(a), thinking of the axes as the corner the flow goes around. ▲

Before leaving this topic, we take a moment to generalize Example 1.2.10 to flows around a corner with angle π/n as it is a particularly nice application of polar coordinates.

Example 1.2.11. *Flows around arbitrary corners*

To motivate the generalization, we first find polar equations for a right-angled flow. Using double-angle formulas from trig, we see

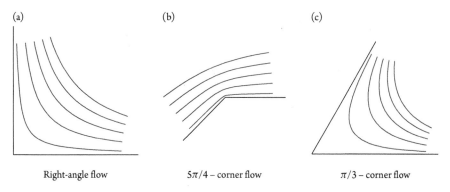

(a) (b) (c)

Right-angle flow $5\pi/4$ – corner flow $\pi/3$ – corner flow

Figure 1.2.12 Streamlines of ideal fluid flow

$$\varphi(x,y) = x^2 - y^2 = r^2(\cos^2\theta - \sin^2\theta) = r^2\cos 2\theta,$$
$$\psi(x,y) = 2xy = r^2 2\cos\theta\sin\theta = r^2\sin 2\theta.$$

(1.2.2)

To generalize from a corner angle of $\pi/2$ to one of π/n, it would be great to change the 2's to n's in Equation 1.2.2. Somewhat miraculously, this is indeed the proper generalization, so that an ideal flow around a corner π/n has velocity potential and stream function given by:

$$\varphi(r,\theta) = r^n\cos n\theta,$$
$$\psi(r,\theta) = r^n\sin n\theta.$$

(1.2.3)

To be more specific, let's sketch some streamlines for an ideal fluid flow around a 60° corner (see Figure 1.2.12(c)). In this case, the angle is $\pi/3$ and the stream function is $\psi = r^3\sin 3\theta$. Note that $\psi = 0$ when θ is 0 or $\pi/3$ (among other values), so those rays are level curves of ψ at level 0. These curves represent the boundary of the flow, and for this reason we only sketch the curves for $0 < \theta < \pi/3$.

For $c > 0$, the streamline equation $\psi = c$ can be solved for r to be

$$r = \sqrt[3]{\frac{c}{\sin 3\theta}}.$$

For θ near 0 and $\pi/3$ the sine approaches 0 and $r \to \infty$, so the streamlines are asymptotic to the boundary of the flow. Additionally, the distance r is minimized when $\sin 3\theta$ is maximized at $3\theta = \pi/2$, or when $\theta = \pi/6$, which is half-way through the domain. These features are illustrated in Figure 1.2.12(c).

Finally, since one might be more comfortable with Cartesian coordinates, we use angle sum identities to translate φ and ψ into Cartesian coordinates. While these may be more familiar, they are definitely less illuminating, thereby illustrating the usefulness of polar coordinates!

$$\varphi = r^3 \cos 3\theta = r^3 (\cos 2\theta \cos\theta - \sin 2\theta \sin\theta)$$
$$= r^3 \left((\cos^2\theta - \sin^2\theta)\cos\theta - 2\sin^2\theta \cos\theta \right)$$
$$= r^3 \cos^3\theta - 3r^3 \sin^2\theta \cos\theta$$
$$= x^3 - 3xy^2,$$
$$\psi = r^3 \sin 3\theta = r^3 (\sin 2\theta \cos\theta + \cos 2\theta \sin\theta)$$
$$= r^3 \left(2\sin\theta \cos^2\theta + (\cos^2\theta - \sin^2\theta)\sin\theta \right)$$
$$= 3r^3 \cos^2\theta \sin\theta - r^3 \sin^3\theta$$
$$= 3x^2 y - y^3. \; \blacktriangle$$

In the above generalization to flows around π/n-corners, n need not be an integer. As an example, the streamlines for a $5\pi/4$-corner and are shown in Figure 1.2.12(b).

Things to know/Skills to have

- Know the definition of the graph of a function of two variables.
- Be able to sketch level curves for a given function and level.
- Solve a constrained optimization problem using level curves.
- Be able to determine maxima, minima, and saddle points for a function from its level curves. More generally, relate the level curves to the geometry of a surface.
- Sketch streamlines given the stream function for an ideal fluid flow.

Exercises

1. Decide whether each point P is a relative maximum, relative minimum, or a saddle point. Give a one-sentence justification for your answer.

(a)

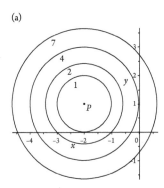

(b)

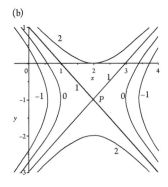

2. Sketch the level curves of $f(x,y) = xy$ at levels $c = -1, 0, 1, 2$. What kind of surface is this?

3. Sketch the level curves of $f(x,y) = x^2 + 4y^2$ at levels $c = 0, 4, 16$. For what levels c does the level curve $f(x,y) = c$ not exist? What does this imply about the intersection of the surface $z = f(x,y)$ and the plane $z = c$ for these levels?

4. Some level curves of the function $f(x,y) = x^3 - 3x + y^2$ are pictured below. Guess whether the points P and Q represent maxima, minima, or saddles. What keeps you from being definite in your answer?

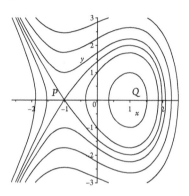

5. Sketch the level curves of $f(x,y) = 3x + 2y + 7$ for levels $c = -2, 1, 8$. Can you guess at the shape of the surface?

6. Sketch the level curves of $f(x,y) = x^2 - y^2 - 4x - 2y$ at levels $c = -4, -3, -2, 0$. What kind of surface do you get?

7. Sketch the level curves of $f(x,y) = \sqrt{16 - x^2 - y^2}$ for levels $c = 0, 1, 2, 3, 4$. Can you describe the surface?

8. Sketch the level curves of $f(x,y) = x^2 + y^2 - 2xy$ for levels $c = 0, 1, 4$. Can you guess at the shape of the surface? (Hint: factor $f(x,y)$ first)

9. Maximize $f(x,y) = x - \sqrt{3}y$ subject to the constraint $x^2 + y^2 = 1$.

10. Sketch the level curves for $f(x,y) = 3x - 3y + 4$ at levels $c = 0, 4, 16$. An ant is crawling on the surface $z = f(x,y)$ above the unit circle in the xy-plane. What are the highest and lowest elevations the ant attains? Sketch the surface in $\mathbb{R}^3$.

11. Sketch the level curves for $f(x,y) = x^2 + y^2 - 6x + 16y$ at levels $c = 0, 4, 16$. An ant is crawling on the surface $z = f(x,y)$ above the unit circle in the xy-plane. What are the highest and lowest elevations the ant attains? Sketch the surface in $\mathbb{R}^3$.

12. Different surfaces can have very similar level curves!
    (a) Sketch the level curves of $f(x,y) = x^2 + y^2$ at levels $c = 0, 1, 2$.
    (b) Sketch the level curves of $f(x,y) = \sqrt{x^2 + y^2}$ at levels $c = 0, 1, 2$.
    (c) The level curves for both surfaces are concentric circles centered at the origin. How can you tell the surfaces apart from their level curves?

13. The saddle $f(x,y) = (x-y)(x+y)$ has two portions sloping down and two sloping up. The monkey saddle $f(x,y) = x(x-y)(x+y)$ has three portions of the surface sloping down (two for the legs and one for the tail) and three up. Can you guess a formula for a surface with four upward and four downward sloping portions? Can you generalize?

14. Sketch some level curves of $h(x,y) = \ln(x^2 + y^2 + 1)$. Describe the surface in a sentence or two.

15. Sketch some level curves of $h(x,y) = \cos(x^2 + y^2)$. Describe the surface in a sentence or two.

16. Sketch some level curves of $h(x,y) = 1/(1 + x^2 + y^2)$, and describe what the surface does near infinity.

17. An ideal fluid flow is modeled with the velocity potential $\varphi = 4x - 3y$ and stream function $\psi = 3x + 4y$. Sketch some streamlines for this flow. Can you describe this flow in a sentence or two?

18. The velocity potential $\varphi = x + 2y$ and stream function $\psi = -2x + y$ model a two-dimensional flow in the plane. Sketch a few streamlines for the flow.

19. Find the velocity potential and stream function for an ideal fluid flow around a $\pi/4$-corner. Sketch some streamlines for the flow.

20. Guess what the streamlines around a π-corner would look like (write your guess down first—AND why you think so!). Now find the velocity potential and stream function for this case in polar and in Cartesian coordinates, and sketch some streamlines. Write a sentence or two comparing your guess to the analysis that followed.

21. A point source in two-dimensional ideal fluid flow emits fluid at a constant rate μ uniformly in all directions (think of a fountain, for example). A point source at the origin can be modeled in polar coordinates with velocity potential and stream function

$$\varphi = \frac{\mu}{2\pi}\ln r, \ \psi = \frac{\mu}{2\pi}\theta.$$

Letting $\mu = 1$, sketch some streamlines for the flow.

22. A vortex in a two-dimensional flow has velocity and stream functions

$$\varphi = K\theta, \ \psi = -K\ln r.$$

Where K is the strength of the vortex. Letting $K = 1$, sketch some streamlines for the flow. Compare the streamlines for a source to those of a vortex.

1.3 Surfaces from Equations

In this section we present a potpourri of surfaces arising as the solution sets of equations, and techniques to study them. We study slicing surfaces with planes, quadric surfaces, generalized cylinders, planes, level surfaces, and equations in other coordinate systems. The basic technique for studying quadric surfaces and level surfaces will be to slice the surfaces with planes parallel to coordinate planes. Generalized cylinders turn out to be easy to recognize from their equations, and easy to sketch. Geometric understanding will help analyze surfaces given by equations in other coordinate systems. Enjoy the variety!

Planes: Lines are some of the first curves you learn to graph in the plane. Analogously, we begin this section on surfaces with planes. The general form for an equation of a line is $Ax + By = C$, so it is no surprise that any plane in $\mathbb{R}^3$ has an equation of the form

$$Ax + By + Cz = D,$$

where A through D are constants and at least one of A, B, or C is not zero. The constants in the equation $Ax + By + Cz = D$ turn out to have significant geometric meaning, which we will see after discussing vectors in Section 2.3... something to look forward to!

Example 1.3.1. *The plane $5x + 2y + z = 10$*

An easy way to visualize the plane $5x + 2y + z = 10$ is to plot the intersections with the coordinate axes, connect the dots to form a triangle. When sketching planes, the triangle might be enough to visualize it as long as you remember that the plane extends without bound in all directions.

The y- and z-coordinates of points on the x-axis in $\mathbb{R}^3$ are both zero. This implies that letting $y = z = 0$ in the equation for the plane gives the x-coordinate, and we get the intercept $(2,0,0)$. Similarly, we get y-intercept $(0,5,0)$ and z-intercept $(0,0,10)$. Plotting these points and connecting the dots gives a triangle that lives in the plane (see Figure 1.3.1(a)). Extending the triangle in all directions gives the plane itself, as in Figure 1.3.1(b). ▲

Example 1.3.2. *Relating a planar equation to the graph*

Let $\mathcal{P}$ be the plane given by $Ax + By + Cz = D$.

1. Suppose $D = 0$, what can you say about the plane?
   If $D = 0$, then the origin $(0,0,0)$ satisfies the equation, so $\mathcal{P}$ goes through the origin. In this case the above technique for visualizing planes will not work, but Section 2.3 will use vectors perpendicular to the plane.

2. Suppose $A \cdot B < 0$ and $D \neq 0$, what can you say about the plane $\mathcal{P}$?
   Since $A \cdot B < 0$, we know A and B have opposite signs. If the x-intercept is positive, then the y-intercept will be negative, and vice versa. Thus we can say the x- and y-intercepts have opposite signs. ▲

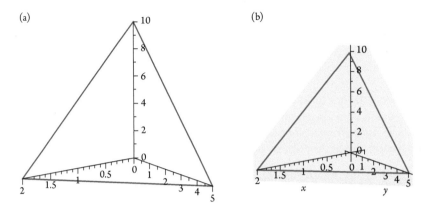

(a)

(b)

Figure 1.3.1 Visualizing the plane $Ax + By + Cz = D$

Generalized Cylinders: We have already seen that the cylindrical coordinate equation $r = 2$ describes a cylinder in space. Translating to Cartesian coordinates for $\mathbb{R}^3$ this corresponds to the equation $\sqrt{x^2 + y^2} = 2$, or $x^2 + y^2 = 4$, which is missing the z-coordinate entirely. One way to think of $r = 2$, then, is that it is the curve $x^2 + y^2 = 4$ in the xy-plane translated up and down along the z-axis. More generally, we'll say that:

Definition 1.3.1. *Any surface whose Cartesian equation is missing one variable is a* generalized cylinder.

To visualize generalized cylinders:

1. Sketch the curve in the appropriate plane.
2. Translate the curve along the axis of the missing variable.

Example 1.3.3. *A hyperbolic cylinder*

Sketch the surface $x^2 - 4z^2 = 1$. We know this is a generalized cylinder since the y-coordinate is missing from the equation. The first step is to sketch the curve $x^2 - 4z^2 = 1$ in the xz-plane. This is a hyperbola with vertices $(\pm 1, 0, 0)$, and asymptotes $z = \pm x/2$ (see the dark curve of intersection in Figure 1.3.2). To get the surface, simply translate the hyperbola back and forth along the y-axis, as in Figure 1.3.2. ▲

Example 1.3.4. *A washboard*

You can sketch the surface $y = \sin z$ in $\mathbb{R}^3$ in two steps. First sketch the curve $y = \sin z$ in the yz-plane, then translate the curve back and forth along the x-axis. This process is illustrated in Figure 1.3.3. ▲

Plane Sections: In Section 1.2 we used level curves to make topographic maps for surfaces arising from the graphs of functions $z = f(x, y)$. Geometrically, this corresponded to considering the intersection of the surface $z = f(x, y)$ with a horizontal plane $z = c$.

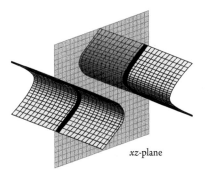

Figure 1.3.2 The generalized cylinder $x^2 - 4z^2 = 1$

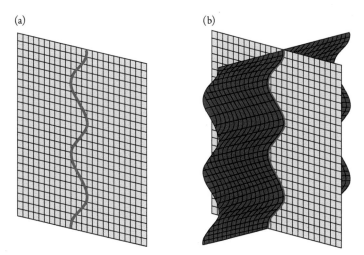

Figure 1.3.3 A "cylinder"

We generalize Section 1.2 in two ways. One generalization is to consider surfaces that need not be graphs of functions. This is analogous to considering curves in the plane defined as solution sets to equations, such as the ellipse $\frac{x^2}{9} + 4y^2 = 1$, which is not the graph of a function. The second generalization is that the planes we slice them with need not be horizontal. Allowing for this flexibility can give greater insight into the structure of some surfaces. Which planes give you greatest insight into a surface is usually dictated by the equation that defines it.

We begin with some examples of slicing graphs of functions with other planes, which could be done using the level curve technique but are easier to understand by slicing with other planes. We then proceed to surfaces whose defining equations are not functions of one of the variables.

Example 1.3.5. *Slicing surfaces for geometric understanding*

Level curves of the function $f(x,y) = x^3 - x + y^2$ are curves in the xy-plane defined by

$$x^3 - x + y^2 = c,$$

for some constant c. Since these curves are hard to visualize, it may be easier to find the intersections with other planes instead. Let's investigate the graph of $f(x,y) = x^3 - x + y^2$ by slicing it with planes parallel to the xz-plane.

First, the curve of intersection of the surface $z = x^3 - x + y^2$ with the plane $y = 0$ is the solution to the system of equations

$$\begin{cases} z = x^3 - x + y^2 \\ y = 0. \end{cases}$$

An equation for the curve is obtained by substituting 0 for y in the first equation, yielding $z = x^3 - x$. One can plot this curve in the xz-plane, as in Figure 1.3.4(b). The intersection with the plane $y = -3$ has equation $z = x^3 - x + (-3)^2 = x^3 - x + 9$, and is pictured in 1.3.4(a). Similarly the $y = 2$ cross-section is in 1.3.4(c), and they are put together in Figure 1.3.5.

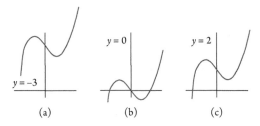

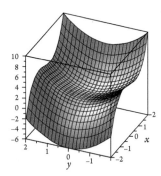

Figure 1.3.4 Plane sections of $f(x,y) = x^3 - x + y^2$

Figure 1.3.5 The surface $z = x^3 - x + y^2$

More generally, to describe the intersection of the surface with the plane $y = c$ analytically we substitute c for y, obtaining $z = x^3 - x + c^2$. This is a vertical translation of the curve $z = x^3 - x$ by c^2 units. So as you move back and forth along the y-axis, the curve $z = x^3 - x$ gets translated up by the appropriate amount, sweeping out the surface as in Figure 1.3.5. The grid lines on the surface that are "parallel" to the highlighted one in the middle are all intersections with planes $y = c$. This technique is convenient because the x's and y's in $f(x,y) = x^3 - x + y^2$ are added together, making the vertical translation easy to see. ▲

Math App 1.3.1. *Surface sections*

This Math App gives more hands-on experience with plane sections. Click on the hyperlink, or print users open manually, to improve your geometric intuition.

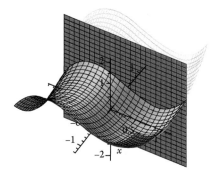

Example 1.3.6. *Fun with trigonometric functions*

Equations for the level curves of $f(x,y) = \cos x + \sin y$ are of the form

$$\cos(x) + \sin(y) = c,$$

which are definitely not fun to work with. To make this example more enjoyable, we'll choose other planes to get some two-dimensional slices with. If we choose to slice the surface $z = \cos x + \sin y$ with planes parallel to the yz-plane, that amounts to replacing x with a constant and analyzing the curves. The curve of intersection with the yz-plane (or $x = 0$) is given by $z = \cos 0 + \sin x = 1 + \sin y$, which is a translation of the sine curve one unit up. In fact, slicing with the plane $x = c$ gives the curve

$$z = \cos c + \sin y,$$

which is a vertical translation of $z = \sin y$ by the value $\cos c$. The surface $z = \cos x + \sin y$ can be thought of, then, as taking the sine curve in the y-direction, and having it ride along a roller coaster in the x-direction. The roller coaster is the cosine curve (see Figure 1.3.6). ▲

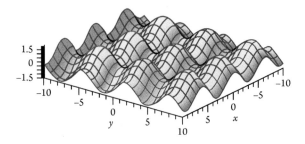

Figure 1.3.6 The surface $z = \cos x + \sin y$

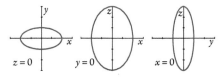

Figure 1.3.7 Planar cross-sections of $\frac{x^2}{4} + y^2 + \frac{z^2}{9} = 1$

The preceeding two examples involved surfaces that were graphs of functions, but didn't have particularly nice level curves. The strategy was to get two-dimensional slices by cutting with planes parallel to one of the coordinate planes, other than the xy-plane. We now consider some surfaces arising from equations that can't be solved for one of the variables. We will find that slicing with a variety of planes gives geometric insight to the surfaces.

A quadric surface is the solution to a quadratic equation in three variables. The quadratic equations we will consider have the form

$$Ax^2 + By^2 + Cz^2 + Dx + Ey + Fz = G$$

(we avoid equations with mixed terms xy, xz, or yz). Our basic strategy for understanding such surfaces will be to sketch a "skeleton" by intersecting the surface with coordinate planes, then "connect the curves". We illustrate this with two examples.

Example 1.3.7. *An ellipsoid*

The solution set to $\frac{x^2}{4} + y^2 = 1$ is an ellipse. Analogously, the solution set to the three-variable equation $\frac{x^2}{4} + y^2 + \frac{z^2}{9} = 1$ is an ellipsoid. To get a feel for what it looks like, consider the intersections with coordinate planes. When first sketching these surfaces, it might help to sketch the intersections in separate planes, as in Figure 1.3.7, then piece them together. To find equations for these curves of intersection, set one of the variables to zero and sketch the resulting curve in the appropriate coordinate plane. The intersection of the surface $\frac{x^2}{4} + y^2 + \frac{z^2}{9} = 1$ with the xy-plane is gotten by setting $z = 0$ (technically, we're solving the *system* of equations consisting of both, and getting the intersection of the surfaces). The intersection with the xz-and yz-planes are the ellipses $\frac{x^2}{4} + \frac{z^2}{9} = 1$ and $y^2 + \frac{z^2}{9} = 1$, respectively. The planes pieced together form a "skeleton" for the whole ellipsoid pictured in Figure 1.3.8. ▲

Example 1.3.8. *The one-sheeted hyperboloid*

The solution set to the equation $x^2 + y^2 - z^2 = 1$ is called a hyperboloid of one sheet. Intersecting it with the coordinate planes yields the circle $x^2 + y^2 = 1$ in the xy-plane, and the hyperbolas $x^2 - z^2 = 1$ and $y^2 - z^2 = 1$ in the xz- and yz-planes, respectively (see Figure 1.3.9).

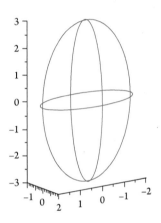 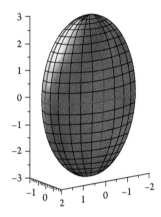

Figure 1.3.8 An ellipsoid and its skeleton

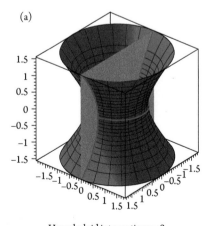 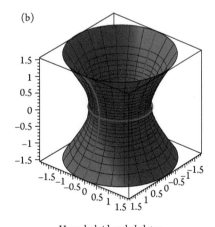

Hyperboloid intersecting $y = 0$ Hyperboloid and skeleton

Figure 1.3.9 A one-sheeted hyperboloid

The one-sheeted hyperboloid can also be thought of as a surface of revolution. To see this, just revolve one component of $x^2 - z^2 = 1$ around the z-axis. You can also see this analytically by changing the equation to cylindrical coordinates. Recall that $x^2 + y^2 = r^2$, so the Cartesian equation for the surface $x^2 + y^2 - z^2 = 1$ becomes the cylindrical equation $r^2 - z^2 = 1$. You notice the equation is independent of θ, which implies that this is a surface of revolution. ▲

Level Surfaces: A level curve is a curve in the plane given by $f(x,y) = c$ for some function of two variables $f(x,y)$. Analogously, a *level surface* is a surface in space obtained by setting a function of three variables equal to a constant, in symbols $g(x,y,z) = c$. Since the result is a three variable equation, level surfaces are examples of surfaces given as the solution set to an equation. We illustrate them with several examples.

Example 1.3.9. *Planar level surfaces*

Let $g(x,y,z) = x - y + 2z$, and sketch the level surfaces at levels $c = -1, 1, 4$. The level surface at level $c = -1$ is the plane $x - y + 2z = -1$. As above we can find the x-, y-, and z-intercepts, plot them and sketch. In this case they are $(-1,0,0)$, $(0,1,0)$, and $(0,0,-1/2)$. The resulting level surface is the lowest plane pictured in Figure 1.3.10. The remaining level surfaces are pictured as well.

As the level increases, then, the level surface remains planar but is translated up the z-axis. In this example, all level surfaces are the same geometric object—a plane. We will soon see that this is not always the case. ▲

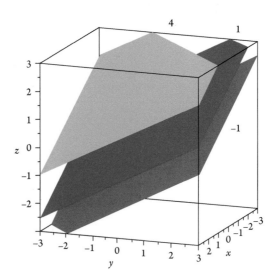

Figure 1.3.10 Level surfaces of $g(x,y,z) = x - y + 2z$

Example 1.3.10. *Quadric level surfaces*

In this example we sketch the level surfaces of $g(x,y,z) = x^2 - y^2 - z^2$ at levels $c = -1, 0, 1, 2$. We treat the cases separately, and it will be interesting to see how different levels can lead to very different surfaces.

The level $c = -1$ yields the equation $-1 = x^2 - y^2 - z^2$. Multiplying by negative one gives $y^2 + z^2 - x^2 = 1$, which is the same equation as in Example 1.3.8 except the roles of the x and z variables are switched. As Example 1.3.8 yielded a one-sheeted hyperboloid with z-axis as core, the surface $y^2 + z^2 - x^2 = 1$ is a one-sheeted hyperboloid with x-axis as core (see Figure 1.3.11(a)).

At level zero, the resulting equation simplifies to $x^2 = y^2 + z^2$. We analyze this surface by first finding intersections with the coordinate planes. Letting $z = 0$ gives $x^2 = y^2$,

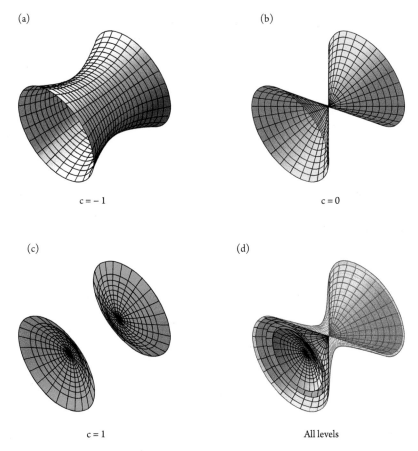

(a)

$c = -1$

(b)

$c = 0$

(c)

$c = 1$

(d)

All levels

Figure 1.3.11 Level surfaces of $g(x,y,z) = x^2 - y^2 - z^2$

which is the pair of lines $y = \pm x$. Similarly, letting $y = 0$ gives the pair of lines $z = \pm x$ in the xz-plane. Further, one sees from the equation $x^2 = y^2 + z^2$ that fixing $x = d$ yields a circle in that plane of radius d. Thus the surface is a cone with x-axis as core, as in Figure 1.3.11(b).

The level surface at level $c = 1$ is the solution to the equation $x^2 - y^2 - z^2 = 1$. Clearly if $x = 0$ there are no solutions to the equation, so this surface misses the yz-plane. The intersection with the xy-plane is the hyperbola $x^2 - y^2 = 1$, and with the xz-plane is $x^2 - z^2 = 1$. Notice that letting $x = d > 1$ yields the curve $y^2 + z^2 = d^2 - 1$. This is a circle in the plane $x = d$ centered on the x-axis. Thus the surface is the two-sheeted hyperboloid of Figure 1.3.11(c).

The level surface at $c = 2$ is similar to $c = 1$, and the results of this discussion are summarized in the following table. The level surfaces are pictured on the same set of axes in Figure 1.3.11(d).

Level	Level Surface	Surface Type
$c = -1$	$-1 = x^2 - y^2 - z^2$	One-sheeted hyperboloid
$c = 0$	$0 = x^2 - y^2 - z^2$	Cone
$c = 1$	$1 = x^2 - y^2 - z^2$	Two-sheeted hyperboloid
$c = 2$	$2 = x^2 - y^2 - z^2$	Two-sheeted hyperboloid

In summary, then, for $c < 0$ the level surface $x^2 - y^2 - z^2 = c$ will be a one-sheeted hyperboloid. As $c \to 0^-$, the "waist" of the hyperboloid shrinks. In the limit, then, one gets a cone when $c = 0$. The cone pinches the waist so that when $c > 0$ the level surfaces are two-sheeted hyperboloids. ▲

Notice that just as level curves of $f(x, y)$ fill up the plane, level surfaces of $g(x, y, z)$ fill up $\mathbb{R}^3$. This is demonstrated quite nicely in the next Math App.

Math App 1.3.2. *Level surfaces*

Click the link below, or print users open manually, to explore how level surfaces fill space as the level changes.

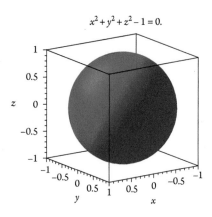

$$x^2 + y^2 + z^2 - 1 = 0.$$

Example 1.3.11. *Temperature near a heat source*

Suppose the function $T(x,y,z) = 20e^{-(x^2+y^2+z^2)}$ gives the temperature (in degrees Celsius) near a heat source at the origin. In this context, the level surface at level $c = 11$ is the surface of constant temperature $11°C$ (this is sometimes called an isothermal surface). Thus, if our original function is temperature, level surfaces represent surfaces of constant temperature. In our case, the surface is the solution set to the equation $11 = 20e^{-(x^2+y^2+z^2)}$. Isolating the variables gives

$$11 = 20e^{-(x^2+y^2+z^2)}$$
$$x^2 + y^2 + z^2 = -\ln(11/20),$$

which is the equation of a sphere centered at the origin of radius $\sqrt{\ln(20/11)}$. More generally, the isothermal surfaces of this example are concentric spheres centered at the origin. ▲

Equations in Other Coordinate Systems: Since we are studying surfaces given by solution sets of an equation, it makes sense to look at the solution set of cylindrical and spherical equations. Even simple equations can give quite interesting surfaces.

Example 1.3.12. *A familiar surface*

Using different coordinate systems can also lead to a better geometric understanding of surfaces. To examine the graph of $f(x,y) = x^2 + y^2$, we can write the equation $z = f(x,y)$ in cylindrical coordinates as follows:

$$z = x^2 + y^2 = (r\cos\theta)^2 + (r\sin\theta)^2 = r^2\cos^2\theta + r^2\sin^2\theta = r^2.$$

Since $z = r^2$ has no θ in it, points on the surface are independent of θ. This really means the surface has rotational symmetry about the z-axis. To sketch it, then, you can sketch the curve in the yz-plane, then rotate it about the z-axis. The yz-plane is given by $x = 0$, so the equation for the curve of intersection is $z = y^2$, a parabola. Revolving the parabola $z = y^2$ around the z-axis yields the surface. ▲

Example 1.3.13. *A cone*

To understand the graph of $f(x,y) = \sqrt{x^2 + y^2}$, we can write the equation $z = \sqrt{x^2 + y^2}$ in spherical coordinates. Using the change of coordinates, we see that $z = \sqrt{x^2 + y^2}$ becomes

$$\rho\cos\phi = \sqrt{(\rho\sin\phi\cos\theta)^2 + (\rho\sin\phi\sin\theta)^2}$$
$$= \sqrt{\rho^2\sin^2\phi}$$
$$= \rho\sin\phi. \qquad\qquad (1.3.1)$$

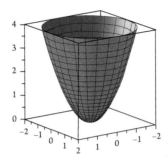

Figure 1.3.12 The paraboloid $z = x^2 + y^2$

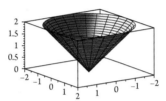

Figure 1.3.13 The cone $z = \sqrt{x^2 + y^2}$

Now $\rho\cos\phi = \rho\sin\phi$ simplifies to $\tan\phi = 1$, which is satisfied by the spherical equation $\phi = \pi/4$. As before, this is a cone that makes an angle of $\pi/4$ with the positive z-axis. ▲

Example 1.3.14. *The surface $z = Ar$*

One way to understand the surface $z = \sqrt{3}r$ is to change to spherical coordinates analytically, as in the previous example. Alternatively, note that the ratio z/r for all points on the surface is $\sqrt{3}$. Secondly, recall that this ratio is merely $\cot\phi$ (see Figure 1.1.6(a)) so the spherical equation for the surface is $\phi = \cot^{-1}(\sqrt{3})$. Thus the surface $z = \sqrt{3}r$ is the cone that makes an angle of $\pi/6$ with the positive z-axis.

Of course, this generalizes to $z = Ar$ for any constant A. Since $z/r = A$, the same reasoning implies $\phi = \cot^{-1}(A)$. So the coefficient of r is the cotangent of the angle the cone $z = Ar$ makes with the z-axis.

Notice that the equation $z = Ar$ is independent of θ. Hence the surface has rotational symmetry. To sketch it, then, you can sketch the intersection with the yz-plane and rotate it about the z-axis. ▲

Example 1.3.15. *A surface of revolution in spherical coordinates*

Sketch the surface $\rho = 1 + \cos\phi$, for $0 \le \phi \le \pi$.

To do so, sketch the portion in the yz-plane, where it looks like a cardioid pointing down (recall that ϕ is the angle with the positive z-axis). Rotating gives the upside-down "apple" of Figure 1.3.14. ▲

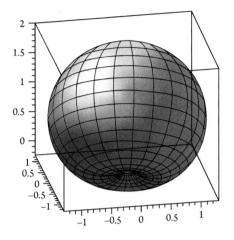
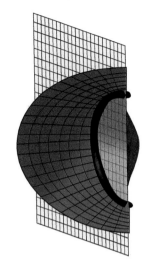

Figure 1.3.14 The surfaces $\rho = 1 + \cos\phi$ and $\rho = \theta$

Example 1.3.16. *A sea shell*

Let's look at the surface described by the spherical equation $\rho = \theta$, for $\theta > 0$. Fixing $\theta = \frac{\pi}{3}$ is equivalent to intersecting the surface with the half-plane. So in the half-plane, we're looking at all points of the form $\rho = \frac{\pi}{3}$. Since ρ is the distance to the origin, the intersection of the surface $\rho = \theta$ with the half-plane $\theta = \frac{\pi}{3}$ is a semicircle of radius $\frac{\pi}{3}$. Generalizing this observation, we notice that as θ increases the intersection with $\rho = \theta$ is a semicircle with increasing radius. The result is $\rho = \theta$ looks like a sea shell. ▲

Gravitational Potential: A mass M is placed at the origin creates a gravitational potential at each point in space, given by the formula

$$U = -\frac{GM}{\rho},$$

where $G = 6.67 \times 10^{-11} \mathrm{m}^3/\mathrm{kg} \cdot \mathrm{s}^2$ is the gravitational constant. This is a spherical equation for the potential, which represents the work per unit mass it would take to bring a unit mass from infinity to a point ρ units from the origin. That's a lot of words which we will ignore for the time being, although work will be described in Section 2.3. For the time being, we satisfy ourselves with sketching a few level surfaces for the function U.

Sketch the level surfaces for U at levels $c = -0.25GM, -GM, -2GM$.

We obtain spherical equations by substituting the levels for U, as in the following table.

Level	Level Surface	Simplified form
$c = -0.25GM$	$-0.25GM = -\frac{GM}{\rho}$	$\rho = 4$
$c = -GM$	$-GM = -\frac{GM}{\rho}$	$\rho = 1$
$c = -2GM$	$-2GM = -\frac{GM}{\rho}$	$\rho = \frac{1}{2}$

Note that the level surfaces are all spheres centered at the origin with different radii. The more negative the level, the smaller the radius of the sphere. ▲

Things to know/Skills to have

In this section we studied surfaces resulting from the solution set of equations. You know you understand the material of this section when you are able to:

- Sketch the plane given by a linear equation in three variables.
- Sketch generalized cylinders.
- Sketch certain quadric surfaces, using their intersection with coordinate planes.
- Sketch level surfaces for a function of three variables.
- Sketch solution sets to equations given in cylindrical and spherical coordinates.

Exercises

1. Sketch the plane $7x - 3y + z = 21$.

2. Sketch the plane $-x + 2y + z = 8$.

3. Sketch the plane $4x - 3y - 3z = 12$.

4. Find coefficients A, B, C so that the plane $Ax + By + Cz = 15$ has intercepts $(1,0,0)$, $(0,-5,0)$, and $(0,0,3)$.

5. Find coefficients A, B, C so that the plane $Ax + By + Cz = 8$ has intercepts $(-4,0,0)$, $(0,2,0)$, and $(0,0,-8)$.

6. Assume $D > 0$, and the plane $Ax + By + Cz = D$ intersects the positive z-axis. What can you say about the coefficient C?

7. Find an equation for the plane with intercepts $(2,0,0)$, $(0,-4,0)$, and $(0,0,6)$.

8. Find an equation for the plane with intercepts $(-3,0,0)$, $(0,3,0)$, and $(0,0,-2)$.

9. Sketch the surface in $\mathbb{R}^3$ whose Cartesian equation is $x = y^2$. Hint: this is a generalized cylinder.

10. Sketch the surface in $\mathbb{R}^3$ whose Cartesian equation is $x^2 + z^2 = 4$.

11. Sketch the surface in $\mathbb{R}^3$ whose Cartesian equation is $z = 4 - x^2$.

12. What is the Cartesian equation for the right circular cylinder of radius 1, with the x-axis as its core?

13. Sketch the intersections of the surface $x^2 + y^2 - z^2 = -1$ with the coordinate planes. Sketch the entire surface. Why might this be called a two-sheeted hyperboloid?

14. Sketch the intersections of the surface $x^2 + 4y^2 + z^2 = 1$ with the coordinate planes. What type of surface is this?

15. Sketch the intersections of the surface $x^2 + y^2 - z^2 = 0$ with the coordinate planes. What type of surface is this?

16. Sketch the intersections of the surface $x^2 - y + z^2 = 1$ with the coordinate planes. What type of surface is this?

17. Let $g(x,y,z) = x + y + z$, and sketch the level surfaces at levels $-1, 1, 2$.

18. Sketch the level surfaces of $g(x,y,z) = x - y^2 - z^2$ at levels $-1, 0, 2$.

19. Sketch the level surfaces of $g(x,y,z) = x^2 + y^2 - z^2$ at levels $-1, 0, 2$.

20. Sketch the level surfaces of $g(x,y,z) = x^2 + y^2 + z^2$ at levels $-1, 0, 4$.

21. Sketch the level surfaces of $g(x,y,z) = 4x^2 + y^2 + 9z^2$ at levels $-1, 1, 9$.

22. Match the following surfaces with their equations.

$$a.\, z = 2y - y^2 \qquad b.\, 4x^2 + y^2 - z^2 = 1 \qquad c.\, z = r^2 \sin^2\theta$$
$$d.\, x^2 + 4y^2 = z^2 \qquad e.\, x^2 + 4y^2 - z^2 = 1 \qquad f.\, 4x^2 + y^2 - z^2 = -1$$

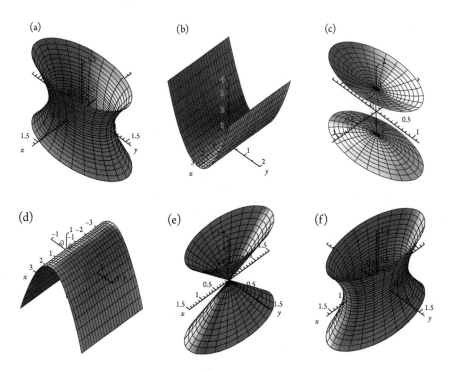

(a)   (b)   (c)   (d)   (e)   (f)

23. A wire runs along the z-axis, and conducts heat so that the temperature near it is given by $T(x,y,z) = 12e^{-x^2-y^2}$ (so temperature is independent of z). Find the isothermal surfaces at temperatures $1°$, $6°$, $10°$.

24. Level surfaces and electric potential: The electric potential V a distance ρ units from a point charge Q is given by

$$V = k\frac{Q}{\rho}.$$

where $k = 9.0 \times 10^9 \text{Nm}^2/\text{C}^2$ is Coulomb's constant. Suppose a point charge of 1 nanocoulomb (1×10^{-9} coulombs) is placed at the origin in $\mathbb{R}^3$.

(a) Find a spherical equation for V.

(b) An *equipotential surface* is one on which the potential function V is constant—a level surface for V! Find the 16 volt equipotential surface.

(c) Describe the equipotential surfaces of V geometrically.

25. Sketch the surface whose cylindrical equation is $z^2 + r^2 = 9$. Hint: change to spherical coordinates.

26. Sketch the surface $z = r$, assume the convention that $r \geq 0$.

27. Find a cylindrical equation for the surface $\phi = \pi/4$, and sketch.

28. Let $0 < \alpha < \pi$ be a constant. Find a cylindrical equation for the surface $\phi = \alpha$.

29. (a) Sketch the curve in $\mathbb{R}^2$ given by the polar equation $r = \theta$.

(b) Sketch the surface in $\mathbb{R}^3$ given by the cylindrical equation $r = \theta$.

1.4 Parametric Curves

Curves in the plane can be described as graphs of functions and as solution sets to equations. In the previous two sections we saw how these methods of describing curves in $\mathbb{R}^2$ generalize naturally to describe surfaces in $\mathbb{R}^3$. It's also common to describe curves in the plane parametrically. We will see that parametric curves in $\mathbb{R}^2$ have two natural generalizations to three dimensions. One generalization leads to parametric curves in $\mathbb{R}^3$, and the other to parametric surfaces. In this section we recall the definition of planar parametric curves and describe several techniques for parameterizing planar curves. A brief foray into some of the flexibility, as well as the structure, that one has when parameterizing curves follows these planar examples. Finally, a few techniques for parameterizing curves in $\mathbb{R}^3$ are introduced. The extension of parameterizations to surfaces will be postponed until Section 1.5.

Recall that parametric curves are given by so-called parametric equations $x = x(t)$, $y = y(t)$. We sometimes use the vector notation $\mathbf{C}(t) = \big(f(t), g(t)\big)$, combining the two parametric equations into a single ordered pair. In this context, both the horizontal and vertical coordinates are functions of a third variable t, called the *parameter*.

Example 1.4.1. *A Fermat spiral*

The curve given by the parametric equations

$$\mathbf{C}(t) = (x(t), y(t)) = (\sqrt{t}\cos t, \sqrt{t}\sin t), \ 0 \le t \le 8\pi$$

is the Fermat spiral of Figure 1.4.1. Recall that parametric curves are implicitly endowed with a direction—that of increasing t. For the Fermat spiral the direction of increasing t is counterclockwise. ▲

We describe two ways of interpreting parametric curves. One way to think of parametric curves is that they describe the position $\mathbf{C}(t)$, at time t, of an ant crawling on the plane.

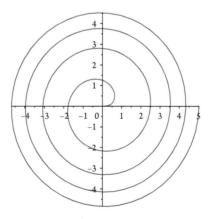

Figure 1.4.1 The Fermat spiral

In the previous example, then, at time $t = \frac{\pi}{2}$ we substitute into the parametric equations to find that the ant is at position

$$C(\pi/2) = \big(x(\pi/2), y(\pi/2)\big)$$
$$= \Big(\sqrt{\pi/2}\cos(\pi/2), \sqrt{\pi/2}\sin(\pi/2)\Big) \approx (0, 1.253).$$

Another way to think of parametric curves is that they tell you how to bend, stretch, and shrink a piece of wire into the plane. In our example, we start with a straight wire from 0 to 8π on the t-number line. The parametric equations $x(t) = \sqrt{t}\cos t$ and $y(t) = \sqrt{t}\sin t$ are the instructions for how to bend and stretch the wire into the plane. Thus the domain of the parametric curve $C(t)$ is the interval $[0, 8\pi]$ on the t-axis, and the image of $C(t)$ is a subset of $\mathbb{R}^2$. Then $C(t)$ is viewed as a mapping from one to the other, giving rise to the notation $C : I \to \mathbb{R}^2$ for parametric curves. In this case $I = [0, 8\pi]$ and the image $C(I)$ is a subset of $\mathbb{R}^2$, namely the curve itself. Conceptually you can think of it as in Figure 1.4.2.

This is a different way of viewing curves than we might be used to. When working with the graphs of functions, the domain and range are in the same space; whereas parametric curves place the domain and range in separate spaces.

Curves in $\mathbb{R}^2$: In this subsection we review parameterizations of several familiar curves in the plane: lines, ellipses, and hyperbolas. We also review techniques for parameterizing the graph of a function and a curve given by a polar equation. Both techniques generalize nicely to parametric surfaces.

Parameterizing Lines: It turns out that any line can parameterized using linear functions of the parameter. For any constants a, b, c, d, with a and c not both zero, the parametric equations

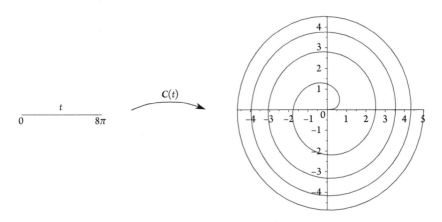

Figure 1.4.2 Parametric curves: bending wire

$$x = at + b$$
$$y = ct + d$$

for $-\infty < t < \infty$ yield a line. One way to see this is to subtract c times the first equation from a times the second. This gives

$$
\begin{aligned}
ay &= act + ad \\
- \quad cx &= cat + cb \\
\hline
ay - cx &= ad - bc,
\end{aligned}
$$

which is of the form $Ax + By = C$, making it a line.

Parametrizing Graphs of Functions: It is easy to find parametric equations for the graph of a function. More precisely, to parameterize the curve $y = f(x)$, simply let $x = t$ and $y = f(t)$. Parametric equations for our hyperbolic sine function are then

$$x = t$$
$$y = \frac{e^t - e^{-t}}{2}$$

for $-\infty < t < \infty$. This seems rather artificial, but it does represent a change of perspective which can sometimes be useful. Similarly, parametric equations for the graph of $y = x^3 - x$ are $C(t) = (t, t^3 - t)$, for $-\infty < t < \infty$. The direction of these parameterizations is the direction of increasing t, which is left to right.

Parameterizing Graphs of Functions

To parameterize the graph of $y = f(x)$ in the plane, let the independent variable be the parameter and use the function to parameterize the dependent variable. Symbolically we have

$$x = t, \; y = f(t), \tag{1.4.1}$$

or more concisely $C(t) = (t, f(t))$. Note that the same strategy works for graphs in any plane using other coordinates, e.g. $x = g(z)$.

Parameterizing Conic Sections: Recall that conic sections are parabolas, ellipses, and hyperbolas. Parabolas in standard position have equations similar to $y = x^2$ or $x = y^2 - 4$, which can be solved for one of the variables. The technique for parameterizing graphs of functions applies, and parabolas are easy to parameterize. We turn our attention to ellipses and hyperbolas, which are most easily parameterized using trigonometric functions.

Example 1.4.2. *The unit circle*

The unit circle has Cartesian equation $x^2 + y^2 = 1$, and the point $(\cos\theta, \sin\theta)$ lies on the unit circle making an angle θ with the positive x-axis. The parametric curve

$$\mathbf{C}(\theta) = (\cos\theta, \sin\theta), \ 0 \le \theta \le 2\pi,$$

is therefore the unit circle! We've used the parameter θ instead of t because it strongly hints that the parameter is just the angle from polar coordinates (which it is). This parameterization starts at $\mathbf{C}(0) = (\cos 0, \sin 0) = (1, 0)$, and goes counterclockwise around the origin. ▲

Example 1.4.3. *Parametrizing ellipses*

The Cartesian equation for an ellipse in standard position is

$$\frac{x^2}{a^2} + \frac{y^2}{b^2} = 1.$$

The ellipse given by this equation is obtained by stretching the unit circle by a factor of a units horizontally and b units vertically. The same rescaling of parametric equations parametrizes the ellipse. More precisely, $\mathbf{C}(t) = (a\cos t, b\sin t)$ for $0 \le t \le 2\pi$ parametrizes the ellipse.

To see this analytically, a simple substitution verifies that the parametric equations satisfy the Cartesian equation of the ellipse:

$$\frac{x^2}{a^2} + \frac{y^2}{b^2} = \frac{(a\cos t)^2}{a^2} + \frac{(b\sin t)^2}{b^2} = \cos^2 t + \sin^2 t = 1. \tag{1.4.2}$$

The given parameterization starts at $(a, 0)$ and traverses the ellipse counterclockwise. One can also parameterize the ellipse in a clockwise fashion. One way to do this is to negate the y-coordinate, giving $\mathbf{C}(t) = (a\cos t, -b\sin t)$. This starts at $(a, 0)$, goes clockwise, and lies on the ellipse. Another clockwise parameterization comes from switching the sine and cosine functions, resulting in $\mathbf{C}(t) = (a\sin t, b\cos t)$ for $0 \le t \le 2\pi$. Indeed, this starts at the point $\mathbf{C}(0) = (0, b)$, and at time $t = \pi/2$ we have $\mathbf{C}(\pi/2) = (a, 0)$. Thus the parameterization has a different starting point, and goes clockwise around the ellipse. A calculation as in Equation 1.4.2 verifies that $\mathbf{C}(t)$ is on the ellipse. ▲

Example 1.4.4. *Parameterizing hyperbolas*

This section started with the graph of $f(x) = \sinh x$. The hyperbolic cosine function is similar, and defined by $\cosh x = (e^x + e^{-x})/2$. Note that $\cosh x \ge 1$ for all x. There is a hyperbolic Pythagorean identity, which is analogous to the familiar $\sin^2 x + \cos^2 x = 1$. It is not hard to verify that $\cosh^2 x - \sinh^2 x = 1$. Using this, one can show that $x = a\cosh t$, $y = b\sinh t$ parameterizes the right half of the hyperbola $\frac{x^2}{a^2} - \frac{y^2}{b^2} = 1$ (we are assuming both a and b are positive).

To verify that points on the parametric curve $\mathbf{C}(t) = (a \cosh t, b \sinh t)$ are on the hyperbola $\frac{x^2}{a^2} - \frac{y^2}{b^2} = 1$, substitute and simplify:

$$\frac{x^2}{a^2} - \frac{y^2}{b^2} = \frac{(a \cosh t)^2}{a^2} - \frac{(b \sinh t)^2}{b^2}$$
$$= \cosh^2 t - \sinh^2 t = 1.$$

Since $\cosh t \geq 1$, the curve $\mathbf{C}(t) = (a \cosh t, b \sinh t)$ for $-\infty < t < \infty$ is the right component of the hyperbola. Since $\sinh t$ is negative for $t < 0$, the parameterization starts at the bottom right, goes to the vertex $(1,0)$ at $t = 0$, and continues up as t increases. ▲

Parameterizing Polar Functions: Curves in $\mathbb{R}^2$ can also be described by polar equations $r = f(\theta)$. For example, the polar function $r = \sqrt{\theta}$ for $\theta > 0$ yields Fermat's spiral of Figure 1.4.1. To find parametric equations for Fermat's spiral, you need to find component functions for Cartesian coordinates x and y. This can be accomplished in two steps.

First let the independent variable θ be the parameter, then use the change of coordinates $x = r \cos \theta$, $y = r \sin \theta$. In our case, substituting $r = f(\theta) = \sqrt{\theta}$ into the change of coordinates yields the parametric equations we had originally:

$$\mathbf{C}(\theta) = (x,y) = \left(\sqrt{\theta} \cos \theta, \sqrt{\theta} \sin \theta \right), \text{ for } \theta \geq 0.$$

Parameterizing Polar Functions

To parameterize a curve given by the polar function $r = f(\theta)$, let θ be the parameter and use the change of coordinates formula. More precisely, a parameterization for the curve $r = f(\theta)$, $a \leq \theta \leq b$, is

$$\mathbf{C}(\theta) = \left(f(\theta) \cos \theta, f(\theta) \sin \theta \right), a \leq \theta \leq b \qquad (1.4.3)$$

Example 1.4.5. *A three-leafed rose*

Parameterize the three-leaf rose $r = \cos 3\theta$ pictured in Figure 1.4.3.

Letting $\theta = t$ we find $r = \cos 3t$. Substituting into the coordinate conversion we get

$$\mathbf{C}(t) = (\cos 3t \cos t, \cos 3t \sin t), \, 0 \leq t \leq \pi.$$

The limits on t come from observing that for $\pi \leq t \leq 2\pi$ the curve traces back over itself. We could also parameterize just the horizontal petal by determining when $r = 0$. These would be the t-values nearest 0 where $\cos 3t = 0$, or $t = \pm \pi/6$. Thus the limits $-\pi/6 \leq t \leq \pi/6$ parameterize the horizontal petal. ▲

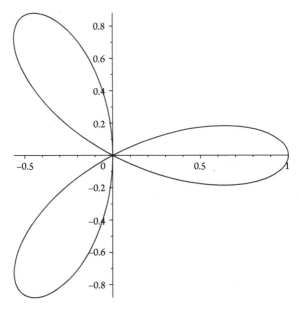

Figure 1.4.3 The curve $r = \cos 3\theta$

This technique for parameterizing planar curves from a polar equation will generalize to parameterizing curves and surfaces in higher dimensions. For example, one can parameterize a surface given by a spherical equation using the appropriate coordinate conversions to Cartesian coordinates.

Geometric Curves and Parameterizations: A parametric curve $\mathbf{C} : I \to \mathbb{R}^2$ can be thought of as giving the position of a particle moving in the plane. As such, the parameterization $\mathbf{C}(t) = (\cos t, \sin t)$, $0 \leq t \leq 2\pi$, contains more information than just the set of points satisfying $x^2 + y^2 = 1$. One can ask for the direction that the particle is traveling, where it starts, or for its speed.

Conversely, a particle traveling in the plane yields a parametric curve by letting $\mathbf{C}(t)$ represent the position of the particle at time t. Two particles traveling the same path at different speeds would give two different parameterizations of that path. A particle traveling the opposite direction would lead to a parameterization of the same path backwards.

A *geometric curve* means a set of points in the plane (think of the unit circle as the solution to $x^2 + y^2 = 1$). A *parametric curve* is a particular parameterization $\mathbf{C}(t)$, $a \leq t \leq b$. By the preceding remarks, each geometric curve has many parameterizations. The image set of a parametric curve $\mathbf{C}(t)$, $a \leq t \leq b$, is sometimes called its trace. In this subsection we investigate the relationship between different parameterizations of the same geometric curve or, equivalently, parametric curves with the same trace. We begin by parameterizing part of the unit circle in several ways.

Example 1.4.6. *The quarter unit circle*

Let C be the quarter unit circle in the first quadrant. The curve C is the geometric curve satisfying $x^2 + y^2 = 1$, with the restrictions $x, y \geq 0$. One way to parameterize C is to restrict the parameterization of Example 1.4 to include just that portion of the unit circle in the first quadrant. This simply amounts to restricting the parameter as in

$$\mathbf{C}_1(t) = (\cos t, \sin t),\ 0 \leq t \leq \pi/2.$$

A second parameterization can be found by thinking of C as the graph of a function, which we already know how to parameterize (see Equation 1.4.1). Solving the Cartesian equation of the unit circle for x, we get that C is the graph of $x = \sqrt{1 - y^2}$ for $0 \leq y \leq 1$. Letting the independent variable be the parameter s we see that

$$\mathbf{C}_2(s) = \left(\sqrt{1 - s^2}, s\right),\ 0 \leq s \leq 1,$$

is a second parameterization of the same geometric curve C.

Both $\mathbf{C}_1$ and $\mathbf{C}_2$ are counterclockwise parameterizations of C. The reader should verify that

$$\mathbf{C}_3(t) = \left(\sqrt{1 - (1 - t)^2}, (1 - t)\right),\ 0 \leq t \leq 1,$$

parameterizes C in a clockwise direction, starting at $(0, 1)$. ▲

From this example it is clear that the same geometric curve can have quite different-looking parameterizations, and our goal is to describe how to obtain all possible parameterizations of a given geometric curve. To do so, we need to make some reasonable assumptions on the types of parameterizations we consider, and we do so with the definition:

Definition 1.4.1. *The parametric curve $\mathbf{C}(t) = (x(t), y(t))$, $a \leq t \leq b$, is regular if $x'(t)$ and $y'(t)$ are not simultaneously zero for $a < t < b$.*

Is the parameterization $\mathbf{C}(t) = (\cos t, \sin t)$, $0 \leq t \leq 2\pi$ regular? How about the other parameterizations of Example 1.4?

The restriction that $x'(t)$ and $y'(t)$ are not both zero is equivalent to the statement that $\sqrt{x'^2(t) + y'^2(t)} > 0$ for all $a \leq t \leq b$. Section 2.5 will show that this means that the tangent vector to the curve never vanishes. If we interpret $\mathbf{C}(t)$ as the position of a particle, this definition implies the particle never stops or changes direction. This fact will allow us to compute the length of a geometric curve from a regular parameterization in Section 2.5.

Regularity of parameterizations, then, is a desirable property. In our efforts to classify all parameterizations of a fixed geometric curve, we will restrict our attention to regular parameterizations. We begin with an example.

Example 1.4.7. *A parabolic arc*

Let C be the portion of the parabola $y = x^2$ for $0 \leq x \leq 1$. One can verify that the following parameterizations all have trace C:

$$\mathbf{C}_1(t) = (t, t^2), \ 0 \leq t \leq 1,$$
$$\mathbf{C}_2(s) = (6s, 36s^2), \ 0 \leq s \leq 1/6, \text{ and}$$
$$\mathbf{C}_3(u) = (u^2, u^4), \ 0 \leq u \leq 1.$$

Indeed, $\mathbf{C}_2$ is just $\mathbf{C}_1$ with the substitution $t = 6s$, and letting $t = u^2$ in $\mathbf{C}_1$ gives $\mathbf{C}_3$. Note that some care is taken to change the limits on the parameters. To get new limits you set the old limits equal to the substitution and solve. For example, the lower limit on s is the solution to $6s = 0$, and you solve $6s = 1$ to get the upper limit. ▲

Pop quiz: Are the parameterizations of Example 1.4.7 regular?

It is tempting to generalize this example and say that given a parameterization $\mathbf{C}_1(t)$, $a \leq t \leq b$, of a geometric curve you can get another one by substituting $t = f(s)$ and taking care with the limits on s. This substitution, of course, is equivalent to composition of maps. Given an onto map $f : I_2 \to I_1$ of intervals and $\mathbf{C}_1 : I_1 \to \mathbb{R}^2$, the composition $\mathbf{C}_2 = \mathbf{C}_1 \circ f : I_2 \to \mathbb{R}^2$ parameterizes the same geometric curve (see Figure 1.4.4).

Recall that both $\mathbf{C}_1(t)$ and the composition $\mathbf{C}_2(s) = \mathbf{C}_1(f(s))$ are required to be regular curves. Certain properties of the function $f(s)$ will guarantee that $\mathbf{C}_2$ is regular, and the chain rule for differentiation sorts this out.

Given a regular parametric curve $\mathbf{C}(t) = (x(t), y(t))$, and a single-variable function $f(s)$, one obtains the parametric equations $\mathbf{C}(f(s)) = \big(x(f(s)), y(f(s))\big)$. Differentiating with respect to s gives

$$\frac{d}{ds}x(f(s)) = x'(f(s))f'(s) \quad \text{and} \quad \frac{d}{ds}y(f(s)) = y'(f(s))f'(s).$$

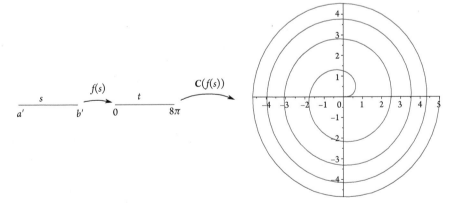

Figure 1.4.4 Reparameterization as composition

If $f'(s) = 0$ at some point, then *both* $\frac{dx}{ds} = \frac{dy}{ds} = 0$ and $\mathbf{C}(f(s))$ is not regular. Conversely, since $\mathbf{C}(t)$ is regular, if $f'(s) \neq 0$ for any s then $\frac{dx}{ds}$ and $\frac{dy}{ds}$ cannot be simultaneously zero (justify that statement—don't just believe me!). For this reason, the definition of a reparameterization is:

Definition 1.4.2. *The composition* $\mathbf{C}_2 = \mathbf{C}_1 \circ f$ *is a* reparameterization *of the parametric curve* $\mathbf{C}_1 : I \to \mathbb{R}^2$.

Here the function $f : J \to I$ *is continuously differentiable with non-vanishing derivative, and maps endpoints of the interval* J *to those of* I *(i.e.* f *is onto).*

Basically, you want to think of a reparameterization as a substitution $t = f(s)$—one where $f'(s) \neq 0$. By the remarks preceding Definition 1.4.2, the non-vanishing derivative restriction guarantees that the reparameterization of a regular curve is again a regular curve.

Example 1.4.8. *A non-reparameterization*

The parametric curve $\mathbf{C}(t) = (3t, \sqrt{36 - 9t^2})$, $-2 \leq t \leq 2$, gives a clockwise parameterization of the top half of $x^2 + y^2 = 36$.
Letting $f(s) = (2s^3 - 5s)/3$, note that $f(\pm 2) = \pm 2$, so that $f : [-2,2] \to [-2,2]$. Despite this, the curve

$$\mathbf{C}_2(s) = \mathbf{C}(f(s)) = \left(2s^3 - 5s, \sqrt{36 - (2s^3 - 5s)^2}\right), \quad -2 \leq s \leq 2,$$

is not a reparameterization of $\mathbf{C}(t)$. The reason it fails to be a reparameterization is that $f'(s)$ vanishes at $s = \pm\sqrt{\frac{5}{8}}$.
The intuition behind this example is easy to grasp, and underscores the reason for requiring that $f'(s) \neq 0$. While starting and stopping at the same endpoints as $\mathbf{C}(t)$, the particle traveling along $\mathbf{C}_2(s)$ changes direction at times $s = \pm\sqrt{\frac{5}{8}}$. See Math App 1.4.1 for further detail. ▲

Math App 1.4.1. *Parameterizing curves*

Click the following hyperlink, or print users open manually, to access a Math App that illustrates different parameterizations of the same curve, as well as the non-reparameterization of the previous example.

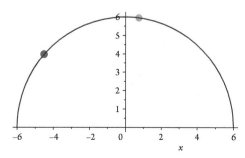

Example 1.4.9. *Linear reparameterization*

Reparameterize the curve $C(t) = (t^2, e^t)$, $-3 \leq t \leq 2$, using the substitution $t = 3s - 4$.

Notice that $\frac{dt}{ds} = 3 > 0$, so the substitution will be a reparameterization and all we have to do is find the limits on s. Substituting the limits for t in $t = 3s - 4$ and solving for s gives $-3 = 3s - 4 \implies s = 1/3$ for the lower limit and $s = 2$ for the upper limit. The reparameterization is

$$C(s) = \left((3s - 4)^2, e^{3s-4}\right), \frac{1}{3} \leq s \leq 2. \ \blacktriangle$$

Example 1.4.10. *A specified range*

Show that for any regular parametric curve $C(t)$, $a \leq t \leq b$, there is a reparameterization with limits $0 \leq s \leq 1$.

To see this we need an increasing function $f(s)$ such that $f(0) = a$ and $f(1) = b$. A linear function $f(s)$ will do, so this is our goal. The t-intercept will be a since $f(0) = a$. Also, the slope is $m = \frac{b-a}{1-0} = (b - a)$, and we have

$$t = (b - a)s + a.$$

By construction this substitution has limits $0 \leq s \leq 1$, and since $f'(s) = (b - a) > 0$ letting $f(s) = (b - a)s + a$ is a valid reparameterization. $\blacktriangle$

Example 1.4.11. *The quarter unit circle—revisited*

The parametric curve $C(t) = (\sqrt{1 - t^2}, t)$, $0 \leq t \leq 1$ is a regular parameterization of the quarter unit circle in the first quadrant.

To reparameterize $C(t)$ we need a function with positive derivative, so we try $f(s) = \sin s$, whose derivative is $f'(s) = \cos s$, which is positive sometimes.

Making the substitution $t = \sin s$ we get the parametric equations

$$C(s) = \left(\sqrt{1 - \sin^2 s}, \sin s\right) = \left(\sqrt{\cos^2 s}, \sin s\right).$$

The limits on s come from solving $\sin s = 0$ and $\sin s = 1$. The s-interval is therefore $0 \leq s \leq \pi/2$. On the interior of that interval $f'(s) = \cos s > 0$, and $\sqrt{\cos^2 s} = \cos s$, which implies that

$$C(s) = (\cos s, \sin s), \ 0 \leq s \leq \pi/2$$

is a reparameterization of $C(t)$. Thus the parameterizations of Example 1.4 are just reparameterizations of each other. $\blacktriangle$

Definition 1.4.2 will be useful in showing that parametric quantities (ones defined on parametric curves) are actually geometric quantities (ones defined on geometric curves). Arclength, for example, is a property of a geometric curve. The arclength of a parametric curve is defined in Section 2.5. Definition 1.4.2 is used to show that all reparameterizations of the same geometric curve have the same arclength, which implies the parametric quantity is really a geometric one! This proof strategy is worth pondering a bit—until it makes sense.

A similar strategy will be implemented in Sections 4.6 and 5.1, to show that certain integrals defined on parametric curves are independent of parameterization—showing that the integrals depend only on the geometric curve being parameterized.

Backwards Parameterizations: Finally, we described how to parameterize curves backwards. To do so, find a function $f(s)$ on the interval $a' \leq s \leq b'$ which satisfies $f(a') = b$, $f(b') = a$, and $f'(s) < 0$ on the interval $a' < s < b'$. The curve $C(f(s))$, $a' \leq s \leq b'$ is a parameterization of the same geometric curve as $C(t)$, $a \leq t \leq b$, but in the opposite direction.

Example 1.4.12. *A parabolic arc—again*

We have already seen that the geometric curve C given by $y = x^2$, $0 \leq x \leq 1$ is parameterized by $C(t) = (t, t^2)$, $0 \leq t \leq 1$. Moreover, the function $f(s) = 2 - s^2$ on the interior of the interval $1 \leq s \leq \sqrt{2}$ has negative derivative, $f(1) = 1$ and $f(\sqrt{2}) = 0$. By the remarks preceding the example, we know that the regular curve

$$C_4(s) = (2 - s^2, (2 - s^2)^2) = (2 - s^2, 4 - 4s^2 + s^4), \ 1 \leq s \leq \sqrt{2}$$

parameterizes C backwards. ▲

Example 1.4.13. *General backwards technique*

Show that letting $t = (b + a) - s$ on the interval $a \leq s \leq b$ parameterizes the curve $C(t)$, $a \leq t \leq b$ backwards. Clearly the function $f(s) = (b + a) - s$ has negative derivative, $f(a) = b$ and $f(b) = a$. Thus, $C(f(s))$, $a \leq s \leq b$ parameterizes C backwards.

For example, parameterizing the quarter circle $C(t) = (\cos t, \sin t)$, $0 \leq t \leq \pi/2$, backwards can be done by substituting $t = \pi/2 - s$ and keeping the same limits on s as were on t. So $C(s) = (\cos(\pi/2 - s), \sin(\pi/2 - s))$, $0 \leq s \leq \pi/2$, does the trick. ▲

Curves in $\mathbb{R}^3$: Sections 1.1, 1.2, and 1.3 all provide examples of curves in $\mathbb{R}^3$ described as the intersection of two surfaces. Thus, they are the solution to a system of two equations in three variables. For example, the solution set of the system $\rho = 3$, $\phi = \pi/4$ is the circle of intersection of the sphere $\rho = 3$ and the cone $\phi = \pi/4$. Using two equations to describe curves, however, would make it difficult (impossible?) to compute arclength, centers of mass, circulation of fluid, and various other quantities related to curves. Using parametric equations, however, allows for such calculations, and is therefore our preferred method for working with curves in three space.

A parametric curve in the plane is given by two component functions $C(t) = (x(t), y(t))$. Recall that we had two intuitive interpretations of parametric curves to aid our understanding. If the parameter t represents time, one can think of $x(t)$ and $y(t)$ as giving the horizontal and vertical position of a bug crawling in the plane at time t. Alternatively, parametric equations can be thought of as instructions for bending, stretching, and shrinking a wire into the plane. The same intuitive interpretations are valid for curves in $\mathbb{R}^3$, the only additional information you need is a third component function $z(t)$. We give a formal definition, and illustrate parametric curves with several examples.

Definition 1.4.3. *A parametric curve is one given by component functions $x(t)$, $y(t)$, and $z(t)$, for $a \le t \le b$. We denote them using ordered triples as in:*

$$\mathbf{C}(t) = \big(x(t), y(t), z(t)\big), \ a \le t \le b.$$

A regular *parametric curve satisfies $\sqrt{x'(t)^2 + y'(t)^2 + z'(t)^2} > 0$ for $a < t < b$.*

Example 1.4.14. *Lines*

Lines in the plane have linear component functions. Analogously, a line in space is given by parametric equations of the form:

$$x = At + x_0$$
$$y = Bt + y_0$$
$$z = Ct + z_0, \ -\infty < t < \infty$$

Letting $t = 0$ in these parametric equations shows that (x_0, y_0, z_0) is a point on the line. After discussing vectors we will see that (A, B, C) gives the direction of the line. Therefore lines in space are determined by two pieces of information: a point and a direction. We look at a couple examples that use parametric equations to determine some geometric properties of lines. ▲

Example 1.4.15. *Lines and coordinate planes*

Let L be the line given parametrically by $\mathbf{L}(t) = (2t + 1, -t + 3, 4t - 2)$. Find the point of intersection of L and the yz-plane.

First we find "when" L intersects the yz-plane by solving $x(t) = 0$. In this case we solve $2t + 1 = 0$ to get $t = -1/2$. Now find the point on L at that time by evaluating $\mathbf{L}(t)$ at $t = -1/2$. The point where L intersects the yz-plane is $\mathbf{L}(-1/2) = (0, 3.5, -4)$. ▲

Example 1.4.16. *Intersecting lines*

Determine if the lines $\mathbf{L}_1(t) = (3t, 2t - 1, -t + 2)$ and $\mathbf{L}_2(t) = (t + 3, t, -t + 3)$ intersect.

If the lines intersect, the component functions must all be equal. There is one subtlety here: if the lines intersect, it may not be at the same time. In other words, bugs flying along these two paths could visit the same point, but not hit each other because they arrive there at different times. To account for this, change one of the parameters to a different letter, say $L_2(s) = (s + 3, s, -s + 3)$. Now equating component functions gives a system of three equations in two variables. A solution to the system corresponds to a point of intersection of the lines. Solving

$$3t = s + 3$$
$$2t - 1 = s$$
$$-t + 2 = -s + 3$$

gives $t = 2$, $s = 3$, which means the lines intersect.

Thus the point of intersection is $L_1(2) = L_2(3) = (6, 3, 0)$. ▲

Parameterizing Surface Intersections: In general the intersection of two surfaces in $\mathbb{R}^3$ will be a curve. In some cases there are effective strategies for parameterizing curves of intersection. We treat three such cases now: where one surface is a plane, where one surface is a generalized cylinder, and where the surfaces are given in cylindrical or spherical coordinates.

Example 1.4.17. *Horizontal slices*

Parameterize the curve of intersection of the elliptical paraboloid $z = x^2 + 4y^2$ and the horizontal plane $z = 36$.

We know we want component functions for the curve. Further, the x- and y-coordinates of the curve of intersection satisfy the equation of the level curve $x^2 + 4y^2 = 36$, which is an ellipse. Thus the x- and y-component functions parameterize the curve

$$\frac{x^2}{36} + \frac{y^2}{9} = 1,$$

and are given by

$$x = 6\cos t$$
$$y = 3\sin t, \ 0 \le t \le 2\pi.$$

Since z is a constant, its component function will be too, and we get the parameterization $C(t) = (6\cos t, 3\sin t, 36)$, $0 \le t \le 2\pi$. ▲

Recall that a generalized cylinder is a surface whose defining Cartesian equation is missing a variable, like $x^2 + z^2 = 9$. Given a second surface, which is the graph of a function $y = f(x, z)$, let C be the curve of intersection with the cylinder. One can parameterize C by first parameterizing the curve $x^2 + z^2 = 9$ in the xz-plane, then substituting the resulting component functions in $f(x, z)$ to find $y(t)$. We illustrate this with some examples.

(a)

(b)

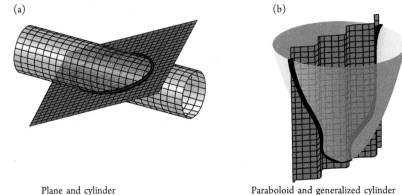

Plane and cylinder

Paraboloid and generalized cylinder

Figure 1.4.5 Parameterizing surface intersections

Example 1.4.18. *A cylinder and a plane*

Let C be the curve of intersection of the cylinder $x^2 + z^2 = 9$ and the plane $x - y + 2z = 10$ as in Figure 1.4.5. We can parameterize C in two steps:

Step 1: Parameterize the curve $x^2 + z^2 = 9$ in the xz-plane.

Step 2: Use the equation $x - y + 2z = 10$ to find $y(t)$.

We now carry out the above steps. The curve $x^2 + z^2 = 9$ can be parameterized using component functions $x(t) = 3\cos t$, $z(t) = 3\sin t$ for $0 \leq t \leq 2\pi$. To accomplish step 2, solve the plane equation for y, giving $y = x + 2z - 10$. Substituting we find the component function $y(t) = 3\cos t + 6\sin t - 10$. Thus the curve C is parameterized by

$$\mathbf{C}(t) = (3\cos t, 3\cos t + 6\sin t - 10, 3\sin t), \text{ for } 0 \leq t \leq 2\pi. \; \blacktriangle$$

Example 1.4.19. *Paraboloid and "washboard"*

We do a second example, employing the above technique. Parameterize the curve of intersection of the generalized cylinder $y = \cos x$ and the paraboloid $z = x^2 + y^2$.

The first step is to parameterize the graph of $y = \cos x$. Recall from Equation 1.4.1 that to parameterize the graph of a function, we let the independent variable be the parameter and use the function to parameterize the dependent variable. In this case we get $x = t$, $y = \cos t$ for $-\infty < t < \infty$. This gives the component functions for x and y, and we use the equation of the surface $z = x^2 + y^2$ to find the component function for z. In this case $z = t^2 + \cos^2 t$, and we summarize:

$$\mathbf{C}(t) = \left(t, \cos t, t^2 + \cos^2 t\right), \text{ for } -\infty < t < \infty. \; \blacktriangle$$

Parameterizing Intersections of Cylinders and Surfaces

Let $g(x,y) = 0$ be a generalized cylinder in $\mathbb{R}^3$, $z = f(x,y)$ a surface, and C be the intersection of the two surfaces. To parameterize C:

1. Parameterize the curve $g(x,y) = 0$ in the xy-plane. This gives component functions $x(t)$ and $y(t)$.

2. Substitute the component functions from step 1 into the function $f(x,y)$ to find the component function $z(t)$.

Note that there is nothing special about the cylinder missing the variable z. The same strategy works for cylinders $g(x,z) = 0$ and surfaces $y = f(x,z)$, etc.

Concept Connection: In some applications the direction of a curve is important. In particular, both Green's theorem and Stokes' theorem require a certain orientation (or direction) on a curve. The reason for this is that many physical quantities have a direction associated with them (velocity, force, etc.), and going the wrong direction gives the negative of the answer you want! We do one example where the direction of the parameterization is specified.

Example 1.4.20. *Paraboloid and cylinder—clockwise orientation*

Parameterize the curve of intersection of the paraboloid $z = -x^2 + 6x - y^2 - 8y$ and the cylinder $x^2 + y^2 = 25$, so that it is oriented clockwise when viewed from above.

The added feature of this example is the requirement that the parameterization be oriented clockwise. To do so, note that when viewing "from above" the z-coordinate disappears and we are looking at the projection in the xy-plane. Thus, the "clockwise from above" requirement just means that our parameterization of the curve $x^2 + y^2 = 25$

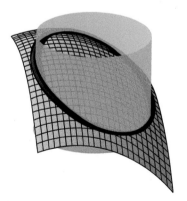

Figure 1.4.6 A clockwise parameterization

in the xy-plane be clockwise. The standard parameterization of the circle is counter-clockwise, but recall that negating the y-component function results in a clockwise parameterization.

Thus step 1 of the process gives $x(t) = 5\cos t$, $y(t) = -5\sin t$, for $0 \le t \le 2\pi$. Substituting these equations in the formula for the paraboloid gives

$$z(t) = -(5\cos t)^2 + 30\cos t - (-5\sin t)^2 + 40\sin t$$
$$= 30\cos t + 40\sin t - 25.$$

Thus $\mathbf{C}(t) = (5\cos t, -5\sin t, 30\cos t + 40\sin t - 25)$, for $0 \le t \le 2\pi$, is a clockwise parameterization of the curve. ▲

We have seen several examples of parameterizing curves of intersection when one of the surfaces is a generalized cylinder. Surfaces given by equations in spherical or cylindrical coordinates can also intersect in curves that are relatively easy to parameterize. The general strategy is to let the independent variable be the parameter and use the conversions between coordinate systems to find component functions.

Example 1.4.21. *Intersecting constant coordinate surfaces*

Parameterize the curve of intersection of the sphere $\rho = 5$ and the half-plane $\theta = \pi/3$.

Parametric curves are given by component functions of a single variable, and there is nothing special about using the letter t for the parameter. Thus substituting the given values for ρ and θ into the conversions gives component functions of the single variable ϕ, which is a parameterization! More explicitly,

$$\mathbf{C}(\phi) = (\rho\sin\phi\cos\theta, \rho\sin\phi\sin\theta, \rho\cos\phi)$$
$$= (5\sin\phi\cos(\pi/3), 5\sin\phi\sin(\pi/3), 5\cos\phi)$$
$$= \left(\frac{5}{2}\sin\phi, \frac{5\sqrt{3}}{2}\sin\phi, 5\cos\phi\right), \ 0 \le \phi \le \pi.$$

If you aren't comfortable using ϕ as a parameter, you can replace it with t for now. However, you do want to recognize that any letter can be used as a parameter. ▲

Example 1.4.22. *An example in cylindrical coordinates*

Parameterize the curve of intersection of $z = r^2$ and $r = \theta$, for $\theta \ge 0$.

Analytically, this is easy to do. Let the independent variable, r, be the parameter. The given equations express all the cylindrical coordinates (r, θ, z) in terms of r, so that $(r, \theta, z) = (r, r, r^2)$. Now substitute these values into the coordinate conversions to get the component functions $x(r)$, $y(r)$, and $z(r)$. Since the result is component functions of one variable, this is a parameterization.

$$\mathbf{C}(r) = (r\cos\theta, r\sin\theta, z)$$
$$= \left(r\cos r, r\sin r, r^2\right), \ 0 \le r$$

(a)   (b)

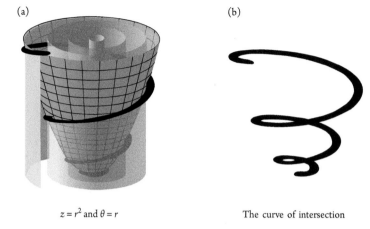

$z = r^2$ and $\theta = r$ The curve of intersection

Figure 1.4.7 Cylindrical coordinate surfaces

It's a little more difficult to see what is going on geometrically. First, the surface $z = r^2$ has Cartesian equation $z = x^2 + y^2$, so is our favorite paraboloid. Second, the *curve* $\theta = r$ in the xy-plane is a spiral since, as θ increases, the distance to the origin does too. Since z is missing from the equation, the surface is obtained by translating the spiral up and down along the z-axis. The resulting surface is a spiral cookie cutter, and the curve of intersection is pictured in Figure 1.4.7. ▲

Parameterizing Curves Using Other Coordinate Systems

To parameterize the intersection of two surfaces given in cylindrical or spherical coordinates:

1. Parameterize the given coordinates (i.e. get them all in terms of one variable).
2. Use conversions to Cartesian coordinates to find the component functions.

We conclude our introduction to parametric curves with the classic example of a helix, describing knots, and determining when a curve lies on a surface. These examples follow.

Example 1.4.23. *Helices*

Everyone who has seen parametric curves has seen the helix, and then shown it to their brother because it was cool. The parametric equations for a helix are

$$C(t) = (\cos t, \sin t, t), \text{ for } -\infty < t < \infty.$$

Note that if you project into the xy-plane (i.e. cover up the z-coordinate), you get parametric equations for the unit circle in the plane. Thus as t increases the bug flying along $\mathbf{C}(t)$ keeps circling the z-axis in a counterclockwise fashion when viewed from above. The third component function $z(t) = t$ means that the bug is increasing in altitude the entire time.

More general helicies have parametric equations $\mathbf{C}(t) = (a\cos t, a\sin t, bt)$ for constants a and b. These all live on the cylinders given by $x^2 + y^2 = a^2$. How might you parameterize a helix on the elliptical cylinder $\frac{x^2}{4} + y^2 = 1$? ▲

Example 1.4.24. *Knots*

Knots are curves that don't intersect themselves, and end where they started. Lissajous knots are knots with particularly easy parameterizations; they are always of the form:

$$x = \cos(At + x_0)$$
$$y = \cos\left(Bt + y_0\right)$$
$$z = \cos(Ct + z_0), \ 0 < t < 2\pi.$$

The Lissajous knot $\mathbf{C}(t) = (\cos(3t + 1.5), \cos(2t + .2), \cos(5t))$ is called a Stevedore knot and is pictured in Figure 1.4.8. You can change the constants in a Lissajous parameterization and try to get different knots. ▲

Example 1.4.25. *Analytically verifying curves on surfaces*

To show that the parametric line $\mathbf{C}(t) = (1, t, t)$ lies on the one-sheeted hyperboloid $x^2 + y^2 - z^2 = 1$, just substitute the parametric equations for the line into the Cartesian coordinates and verify that the equation is satisfied. In this case we see $x^2 + y^2 - z^2 = 1 + t^2 - t^2 = 1$ satisfies the equation. To see that the line $\mathbf{C}(t) = (t + 1, 0, -t)$ doesn't lie on the surface, substitute and simplify to get $x^2 + y^2 - z^2 = (1 + t)^2 + 0^2 - (-t)^2 = 1 + 2t + t^2 - t^2 = 1 + 2t$. Since the result is not identically 1, this does not lie on the surface. In fact, the only value of t which makes this one is $t = 0$. This means that the only point where the line intersects the surface is $\mathbf{C}(0) = (1, 0, 0)$. These lines and the surface are pictured in Figure 1.4.9.

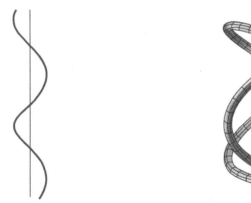

Figure 1.4.8 A helix and Stevedore's knot

Figure 1.4.9 Curves on surfaces

To show that the curve $\mathbf{C}(t) = (\sin t \cos t, \sin^2 t, \cos t)$ lies on the unit sphere $x^2 + y^2 + z^2 = 1$, we substitute and simplify.

$$x^2 + y^2 + z^2 = (\sin t \cos t)^2 + \sin^4 t + \cos^2 t$$
$$= \sin^2 t \left(\cos^2 t + \sin^2 t\right) + \cos^2 t$$
$$= \sin^2 t + \cos^2 t = 1. \; \blacktriangle$$

Things to know/Skills to have

In this section we studied parametric curves in $\mathbb{R}^2$ and $\mathbb{R}^3$. You know you understand the material of this section when you are able to:

- Parameterize ellipses clockwise and counterclockwise.
- Parameterize graphs of $y = f(x)$.
- Parameterize graphs of polar equations $r = f(\theta)$.
- Reparameterize a given curve.
- Parameterize curves backwards.
- Parameterize lines in $\mathbb{R}^3$.
- Parameterize curves of intersection of surfaces whether they are planes, generalized cylinders, or surfaces in other coordinate systems.
- Determine if a parametric curve lies on a given surface.

Exercises

1. Parameterize the circle $x^2 + y^2 = 9$ clockwise. Now parameterize it counterclockwise.

2. Parameterize the circle $(x-2)^2 + y^2 = 4$. (Hint: how should you change the parameterization of $x^2 + y^2 = 4$ to get the desired result?)

3. Parameterize the circle $(x+3)^2 + (y-1)^2 = 25$.

4. Find parametric equations for the circle in the xy-plane with center (h,k) and radius r (i.e. the circle $(x-h)^2 + (y-k)^2 = r^2$).

5. Parameterize the ellipse $\frac{x^2}{16} + \frac{y^2}{9} = 1$ counterclockwise.

6. Parameterize the ellipse $4x^2 + 9y^2 = 1$ clockwise.

7. Parameterize the ellipse $9x^2 + z^2 = 1$ in the xz-plane.

8. Parameterize the curve in $\mathbb{R}^2$ given by the function $y = x^4 + 8x$.

9. Parameterize the curve in $\mathbb{R}^2$ given by the function $y = \tan^{-1} x$.

10. Parameterize $y = \dfrac{x+1}{x^2+4}$.

11. Parameterize the graph of $y = \sqrt{1-x^2}$, be sure to include limits on your parameter.

12. Parameterize one component of the hyperbola $y^2 - x^2 = 1$ using the tangent and secant functions.

13. Parameterize one component of $x^2 - y^2 = 4$ using the cotangent and cosecant functions.

14. Parameterize the left component of $\frac{x^2}{4} - \frac{y^2}{9} = 1$.

15. Find parametric equations for the cardioid $r = 1 + \cos\theta$.

16. Parameterize the four-leaf rose $r = \cos 2\theta$. Now parameterize just the petal pointing straight up.

17. Recall that the stream function for an ideal fluid flow around a $\pi/3$-corner in polar coordinates is $\psi = r^3 \sin 3\theta$, and that level curves are streamlines of the flow (paths that particles in the flow follow). Parameterize the streamline $1 = r^3 \sin 3\theta$.

18. The stream function for an ideal fluid flow around a $\pi/2$-corner is $\psi = 2xy$, so level curves are of the form $c = 2xy$.
    (a) Show that $\mathbf{C}(t) = \sqrt{c/2}\,\langle 1/t, t \rangle$, $t > 0$ parameterizes the streamlines. (Don't forget the scalar multiplication when finding component functions!)
    (b) Show that $\mathbf{C}(t) = \sqrt{c/2}\,\langle e^{-t}, e^t \rangle$, $-\infty < t < \infty$ parameterizes the streamlines as well.

19. Parameterize the spiral $r = \frac{\theta}{2\pi}$, for $\theta \geq 0$. Where does this curve intersect the positive x-axis? The negative x-axis?

20. Parameterize the infinite spiral $r = e^\theta$.

21. Reparameterize the curve $\mathbf{C}(t) = (2\cos t, 3\sin t)$, $0 \leq t \leq 2\pi$, to travel twice as fast.

22. Reparameterize $\mathbf{C}(t) = (\cos t, \sin t, t)$, $0 \leq t \leq 2\pi$, to travel one-third as fast.

23. Reparameterize $\mathbf{C}(t) = (\sec t, \tan t)$, $0 \leq t \leq \pi/4$:
    (a) so that t-limits are $0 \leq t \leq 1$

(b) using $f(s) = \tan^{-1} s$

(c) backwards.

24. Show that the substitution $t = s^u$, for $u > 0$, reparameterizes the curve $C(t) = (x(t), y(t))$, $0 \le t \le 1$ with the limits remaining $0 \le s \le 1$.

25. Two particles travel along $C_1(t) = (t, \sqrt{4 - t^2}), 0 \le t \le 2$ and $C_2(t) = (2t, 2\sqrt{1 - t^2})$, $0 \le t \le 1$. Show that the particles travel the same path, and describe the difference between their travels.

26. Where do the following lines intersect the coordinate planes?
    (a) $C(t) = (2t + 4, -t - 5, 3t + 1)$
    (b) $C(t) = (2t, -4t, 7t)$
    (c) $C(t) = (3t + 1, 2, -t)$

27. Determine if the lines $C_1(t) = (t + 1, -2t, t - 2)$ and $C_2(t) = (t, 2 - t, t - 3)$ intersect.

28. Determine if the lines $C_1(t) = (3t, 2t + 1, 2 - t)$ and $C_2(t) = (t + 2, 2 - t, 2t - 1)$ intersect.

29. Without doing any algebra, determine whether or not the lines $C_1(t) = (4t, 3, -t + 1)$ and $C_2(t) = (1 + t, -1, t + 2)$ intersect.

30. At what point does the line $C(t) = (4t + 1, 2t - 3, t + 4)$ intersect the plane $x - y + 2z = 8$?

31. At what point does the helix $C(t) = (\cos t, \sin t, t)$ intersect the paraboloid $z = x^2 + y^2$?

32. Parameterize the curve of intersection of the cylinder $x^2 + y^2 = 4$ and the plane $2x + y - 6z = 6$ clockwise when viewed from above.

33. Parameterize the curve of intersection of the cylinder $r = 4$ and the paraboloid $z = 4 - x^2 - 4x - y^2$ clockwise when viewed from above.

34. Parameterize the intersection of the generalized cylinder $z = x^2$ and the plane $2x - y + 3z = 4$.

35. Parameterize the intersection of the generalized cylinder $9y^2 + z^2 = 36$ and the surface $x = y^2 + z^2 + 2y$.

36. Parameterize the curve of intersection of the generalized cylinder $x = \sin z$ and the surface $y = x^2 - z^2$.

37. Parameterize the curve of intersection of the plane $z = 6$ and the half-plane $\theta = \pi/6$. Be careful with the limits on your parameterization.

38. Parameterize the curve of intersection of the cylinder $r = 3$ and the half-plane $\theta = -\pi/4$.

39. Parameterize the curve of intersection of the plane $z = -3$ and the cylinder $r = 2$, clockwise when viewed from above.

40. Parameterize, counterclockwise when viewed from above, the curve of intersection of the cone $\phi = \pi/3$ and the sphere $\rho = 2$.

41. Parameterize the curve of intersection of the sphere $\rho = 3$ and the half-plane $\theta = 2\pi/3$.

42. Parameterize the curve of intersection of the cone $\phi = \pi/4$ and the half-plane $\theta = 5\pi/4$.

43. Show that the curve $\mathbf{C}(t) = (t\cos t, t\sin t, t)$, for $t \geq 0$, lies on the cone $z = \sqrt{x^2 + y^2}$.

44. Show that the curve $\mathbf{C}(t) = (t\cos t, t^2, t\sin t)$, for $t \geq 0$, lies on the paraboloid $y = x^2 + z^2$.

1.5 Parametric Surfaces

A curve in space is one-dimensional, so you need one parameter to describe it. A surface is two-dimensional, so the parametric equations will be functions of two variables: $S(s,t) = (x(s,t), y(s,t), z(s,t))$. More formally we have

Definition 1.5.1. *A parametric surface is one defined by two-variable component functions $S(s,t) = (x(s,t), y(s,t), z(s,t))$, together with limits on the parameters $a \leq s \leq b$, $c \leq t \leq d$.*

It should be noted that the individual component functions could be constant or have just one variable, as long as the component functions combined involve two variables. Thus $S(s,t) = (s^3 - 2, t^2 + 1, 5)$, $0 \leq s,t \leq 1$ is a parametric surface, but $C(s,t) = (t^3 - 2t, \cos t, t^2)$ and $X(s,t,u) = (t^2, s^2 u^3, t - s + u)$ aren't (one has too few variables, the other too many).

The "bug flying" interpretation that works for parametric curves doesn't work so well for parametric surfaces. The bending wire interpretation, however, can be generalized nicely to this setting. Instead of wire, imagine starting with a sheet of aluminum foil. The parametric equations are then instructions for how to bend it into space (foil doesn't stretch, so maybe elastic fabric is a better choice). More precisely, the domain of a parametric surface $S(s,t)$ is a subset of the st-plane, while the range lives in $\mathbb{R}^3$. In this section we give several examples and strategies for parameterizing surfaces.

Example 1.5.1. *Planes*

As parametric equations for lines are linear functions in t, parametric equations for planes are linear functions in two variables. They have the form:

$$x = As + Dt + x_0$$
$$y = Bs + Et + y_0$$
$$z = Cs + Ft + z_0, \quad -\infty < s, t < \infty$$

Letting $s = t = 0$ shows that (x_0, y_0, z_0) is a point on the plane. We'll see in Chapter 2 that the coefficients of s determine a direction in the plane, as do the coefficients of t. ▲

Constant Coordinate Surfaces: Parameterizing constant coordinate surfaces is straightforward. An immediate substitution parameterizes the plane $y = -3$:

$$S(x,z) = (x, -3, z), \quad -\infty < x, z < \infty.$$

Parameterizations of the planes $x = c$ and $z = d$ are obtained similarly.

We will use parameters s and t for many parameterizations in this book; however, that is not necessary as long as there are two variables. In this case it's easier, and more illuminating, to use the variables x and z as parameters.

To parameterize constant coordinate surfaces in other coordinate systems, you again substitute the given constant. Now, however, you substitute it in the relevant coordinate system. For example, to parameterize the sphere $\rho = 5$ you get:

$$S(\theta,\phi) = (5\sin\phi\cos\theta, 5\sin\phi\sin\theta, 5\cos\phi),\ 0 \le \phi \le \pi, 0 \le \theta \le 2\pi.$$

We point out that when $\phi = 0$ we have $S(\theta,0) = (0,0,5)$ for all θ. Similarly, $S(\theta,\pi) = (0,0,-5)$ for all θ. The fact that many points in the domain get mapped to the same point in the image is not desireable. In this case it makes little difference because they are isolated points on the sphere, but it is something to be aware of when determining limits of a parameterization. For example, if we let ϕ range from 0 to 2π every point on the sphere would be hit twice since $S(\theta,\phi) = S(\theta + \pi, 2\pi - \phi)$.

As in the case of parametric curves, there is a notion of a *regular* parametric surface that avoids difficulties such as these. Regularity for surfaces, however, cannot be defined until after discussing cross-products and partial derivatives.

Parameterizing Constant Coordinate Surfaces

To parameterize a constant coordinate surface, substitute the constant into the conversion to Cartesian coordinates.

One last example will round out this paragraph: A parameterization of $r = 7$ is

$$S(s,t) = (7\cos t, 7\sin t, s),\ 0 \le t \le 2\pi,\ -\infty < s < \infty.$$

Parameterizing Graphs: In Section 1.2 we saw that the graph of a function of two variables is a surface in $\mathbb{R}^3$. Parametric equations for the surface $z = f(x,y)$ are obtained in the same way that we parameterize the curve $y = f(x)$ in the plane. Simply let the independent variables be the parameters, and use the function $f(x,y)$ to parameterize z.

Example 1.5.2. *Parameterizing a Saddle*

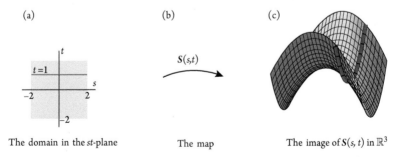

(a)   (b)   (c)

The domain in the *st*-plane   The map   The image of $S(s,t)$ in $\mathbb{R}^3$

Figure 1.5.1 A parameterized saddle

Using the above strategy we see that one way to parametrize the saddle $z = x^2 - y^2$ is $S(s,t) = (s,t,s^2 - t^2)$. Again, this seems artificial, but represents thinking of the domain and surface as living in separate spaces. The idea is illustrated in Figure 1.5.1 for the restrictions $-2 \leq s, t \leq 2$.

We note that parametric equations for $S(s,t)$ map curves in the domain to curves on the surface. For example, letting $t = 1$ in the domain gives a horizontal line in the st-plane which $S(s,t)$ maps to the thickened curve on the surface given parametrically by $S(s,1) = (s,1,s^2 - 1)$. ▲

Parameterizing Graphs

To parameterize $z = f(x,y)$ let the independent variables be the parameters and use the function to parameterize the dependent variable.

Example 1.5.3. *Using different coordinates*

This approach to parameterizing surfaces works for functions of any two Cartesian coordinates, not just x and y. We illustrate this by parameterizing the surface $y = \sin(x^2 - 2x + z^2)$.

In this case, the independent variables are x and z so they become the parameters, and we have $S(x,z) = (x, \sin(x^2 - 2x + z^2), z)$ for $-\infty < x, z < \infty$. Recall that parameters can be any variables we choose, so long as the component functions use two variables combined, thus we can leave them as x and z. The surface is pictured in Figure 1.5.2 for $-2 \leq x, z \leq 2$. ▲

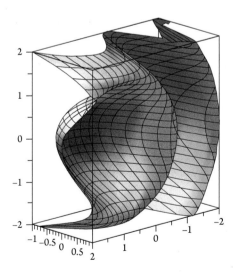

Figure 1.5.2 The surface $y = \sin(x^2 - 2x + z^2)$

Math App 1.5.1. *A parametric paraboloid*

In this Math App you view the paraboloid $z = x^2 + y^2$ parameterized by

$$S(s,t) = (s\cos t, s\sin t, s^2), \ 0 \le s \le 2, \ 0 \le t \le 2\pi.$$

Direct substitution verifies that the parametric equations satisfy the Cartesian equation:

$$\begin{aligned} x^2 + y^2 &= (s\cos t)^2 + (s\sin t)^2 & \text{substitution}\\ &= s^2 \left(\cos^2 t + \sin^2 t\right) = s^2 & \text{algebra}\\ &= z. & \text{un-substitution} \end{aligned}$$

To develop an intuition for how the parametric equations bend, twist, and stretch the rectangle $0 \le s \le 2$, $0 \le t \le 2\pi$ onto the paraboloid $z = x^2 + y^2$, click on the hyperlink below or print users open manually.

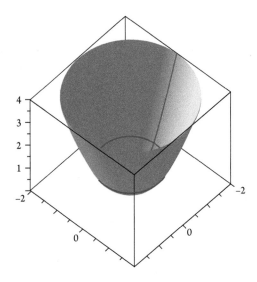

Concept Connection: It turns out that thinking of graphs parametrically will be useful when using calculus to study surfaces. In particular, any formula we derive for parametric surfaces will yield a formula for the graph $z = f(x,y)$ by thinking of it as the parametric surface $S(x,y) = (x,y,f(x,y))$. Such formulas include finding tangent planes to surfaces, determining surface area, and much more. This becomes a powerful tool for remembering the myriad of formulas we encounter in this course!

Parameterizing Generalized Cylinders: Recall that generalized cylinders are surfaces whose Cartesian equations are missing a variable. We can think of them as a planar curve that gets translated back and forth along a perpendicular axis. The only way you know it's a surface, rather than a curve, is by the context. To parameterize a cylinder, first parameterize the planar curve using a single parameter, then let the missing variable be the second parameter. We illustrate the process with an example.

Example 1.5.4. *Parametric generalized cylinders*

Parameterize the cylinder $z = x^2 - 3x - 10$. First we parameterize the parabola in the xz-plane. Since z is a function of x in this case, we let $x = s$ and $z = s^2 - 3s - 10$. To finish parameterizing the surface let $y = t$, yielding the parametric surface $S(s,t) = (s, t, s^2 - 3s - 10)$ for $-\infty < s, t < \infty$.

Similarly, to parameterize the cylinder $y^2 + z^2 = 1$ start by parameterizing the unit circle in the yz-plane then let x be the second parameter. One parameterization is $S(s,t) = (s, \cos t, \sin t)$, with $0 \leq t \leq 2\pi$, $-\infty < s < \infty$. ▲

Parameterizing Generalized Cylinders

To parameterize a generalized cylinder, parameterize the curve given by the equation and let the remaining variable be the second parameter.

Other Coordinate Systems: We have already seen that one can use the change from cylindrical or spherical to Cartesian coordinates to parameterize constant coordinate surfaces. Similar techniques can be used to parameterize surfaces given by equations in other variables. We illustrate with examples.

Example 1.5.5. *A corkscrew*

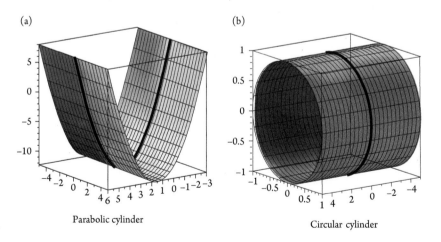

(a) Parabolic cylinder

(b) Circular cylinder

Figure 1.5.3 Two cylinders

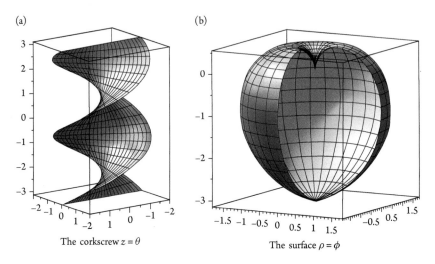

(a)

(b)

The corkscrew $z = \theta$

The surface $\rho = \phi$

Figure 1.5.4 Parameterizing cylindrical and spherical equations

The surface given by the cylindrical equation $z = \theta$ can be parameterized by replacing θ with z in the change of coordinates formula to get $\mathbf{S}(r,z) = (r\cos z, r\sin z, z)$. This surface looks better if we allow r to be negative, and it is plotted in Figure 1.5.4(a) for $-2 \le r \le 2$ and $-3 \le z \le 3$. ▲

Parameterizing Surfaces: Equations in other coordinate systems

Given an equation in cylindrical or spherical coordinates:

1. Use the equation to express one of the coordinates in terms of the other two, then

2. Use the conversion to Cartesian to parameterize the surface.

Example 1.5.6. *Parametric peach*

The surface $\rho = \phi$, for $0 \le \phi \le \pi$, looks somewhat like an apricot or peach.

Using the equation we can replace ρ with ϕ, and let θ be the other parameter. We get the parameterization

$$\mathbf{S}(\theta,\phi) = (\phi\sin\phi\cos\theta, \phi\sin\phi\sin\theta, \phi\cos\phi),\ 0 \le \phi \le \pi,\ 0 \le \theta \le 2\pi.$$

Recall that $x = \rho\sin\phi\cos\theta$, $y = \rho\sin\phi\sin\theta$, and $z = \rho\cos\phi$). In Figure 1.5.4 we plotted the peach for $0 \le \theta \le 3\pi/2$, $0 \le \phi \le \pi$, to highlight the inside. ▲

Surfaces of Revolution: A surface of revolution is obtained by revolving a curve around an axis. It turns out that such surfaces have nice parameterizations. The first step is to parameterize the curve that is being revolved, which requires one parameter. The second parameter is used to carry out the revolution. Below is a procedure that works for the case where the z-axis is the axis of revolution.

Parameterizing Surfaces of Revolution

Let S be a surface of revolution about the z-axis. To parameterize S:

1. Parameterize the curve C of S in the half-plane $\theta = 0$ by $(x(s), 0, z(s))$ for $a \le s \le b$.

2. Parametric equations for S are

$$\mathbf{S}(s,t) = (x(s)\cos t, x(s)\sin t, z(s)), \; a \le s \le b, \; 0 \le \theta \le 2\pi.$$

We remark that one could use different half-planes to find curves C in step one. Another natural choice would be the half-plane $\theta = \pi/2$. We now describe why this process works.

Let S be a surface obtained by revolving a curve around the z-axis. We will use the technique of parameterizing surfaces in other coordinate systems to parameterize S. Thus we need to find component functions for r, θ, and z, then use coordinate conversion to get the parameterization. First note that θ should be one of the parameters. Indeed, changing the θ coordinate of a point on the surface gives another point on the surface since S is a surface of revolution. Thus we let $\theta = t$ be one of the parameters.

To parameterize r and z, note that the half-plane $\theta = 0$ intersects S in a curve C. Since C lies in $\theta = 0$, the y-coordinate of points on C is 0 and one can use curve techniques to find component functions $x(s)$ and $z(s)$. This obviously gives us $z(s)$, but we also get $r(s)$

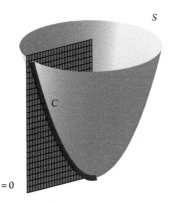

Figure 1.5.5 Parameterizing a surface of revolution

for free! Recall that the cylindrical coordinate r is the distance to the z-axis. In the half-plane $\theta = 0$ the x-coordinate gives the distance to the z-axis: so in this half-plane $r = x$. Therefore $r = x(s)$ is the desired parameterization for r. Now that we have parametric equations for r, θ, and z, using coordinate conversion parameterizes the surface, as in step 2 above. This discussion justifies the above procedure.

We briefly note that this technique is more general than we've described. For example, if the half-plane $\theta = \pi/2$ is used in step one, the only difference is that the y-coordinate gives r. More precisely, suppose $\mathbf{C}(s) = (0, y(s), z(s))$, $a \leq s \leq b$, parametrizes a curve in the yz-plane. One parameterization of the surface obtained by revolving $\mathbf{C}(s)$ around the z-axis is

$$\mathbf{S}(s,t) = \big(y(s)\cos t, y(s)\sin t, z(s)\big), \ a \leq s \leq b, \ 0 \leq \theta \leq 2\pi.$$

Additionally, if the x- or y-axes are used for the revolution, one uses similar techniques with "cylindrical" coordinates defined appropriately.

Example 1.5.7. *Revolving the hyperbolic cosine curve*

The hyperbolic cosine function is the even function defined by $\cosh x = (e^x + e^{-x})/2$. To get a surface, rotate the curve $z = \cosh x$, for $x \geq 0$, about the z-axis.

Our goal is to find parametric equations for this surface, and a first step is parameterizing the curve. To parameterize the function, let the independent variable be the parameter and use the function to parameterize the dependent variable. Thus the given curve has parametric equations $(x, y, z) = (s, 0, \cosh s)$, for $s \geq 0$, and is the darkened curve in Figure 1.5.6(a). Applying step 2 of the process gives the parameterization

$$\mathbf{S}(s,t) = (s\cos t, s\sin t, \cosh s), \ s \geq 0, \ 0 \leq t \leq 2\pi. \quad \blacktriangle$$

(a)   (b)

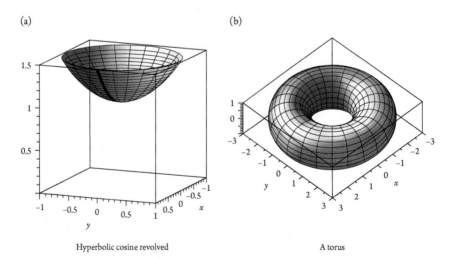

Hyperbolic cosine revolved   A torus

Figure 1.5.6 Surfaces of revolution

Example 1.5.8. *Parameterizing a torus*

The curve $z^2 + (y-2)^2 = 1$ is a circle in the yz-plane with center $(2,0)$ and radius 1 (note that we are not thinking of this as a "generalized cylinder" here). It has parametric equations $y = \cos s + 2$, $z = \sin s$, for $0 \le s \le 2\pi$. To rotate it around the z-axis, use cylindrical coordinates as above with $r = \cos s + 2$, $\theta = t$, and $z = \sin s$. The change to Cartesian coordinates gives the parametric surface

$$S(s,t) = ((\cos s + 2)\cos t, (\cos s + 2)\sin t, \sin s), \ 0 \le s, t \le 2\pi. \ \blacktriangle$$

The torus is interesting enough that we'll take a closer look at it. The grid lines in Figure 1.5.6(*b*) come about by fixing one of the parameters, and letting the other run. In the case of the torus, the grid lines that go through the hole are called meridians and those going around the hole are longitudes. To test your understanding, determine whether the grid line $s = \pi$ is a meridian or longitude!

The torus in Figure 1.5.6(*b*) is the image of the square $0 \le s, t \le 2\pi$ in the st-plane. Why do the parametric equations imply that the left and right sides of the square map to the same grid line, where the left side is $s = 0$, and the right is $s = 2\pi$? They also show that the top and bottom of the square map to the same longitude. You can get more interesting curves by letting $t = 2u$, $s = 3u$ for a new parameter u with $0 \le u \le 2\pi$. The result is a curve that goes twice around the hole and three times through it. This is equivalent to considering the image of the line segments below, each of which has slope $3/2$ and the ends get glued when identifying top and bottom, and left and right. In fact, you can choose any two relatively prime integers p, q (not just 2 and 3), and do the same thing. The resulting curve is called a pq-torus knot because it lives on the surface of a torus.

(a)

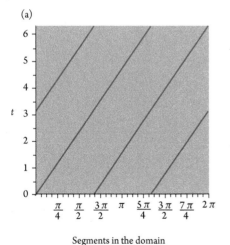

Segments in the domain

(b)

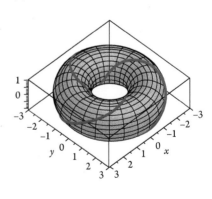

A torus knot: the image of the segments

Figure 1.5.7 A torus knot

Parameterizing Surfaces with Restricted Domain: In many applications we will be concerned with only portions of surfaces. For example, we may want to consider the portion of a plane that lies over a disk in the xy-plane, or that part of a paraboloid over a rectangle. The restrictions on the domain usually change the limits of your parameterization, leaving the component functions the same. To find the limits on the parameters, one needs to describe regions in the plane using systems of inequalities. This topic is discussed more fully in section 1.6, but we illustrate here with a few examples.

Example 1.5.9. *Portions of a plane*

Parameterize the portion of the plane $3x - 2y - 4z = -8$ over the rectangle $0 \le x \le 4, -2 \le y \le 3$.

To parameterize the plane, we can solve for z and think of it as the graph of $z = (3x - 2y + 8)/4$. Letting x and y be parameters, and using the function to find the third component function, we get the parameterization of the entire plane:

$$S(x,y) = \left(x, y, \frac{3x - 2y + 8}{4} \right), \quad -\infty < x, y < \infty.$$

The restriction to the given rectangle changes the limits of the parameterization, but not the component functions, so we have (see Figure 1.5.8(a))

$$S(x,y) = \left(x, y, \frac{3x - 2y + 8}{4} \right), \quad 0 \le x \le 4, -2 \le y \le 3.$$

Now suppose we want to parameterize the portion of the same plane that lies above the disk radius 3 in the xy-plane (i.e. all points satisfying $x^2 + y^2 \le 9$). One way to do this is to describe the disk using a system of inequalities involving x and y. For example, the y-coordinate is always between the top and bottom semicircles, so $-\sqrt{9 - x^2} \le y \le \sqrt{9 - x^2}$. Since x is between ± 3, one parameterization is

$$S(x,y) = \left(x, y, \frac{3x - 2y + 8}{4} \right),$$

(a)   (b)   (c)

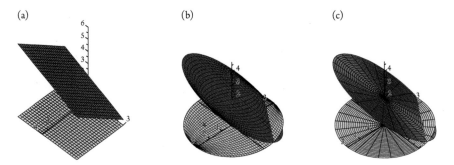

Figure 1.5.8 Restricted domain parameterization

with the limits on the parameterization being

$$-3 \le x \le 3, \ -\sqrt{9-x^2} \le y \le \sqrt{9-x^2}.$$

The disadvantage to this parameterization is that the limits on y are ugly (see Figure 1.5.8(b)). We can use polar coordinates instead to make the limits more appealing.

In polar coordinates, the disk is described by the system of inequalities $0 \le r \le 3$, $0 \le \theta \le 2\pi$. If we use r and θ as our parameters, the limits will be nice. To do so, we just use the conversions to Cartesian coordinates $x = r\cos\theta$, $y = r\sin\theta$. Making the substitution we get the alternate parameterization (see Figure 1.5.8(c))

$$\mathbf{S}(r,\theta) = \left(r\cos\theta, r\sin\theta, \frac{3r\cos\theta - 2r\sin\theta + 8}{4} \right),$$

with limits

$$0 \le r \le 3, \ 0 \le \theta \le 2\pi.$$

Take a moment to compare the grid lines in Figures 1.5.8(b) and 1.5.8(c). Why are they different? Which is more aesthetically pleasing? ▲

Example 1.5.10. *Parameterizing a paraboloid portion*

Parameterize the portion of the paraboloid $z = x^2 + y^2 - 2x$ lying above the square $-1 \le x, y \le 1$, and above the unit disk $x^2 + y^2 \le 1$. The different parameterizations are pictured in Figure 1.5.9.

As in the previous example we use Cartesian coordinates to parameterize the paraboloid over the square, and polar over the disk. Over the square we have

$$\mathbf{S}(x,y) = (x,y,x^2+y^2-2x), \ -1 \le x, y \le 1.$$

(a)

(b)

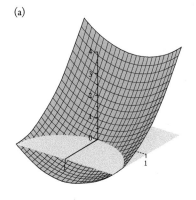
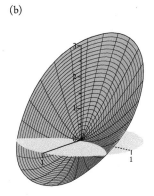

Figure 1.5.9 Paraboloid parameterizations

Substituting coordinate conversions, and using appropriate limits, gives

$$S(s,t) = (s\cos t, s\sin t, s^2\cos^2 t + s^2\sin^2 t - 2s\cos t)$$
$$= (s\cos t, s\sin t, 1 - 2\cos t),\ 0 \le s \le 1,\ 0 \le t \le 2\pi.\ \blacktriangle$$

Things to Know/Skills to have

- A parametric surface is one given by 2-variable component functions

$$S(s,t) = \big(x(s,t), y(s,t), z(s,t)\big),\ a \le s \le b,\ c \le t \le d$$

- **Constant coordinate surfaces** are parameterized using conversions to Cartesian coordinates.
- One parameterization of the **graph** $z = f(x,y)$ is obtained by letting the independent variables be the parameters and using the function to parameterize the dependent variable.
- Parameterize **generalized cylinders** by parameterizing the corresponding curve with one parameter, and letting the other parameter be the missing variable.
- To parameterize surfaces given in **other coordinate systems**, first parameterize the given coordinates, then convert to Cartesian using coordinate conversions.
- To parameterize a **surface of revolution**, first parameterize the curve in a plane, then use rotational symmetry to finish the parameterization.
- Parameterize surfaces with a restricted domain.

Exercises

1. Parameterize the following constant-coordinate surfaces. Be sure to include limits on your parameters.
   (a) The cylinder $r = 3$.
   (b) The half-plane $\theta = 5\pi/6$.
   (c) The sphere $\rho = 5$.
   (d) The cone $\phi = \pi/4$.

2. Parameterize the ellipsoid $\frac{x^2}{4} + y^2 + \frac{z^2}{9} = 1$. (Hint: Stretch the parameterization of the unit sphere, in the same way ellipses are stretched circles.)

3. Parameterize the ellipsoid $x^2 + 9y^2 + 4z^2 = 36$.

4. Parameterize the plane $3x - y + 2z = 6$ (Hint: Think of y as a function of x and z.)

5. Parameterize the plane with intercepts $(-2,0,0)$, $(0,3,0)$, $(0,0,6)$.

6. (a) Parameterize the surface $z = x^2 + y^2$ by thinking of z as a function of the other variables.

   (b) Show that $S(s,t) = (s\cos t, s\sin t, s^2)$ parameterizes $z = x^2 + y^2$ as well.

7. Verify that $S(s,t) = (\cosh s \cos t, \cosh s \sin t, \sinh s)$ parameterizes the one-sheeted hyperboloid $x^2 + y^2 - z^2 = 1$.

8. Parameterize the following generalized cylinders. Be sure to include limits on your parameters.

   (a) $z = e^y$.

   (b) $y = \sin x$.

   (c) $y^2 + z^2 = 1$.

9. Parameterize the surface obtained by revolving the line $z = 2x + 1$ around the z-axis.

10. Parameterize the surface obtained by revolving the curve $z = e^{y^2}$ for $y \geq 0$ about the z-axis.

11. Parameterize the torus obtained by revolving the circle $(y - 3)^2 + z^2 = 4$ around the z-axis.

12. Parameterize the one-sheeted hyperboloid $x^2 + y^2 - z^2 = 1$, by thinking of it as one component of the curve $x^2 - z^2 = 1$ revolved around the z-axis.

13. Parameterize the surface given by the cylindrical equation $z = \cos r$. Sketch it using computer software.

14. Parameterize the surface given by $\rho = 5 + \cos 10\phi$ for $0 \leq \phi \leq \pi$. Sketch it using computer software.

15. Parameterize the surface given by the spherical equation $\theta = 2\phi$. Sketch it using computer software.

16. Parameterize the portion of the plane $3x + 2y + 5z = 30$ above the disk $x^2 + y^2 \leq 9$.

17. Parameterize the portion of the paraboloid $z = x^2 + 4x + y^2$ that lies above the unit square with vertices $(0,0)$, $(1,0)$, $(1,1)$, and $(0,1)$.

18. Parameterize the portion of the plane $x - 2y - 3z = 4$ that lies inside the cylinder $r = 2$.

19. Parameterize the portion of the sphere $x^2 + y^2 + z^2 = 9$ that lies above the unit disc.

20. Parameterize the top portion of the sphere $\rho = 2$ that lies inside the cone $z = r$.

1.6 Describing Regions

In this section we develop the skill of describing regions using systems of inequalities, which will be necessary for setting up limits of integration in multiple integrals. We begin with a few straightforward examples of regions in the plane. These are followed by solids in space, which also serve as a nice review of surfaces in space.

Planar Regions: Double integrals are introduced in Section 4.1, while techniques for calculating them appear in Section 4.2. In order to find limits of integration in double integrals, we will have to describe planar regions using systems of inequalities. At this stage most students are familiar with the plane, so we move directly to examples.

Example 1.6.1. *A region in the plane*

Find a system of inequalities that describes the region R in the first quadrant under $y = 4 - x^2$ (see Figure 1.6.1).

There are two approaches to this problem, depending on how you view the region, and we illustrate both.

Thinking of the x-axis as the bottom of the region and the parabola as the top, we see that the y-coordinate of any point in the region satisfies $0 \le y \le 4 - x^2$. After making this observation, look at the shadow of the region in the x-axis to find limits on x. The restrictions on y are valid for $0 \le x \le 2$. Thus R can be described by the system of inequalities:

$$0 \le y \le 4 - x^2 \qquad (1.6.1)$$
$$0 \le x \le 2$$

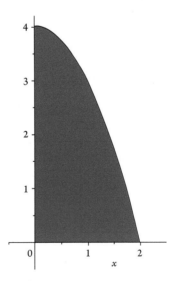

Figure 1.6.1 Parabolic region

The region R can also be thought of as being delimited by the y-axis on the left and the parabola on its right. This point of view restricts the x-coordinate of points in R. To find the system of inequalities on x we have to solve the parabola's equation for x, giving $x = \sqrt{4-y}$. Thus the x-coordinate of any point in the region satisfies $0 \le x \le \sqrt{4-y}$. Now consider the shadow of the region on the y-axis to see that this inequality is valid for $0 \le y \le 4$. We end up with the system

$$0 \le x \le \sqrt{4-y} \qquad\qquad (1.6.2)$$
$$0 \le y \le 4. \ \blacktriangle$$

In Chapter 4 switching perspectives between the systems in Equations 1.6.1 and 1.6.2 will correspond to changing the order in which you integrate. It will be beneficial, then, to be comfortable with both methods.

The above technique generalizes. If R is a region in the plane with top $y = f(x)$ and bottom $y = g(x)$, and its shadow on the x-axis is the interval $a \le x \le b$, then R is defined by the system of inequalities

$$g(x) \le y \le f(x)$$
$$a \le x \le b.$$

If R is a region in the plane with right side $x = f(y)$ and left side $x = g(y)$, and the shadow on the y-axis is the interval $c \le y \le d$, then R is defined by the system of inequalities

$$g(y) \le x \le f(y)$$
$$c \le y \le d.$$

Example 1.6.2. *A more complicated region*

Let R be the region in the first quadrant bounded by $y = 2$ and $y = 4 - 2x$. Describe R using a system of inequalities (see Figure 1.6.2(a)).

We illustrate this using both techniques outlined above. If we want to use the "top/bottom" approach, you notice that the top function changes from $y = 2$ to $y = 4 - 2x$ at $x = 1$. To accurately describe R we have to use two systems of inequalities. The region R consists of two regions R_1 and R_2, described by the two systems of inequalities,

Region R_1	Region R_2
$0 \le y \le 2$	$0 \le y \le 4 - 2x$
$0 \le x \le 1$	$1 \le x \le 2.$

On the other hand, the region R has just one left and right side. Therefore it can be described using a single system of inequalities using a "left/right" approach. To do this,

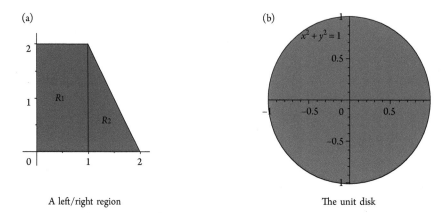

(a)

R₁

R₂

A left/right region

(b)

$x^2 + y^2 = 1$

The unit disk

Figure 1.6.2 Two regions

the equations for the left and right curves must be solved for x. Solving the right side for x gives $x = 2 - \frac{1}{2}y$, and we get the system

$$0 \le x \le 2 - \frac{1}{2}y$$
$$0 \le y \le 2.$$

The fact that the "left/right" approach is easier to use than the "top/bottom" indicates that the shape of the region can dictate which approach to use. ▲

Example 1.6.3. *A disk*

The unit disk in the plane is all points inside the unit circle, and is defined to be all points satisfying $x^2 + y^2 \le 1$ (see Figure 1.6.2(b)). The top of the region is the upper semicircle, and the bottom is the lower. Solving the equation $x^2 + y^2 = 1$ of the boundary curve for y gives $y = \pm\sqrt{1 - x^2}$. You could also use the left and right semicircle, solving the equation for x. Therefore the disk is given by

Top/bottom	Left/right
$-\sqrt{1 - x^2} \le y \le \sqrt{1 - x^2}$	$-\sqrt{1 - y^2} \le x \le \sqrt{1 - y^2}$
$-1 \le x \le 1$	$-1 \le y \le 1$

Both of these descriptions are cumbersome, and the unit disk is much more easily described in polar coordinates. We see that r is between 0 and 1, and there is no restriction on θ, therefore the unit disk is described by the system of polar inequalities

$$0 \le r \le 1$$
$$0 \le \theta \le 2\pi. \;▲$$

Describing Solids: Limits of integration in the triple integrals of Section 4.3 require describing solids using systems of inequalities. The solids in question will have bounding surfaces, so a good understanding of the relative position of surfaces is necessary. We briefly review some surfaces before describing some solids.

Example 1.6.4. *Paraboloids*

Be able to describe the vertex and direction of a paraboloid.

The classic paraboloid is $z = x^2 + y^2$. Its vertex is at the origin, and it points up along the positive z-axis. Adding a constant just changes the vertex, while negating the $x^2 + y^2$ changes the direction. The surfaces with equations $z = x^2 + y^2 - 3$ and $z = x^2 + y^2 + 5$ point up along the positive z-axis with vertices $(0, 0, -3)$ and $(0, 0, 5)$ respectively, while $z = -x^2 - y^2$ and $z = 4 - x^2 - y^2$ point down with vertices at the origin and $(0, 0, 4)$.

Of course, paraboloids can have other axes than the z-axis, and their equations are similar. Therefore $x = y^2 + z^2 + 2$ points along the positive x-axis with vertex $(2, 0, 0)$ and $y = 1 - x^2 - z^2$ points along the negative y-axis with vertex at $(0, 1, 0)$. ▲

Example 1.6.5. *Cones*

Be able to recognize cones from their equations.

The idea is the same as for paraboloids. Pick the vertex and direction from the equation. The standard cone has Cartesian equation $z = \sqrt{x^2 + y^2}$ and cylindrical equation $z = r$ (how about spherical?). Its vertex and axis are the origin and the positive z-axis. Similarly, $x = 4 - \sqrt{y^2 + z^2}$ has vertex $(4, 0, 0)$ and points along the negative x-axis. ▲

Example 1.6.6. *Cylinders*

Cylinders have equations that look like equations for circles, but they are surfaces. It's just that their equations are missing a variable. For example, $x^2 + y^2 = 4$ is a cylinder radius 2 with axis the z-axis, while $y^2 + z^2 = 9$ has radius 3 with core axis the x-axis. ▲

After this quick review of surfaces and their equations, we return to the task of describing solids. In each case equations for the boundary surfaces of the solids will provide limits on the variables. The first example is a solid with planes for boundary surfaces.

Example 1.6.7. *A tetrahedron*

A tetrahedron is the three-dimensional analog of a triangle. To construct a triangle, connect three vertices with line segments and take all the points inside. The vertices can't lie on a line, or it doesn't work. For a tetrahedron, connect four non-coplanar points in space and take all the points inside (as with the triangle, we also mean the points on the edge, or boundary).

Consider the tetrahedron T determined by the origin and the intercepts of the plane $2x + y - 3z = 6$. The intercepts are $(3, 0, 0)$, $(0, 6, 0)$, and $(0, 0, -2)$, so the triangle they determine lies below the xy-plane, and is the bottom of the tetrahedron T. The top of the tetrahedron is the xy-plane. Therefore the z-coordinate of any point in T is between the equations for those surfaces solved for z. Since the xy-plane is $z = 0$, we have $(2x + y - 6)/3 \leq z \leq 0$. To get inequalities on x and y, we reduce to a two-dimensional picture

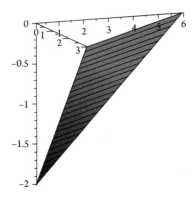

Figure 1.6.3 A tetrahedron

by taking the shadow of T in the xy-plane (the technical term is the *projection* of T into the xy-plane). Analytically, that amounts to letting $z = 0$ in the boundary surfaces. In this case you get the triangle in the first quadrant bounded by the line $2x + y = 6$. Now find limits on x and y using the two-dimensional techniques described above. This gives that T is described by the system

$$(2x + y - 6)/3 \le z \le 0$$
$$0 \le y \le 2x - 6$$
$$0 \le x \le 3.$$

Alternatively you can consider the plane as describing the right side of T, and the triangle in the xz-plane as the left side. Solving equations for y gives limits on y, then project into the xz-plane and use two-dimensional techniques to find the remaining limits. Thinking of the plane as the front, solving equations for x, and projecting into the yz-plane yields the final system of inequalities describing T.

Right/left	Front/back
$0 \le y \le 6 - 2x + 3z$	$0 \le x \le (6 - y + 3z)/2$
$(2x - 6)/3 \le z \le 0$	$0 \le y \le 3z + 6$
$0 \le x \le 3$	$-2 \le z \le 0.$ ▲

To describe a region in $\mathbb{R}^3$ using a system of inequalities in Cartesian coordinates, first decide if it's a top/bottom, right/left, or front/back solid region. If it's a front/back solid, solve all equations for x getting the front surface is $x = f(y,z)$ and the back $x = g(y,z)$. The limits on x are then $g(y,z) \le x \le f(y,z)$. Now project the solid into the yz-plane and use two-dimensional techniques to find the remaining limits. For a top/bottom approach:

1. Solve equations of surfaces for z: suppose the top surface is $z = f(x,y)$ and the bottom $z = g(x,y)$.

2. The limits on z are $g(x,y) \le z \le f(x,y)$.

3. Project the solid into the xy-plane, and use two-dimensional techniques.

Math App 1.6.1. *Describing tetrahedra*

This Math App allows you to experiment with using systems of inequalities to describe tetrahedra. The tetrahedra will have the origin as one vertex, and the other three vertices on the coordinate axes.

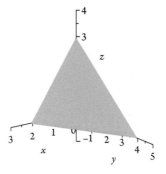

Example 1.6.8. *A tubular solid*

Describe the solid inside $y^2 + z^2 = 1$ and between the *yz*-plane and $x + y + z = 5$ using a system of inequalities.

The first surface is a cylinder along the *x*-axis that intersects the *yz*-plane in its unit circle. Because the plane $x + y + z = 5$ intersects each coordinate axis at 5, the plane intersects the cylinder $y^2 + z^2 = 1$ *in front of* the *yz*-plane (see Figure 1.6.4(a)).

The front of the solid, then, is $x + y + z = 5$ while the back is the *yz*-plane. Solving these equations for *x* gives the limits

$$0 \le x \le 5 - y - z.$$

Projecting the solid in the *yz*-plane gives the unit disk, and Example 1.6.3 can be adapted to give the limits

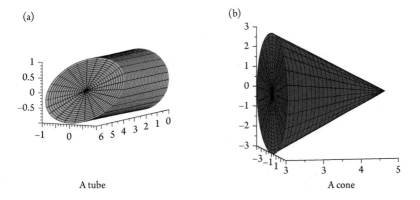

(a)

(b)

A tube

A cone

Figure 1.6.4 Two solids

$$-\sqrt{1-z^2} \le y \le \sqrt{1-z^2} \quad \text{and } -1 \le z \le 1. \ \blacktriangle$$

Example 1.6.9. *Left-right solid*

Let W be the solid between $y = 2$ and $y = 5 - \sqrt{x^2 + z^2}$ (see Figure 1.6.4(b)). Find a system of inequalities that describe W.

In this case, $y = 5 - \sqrt{x^2 + z^2}$ is a cone opening along the negative y-axis with vertex at $(0,3,0)$. Thus the plane $y = 2$ forms the left side of W and the cone is the right, giving the inequalities

$$2 \le y \le 5 - \sqrt{x^2 + z^2}.$$

The projection of W onto the xz-plane is a region with boundary curve. You can determine the boundary curve by equating the limits on y, which gives $2 = 5 - \sqrt{x^2 + z^2}$. Simplifying gives the equation $x^2 + z^2 = 9$, which gives the system

$$-\sqrt{9 - z^2} \le x \le \sqrt{9 - z^2} \quad \text{and } -3 \le x \le 3.$$

As in Example 1.6.3 polar coordinates might make for a simpler system, but that is polar coordinates in the xz-plane. First we choose coordinates $x = r\cos\theta, z = r\sin\theta$, and note that the equation for the cone becomes $y = 5 - r$. Since the projection into the xz-plane is the disk centered at the origin and radius 3, we get the system of inequalities

$$2 \le y \le 5 - r$$
$$0 \le r \le 3$$
$$0 \le \theta \le 2\pi. \ \blacktriangle$$

We conclude this section with an example containing inequalities in each of the coordinate systems.

Example 1.6.10. *A snowcone*

In this example we describe a region in space as a system of inequalities in Cartesian, cylindrical, and spherical coordinates. The region R is above the cone $z = r$ and inside the unit sphere, which looks like a snowcone.

Cylindrical Coordinates

We begin by describing all surfaces involved in cylindrical coordinates. The equation for the cone is given in cylindrical coordinates, while the easiest equation for the sphere is $\rho = 1$. Since $\rho^2 = z^2 + r^2$, we have that R is above $z = r$ and below $z = \sqrt{1 - r^2}$. This determines the restrictions on z, and we must consider the shadow in the xy-plane to get limits on r and θ.

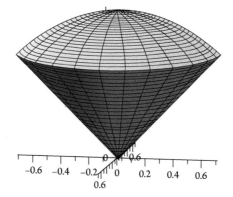

-0.6 -0.4 -0.2 0 0.2 0.4 0.6
0.6

θ 0.6
0.6

Figure 1.6.5 A snowcone

It is clear that this shadow is a disk centered at the origin. To find the radius of the disk, set the cylindrical equations equal to each other and solve for r. This gives $r = \sqrt{1 - r^2}$, which implies $r = \sqrt{2}/2$. The limits on r and θ are now clear, and we see that the solid region R is described in cylindrical coordinates by

$$r \leq z \leq \sqrt{1 - r^2}$$
$$0 \leq \theta \leq 2\pi$$
$$0 \leq r \leq \sqrt{2}/2.$$

Spherical Coordinates

In spherical coordinates, the equations of the boundary surfaces are particularly straightforward. The sphere is $\rho = 1$ and the cone $\phi = \pi/4$. To get limits in spherical coordinates requires a slightly different approach than we've encountered so far. The lower limit on ρ is the equation of the surface where a ray from the origin enters the region. Any ray entering R starts in R, since the origin is in R, so the lower limit is 0. The upper limit is the equation of the surface where the ray leaves the region, which is $\rho = 1$ in our case. Thus we have $0 \leq \rho \leq 1$. Notice that θ is unrestricted, and ϕ is between 0 and $\pi/4$, so the snow cone is

$$0 \leq \rho \leq 1$$
$$0 \leq \theta \leq 2\pi$$
$$0 \leq \phi \leq \pi/4.$$

Cartesian Coordinates

In Cartesian coordinates the sphere is $x^2 + y^2 + z^2 = 1$, and the cone $z = \sqrt{x^2 + y^2}$. Since the sphere is the top and the cone the bottom of the solid we'll use a top/bottom approach. The shadow in the *xy*-plane is the disk with boundary $x^2 + y^2 = 1/2$. Using the same strategy as for the unit disk above gives

$$\sqrt{x^2 + y^2} \leq z \leq \sqrt{1 - x^2 - y^2}$$
$$-\sqrt{1/2 - x^2} \leq y \leq \sqrt{1/2 - x^2}$$
$$-\sqrt{2}/2 \leq x \leq \sqrt{2}/2. \ \blacktriangle$$

Exercises

1. Find a system of inequalities that describes the region in the first quadrant bounded by the line $y = 6 - 2x$.

2. Find a system of inequalities that describes the region in $\mathbb{R}^2$ between the parabolas $y = 1 - x^2$ and $y = x^2 - 1$.

3. Describe the region of $\mathbb{R}^2$ between $x = 0$ and $x = 2$, above $y = x$ and below $y = 4$ using a system of inequalities.

4. Let R be the triangle in the plane with vertices $(-1, 0)$, $(0, 1)$, and $(1, 0)$. Describe R.

5. Use inequalities to describe the finite regions between $y = x^3 - x$ and the x-axis (ignore the unbounded regions).

6. Use polar coordinates to describe the quarter of the unit circle in the second quadrant.

7. Find a system of inequalities in polar coordinates defining the region between circles of radii 3 and 5 centered at the origin.

8. Determine a system of inequalities describing the tetrahedron in the first octant bounded by the plane $2x + y + 3z = 6$.

9. Determine a system of inequalities describing the tetrahedron bounded by coordinate planes and the plane $x - y + 3z = 9$.

10. Use a system of inequalities to describe the solid between $y = x^2 + z^2 - 4$ and the xz-plane.

11. Describe the solid between $x = 4 - y^2 - z^2$ and $x = y^2 + z^2 + 2$ using a system of inequalities.

12. Use a system of inequalities to describe the solid inside $y^2 + z^2 = 4$ and between the xz-plane and $-x + y + z = 4$.

13. Describe the solid under $z = 8 - \sqrt{x^2 + y^2}$ and above $z = 4$ using a system of inequalities.

14. Find a system of inequalities describing the solid between $x = y^2 + z^2$ and $x = 2 - \sqrt{y^2 + z^2}$.

15. Use a system of inequalities to describe the solid under the cone $z = 5 - r$ and above the xy-plane.

16. Describe the solid inside the cylinder $r = 1$, above the xy-plane, and under the unit sphere using a system of inequalities.

17. Describe the region in space between spheres of radius 1 and 2 using inequalities in spherical coordinates.

18. The region R is inside the cylinder $y^2 + z^2 = 1$ and above the xy-plane, for $-1 \leq x \leq 2$. Describe R using a system of inequalities on Cartesian coordinates.

2 Introduction to Vectors

In this chapter we are introduced to vectors, which are directed line segments, or arrows. Vectors are a very powerful tool for studying geometric properties of surfaces, and have many applications in physics. For these reasons, this chapter is fundamental for the rest of the book.

Vectors are defined in Section 2.1, along with vector addition and scalar multiplication. Both of these operations on vectors are defined geometrically and algebraically. An application to parameterizations is given in Section 2.2 to illustrate the utility of vector analysis. The next sections introduce two vector products: the dot products and cross products. Both are easily defined algebraically, and both have considerable geometric significance. The dot product, defined in Section 2.4, is related to angles between vectors, and can be used to project one vector onto another. The cross product is studied in Section 2.4, and can be used to find areas of parallelograms and normal vectors to planes. Vectors are used to study calculus on parametric curves in Section 2.5 and vector fields in Section 2.6. Vector fields will be used extensively in Chapter 5.

2.1 Geometry and Algebra of Vectors

The Geometry of Vectors We begin by defining what we mean by a vector, and when two vectors are the same. This is followed by geometric definitions of vector addition and scalar multiplication which are quite natural. Decomposing vectors into components will connect the geometric definitions to algebraic manipulation, facilitating vector analysis.

Definition 2.1.1. *A vector is a quantity that has a magnitude and direction. Thus a vector is a directed line segment.*

A vector can be thought of as an arrow, as in 2.1.1. Vectors will be denoted by bold face letters, such as **v**. The magnitude of the vector will be denoted $\|\mathbf{v}\|$. The number $\|\mathbf{v}\|$ is also called the *norm*, or length, of **v**. There is one special vector, the zero vector **0**. It is the vector with magnitude zero, which symbolically looks like $\|\mathbf{0}\| = 0$. The bold-faced zero represents a vector, while the zero on the right represents a real number.

Note that polar coordinates are similar in flavor to vectors since r gives the length and θ the direction. In spherical coordinates, ρ gives the length and θ and ϕ combine to provide the direction. Thus we've already been thinking of points in vector fashion and we'll now formalize that a little more.

Geometrically, the vector **v** is represented by an arrow pointing in the direction of **v** with length $\|\mathbf{v}\|$. When thinking of them in this way, we call the tail of the arrow the *initial point*, and the *terminal point* is its tip.

Example 2.1.1. *Vectors*

The initial point of vector **v** in Figure 2.1.1 is $(0,2)$ and its terminal point is $(3,-1)$. The length $\|\mathbf{v}\|$ of the vector **v** is the distance from its initial to its terminal point. Using the distance formula in the plane we have

$$\|\mathbf{v}\| = \sqrt{(3-0)^2 + (-1-2)^2} = 3\sqrt{2}. \ \blacktriangle$$

Definition 2.1.2. *Two vectors are the same vector if they have the same magnitude and direction.*

This means it doesn't matter where the vectors start: if they have the same magnitude and direction, they are the same vector! So all the vectors pictured in Figure 2.1.1 are the same. This sometimes causes confusion at first, but just remember that you can translate vectors around and they don't change (provided you keep the same length and direction).

A *position vector* for the point P will be the vector **P** with its initial point at the origin and its terminal point at P. In Figure 2.1.1, the position vector representing **v** is labeled **P**, and is the position vector of the point $P = (3,-3)$. We'll denote position vectors using angled parentheses, as in $\mathbf{P} = \langle 3,-3 \rangle$. Points will frequently be thought of as position vectors, and then as points again. It is important to be versatile enough to switch between the two interpretations with ease.

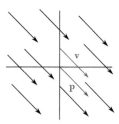

Figure 2.1.1 Many instances of the same vector

Example 2.1.2. *Length of a position vector*

Find the length of the vector $\mathbf{P} = \langle 3, 7, -2 \rangle$.

Since the initial point of $\mathbf{P}$ is the origin and the terminal point is $(3, 7, -2)$, the length of $\mathbf{P}$ is just the distance from the point P to the origin. We have $\|\mathbf{P}\| = \sqrt{3^2 + 7^2 + (-2)^2} = \sqrt{62}$. ▲

More generally, given a position vector $\mathbf{P} = \langle a, b, c \rangle$, then

$$\|\langle a, b, c \rangle\| = \sqrt{a^2 + b^2 + c^2}. \tag{2.1.1}$$

Example 2.1.3. *Position vector from initial and terminal points*

Find a position vector $\mathbf{P}$ equal to the vector $\mathbf{v}$ with initial point $(2, -1, 1)$ and terminal point $(-3, 2, -4)$.

Since $\mathbf{P}$ equals $\mathbf{v}$, the distance from initial to terminal points in each coordinate direction must be the same. Now $\mathbf{v}$ goes -5 units in the x-direction, 3 units in the y-direction, and -5 units vertically, so $\mathbf{P}$ does too. Since $\mathbf{P}$ starts at the origin, these distances are just its components, so $\mathbf{P} = \langle -5, 3, -5 \rangle$. ▲

Thus far we've introduced the notions of a vector, equal vectors, and their lengths. Position vectors turn out to be particularly nice vector representatives. We now describe two important operations on vectors: vector addition and scalar multiplication.

Vector Addition: Let $\mathbf{v}, \mathbf{w}$ be vectors, we define their sum $\mathbf{v} + \mathbf{w}$ as follows. The vector $\mathbf{v} + \mathbf{w}$ is obtained by concatenating $\mathbf{v}$ and $\mathbf{w}$. In other words, placing $\mathbf{w}$ on the terminal point of $\mathbf{v}$, then going from tail to tip (see Figure 2.1.2(a)).

This moving $\mathbf{w}$ around makes sense, since vectors are not tied to initial points. Notice that vector addition is commutative! Technically the sum $\mathbf{v} + \mathbf{w}$ is obtained by placing $\mathbf{w}$ on the tip of $\mathbf{v}$, rather than placing $\mathbf{v}$ on the tip of $\mathbf{w}$. Observe that the two constructions form a parallelogram, and both $\mathbf{w} + \mathbf{v}$ and $\mathbf{v} + \mathbf{w}$ are the diagonal of that parallelogram. Algebraically, then $\mathbf{w} + \mathbf{v} = \mathbf{v} + \mathbf{w}$. One can also see geometrically that vector addition is associative, but the picture is not very enlightening.

Math App 2.1.1. *Vector addition is commutative*

The following Math App verifies that vector addition is commutative using a more dynamic approach. Take a moment to investigate it.

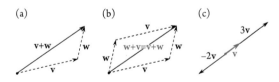

Figure 2.1.2 Vector addition and scalar multiplication

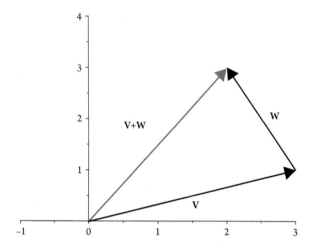

Example 2.1.4. *Vector addition and position vectors*

Let **v** be the position vector $\langle 3,1 \rangle$, and $\mathbf{w} = \langle 1,2 \rangle$. Then $\mathbf{v} + \mathbf{w}$ is obtained by placing **w** on the tip of **v**, and going from tail to tip (see Figure 2.1.2(a)). Since **v** is a position vector, the initial point of $\mathbf{v} + \mathbf{w}$ is the origin. To find the terminal point start at the origin, walk along **v** then walk along **w**. Thus the *total horizontal* distance traveled is the horizontal distance from **v** plus the horizontal distance from **w**. Similarly the *total vertical* distance for $\mathbf{v} + \mathbf{w}$ is the sum of the vertical distances for **v** and **w**. The tip of $\mathbf{v} + \mathbf{w}$, then, is the point whose coordinates are the sums $(3 + 1, 1 + 2)$, and $\mathbf{v} + \mathbf{w}$ is the position vector $\langle 4,3 \rangle$. Take a moment to notice what this equality looks like using position vector notation:

$$\langle 3,1 \rangle + \langle 1,2 \rangle = \langle 3+1, 1+2 \rangle = \langle 4,3 \rangle.$$

This is not a coincidence, and provides an algebraic method for summing two position vectors—add them coordinate-wise! ▲

This observation generalizes to vector addition in arbitrary dimensions, so we see

---

Adding Position Vectors

Let $\mathbf{v} = \langle a,b,c \rangle$ and $\mathbf{w} = \langle d,e,f \rangle$ be position vectors, then

$$\mathbf{v} + \mathbf{w} = \langle a+d, b+e, c+f \rangle \qquad (2.1.2)$$

Scalar Multiplication: We can multiply vectors by real numbers. Intuitively, multiplying the vector **v** by the number α just stretches it by a factor of α. Since this multiplication results in a rescaling, real numbers will also be called *scalars*. Moreover, this multiplication will be called *scalar multiplication*. If α is negative, then $\alpha\mathbf{v}$ goes in the opposite direction from $\mathbf{v}$. Symbolically we see that $\|\alpha\mathbf{v}\| = |\alpha|\,\|\mathbf{v}\|$, where $|\alpha|$ is the absolute value of a number while both $\|\alpha\mathbf{v}\|$ and $\|\mathbf{v}\|$ represent lengths of vectors. If $\alpha = 0$, we have that $0 \cdot \mathbf{v} = \mathbf{0}$ is the zero vector.

Convince yourself that $\alpha\mathbf{v} + \beta\mathbf{v} = (\alpha + \beta)\mathbf{v}$ for scalars $\alpha, \beta \in \mathbb{R}$ and any vector $\mathbf{v}$.

Example 2.1.5. *Scalar multiplication of position vectors*

In Figure 2.1.2(c) we see $\mathbf{v} = \langle 1, 3/4 \rangle$ and $\|\mathbf{v}\| = 5/4$. Then $4\mathbf{v}$ should be in the same direction but four times as long. The definition of scalar multiplication just described implies that $4\mathbf{v}$ should be four times as long as $\mathbf{v}$ and in the same direction. Since rescaling position vectors corresponds algebraically to multiplying coordinates, we see that

$$4\langle 1, 3/4 \rangle = \langle 4 \cdot 1, 4 \cdot 3/4 \rangle = \langle 4, 3 \rangle. \blacktriangle$$

Thus scalar multiplication with position vectors can be done coordinate-wise as well! We summarize with

---

Scalar Multiplication of Position Vectors

Let $\alpha \in \mathbb{R}$ be a scalar, and let $\mathbf{v} = \langle a, b, c \rangle$ be a position vector, then

$$\alpha\mathbf{v} = \langle \alpha a, \alpha b, \alpha c \rangle. \qquad (2.1.3)$$

Length and Vector Operations: Now that we have the operations of vector addition and scalar multiplication, we take a moment to discuss how the length of a vector can change under these operations. If $\mathbf{v} = \langle a, b, c \rangle$ and α is a scalar, we have

$$\|\alpha\mathbf{v}\| = \|\langle \alpha a, \alpha b, \alpha c \rangle\| = \sqrt{(\alpha a)^2 + (\alpha b)^2 + (\alpha c)^2} = |\alpha|\sqrt{a^2 + b^2 + c^2} = |\alpha|\,\|\mathbf{v}\|.$$

Algebraically, then, $\|\alpha\mathbf{v}\| = |\alpha|\,\|\mathbf{v}\|$. Knowing what you know about scalar multiplication, can you give a one-sentence justification of this fact?

Also note that

$$\|\mathbf{v} + \mathbf{w}\| \neq \|\mathbf{v}\| + \|\mathbf{w}\|$$

in general. Can you find examples where $\|\mathbf{v} + \mathbf{w}\|$ doesn't equal $\|\mathbf{v}\| + \|\mathbf{w}\|$? How about examples where they are equal? Can you use Figure 2.1.2(a) to argue which is bigger in general, $\|\mathbf{v} + \mathbf{w}\|$ or $\|\mathbf{v}\| + \|\mathbf{w}\|$?

Unit Vectors: Vectors that are one unit long enjoy many nice properties, so we give them a name. A *unit vector* is any vector of length one.

Example 2.1.6. *Unit vector*

The vector $\mathbf{u} = \left\langle \frac{1}{\sqrt{3}}, -\frac{1}{\sqrt{3}}, \frac{1}{\sqrt{3}} \right\rangle$ is a unit vector since

$$\|\mathbf{u}\| = \left\| \left\langle \frac{1}{\sqrt{3}}, -\frac{1}{\sqrt{3}}, \frac{1}{\sqrt{3}} \right\rangle \right\|$$
$$= \sqrt{\left(\frac{1}{\sqrt{3}}\right)^2 + \left(-\frac{1}{\sqrt{3}}\right)^2 + \left(\frac{1}{\sqrt{3}}\right)^2} = 1. \; \blacktriangle$$

There are many instances in which a vector is given and your task is to find a unit vector in the same direction (see Example 2.1.14 for one instance involving electric force and Coulomb's law). The process of finding a unit vector parallel to a given one is called *normalizing* the vector. Before reading the next example, guess what you would do to a vector 5 units long to normalize it—to make it one unit long.

Example 2.1.7. *Normalizing vectors*

Normalize the vector $\mathbf{v} = 2\mathbf{i} + 1\mathbf{j} - 3\mathbf{k}$.

First note that the length is $\|\mathbf{v}\| = \sqrt{2^2 + 1^2 + (-3)^2} = \sqrt{14}$. Dividing the vector $\mathbf{v}$ by this scalar gives

$$\mathbf{u} = \frac{\mathbf{v}}{\|\mathbf{v}\|} = \left\langle \frac{2}{\sqrt{14}}, \frac{1}{\sqrt{14}}, -\frac{3}{\sqrt{14}} \right\rangle.$$

One can compute directly that $\|\mathbf{u}\| = 1$, but hopefully it is more intuitive than that. Indeed, to normalize a vector, merely divide $\mathbf{v}$ by its length $\|\mathbf{v}\|$. Is that what you guessed? $\blacktriangle$

We now use what we know about the unit circle in $\mathbb{R}^2$ and the unit sphere in $\mathbb{R}^3$ to find forms of unit vectors more generally. Any vector of the form $\mathbf{u} = \langle \cos\theta, \sin\theta \rangle$ is a unit vector in $\mathbb{R}^2$. Moreover, any unit length position vector in $\mathbb{R}^3$ has its terminal point on the unit sphere $\rho = 1$. Letting $\rho = 1$ in the coordinate transformation shows that any unit vector in $\mathbb{R}^3$ is of the form $\mathbf{u} = \langle \sin\phi \cos\theta, \sin\phi \sin\theta, \cos\phi \rangle$ for some values of θ and ϕ.

Algebra with Vectors: The operations of vector addition and scalar multiplication have convenient algebraic interpretations when we work with position vectors. Sometimes, however, vectors are not given as position vectors; rather they are given in terms of their initial and terminal points, or by a length and direction. Before we can manipulate such vectors algebraically it is convenient to decompose them into their components, which is a fancy way of saying we find a position vector representative. Given an arbitrary vector, then, we now define its components and show how to find them. We can then use the rules of Equations 2.1.2 and 2.1.3 to engage in vector analysis.

To begin with, we introduce special vectors in $\mathbb{R}^2$ and $\mathbb{R}^3$. The vector $\mathbf{i}$ is a unit vector in the positive x-direction. Another way to say this is that $\mathbf{i}$ is the position vector of the point $P = (1,0,0)$, so $\mathbf{i} = \langle 1,0,0 \rangle$. Similarly, $\mathbf{j} = \langle 0,1,0 \rangle$ and $\mathbf{k} = \langle 0,0,1 \rangle$, are the position

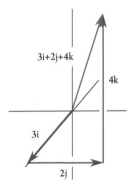

Figure 2.1.3 A vector as a sum of components

vectors of $(0,1,0)$ and $(0,0,1)$, respectively. When considering vectors in $\mathbb{R}^2$, we still use **i** and **j** to represent the position vectors of $(1,0)$ and $(0,1)$. These vectors are so special, they get a name.

Definition 2.1.3. *The vectors* **i**, **j**, **k** *are called the* standard basis vectors *for* $\mathbb{R}^3$.

The standard basis vectors will give us an easy way to describe all vectors in $\mathbb{R}^3$, as well as calculate with them. Since scalar multiplication is just stretching, the vector $a\mathbf{i}$ can be described as the vector with initial point $(0,0,0)$ and terminal point $(a,0,0)$ (i.e. $a\mathbf{i} = \langle a,0,0\rangle$). Similarly we obtain the vectors $b\mathbf{j} = \langle 0,b,0\rangle$ and $c\mathbf{k} = \langle 0,0,c\rangle$. Moreover, when we start at the origin and concatenate the vectors $a\mathbf{i}$, $b\mathbf{j}$, and $c\mathbf{k}$, we travel a units in the x-direction, b in the y-direction, and c in the z-direction. At the end of it all we are at the terminal point $(a,b,c) \in \mathbb{R}^3$. Algebraically we have $a\mathbf{i} + b\mathbf{j} + c\mathbf{k} = \langle a,b,c\rangle$ (see Figure 2.1.3).

Definition 2.1.4. *The* **i**,**j**, *and* **k** components *of the vector* $\mathbf{v} = a\mathbf{i} + b\mathbf{j} + c\mathbf{k} = \langle a,b,c\rangle$ *are the scalars a, b, c.*

More precisely a is the component of $\mathbf{v}$ in the **i** direction, b is the component of $\mathbf{v}$ in the **j** direction, and c is the **k**-component of $\mathbf{v}$.

Writing a given vector as a sum of scalar multiples of other vectors, like **i**, **j**, and **k**, is an important skill. How you do it depends on how the vector is given to you initially, and we illustrate it with a couple examples below. First we define a *linear combination* of vectors.

Definition 2.1.5. *A* linear combination *of vectors is any sum of scalar multiples of them.*

Example 2.1.8. *A linear combination*

The vector $\mathbf{v} = 3\mathbf{i} + 2\mathbf{j} + 4\mathbf{k}$ is a linear combination of **i**, **j**, and **k**. Its **i**-component is 3, its **j**-component is 2, and 4 is its component in the **k** direction.

The operations of vector addition and scalar multiplication yield a position vector for **v** as follows

$$\mathbf{v} = 3\mathbf{i} + 2\mathbf{j} + 4\mathbf{k} = 3\langle 1,0,0 \rangle + 2\langle 0,1,0 \rangle + 4\langle 0,0,1 \rangle$$
$$= \langle 3,0,0 \rangle + \langle 0,2,0 \rangle + \langle 0,0,4 \rangle = \langle 3,2,4 \rangle.$$

Thus the process of writing a vector as a linear combination of standard basis vectors is equivalent to finding a position vector representative. Both make for ease in computation. ▲

Example 2.1.9. *Points in the plane*

Show that any position vector in the plane is a linear combination of the standard basis vectors $\mathbf{i}, \mathbf{j}$.

Let $\mathbf{P} = \langle x, y \rangle$ be any position vector in the plane. Then

$$\mathbf{P} = \langle x,y \rangle = \langle x,0 \rangle + \langle 0,y \rangle = x\langle 1,0 \rangle + y\langle 0,1 \rangle = x\mathbf{i} + y\mathbf{j}.$$

Now show that any position vector in the plane is a linear combination of $\mathbf{v} = \langle 1,2 \rangle$ and $\mathbf{w} = \langle -1,1 \rangle$.

Again, let $\mathbf{P} = \langle x,y \rangle$ be any position vector in the plane. Our goal is to write $\mathbf{P}$ as a sum of scalar multiples of $\mathbf{v}$ and $\mathbf{w}$, so we solve the following vector equation for a, b in terms of x and y:

$$\mathbf{P} = a\mathbf{v} + b\mathbf{w}$$
$$\langle x,y \rangle = a\langle 1,2 \rangle + b\langle -1,1 \rangle$$
$$\langle x,y \rangle = \langle a - b, 2a + b \rangle.$$

This one vector equation is, of course, the system of two scalar equations

$$x = a - b$$
$$y = 2a + b,$$

Whose solution is $a = (x+y)/3, b = (-2x+y)/3$, so we conclude that

$$\mathbf{P} = \frac{x+y}{3}\mathbf{v} + \frac{-2x+y}{3}\mathbf{w}.$$

Since $\mathbf{P}$ is arbitrary, this shows that all position vectors are linear combinations of $\mathbf{v}$ and $\mathbf{w}$. For example, to write $\langle -4,3 \rangle$ as a linear combination of $\mathbf{v}$ and $\mathbf{w}$, substitute into the general solution

$$\langle -4,3 \rangle = \frac{-4+3}{3}\mathbf{v} + \frac{-2(-4)+3}{3}\mathbf{w} = \frac{-1}{3}\mathbf{v} + \frac{11}{3}\mathbf{w},$$

which is easily verified by letting $\mathbf{v} = \langle 1,2 \rangle$ and $\mathbf{w} = \langle -1,1 \rangle$. ▲

The vectors $\{\mathbf{i}, \mathbf{j}\}$ are said to *span* the plane since any point is a linear combination of them. Similarly, the set of vectors $\{\mathbf{v}, \mathbf{w}\}$ of the previous example spans the plane. We will encounter this again when parameterizing planes. Moreover, the notion of a spanning set is an important one in linear algebra.

Finding Vector Components: We have mentioned that vectors aren't always given as a linear combination of the standard basis vectors. We may be given initial and terminal points, or a distance and direction. In the next few examples we will illustrate writing vectors as linear combinations of the standard basis vectors. The strategies for doing so depend on how the vector is initially given to you.

Example 2.1.10. *Given terminal and initial points*

The vector $\mathbf{v}$ has initial point $(3, -4, 2)$, and terminal point $(-2, 3, 1)$. Write $\mathbf{v}$ as a linear combination of the standard basis vectors.

To do so, note that the coefficent of $\mathbf{i}$ is the distance in the x-direction from initial to terminal point of $\mathbf{v}$. In this case, that is from 3 to -2, so the component is $(-2) - 3 = -5$. Using similar reasoning in the other coordinate directions, we have

$$\mathbf{v} = ((-2) - 3)\mathbf{i} + (3 - (-4))\mathbf{j} + (1 - 2)\mathbf{k} = -5\mathbf{i} + 7\mathbf{j} - 1\mathbf{k}.$$

We see that the vector $\mathbf{v}$, which is given as having initial point $(3, -4, 2)$ and terminal point $(-2, 3, 1)$, is the same as the position vector $\langle -5, 7, -1 \rangle$. ▲

This example generalizes, and we state it in an intuitive way: To write $\mathbf{v}$ as a linear combination of the standard basis vectors, take the terminal point minus the initial point. Technically we aren't "subtracting points". We *are* subtracting the position vectors of the points. Analytically the two processes are equivalent.

Terminal Minus Initial

The vector $\mathbf{v}$ from $P(a, b, c)$ to $Q(d, e, f)$ is the position vector

$$\mathbf{v} = \mathbf{Q} - \mathbf{P} = \langle d - a, e - b, f - c \rangle.$$

Geometrically we see that $\mathbf{v}$ and $\mathbf{Q} - \mathbf{P}$ are opposite sides of a parallelogram, and are therefore the same vector.

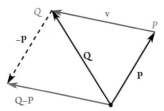

Example 2.1.11. *Given magnitude and direction*

Let **v** be the position vector in $\mathbb{R}^2$ that makes an angle of $\frac{2\pi}{3}$ with the positive *x*-axis and is 3 units long. Write **v** as a linear combination of standard basis vectors.

Notice that this vector is defined by specifying a direction and magnitude. In polar coordinates, the endpoint of the position vector is at $r = 3$ and $\theta = \frac{2\pi}{3}$. Using the change of coordinates from polar to Cartesian, the endpoint is $(3\cos\frac{2\pi}{3}, 3\sin\frac{2\pi}{3}) = (-3/2, 3\sqrt{3}/2)$. Since the coordinates of the terminal point of a position vector are the components, we have

$$\mathbf{v} = -\frac{3}{2}\mathbf{i} + \frac{3\sqrt{3}}{2}\mathbf{j} = \left(-3/2, 3\sqrt{3}/2\right). \ \blacktriangle$$

Example 2.1.12. *Spherical coordinates and vectors in* $\mathbb{R}^3$

In $\mathbb{R}^2$ you can specify a direction by fixing θ. In $\mathbb{R}^3$ you must fix both θ and ϕ to determine a direction. The coordinate transformations $(x, y, z) = (\rho\sin\phi\cos\theta, \rho\sin\phi\sin\theta, \rho\cos\phi)$ determine components for a given vector. Let **v** be the vector 2 units long in the direction determined by $\theta = \frac{2\pi}{3}$ and $\phi = \frac{\pi}{4}$. Think of **v** with its initial point at the origin. Then the length of **v** is ρ, and the specifications determine the terminal point: $(x, y, z) = (2\sin\frac{\pi}{4}\cos\frac{2\pi}{3}, 2\sin\frac{\pi}{4}\sin\frac{2\pi}{3}, 2\cos\frac{\pi}{4}) = (-\frac{\sqrt{2}}{2}, \frac{\sqrt{6}}{2}, \sqrt{2})$. In vector notation we have

$$\mathbf{v} = -\frac{\sqrt{2}}{2}\mathbf{i} + \frac{\sqrt{6}}{2}\mathbf{j} + \sqrt{2}\mathbf{k} = \left(-\sqrt{2}/2, \sqrt{6}/2, \sqrt{2}\right). \ \blacktriangle$$

Our original definitions of vector addition and scalar multiplication involved geometric interpretations of vectors. Expressing vectors as linear combinations of the standard basis vectors makes working with them much easier algebraically. We summarize the algebra of vectors in terms of standard basis vectors as well as position vectors.

Algebra with Position Vectors

$$\langle a, b, c\rangle + \langle d, e, f\rangle = \langle a + d, b + e, c + f\rangle$$
$$t\langle a, b, c\rangle = \langle ta, tb, tc\rangle$$
$$\|\langle a, b, c\rangle\| = \sqrt{a^2 + b^2 + c^2}$$

Algebra using Standard Basis Vectors

If $\mathbf{v} = \langle a, b, c\rangle$ and $\mathbf{w} = \langle d, e, f\rangle$ and $t \in \mathbb{R}$, we have

$$\mathbf{v} + \mathbf{w} = (a\mathbf{i} + b\mathbf{j} + c\mathbf{k}) + (d\mathbf{i} + e\mathbf{j} + f\mathbf{k})$$
$$= (a + d)\mathbf{i} + (b + e)\mathbf{j} + (c + f)\mathbf{k}$$
$$t\mathbf{v} = t(a\mathbf{i} + b\mathbf{j} + c\mathbf{k}) = ta\mathbf{i} + tb\mathbf{j} + tc\mathbf{k}$$
$$\|\mathbf{v}\| = \|a\mathbf{i} + b\mathbf{j} + c\mathbf{k}\| = \sqrt{a^2 + b^2 + c^2},$$

This section concludes with looking at a couple applications of vector algebra. These demonstrate the usefulness of vectors in many applied disciplines.

Force: Force in physics is an influence (a push or pull) that one object has on another. This could be the force of gravity on us, resulting from our interaction with the earth, or the force of friction a table applied to a glass of water sliding across it. Force is a vector quantity, having both magnitude and direction. Its magnitude is measured in newtons, which is the force needed to cause a 1m/s^2 acceleration on a 1kg object. If several forces are acting on an object, the *net force* is just the vector sum of them.

Many of us have an intuitive notion of what force is, but we take a moment to discuss force from the perspective of Newton's laws of motion. Newton's first law says that an object's motion will remain uniform unless it is acted upon by an external force. Therefore, forces cause *changes* in velocity—also known as acceleration. Indeed, a car traveling at a constant rate of 65 mph on a straight road has zero net force acting on it—even though it's moving at a good clip. Approaching a curve the driver turns the wheel, resulting in an external force being applied, and a subsequent acceleration. Note that acceleration exists even if the car's speed remains constant because velocity is a vector quantity which changes direction as the car navigates a turn.

Newton's second law quantifies the relationship between net force and acceleration, stating that $\mathbf{F} = m\mathbf{a}$, or force is mass times acceleration. This is a *vector* equation since both net force $\mathbf{F}$ and acceleration $\mathbf{a}$ are vector quantities and mass is a scalar. From this law it is easy to see that no acceleration implies zero net force. For the car traveling along the straight highway at constant speed, the car's force must exactly negate all other forces acting on it (gravity, friction, wind, etc.). These observations emphasize that forces do not cause velocity, they cause acceleration.

We continue with examples from physical situations.

Example 2.1.13. *Calculating net force*

A simple pendulum has a 10kg mass, so that the constant force due to gravity is $\mathbf{F}_g = 10\langle 0, -9.8\rangle$ with magnitude $\|\mathbf{F}_g\| = 98$N. There is a second force acting on the pendulum, that of tension due to the string (we ignore air resistance). The force due to tension $\mathbf{F}_t$ is not constant, but let's consider a particular moment in time.

Assume that when the pendulum is at its maximum height, it makes an angle of $60°$ below the horizontal. At this height it is a fact that $\mathbf{F}_t = \frac{1}{2}\langle -49\sqrt{3}, 147\rangle$ (take my word for it at this point). Find the net force $\mathbf{F}_n$ on the pendulum at that moment, together with its magnitude.

The net force is simply the vector sum of the forces acting on the pendulum, so

$$\mathbf{F}_n = \mathbf{F}_g + \mathbf{F}_t = 10\langle 0, -9.8\rangle + \frac{1}{2}\langle -49\sqrt{3}, 147\rangle$$

$$= \langle 0, -98\rangle + \langle -49\sqrt{3}/2, 147/2\rangle = \langle -49\sqrt{3}/2, -49/2\rangle.$$

Further, we compute the magnitude of the net force to be

$$\|\mathbf{F}_n\| = \sqrt{\left(-49\sqrt{3}/2\right)^2 + (-49/2)^2} = 49\text{N}.$$

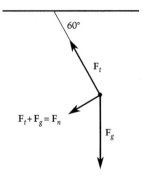

Figure 2.1.4 Forces on a pendulum

In Section 2.3 we will define the *component* of one vector in the direction of another. Since the pendulum does not accelerate in the direction of the string, which is always taught, there is no net force in that direction. This implies that the force due to tension cancels with the component of $\mathbf{F}_g$ along the string. In the notation of Section 2.3 this is written $\mathbf{F}_t = -comp_{F_t}\mathbf{F}_g$, which is how we calculated $\mathbf{F}_t = \frac{1}{2}\left(-49\sqrt{3}, 147\right)$ above. ▲

Electric Force: There is an electric force between any two charged objects. As an example, blow up two balloons and dangle them from threads held in the same hand. They contentedly bump against each other as they hang. Now rub them against your hair, placing negative charge on both. Dangle them next to each other now, and they will repel each other. This is an instance of a more general fact that like charges repel each other. Coulomb's law provides a more quantitative description of electric force.

One charged object at point A, with charge Q coulombs, exerts an electric force on object at point B with charge q coulombs described by

$$\mathbf{F} = k\frac{Q \cdot q}{\rho^2}\hat{\boldsymbol{\rho}},$$

where ρ is the distance between the objects and $\hat{\boldsymbol{\rho}}$ is the unit vector pointing from A to B, and $k = 9.0 \times 10^9 \frac{\text{N·m}^2}{\text{C}^2}$ is Coulomb's constant.

The principle of superposition allows us to extend Coulomb's law to systems of point charges. It states that the interaction between two charges is independent of the presence of other charges. In practice this allows us to calculate the net electric force that a system of charges $Q_1, \ldots, Q_n$ exerts on a test charge q. In fact, the net force on q is just the sum of the forces due to the individual charges. Analytically this means that the net force $\mathbf{F}$ on q is

$$\mathbf{F} = \sum k\frac{Q_i \cdot q}{\rho_i^2}\hat{\boldsymbol{\rho}}_i, \tag{2.1.4}$$

where ρ_i is the distance between the charges q and Q_i, and $\hat{\rho}_i$ is the unit vector pointing from charge Q_i to q, and k is Coulomb's constant. The following examples illustrate Coulomb's law and the principle of superposition.

Example 2.1.14. *Coulomb's law*

Find the force that a 3 coulomb charge A at the origin exerts on a $-2C$ charge B at the point $(3, 2, -1)$.

To find ρ and $\hat{\rho}$, note that the position vector $\langle 3, 2, -1 \rangle$ points from the A to B. Then $\rho = \|\langle 3, 2, -1 \rangle\| = \sqrt{14}$ and normalizing the vector $\langle 3, 2, -1 \rangle$ we find $\hat{\rho} = \langle 3, 2, -1 \rangle / \sqrt{14}$, and Coulomb's law gives

$$\mathbf{F} = k \frac{3 \cdot -2}{14} \cdot \frac{1}{\sqrt{14}} \langle 3, 2, -1 \rangle = \frac{-6k}{14^{3/2}} \langle 3, 2, -1 \rangle.$$

Because $\mathbf{F}$ represents the force of Q on q we consider its initial point on B, the charge being acted on. Note that $\mathbf{F}$ points from B toward A, which is consistent with the fact that opposite charged objects attract. This is because of choice of the unit vector $\hat{\rho}$. Letting $\hat{\rho}$ point from A to B makes the vector in Coulomb's law point in the correct direction when the signs of the charges are included. Indeed, if Q and q are the same sign, then Coulomb's law yields a repulsive force while if they are opposite signs $\mathbf{F}$ represents an attractive one. ▲

Example 2.1.15. *The principle of superposition*

Suppose a positive unit charge is placed at the point $A = (-1, 0)$ in the plane, and a negative unit charge at $B = (1, 0)$. Determine the electric force this system exerts on a unit positive test charge at the point $C = (-1, 2)$.

Figure 2.1.5 illustrates the individual forces $\mathbf{F}_A, \mathbf{F}_B$ that the particles at A and B exert on C, together with their sum, the net force $\mathbf{F} = \mathbf{F}_A + \mathbf{F}_B$. We now compute $\mathbf{F}$ analytically.

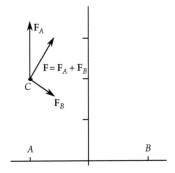

Figure 2.1.5 Net electric force **F** on C

Using the distance formula in the plane, $\rho_1 = 2$ and $\rho_2 = 2\sqrt{2}$. To find the $\hat{\rho}_i$, just normalize the vectors from A to C and from B to C. We get

$$\hat{\rho}_1 = \frac{C - A}{\|C - A\|} = \frac{\langle -1, 2 \rangle - \langle -1, 0 \rangle}{2} = \langle 0, 1 \rangle, \text{ and}$$

$$\hat{\rho}_2 = \frac{C - B}{\|C - B\|} = \frac{\langle -1, 2 \rangle - \langle 1, 0 \rangle}{2\sqrt{2}} = \left\langle -\sqrt{2}/2, \sqrt{2}/2 \right\rangle.$$

Using the above calculations in equation 2.1.4 gives the electric force

$$\mathbf{F} = k\frac{1 \cdot 1}{2^2} \langle 0, 1 \rangle + k\frac{-1 \cdot 1}{(2\sqrt{2})^2} \left\langle -\sqrt{2}/2, \sqrt{2}/2 \right\rangle$$

$$= k \langle 0, 1/4 \rangle + k \left\langle \sqrt{2}/16, -\sqrt{2}/16 \right\rangle$$

$$= k \left\langle \sqrt{2}/16, \left(4 - \sqrt{2} \right)/16 \right\rangle = \frac{k}{16} \left\langle \sqrt{2}, 4 - \sqrt{2} \right\rangle.$$

Since both components of $\mathbf{F}$ in this case are positive, a unit test charge released at $C = (-1, 2)$ would begin to travel up and to the right. ▲

Example 2.1.16. *Three equal point charges*

Suppose three equal point charges of Q coulombs are placed on the vertices of an equilateral triangle. Let q be a test charge placed at the centroid of the triangle. Show that the net force of this system on q is zero.

By Coulomb's law the force of Q on q is the vector $k\frac{Q \cdot q}{\rho^2} \hat{\rho}$, where $\hat{\rho}$ is the unit vector from a vertex to the centroid. By the symmetry of the problem, these all have the same length, and meet at an angle $2\pi/3$ at the centroid (see Figure 2.1.6(a) for the case where Q and q have opposite signs). Adding them, as in Figure 2.1.6(b), results in an equilateral triangle, so the sum of the vectors is the zero vector. ▲

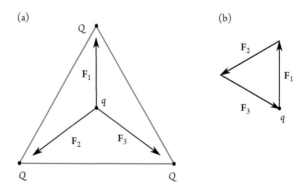

Figure 2.1.6 Zero force at centroid

Exercises

1. Which of the following vectors are the same?
   (a) $\mathbf{v} = \langle 3, -3 \rangle$
   (b) $\mathbf{v}$ has initial point: $(3, -2)$ and terminal point: $(2, -3)$
   (c) $\|\mathbf{v}\| = 3\sqrt{2}$ and $\mathbf{v}$ makes an angle of $-\pi/4$ with the positive x-axis.
   (d) $\|\mathbf{v}\| = \sqrt{2}$ and $\mathbf{v}$ makes an angle of $5\pi/4$ with the positive x-axis.
   (e) $\mathbf{v} = \langle -1, -1 \rangle$

2. Which of the following vectors are the same?
   (a) $\mathbf{v} = \langle 4, 3 \rangle$
   (b) $\mathbf{v}$ has initial point: $(5, -3)$ and terminal point: $(9, 0)$
   (c) $\|\mathbf{v}\| = 5$ and $\mathbf{v}$ makes an angle of $-\pi/2$ with the positive x-axis.
   (d) $\|\mathbf{v}\| = 5$ and $\mathbf{v}$ makes an angle of $\tan^{-1}(3/4)$ with the positive x-axis.
   (e) $\mathbf{v} = \langle 0, -5 \rangle$

3. Find the components of a vector in the plane that is 4 units long and makes an angle of $\frac{4\pi}{3}$ with the positive x-axis.

4. Normalize the vector $\langle 3, 4, 7 \rangle$.

5. Find a unit vector that points in the same direction as the vector from $(2, 1, 5)$ to $(3, -1, 2)$.

6. Normalize the vector in $\mathbb{R}^2$ given by $\|\mathbf{v}\| = 7$ and $\mathbf{v}$ makes an angle of $5\pi/6$ with the positive x-axis.

7. Find the components of the unit vector in the direction $\phi = \frac{\pi}{4}, \theta = \frac{3\pi}{4}$.

8. Find the vector length 5 in the direction $\phi = \frac{3\pi}{4}, \theta = -\frac{\pi}{3}$.

9. Normalize the vector $\langle 13, -1, 4 \rangle$.

10. Find a vector 3 units long in the opposite direction from $\mathbf{v} = \langle 3, 1, -4 \rangle$.

11. What scalar multiple of $\mathbf{v} = \langle 3, 1, -2 \rangle$ must be added to the vector $\mathbf{u} = \langle 4, 2, 1 \rangle$ to result in the vector $\mathbf{w} = \langle -5, -1, 7 \rangle$.

12. Describe, in English, why $\alpha \mathbf{v} + \beta \mathbf{v} = (\alpha + \beta)\mathbf{v}$ for scalars $\alpha, \beta \in \mathbb{R}$ and any vector $\mathbf{v}$.

13. Use Figure 2.1.2(a) to determine which of $\|\mathbf{v} + \mathbf{w}\|$ or $\|\mathbf{v}\| + \|\mathbf{w}\|$ is larger in general.

14. A boy pulls a sled along a horizontal path with a force of 40 newtons. If the rope makes and angle of $60°$ with the horizontal, find the component of force in the direction of movement.

15. One point charge Q_1 of 2C is placed at the point $A = (-1, 0)$ and a second Q_2 of $-3C$ is placed at $B = (1, 2)$. Find the force that Q_1 exerts on Q_2. Is this an attracting or repelling force?

16. A charge of 3C at $(1, 0)$ exerts a force of $\mathbf{F} = k\langle -6, 0 \rangle$ on a charge of Q coulombs at the origin. Find Q.

17. A unit positive charge is placed at $(-1,0)$, and a unit negative charge at $(1,0)$ in the plane. What is the net force of the system on a unit positive charge placed at $(0,1)$? $(\cos\theta, \sin\theta)$?

18. Four equal charges are placed at the vertices of a square. Argue, using a sketch, that the force of this system on a charge at the center of the square is **0**.

19. Suppose three equal point charges of Q coulombs are placed on the vertices of an equilateral triangle. Let q be a test charge placed on an altitude of the triangle. Show that q moves along the altitude.

2.2 Vectors and Parameterizations

The operations of vector addition and scalar multiplication make vectors particularly useful in a variety of settings, and geometry is one of those settings. In this section we see how these operations shed light on the parameterizations of lines and planes that we've already encountered in Sections 1.4 and 1.5. Vector addition and scalar multiplication provide geometric interpretations of different aspects of parametric component functions. We also introduce Bézier curves in this section. Bézier curves are parameterized by choosing some fixed points and using vector algebra. Moreover, they retain their shape under rescaling, making them particularly useful in computer graphics.

Parametric Lines: The operations of vector addition and scalar multiplication demonstrate that lines (even in $\mathbb{R}^3$) are determined by two things: a point P on the line and a vector $\mathbf{v}$ in the direction of the line. We will derive this by first considering lines through the origin.

Suppose ℓ is a line through the origin in $\mathbb{R}^3$ and $\mathbf{v} = \langle A, B, C \rangle$ is a position vector that lies on ℓ. Then any scalar multiple $t\mathbf{v}$ also lies on ℓ, since it's just a rescaling of $\mathbf{v}$. In fact, since the set of all scalar multiples of $\mathbf{v}$ forms a line, it must be true that $C(t) = t\mathbf{v}$, $-\infty < t < \infty$, parameterizes ℓ!

We now consider lines that don't go through the origin. Let ℓ be a line through the point P, and in the direction of the vector $\mathbf{v}$. Then $C(t) = t\mathbf{v}$ parameterizes the line through the origin parallel to ℓ. To get a parameterization for ℓ we just have to translate it so that it goes through P. Adding the position vector $\mathbf{P}$ to each point on the line $t\mathbf{v}$ moves the origin to P, and the rest of the line $t\mathbf{v}$ to ℓ. Thus parametric equations for ℓ are $C(t) = \mathbf{P} + t\mathbf{v}$ (see Figure 2.2.1(a)). Notice that the right-hand side of this equality is a linear combination of the vectors $\mathbf{P}$ and $\mathbf{v}$, which is another vector. However, the left-hand side we are considering to be a point on the line, not a vector. Thus we are blurring the distinction between position vectors and points here.

Parameterizing Lines

One parameterization of the line ℓ through the point P and in the direction of the vector $\mathbf{v}$ is

$$C(t) = \mathbf{P} + t\mathbf{v}, \quad -\infty < t < \infty.$$

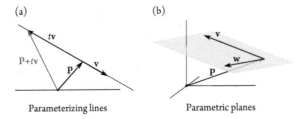

(a)

(b)

Parameterizing lines

Parametric planes

Figure 2.2.1 Parametric lines and planes

Example 2.2.1. *Given point and direction*

Parameterize the line through $(1, -2, 1)$ parallel to the vector $\mathbf{v} = 2\mathbf{i} + 3\mathbf{j} - \mathbf{k}$. Using the above strategy we get

$$\mathbf{C}(t) = \langle 1, -2, 1 \rangle + t \langle 2, 3, -1 \rangle = \langle 1, -2, 1 \rangle + \langle 2t, 3t, -t \rangle$$
$$= \langle 1 + 2t, -2 + 3t, 1 - t \rangle, \text{ for } -\infty < t < \infty. \ \blacktriangle$$

Example 2.2.2. *Line given two points*

Parameterize the line through the points $(3, -1, 3)$ and $(2, 4, -1)$.

We need a point on the line and a vector in the line. Notice that the vector from $(3, -1, 3)$ to $(2, 4, -1)$ lies in the line. Using the "terminal minus initial" idea we see that a vector in the line is $\mathbf{v} = (2 - 3)\mathbf{i} + (4 - (-1))\mathbf{j} + (-1 - 3)\mathbf{k} = -1\mathbf{i} + 5\mathbf{j} - 4\mathbf{k}$. This gives the parametric equations

$$\mathbf{C}(t) = \langle 3, -1, 3 \rangle + t \langle -1, 5, -4 \rangle = \langle 3 - t, -1 + 5t, 3 - 4t \rangle,$$
$$\text{for } -\infty < t < \infty. \ \blacktriangle$$

In many applications we are interested in parameterizing the line segment between two points. This occurs frequently enough that we highlight one way to do it. Suppose we want to parameterize the directed line segment from point P to point Q. As in Example 2.2.2, we can parameterize the line by noticing that the vector $\mathbf{Q} - \mathbf{P}$ is in the direction of the line. By our strategy for parameterizing lines the entire line ℓ is given by $\mathbf{C}(t) = \mathbf{P} + t(\mathbf{Q} - \mathbf{P})$, $-\infty < t < \infty$, and we have to determine the range of t values that gives the segment from P to Q. Notice that when $t = 0$ we have $\mathbf{C}(0) = \mathbf{P}$. Similarly $\mathbf{C}(1) = \mathbf{P} + \mathbf{Q} - \mathbf{P} = \mathbf{Q}$. Therefore the segment is given by $\mathbf{C}(t) = \mathbf{P} + t(\mathbf{Q} - \mathbf{P})$ for $0 \le t \le 1$. Combining like terms on $\mathbf{C}(t)$ we get the parameterization $\mathbf{C}(t) = \mathbf{P} + t(\mathbf{Q} - \mathbf{P}) = (1 - t)\mathbf{P} + t\mathbf{Q}$ for $0 \le t \le 1$.

It is worth noting that the parameter t has a nice geometric interpretation when using this parameterization of a line segment. Using the original form of the parameterization $\mathbf{C}(t) = \mathbf{P} + t(\mathbf{Q} - \mathbf{P})$ one sees that t is just the fraction of the vector $\mathbf{Q} - \mathbf{P}$ you are from the initial point P. Thus t represents the fraction of the distance from P to Q that $\mathbf{C}(t)$ has traveled! In symbols, letting $\overline{PC(t)}$ and $\overline{PQ}$ be the distances of the corresponding segments, we have the equality $\overline{PC(t)} = t\overline{PQ}$.

Parameterizing Line Segments

One parameterization of the line segment from P to Q is

$$\mathbf{C}(t) = (1 - t)\mathbf{P} + t\mathbf{Q}, \text{ for } 0 \le t \le 1. \qquad (2.2.1)$$

In this setting, t represents the proportion of the segment PQ that $\mathbf{C}(t)$ has traveled.

We illustrate the usefulness of this technique with several examples.

Example 2.2.3. *Line segments*

Parameterize the line segment from $(3,1,3)$ to $(-2,3,5)$. According to Equation 2.2.1 we have

$$\begin{aligned}
\mathbf{C}(t) &= (1-t)\langle 3,1,3\rangle + t\langle -2,3,5\rangle \\
&= \langle (1-t)3 + t(-2), (1-t)+3t, (1-t)3+5t\rangle \\
&= \langle 3-5t, 1+2t, 3+2t\rangle, \ 0 \le t \le 1. \ \blacktriangle
\end{aligned}$$

Example 2.2.4. *Proportions of segments*

Find the point one quarter of the way from $(1,2,-2)$ to $(3,-1,2)$.

By the above remarks, this is the point $\mathbf{C}(1/4)$, where $\mathbf{C}(t)$ is the parameterization of Equation 2.2.1. We have

$$\mathbf{C}(t) = (1-t)\langle 1,2,-2\rangle + t\langle 3,-1,2\rangle = (1+2t, 2-3t, -2+4t), \ 0 \le t \le 1.$$

The point one-quarter of the way is $\mathbf{C}(1/4) = (3/2, 5/4, -1)$. $\blacktriangle$

Example 2.2.5. *Piecewise linear curves*

Parameterize the triangle with vertices $P(1,3,-4)$, $Q(5,-1,6)$, and $R(-2,1,3)$ counterclockwise when viewed from above.

In order to parameterize a triangle, you parameterize each side separately. Each side is a line segment, so we can use Equation 2.2.1. We now consider how to parameterize counterclockwise when viewed from above. When viewing from above one cannot distinguish heights, so we project P, Q, and R into the xy-plane as in Figure 2.2.2. To ensure a counterclockwise orientation, we go from P to R to Q, and back to P again.

Let C_1 be the line segment from P to R, C_2 from R to Q, and C_3 from Q to P. Using Equation 2.2.1 we get the parameterizations:

$$\begin{aligned}
\mathbf{C}_1(t) &= (1-t)\langle 1,3,-4\rangle + t\langle -2,1,3\rangle = (1-3t, 3-2t, -4+7t) \\
\mathbf{C}_2(t) &= (1-t)\langle -2,1,3\rangle + t\langle 5,-1,6\rangle = (-2+7t, 1-2t, 3+3t) \\
\mathbf{C}_3(t) &= (1-t)\langle 5,-1,6\rangle + t\langle 1,3,-4\rangle = (5-4t, -1+4t, 6-10t)
\end{aligned}$$

where the limits on each parameterization is $0 \le t \le 1$. $\blacktriangle$

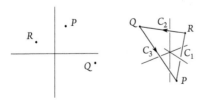

Figure 2.2.2 Parameterizing a triangle

Parametric Planes: Parametric equations for planes can be derived in much the same way as those for lines. As with lines, we begin with planes through the origin, and recall the notion of a spanning set from Example 2.1.9. In that example we showed that any vector in the plane could be written as a linear combination $s\mathbf{v} + t\mathbf{w}$ of the two (non-parallel) vectors $\mathbf{v} = \langle 1, 2 \rangle$ and $\mathbf{w} = \langle -1, 1 \rangle$. Equivalently one says that the set of all linear combinations of $\mathbf{v}$ and $\mathbf{w}$ is the plane. Of course, there is nothing that says the vectors must lie in $\mathbb{R}^2$, and the analogous statement holds in $\mathbb{R}^3$ (or even $\mathbb{R}^n$ for that matter):

If $\mathbf{v}$ and $\mathbf{w}$ are non-parallel position vectors, then all linear combinations $s\mathbf{v} + t\mathbf{w}$ of them form a plane through the origin.

Another way to see this is: we already know that all scalar multiples of one vector form a line through the origin, so linear combinations of two form the plane through the origin containing them.

We take a moment to describe some terminology, and make use of the fact that a plane in $\mathbb{R}^3$ has one direction, denoted by a vector $\mathbf{n}$, that is perpendicular to it. If $\mathbf{v}$ is perpendicular to $\mathbf{n}$, we will say the vector $\mathbf{v}$ is in the plane (or the plane contains $\mathbf{v}$). For example, the vector $\mathbf{j}$ is perpendicular to the xz-plane. The vector $\langle 1, 0, 1 \rangle$ lies in the xz-plane, and is perpendicular to $\mathbf{j}$.

To parameterize a plane through the origin, find two vectors in the plane $\mathbf{v}$ and $\mathbf{w}$ and set

$$S(s,t) = s\mathbf{v} + t\mathbf{w}, \quad -\infty < s, t < \infty.$$

As with lines, if the plane doesn't go through the origin find a point P on it and parameterize it by adding the position vector $\mathbf{P}$ to this. We summarize:

Parameterizing Planes

One parameterization of the plane through the point P and containing the vectors $\mathbf{v}, \mathbf{w}$ is

$$S(s,t) = \mathbf{P} + s\mathbf{v} + t\mathbf{w}, \quad -\infty < s, t < \infty. \tag{2.2.2}$$

We apply these observations to several examples.

Example 2.2.6. *Vectors in a plane*

Find four vectors in the plane

$$S(s,t) = (3 - 2s + t, -4 + s - 2t, 2 + 3s + 2t), \quad -\infty < s, t < \infty.$$

Notice that the parameters in Equation 2.2.2 are coefficients of vectors in the plane. By vector algebra we have

$$S(s,t) = (3 - 2s + t, -4 + s - 2t, 2 + 3s + 2t) = \langle 3, -4, 2 \rangle + s\langle -2, 1, 3 \rangle + t\langle 1, -2, 2 \rangle,$$

so that $\langle -2,1,3 \rangle$ and $\langle 1,-2,2 \rangle$ are vectors in the plane. In the future the vector algebra is unnecessary, it suffices to write the vector whose components are the coefficients of one of the parameters.

To find others, note that the vector between any two points in the plane will lie in the plane. We compute

$$\mathbf{S}(1,1) - \mathbf{S}(0,0) = \langle 2,-5,7 \rangle - \langle 3,-4,2 \rangle = \langle -1,-1,5 \rangle, \text{ and}$$
$$\mathbf{S}(1,-1) - \mathbf{S}(0,0) = \langle 0,-1,3 \rangle - \langle 3,-4,2 \rangle = \langle -3,3,1 \rangle.$$

Alternatively, any linear combination of $\langle -2,1,3 \rangle$ and $\langle 1,-2,2 \rangle$ is a vector in the plane. ▲

Example 2.2.7. *Parametric planes*

Find parametric equations for the plane containing points $(3,-1,2)$, $(-3,2,2)$, and $(1,4,3)$.

First we find two vectors in the plane, thinking of $(3,-1,2)$ as the initial point of both and the remaining points as terminal points. We have

$$\mathbf{v} = (-3-3)\mathbf{i} + (2-(-1))\mathbf{j} + (2-2)\mathbf{k} = -6\mathbf{i} + 3\mathbf{j} = \langle -6,3,0 \rangle$$
$$\mathbf{w} = (1-3)\mathbf{i} + (4-(-1))\mathbf{j} + (3-2)\mathbf{k} = -2\mathbf{i} + 5\mathbf{j} + \mathbf{j} = \langle -2,5,1 \rangle$$

Using Equation 2.2.2 we get

$$\mathbf{S}(s,t) = \langle 3,-1,2 \rangle + s\langle -6,3,0 \rangle + t\langle 2,5,1 \rangle$$
$$= \langle 3-6s+2t, -1+3s+5t, 2+t \rangle, \quad -\infty < s,t < \infty. \text{ ▲}$$

By putting limits on the parameters, we can parameterize portions of planes as well. In particular, we can easily parameterize the parallelogram determined by two position vectors. Let $\mathbf{v} = \langle a,b,c \rangle$ and $\mathbf{w} = \langle d,e,f \rangle$ be two position vectors in $\mathbb{R}^3$. As long as $\mathbf{v}$ and $\mathbf{w}$ are not parallel they determine a parallelogram $\mathcal{P}$, with vertices $(0,0,0)$, (a,b,c), (d,e,f), and whose final vertex is the endpoint of $\mathbf{v} + \mathbf{w} = \langle a+d, b+e, c+f \rangle$. Any point P in $\mathcal{P}$ has a position vector that is a linear combination of $\mathbf{v}$ and $\mathbf{w}$, so that $\mathbf{P} = s\mathbf{v} + t\mathbf{w}$ (see Figure 2.2.3). In fact, the scalars s and t are both between 0 and 1 since $\mathbf{P}$ goes only part way in both the $\mathbf{v}$ and $\mathbf{w}$ directions. Thus a parameterization of the parallelogram $\mathcal{P}$ is

$$\mathbf{S}(s,t) = s\mathbf{v} + t\mathbf{w}, \ 0 \le s,t \le 1.$$

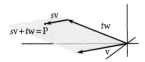

Figure 2.2.3 Parameterizing a parallelogram

Example 2.2.8. *Parameterizing a parallelogram*

Parameterize the parallelogram determined by the position vectors $\langle 3, -4, 2 \rangle$ and $\langle 2, 2, 1 \rangle$.

By the above remarks, one parameterization is

$$S(s,t) = (3s + 2t, -4s + 2t, 2s + t), \ 0 \le s, t \le 1. \ \blacktriangle$$

Concept Connection: Parallelograms and their areas will be vital in deriving formulas for computing surface area and for integrating functions over surfaces. Unfortunately this will have to wait until Chapter 4.

Math App 2.2.1. *Parametric parallelogram*

This Math App allows you to interact with linear combinations of vectors to see how changing scalars changes the sum.

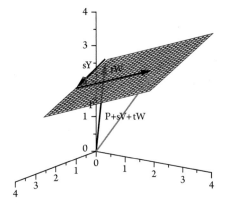

Bézier Parameterizations: We turn our attention to Bézier curves. The French auto industry seems to be where Bézier curves first gained popularity because of their applicability. Paul de Casteljau, who worked for Citroën, and Pierre Bézier, employed by Renault, both took advantage of what are now called Bézier curves in designing automobiles. Since then, Bézier curves have become an integral part of computer graphics programming. In fact, Bézier curves are used to create the fonts you are reading right now!

Before describing them, we mention a few of the desirable properties of Bézier curves that make them so amenable to computing. First, their parameterizations are straightforward and only require knowing a few "control" points, making them easy to store in a computer. Bézier curves can be subdivided into a collection of shorter Bézier curves which are nearly linear. Thus a smooth curve can easily (for a computer) be approximated by line segments after sufficiently subdividing it into smaller Bézier curves. These curves also behave very nicely under linear transformations (to be defined in Section 2.4) and translations—operations that are necessary, for example, when rendering three-dimensional graphics. These properties make Bézier curves quite useful in technological fields.

It turns out that we have already encountered linear Bézier curves when parameterizing the line segment between two points. Recall that the line segment from P to Q is parameterized by thinking of the points as position vectors and using the formula

$$\mathbf{C}(t) = (1-t)\mathbf{P} + t\mathbf{Q}, \ 0 \le t \le 1.$$

This is a linear Bézier curve, and the points $\mathbf{P}, \mathbf{Q}$ are called *control points* in this setting. A quadratic Bézier curve requires three control points, $\mathbf{P}_0, \mathbf{P}_1, \mathbf{P}_2$, and is defined by the parameterization

$$\mathbf{C}(t) = (1-t)^2\mathbf{P}_0 + 2(1-t)t\mathbf{P}_1 + t^2\mathbf{P}_2, \ 0 \le t \le 1. \qquad (2.2.3)$$

The four control points $\mathbf{P}_0, \mathbf{P}_1, \mathbf{P}_2, \mathbf{P}_3$ define the cubic Bézier curve via

$$\mathbf{C}(t) = (1-t)^3\mathbf{P}_0 + 3(1-t)^2t\mathbf{P}_1 + 3(1-t)t^2\mathbf{P}_2 + t^3\mathbf{P}_3, \ 0 \le t \le 1. \qquad (2.2.4)$$

Note that the limits on t are always from zero to one, and that the curve starts and ends at the first and last control point, respectively (just evaluate $\mathbf{C}(0)$ and $\mathbf{C}(1)$). Bézier curves do not go through the intermediate control points, but we will sketch a Bézier curve along with its *control polygon* (the polygon obtained by connecting control points). Before describing how to generate these formulas we illustrate the ideas with some examples.

Example 2.2.9. *Familiar Bézier curves*

(a) Find a Cartesian equation for the quadratic Bézier curve with control points $(-1, 1), (0, -1), (1, 1)$.

Thinking of the control points as position vectors, the parameterization is

$$\begin{aligned}
\mathbf{C}(t) &= (1-t)^2\mathbf{P}_0 + 2(1-t)t\mathbf{P}_1 + t^2\mathbf{P}_2 \\
&= (1-t)^2\langle -1, 1 \rangle + 2(1-t)t\langle 0, -1 \rangle + t^2\langle 1, 1 \rangle \\
&= \langle -(1-t)^2 + t^2, (1-t)^2 - 2(1-t)t + t^2 \rangle = \langle 2t - 1, 4t^2 - 4t + 1 \rangle.
\end{aligned}$$

Since the second component function is the square of the first, this parameterizes the section of $y = x^2$ for $-1 \le x \le 1$. Figure 2.2.4(a) depicts this curve together with its control polygon.

(b) Find a Cartesian equation for the cubic Bézier curve with control points $(-1, -1)$, $(-1/3, 1), (1/3, -1), (1, 1)$.

Utilizing Equation 2.2.4 we have

$$\begin{aligned}
\mathbf{C}(t) &= (1-t)^3\langle -1, -1 \rangle + 3(1-t)^2t\langle -1/3, 1 \rangle + 3(1-t)t^2\langle 1/3, -1 \rangle + t^3\langle 1, 1 \rangle \\
&= \langle 2t - 1, 8t^3 - 12t^2 + 6t - 1 \rangle = \langle (2t - 1), (2t - 1)^3 \rangle, \ 0 \le t \le 1.
\end{aligned}$$

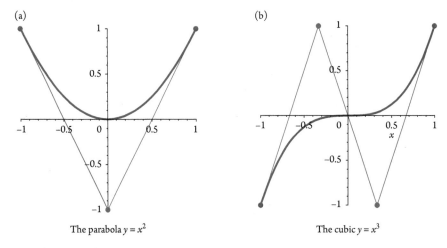

(a)   (b)

The parabola $y = x^2$ The cubic $y = x^3$

Figure 2.2.4 Bézier parameterizations of familiar curves

Since the cube of the x component function is the y component function, this parameterizes $y = x^3$ for $-1 \le x \le 1$. How would you check the limits on x?

See Figure 2.2.4(b) for an illustration of this curve together with its control polygon. Note, in both parts of Figure 2.2.4, that Bézier curves don't go through intermediate control points. ▲

General Bézier Curves The quadratic and cubic Bézier curves given by Equations 2.2.3 and 2.2.4 can be generalized to an arbitrary degree. A degree n Bézier curve $\mathbf{C}(t)$ is determined by $n+1$ points $\mathbf{P}_0, \ldots, \mathbf{P}_n$. The parameterization of $\mathbf{C}(t)$ is a linear combination of its control points, where the coefficients are Bernstein polynomials which arise from a clever application of the binomial theorem. Notice the pattern below:

$$1^1 = (1-t)+t$$
$$1^2 = ((1-t)+t)^2 = (1-t)^2 + 2(1-t)t + t^2$$
$$1^3 = ((1-t)+t)^3 = (1-t)^3 + 3(1-t)^2 t + 3(1-t)t^2 + t^3$$
$$1^4 = ((1-t)+t)^4 = (1-t)^4 + 4(1-t)^3 t + 6(1-t)^2 t^2 + 4(1-t)t^3 + t^4$$
$$\cdots \tag{2.2.5}$$

Notice that the polynomials occurring in the expansion of 1^2 and 1^3 are the coefficients of the control points in Equations 2.2.3 and 2.2.4, respectively. It stands to reason, then, that the quartic Bézier curve determined by control points $\mathbf{P}_0, \ldots, \mathbf{P}_4$ has the parameterization

$$\mathbf{C}(t) = (1-t)^4 \mathbf{P}_0 + 4(1-t)^3 t \mathbf{P}_1 + 6(1-t)^2 t^2 \mathbf{P}_2 + 4(1-t)t^3 \mathbf{P}_3 + t^4 \mathbf{P}_4, \ 0 \le t \le 1.$$

To complete the generalization, observe that the constant coefficients in Equations 2.2.5 are the familiar binomial coefficients $\binom{n}{r} = \frac{n!}{r!(n-r)!}$. The 6 in front of $(1-t)^2 t^2$, for example, is simply $\binom{4}{2}$. One way to describe the pattern is that the ith term in the expansion of $((1-t)+t)^n$ is the Bernstein polynomial $B_{i,n}(t) = \binom{n}{i}(1-t)^{n-i}t^i$, for $i=0,\ldots,n$. Using this description, the degree n Bézier curve with control points $\mathbf{P}_0,\ldots,\mathbf{P}_n$ has the parameterization

$$\mathbf{C}(t) = \sum_{i=0}^{n} \binom{n}{i}(1-t)^{n-i}t^i\mathbf{P}_i, \quad 0 \le t \le 1. \tag{2.2.6}$$

Can you describe why the sum of the coefficients in a Bézier parameterization is always one (Hint: see Equations 2.2.5)? Can you determine $\mathbf{C}(0)$ and $\mathbf{C}(1)$ for any Bézier curve?

There is also a recursive way to define Bézier curves, which we now mention. Let $\mathbf{P}_0,\ldots,\mathbf{P}_n$ be $n+1$ control points, $\mathbf{C}_0(t)$ the Bézier parameterization using $\mathbf{P}_0,\ldots,\mathbf{P}_{n-1}$, and $\mathbf{C}_1(t)$ the Bézier parameterization using $\mathbf{P}_1,\ldots,\mathbf{P}_n$. The Bézier curve determined by $\mathbf{P}_0,\ldots,\mathbf{P}_n$ is

$$\mathbf{C}(t) = (1-t)\mathbf{C}_0(t) + t\mathbf{C}_1(t).$$

Example 2.2.10. *Recursive cubic*

Let $\mathbf{C}_0(t)$ be the quadratic Bézier curve with control points $(1,0),(1,1),(0,3)$, and $\mathbf{C}_1(t)$ with control points $(1,1),(0,3),(-1,2)$. Show that

$$\mathbf{C}(t) = (1-t)\mathbf{C}_0(t) = t\mathbf{C}(t)$$

is the cubic Bézier curve with control points $(1,0),(1,1),(0,3),(-1,2)$.

The strategy is to use Equation 2.2.3 to compute $\mathbf{C}_0$ and $\mathbf{C}_1$, then combine like terms to obtain Equation 2.2.4. Without further ado, we have

$$\mathbf{C}_0(t) = (1-t)^2\langle 1,0\rangle + 2t(1-t)\langle 1,1\rangle + t^2\langle 0,3\rangle$$
$$\mathbf{C}_1(t) = (1-t)^2\langle 1,1\rangle + 2t(1-t)\langle 0,3\rangle + t^2\langle -1,2\rangle$$

from which it follows that $\mathbf{C}(t)$ is given by

$$\begin{aligned}
\mathbf{C}(t) &= (1-t)\mathbf{C}_0(t) + t\mathbf{C}(t)\\
&= (1-t)^3\langle 1,0\rangle + 2t(1-t)^2\langle 1,1\rangle + (1-t)t^2\langle 0,3\rangle\\
&\quad + t(1-t)^2\langle 1,1\rangle + 2t^2(1-t)\langle 0,3\rangle + t^3\langle -1,2\rangle\\
&= (1-t)^3\langle 1,0\rangle + \big(2t(1-t)^2 + t(1-t)^2\big)\langle 1,1\rangle\\
&\quad + \big((1-t)t^2 + 2t^2(1-t)\big)\langle 0,3\rangle + t^3\langle -1,2\rangle\\
&= (1-t)^3\langle 1,0\rangle + 3t(1-t)^2\langle 1,1\rangle + 3t^2(1-t)\langle 0,3\rangle + t^3\langle -1,2\rangle.
\end{aligned}$$

This is the cubic Bézier parameterization given in Equation 2.2.4. ▲

De Casteljau's algorithm: We've seen that Bézier curves require very little storage space, simply maintain the control points and the appropriate Bernstein polynomials. This feature simplifies some of the calculations necessary in computer graphics, but is not optimal for others. Rendering a final curve, for example, is accomplished using (many) straight line segments instead of smooth curves. It turns out that Bézier curves have another property that allows them to be useful in replacing smooth curves with piecewise linear ones, and that is the ability to quickly subdivide them into two "straighter" Bézier curves. De Casteljau's algorithm for subdividing Bézier curves is quite general, and we only illustrate it when subdividing a cubic Bézier curve into two at its midpoint.

Start with a cubic Bézier curve with control points $\mathbf{P}_0, \ldots, \mathbf{P}_3$. The control polygon has three edges, and let $\mathbf{Q}_0, \mathbf{Q}_1, \mathbf{Q}_2$ be the midpoints of those edges. The midpoints $\mathbf{Q}_0, \mathbf{Q}_1, \mathbf{Q}_2$ determine a control polygon with two edges, and let $\mathbf{R}_0, \mathbf{R}_1$ be the midpoints of those edges. Finally, let $\mathbf{S}_0$ be the midpoint of the edge $\mathbf{R}_0\mathbf{R}_1$. This situation is depicted in Figure 2.2.5. We illustrate using control points $\langle 0,0\rangle, \langle 4,-8\rangle, \langle 0,16\rangle, \langle -8,8\rangle$, and recall that the midpoint of an edge is the average of its endpoints.

P	Q	R	S
$\mathbf{P}_0 = \langle 0,0 \rangle$			
	$\mathbf{Q}_0 = \langle 2,-4 \rangle$		
$\mathbf{P}_1 = \langle 4,-8 \rangle$		$\mathbf{R}_0 = \langle 2,0 \rangle$	
	$\mathbf{Q}_1 = \langle 2,4 \rangle$		$\mathbf{S}_0 = \langle 0.5,4 \rangle$
$\mathbf{P}_2 = \langle 0,16 \rangle$		$\mathbf{R}_1 = \langle -1,8 \rangle$	
	$\mathbf{Q}_2 = \langle -4,12 \rangle$		
$\mathbf{P}_3 = \langle -8,8 \rangle$			

It turns out that $\mathbf{S}_0 = \mathbf{C}(0.5)$. Yes, always. Moreover, this process splits the original cubic into two—a mild combinatorial miracle. More precisely, if $\mathbf{C}(t), 0 \le t \le 1$ is determined by the control points $\mathbf{P}_0, \ldots, \mathbf{P}_3$, then the cubic $\mathbf{C}_0(t)$, $0 \le t \le 1$ determined by

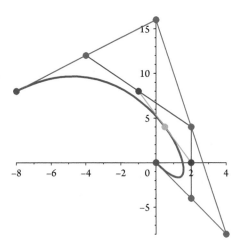

Figure 2.2.5 Subdividing a cubic

P_0, Q_0, R_0, S_0 parameterizes the first half of $C(t)$. More precisely, $C_0(t) = C(t/2)$. The second half of $C(t)$ is the cubic $C_1(t), 0 \le t \le 1$ with control points S_0, R_1, Q_2, P_3. The verification of these facts is not difficult, but is somewhat laborious. They involve writing each of the points in terms of the original control points and lots of algebra.

As an illustration we demonstrate that $C_0(t) = C(t/2)$ for a quadratic Bézier curve. The corresponding proof for higher degree Bézier curves is just uglier algebra, but this shows the elementary nature of the proof. The first step is rewriting the control points involved, as in the table:

P	Q	R
P_0		
	$Q_0 = \dfrac{P_0 + P_1}{2}$	
P_1		$R_0 = \dfrac{P_0 + 2P_1 + P_2}{4}$
	$Q_1 = \dfrac{P_1 + P_2}{2}$	
P_2		

In terms of P_0, P_1, P_2, we have

$$C_0(t) = (1-t)^2 P_0 + 2(1-t)t Q_0 + t^2 R_0$$
$$= \left((1-t)^2 + (1-t)t + \frac{t^2}{4}\right) P_0 + \left((1-t)t + \frac{t^2}{2}\right) P_1 + \frac{t^2}{4} P_2$$
$$= \left(1 - \frac{t}{2}\right)^2 P_0 + \left(1 - \frac{t}{2}\right) t P_1 + \frac{t^2}{4} P_2 = C(t/2).$$

Example 2.2.11. *Subdividing a cubic Bézier curve*

Let $C(t)$ be given by control points $\langle 0,0 \rangle$, $\langle 4,-8 \rangle$, $\langle 0,16 \rangle$, $\langle -8,8 \rangle$. Subdivide $C(t)$ into two cubic Bézier curves using de Casteljau's algorithm.

We begin by recursively finding midpoints of edges, and record this in a table as above.

P	Q	R	S
$\langle 0,0 \rangle$			
	$\langle 2,-4 \rangle$		
$\langle 4,-8 \rangle$		$\langle 2,0 \rangle$	
	$\langle 2,4 \rangle$		$\langle 0.5,4 \rangle$
$\langle 0,16 \rangle$		$\langle -1,8 \rangle$	
	$\langle -4,12 \rangle$		
$\langle -8,8 \rangle$			

Applying Equation 2.2.4, the first half of $C(t)$ is

$$C_0(t) = (1-t)^3 \langle 0,0 \rangle + 3(1-t)^2 t \langle 2,-4 \rangle + 3(1-t)t^2 \langle 2,0 \rangle + t^3 \langle 0.5,4 \rangle.$$

The second half is

$$C_1(t) = (1-t)^3 \langle 0.5, 4 \rangle + 3(1-t)^2 t \langle -1, 8 \rangle + 3(1-t)t^2 \langle -4, 12 \rangle + t^3 \langle -8, 8 \rangle. \; \blacktriangle$$

Exercises

1. Find parametric equations for the line through $(-3, 1, 4)$ and $(2, 5, 1)$.
2. Parameterize the line through $(2, -7, 3)$ and in the direction of $\langle -2, 3, 1 \rangle$.
3. Parameterize the line through $(0, 1, 4)$ and in the direction of $\langle 6, -2, -4 \rangle$.
4. Find a vector parallel to the line $C(t) = (2 - 3t, 3t - 7, 2t + 1)$.
5. Parameterize the line through $(2, 0, 5)$ and parallel to the line $C(t) = (2 + t, t + 5, 2t - 6)$.
6. Parameterize the line segment from $(2, 3, 2)$ to $(7, -6, 3)$.
7. Parameterize the line segment from $(-3, 4, 2)$ to $(6, 5, 7)$.
8. Parameterize the triangle with vertices $(2, -3, 2)$, $(0, -2, -1)$, and $(-4, -7, 3)$, counterclockwise when viewed from above.
9. Parameterize the triangle with vertices $(2, 2, 5)$, $(-5, 2, 1)$, and $(0, 7, -3)$, counterclockwise when viewed from above.
10. Parameterize the triangle with vertices $(-4, 5, 6)$, $(3, 1, 2)$, and $(4, 7, 7)$, clockwise when viewed from above.
11. Find parametric equations for the plane through $(0, 1, 0)$, $(-1, 2, 4)$, and $(3, -2, 5)$.
12. Parameterize the parallelogram determined by the position vectors $\mathbf{u} = \langle 2, 1, 3 \rangle$ and $\mathbf{v} = \langle -1, 4, 4 \rangle$.
13. Find two vectors on the boundary of the parallelogram given by the parameterization $S(s, t) = \langle 3t + 4s, t - s, 4t + 2s \rangle$, $0 \le s, t \le 1$.
14. Parameterize the quadratic Bézier curve with control points $(0, 0), (3, 0), (1, 2)$.
15. Find a Bézier parameterization for the portion of $y = x^2$ with $-2 \le x \le 0$.
16. This problem investigates the effects of translating Bézier curves. Recall from Example 2.2.9 that $y = x^2, 0 \le x \le 2$ has a Bézier parameterization with control points $(-1, 1), (0, -1), (1, 1)$.
    (a) Find control points that parameterize $y = (x + 1)^2, -2 \le x \le 0$.
    (b) How about $y = x^2 + 2, -1 \le x \le 1$?
    (c) And $y = (x + 1)^2 + 2, -2 \le x \le 0$?
    (d) Can you generalize to $y = (x - h)^2 + k$? What interval of x-values would your parameterization use?
17. Parameterize the quadratic Bézier curve with control points $(0, 0), (-2, 0), (2, 6)$. Now subdivide it into two quadratic Bézier curves using de Casteljau's algorithm.

18. Figure 2.2.5 subdivides a cubic into two. The first starts at the origin, the second comprises the last half. If your goal is to build the curve out of cubic Bézier curves which are fairly straight, which half needs more subdivision? Why? Write a computer program that implements it.

19. We presented de Casteljau's algorithm for midpoints. Do some research and find out how this can be generalized.

2.3 The Dot Product

There are several ways to "multiply" vectors. One method is called the *dot product*, denoted $\mathbf{v} \cdot \mathbf{w}$, and it multiplies two vectors to get a scalar. After giving an (unmotivated) algebraic definition of the dot product, we notice some of its algebraic properties in Theorem 2.3.1. These will be useful in applications, calculations, and proving further theorems. Then we prove Theorem 2.3.2, which gives a geometric interpretation of the dot product—relating it to the angle between the vectors. The interplay between the algebraic and geometric interpretations of the dot product provides the computational tools necessary to solve several geometric problems. In particular, we obtain insight into the geometry behind Cartesian equations for planes, as well as the ability to project one vector onto another (a skill necessary for many applications). We define the dot product.

Definition 2.3.1. *The dot product of the vectors* $\mathbf{v} = \langle a, b, c \rangle$ *and* $\mathbf{w} = \langle d, e, f \rangle$ *is the scalar*

$$\mathbf{v} \cdot \mathbf{w} = ad + be + cf.$$

Example 2.3.1. *Calculating dot products*

(a) Let $\mathbf{v} = \langle 1, -2, 1 \rangle$, and $\mathbf{w} = \langle -3, -5, 4 \rangle$. Then

$$\mathbf{v} \cdot \mathbf{w} = \langle 1, -2, 1 \rangle \cdot \langle -3, -5, 4 \rangle = 1 \cdot (-3) + (-2) \cdot (-5) + 1 \cdot 4 = 11.$$

(b) If $\mathbf{v} = \langle 1, 2 \rangle$, find a vector $\mathbf{w} = \langle a, b \rangle$ such that $\mathbf{v} \cdot \mathbf{w} = 0$. The thing that makes this example difficult is that there are infinitely many answers, and you have a choice. We have to solve the equation

$$\mathbf{v} \cdot \mathbf{w} = 1 \cdot a + 2 \cdot b = 0.$$

Perhaps the easiest solution is to interchange coordinates and negate one of them. Thus $\mathbf{w} = \langle 2, -1 \rangle$ is a solution. Notice that any scalar multiple $t\mathbf{w} = \langle 2t, -t \rangle$ is also a solution! In fact, $t\mathbf{w}$ is a parametric line in the plane which represents all solutions. ▲

The dot product has many convenient algebraic properties that follow directly from the definition. It will be useful to know how it interacts with vector addition and scalar multiplication. We list some properties in the following theorem.

Theorem 2.3.1. *Let* $\mathbf{v}$, $\mathbf{w}$, *and* $\mathbf{x}$ *be vectors, and* α *a scalar. Then*

1. $\mathbf{v} \cdot \mathbf{w} = \mathbf{w} \cdot \mathbf{v}$
2. $(\alpha \mathbf{v}) \cdot \mathbf{w} = \alpha (\mathbf{v} \cdot \mathbf{w})$
3. $(\mathbf{v} + \mathbf{w}) \cdot \mathbf{x} = \mathbf{v} \cdot \mathbf{x} + \mathbf{w} \cdot \mathbf{x}$
4. $\|\mathbf{v}\|^2 = \mathbf{v} \cdot \mathbf{v}$

Proof. We prove the first statement, and leave the remaining verifications as exercises. Let $\mathbf{v} = \langle a, b, c \rangle$ and $\mathbf{w} = \langle d, e, f \rangle$ be vectors. A straightforward calculation shows

$$\mathbf{v} \cdot \mathbf{w} = a \cdot d + b \cdot e + c \cdot f = d \cdot a + e \cdot b + f \cdot c = \mathbf{w} \cdot \mathbf{v}.$$

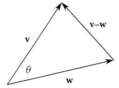

Figure 2.3.1 The Vector form of the law of cosines

The first equality follows from the definition of the dot product, the second is justified because multiplication in $\mathbb{R}$ is commutative, and the last is another application of the definition of the dot product. $\square$

In addition to algebraic properties, the dot product has an important geometric interpretation. It comprises the next theorem:

Theorem 2.3.2. *Let* **v** *and* **w** *be non-zero vectors, and let* θ *be the angle between them. Then*

$$\mathbf{v} \cdot \mathbf{w} = \|\mathbf{v}\| \, \|\mathbf{w}\| \cos\theta.$$

Proof. The result follows from an application of the law of cosines, which states that $C^2 = A^2 + B^2 - 2AB\cos\theta$, where A, B, C are lengths of sides of a triangle and θ is the angle opposite C. If $\mathbf{v} = \langle a, b, c \rangle$ and $\mathbf{w} = \langle d, e, f \rangle$, then $\mathbf{v} - \mathbf{w} = \langle a - d, b - e, c - f \rangle$. Moreover, the vectors $\mathbf{v}$, $\mathbf{w}$, and $\mathbf{v} - \mathbf{w}$ form a triangle in which the angle θ between $\mathbf{v}$ and $\mathbf{w}$ is opposite side $\mathbf{v} - \mathbf{w}$. Applying the law of cosines, and almost all of Theorem 2.3.1, to this situation gives

$$\|\mathbf{v} - \mathbf{w}\|^2 = \|\mathbf{v}\|^2 + \|\mathbf{w}\|^2 - 2\|\mathbf{v}\| \, \|\mathbf{w}\| \cos\theta$$

$$(\mathbf{v} - \mathbf{w}) \cdot (\mathbf{v} - \mathbf{w}) = \mathbf{v} \cdot \mathbf{v} + \mathbf{w} \cdot \mathbf{w} - 2\|\mathbf{v}\| \, \|\mathbf{w}\| \cos\theta$$

$$\mathbf{v} \cdot \mathbf{v} - 2\mathbf{v} \cdot \mathbf{w} + \mathbf{w} \cdot \mathbf{w} = \mathbf{v} \cdot \mathbf{v} + \mathbf{w} \cdot \mathbf{w} - 2\|\mathbf{v}\| \, \|\mathbf{w}\| \cos\theta$$

$$-2\mathbf{v} \cdot \mathbf{w} = -2\|\mathbf{v}\| \, \|\mathbf{w}\| \cos\theta$$

proving the desired result. Which parts of Theorem 2.3.1 justify the transition from the second to the third line? $\square$

There are many consequences of Theorem 2.3.2—the geometry of the dot product—that we will encounter. We start by calculating an angle, then show how to find Cartesian equations of planes. We also see how to use the dot product to find projections of one vector onto another.

Example 2.3.2. *Finding angles between vectors*

Find the angle between $\mathbf{v} = \langle 5, 1, -2 \rangle$, and $\mathbf{w} = \langle -3, 3, 4 \rangle$. By the above formula we have

$$\theta = \cos^{-1}\left(\frac{\langle 5, 1, -2 \rangle \cdot \langle -3, 3, 4 \rangle}{\sqrt{5^2 + 1^2 + (-2)^2}\sqrt{(-3)^2 + 3^2 + 4^2}} \right) = \cos^{-1}\left(-\frac{20}{\sqrt{1020}} \right). \blacktriangle$$

Theorem 2.3.2 also provides us with an easy way to check if two vectors are perpendicular. We take a moment to mention that *perpendicular, orthogonal,* and *normal* are three words for the same thing: meeting at $90°$. We have the following corollary:

Corollary 2.3.1. *The non-zero vectors* **v** *and* **w** *are orthogonal if and only if their dot product is zero, i.e.* $\mathbf{v} \cdot \mathbf{w} = 0$.

Proof. Since **v** and **w** are non-zero vectors, their lengths are positive. This means the only way $\|\mathbf{v}\| \|\mathbf{w}\| \cos\theta$ can be zero is if $\cos\theta = 0$. Thus, by Theorem 2.3.2, we have $\mathbf{v} \cdot \mathbf{w} = 0$ if and only if $\cos\theta = 0$. For us, θ is always between 0 and π, so this occurs if and only if $\theta = \pi/2$ and **v** meets **w** orthogonally. □

Cartesian Equations for Planes: We introduced Cartesian equations for planes in Section 1.3. At that time we asked you to believe that solutions to linear equations in three variables (i.e. equations of the form $Ax + By + Cz = D$) yield planes in $\mathbb{R}^3$ by analogy with the general equation $Ax + By = C$ of a line in $\mathbb{R}^2$. With the help of Theorem 2.3.2 we can now justify what we've taken for granted, and shed light on information that the constants A, B, C, D give.

In Section 2.2 we showed that lines in $\mathbb{R}^3$ are determined by two pieces of information: a point on it and a vector lying in the line. Similarly we will now see that planes are determined by two pieces of information: a point on it and a vector *orthogonal* to it. We begin with an example.

Example 2.3.3. *Orthogonal vectors*

Describe all non-zero vectors that are orthogonal to $\mathbf{v} = \langle 2, -3, 1 \rangle$.

To accomplish this, let $\mathbf{w} = \langle x, y, z \rangle$ be any vector orthogonal to **v**. Corollary 2.3.1 implies that the set of all vectors orthogonal to $\mathbf{v} = \langle 2, -3, 1 \rangle$ is all vectors $\mathbf{w} = \langle x, y, z \rangle$ satisfying the equation

$$\mathbf{v} \cdot \mathbf{w} = \langle 2, -3, 1 \rangle \cdot \langle x, y, z \rangle = 2x - 3y + z = 0.$$

Recall that this is the equation for a plane! Note also that it goes through the origin.▲

Example 2.3.3 uses the fact that two non-zero vectors are orthogonal if and only if their dot product is zero. We extend this use to find Cartesian equations of planes. More precisely, let $\mathcal{P}$ be a plane with $\mathbf{n} = \langle A, B, C \rangle$ a normal vector to $\mathcal{P}$ and $P = (x_0, y_0, z_0)$ a point on $\mathcal{P}$. If $X = (x, y, z)$ is any other point in $\mathcal{P}$, then the vector $\mathbf{v} = \mathbf{X} - \mathbf{P}$ lies in the plane and is therefore orthogonal to **n** (see Figure 2.3.2). This forces the dot product to be zero, and we get the equation

$$0 = \mathbf{n} \cdot \mathbf{v} = \mathbf{n} \cdot (\mathbf{X} - \mathbf{P}) = \mathbf{n} \cdot \mathbf{X} - \mathbf{n} \cdot \mathbf{P},$$

which has the more symmetric form

$$\mathbf{n} \cdot \mathbf{X} = \mathbf{n} \cdot \mathbf{P}.$$

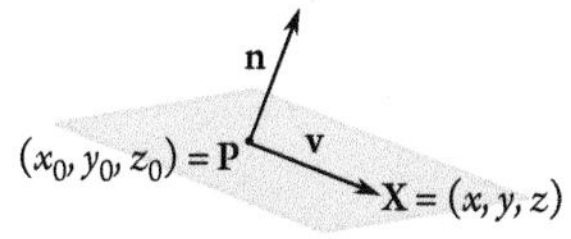

Figure 2.3.2 The vector derivation of Cartesian equations for planes

Replacing $\mathbf{X}$, $\mathbf{P}$, and $\mathbf{n}$ with their coordinates gives the following format for an equation of a plane:

Equations for planes

A Cartesian equation for the plane orthogonal to $\mathbf{n} = \langle A, B, C \rangle$ and through $P(x_0, y_0, z_0)$ is

$$Ax + By + Cz = D \qquad (2.3.1)$$

where $D = Ax_0 + By_0 + Cz_0$.

In vector notation, letting $\mathbf{X} = \langle x, y, z \rangle$, we have the formula

$$\mathbf{n} \cdot \mathbf{X} = \mathbf{n} \cdot \mathbf{P}$$

Example 2.3.4. *Cartesian equations for planes*

(a) Find a Cartesian equation for the plane through $(2, -1, -3)$ with normal vector $\mathbf{n} = \langle 3, 5, 2 \rangle$.

Since the components of the normal vector $\mathbf{n}$ are the coefficients of x, y, and z, and the constant is the dot product of $\mathbf{n}$ with the point, we have that

$$3x + 5y + 2z = \langle 2, -1, -3 \rangle \cdot \langle 3, 5, 2 \rangle$$
$$3x + 5y + 2z = -5$$

is an equation for the plane.

(b) Find a unit normal vector to the plane $2x + y - z = 4$.

The coefficients tell us that $\langle 2, 1, -1 \rangle$ is a normal vector to the plane, so normalizing it gives the answer

$$\mathbf{u} = \frac{\mathbf{n}}{\|\mathbf{n}\|} = \frac{\langle 2, 1, -1 \rangle}{\sqrt{6}} = \left\langle \frac{2}{\sqrt{6}}, \frac{1}{\sqrt{6}}, \frac{-1}{\sqrt{6}} \right\rangle. \; \blacktriangle$$

Math App 2.3.1. *Plane equations*

Now that you know a little about Cartesian equations for planes, the following Math App adds geometric intuition to your knowledge.

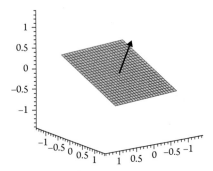

Projections: Vector projections are a second application of Theorem 2.3.2. The projection of **v** onto **w** is obtained in the following way. Pick representatives of **v** and **w** with the same initial point, and drop a perpendicular line from the tip of **v** to **w**. The resulting vector in the direction of **w** is the projection of **v** onto **w**, and is denoted $proj_\mathbf{w}\mathbf{v}$ (see Figure 2.3.3).

The *component* of **v** in the direction of **w** is the signed length of $proj_\mathbf{w}\mathbf{v}$. It is denoted $comp_\mathbf{w}\mathbf{v}$, and is positive if $proj_\mathbf{w}\mathbf{v}$ points in the same direction as **w** and negative if it points in the opposite direction. Our goal is to see how the dot product allows us to easily calculate both $proj_\mathbf{w}\mathbf{v}$ and $comp_\mathbf{w}\mathbf{v}$. We derive the formulas both algebraically and using trigonometry. The derivations make use of the geometry of the dot product given in Theorem 2.3.2.

Algebraic Derivation: To find a formula for $proj_\mathbf{w}\mathbf{v}$, first note that it is a scalar multiple of **w** so there is a scalar α with $proj_\mathbf{w}\mathbf{v} = \alpha\mathbf{w}$. Our task, therefore, is reduced to finding α. Second, notice that the vector $(\mathbf{v} - proj_\mathbf{w}\mathbf{v})$ is perpendicular to **w**, so their dot product vanishes (sketch $(\mathbf{v} - proj_\mathbf{w}\mathbf{v})$ using Figure 2.3.3 as a starting point). The dot product simplifies to

$$\left(\mathbf{v} - proj_\mathbf{w}\mathbf{v}\right) \cdot \mathbf{w} = (\mathbf{v} - \alpha\mathbf{w}) \cdot \mathbf{w} = \mathbf{v} \cdot \mathbf{w} - \alpha\mathbf{w} \cdot \mathbf{w},$$

(a)

θ acute

(b)

θ obtuse

Figure 2.3.3 Projecting vectors

then setting it to zero and solving yields

$$\alpha = \frac{\mathbf{v} \cdot \mathbf{w}}{\mathbf{w} \cdot \mathbf{w}}.$$

Having calculated α, we have the formula for the projection of $\mathbf{v}$ onto $\mathbf{w}$:

$$proj_{\mathbf{w}}\mathbf{v} = \frac{\mathbf{v} \cdot \mathbf{w}}{\mathbf{w} \cdot \mathbf{w}}\mathbf{w}.$$

The formula for the projection vector can be used to find a formula for computing $comp_{\mathbf{w}}\mathbf{v}$. First, using the fact that $\|\mathbf{w}\|^2 = \mathbf{w} \cdot \mathbf{w}$ we see

$$proj_{\mathbf{w}}\mathbf{v} = \frac{\mathbf{v} \cdot \mathbf{w}}{\mathbf{w} \cdot \mathbf{w}}\mathbf{w} = \frac{\mathbf{v} \cdot \mathbf{w}}{\|\mathbf{w}\|^2}\mathbf{w} = \frac{\mathbf{v} \cdot \mathbf{w}}{\|\mathbf{w}\|}\frac{\mathbf{w}}{\|\mathbf{w}\|}.$$

Now $\frac{\mathbf{w}}{\|\mathbf{w}\|}$ is a unit vector, so the signed length of $proj_{\mathbf{w}}\mathbf{v}$ is just its coefficient $\frac{\mathbf{v} \cdot \mathbf{w}}{\|\mathbf{w}\|}$ (pause, take a sip of coffee, and make sure you agree). Since the component of $\mathbf{v}$ in the $\mathbf{w}$ direction *is* the signed length, we've justified

$$comp_{\mathbf{w}}\mathbf{v} = \frac{\mathbf{v} \cdot \mathbf{w}}{\|\mathbf{w}\|}.$$

Take a moment to review the preceding discussion. The key observations in computing $proj_{\mathbf{w}}\mathbf{v}$ were that $proj_{\mathbf{w}}\mathbf{v}$ is a scalar multiple of $\mathbf{w}$ and that $\mathbf{v} - proj_{\mathbf{w}}\mathbf{v}$ is perpendicular to $\mathbf{w}$. To compute $comp_{\mathbf{w}}\mathbf{v}$ we used the general observation that if $\mathbf{u}$ is a unit vector, then $\alpha\mathbf{u}$ has signed length α.

Trigonometric Derivation: Trigonometry provides a derivation of these formulas which is perhaps more familiar, so we include it here. Let θ be the angle between $\mathbf{v}$ and $\mathbf{w}$, and consider the right triangle created by $\mathbf{v}$, $proj_{\mathbf{w}}\mathbf{v}$, and the dotted line in Figure 2.3.3. Recall that $comp_{\mathbf{w}}\mathbf{v}$ is the signed length of $proj_{\mathbf{w}}\mathbf{v}$. Since the cosine of an obtuse angle is negative, the right triangles in Figure 2.3.3 show that $\cos\theta$ is

$$\cos\theta = \frac{comp_{\mathbf{w}}\mathbf{v}}{\|\mathbf{v}\|},$$

whether θ is acute or obtuse. Solving for $comp_{\mathbf{w}}\mathbf{v}$ gives $comp_{\mathbf{w}}\mathbf{v} = \|\mathbf{v}\|\cos\theta$. Now use the result of Theorem 2.3.2 to compute

$$comp_{\mathbf{w}}\mathbf{v} = \|\mathbf{v}\|\cos\theta = \|\mathbf{v}\|\frac{\mathbf{v} \cdot \mathbf{w}}{\|\mathbf{v}\|\|\mathbf{w}\|} = \frac{\mathbf{v} \cdot \mathbf{w}}{\|\mathbf{w}\|}.$$

By Theorem 2.3.1, the length $\|\mathbf{w}\|$ can slide through the dot product until it's under $\mathbf{w}$, giving

$$comp_{\mathbf{w}}\mathbf{v} = \mathbf{v} \cdot \frac{\mathbf{w}}{\|\mathbf{w}\|}.$$

Thus we can say that the dot product of $\mathbf{v}$ with *normalized* $\mathbf{w}$ is the component of $\mathbf{v}$ in the $\mathbf{w}$ direction.

Recall that $proj_{\mathbf{w}}\mathbf{v}$ is the vector that goes $comp_{\mathbf{w}}\mathbf{v}$ units in the direction of $\mathbf{w}$, and the direction of $\mathbf{w}$ is the unit vector $\frac{\mathbf{w}}{\|\mathbf{w}\|}$. Hence

$$proj_{\mathbf{w}}\mathbf{v} = comp_{\mathbf{w}}\mathbf{v}\frac{\mathbf{w}}{\|\mathbf{w}\|} = \frac{\mathbf{v}\cdot\mathbf{w}}{\|\mathbf{w}\|}\frac{\mathbf{w}}{\|\mathbf{w}\|} = \frac{\mathbf{v}\cdot\mathbf{w}}{\|\mathbf{w}\|^2}\mathbf{w} = \frac{\mathbf{v}\cdot\mathbf{w}}{\mathbf{w}\cdot\mathbf{w}}\mathbf{w}.$$

No matter how you slice it, you arrive at the same conclusion. We summarize this discussion below.

The projection $proj_{\mathbf{w}}\mathbf{v}$ of the vector $\mathbf{v}$ onto the vector $\mathbf{w}$ is given by:

$$proj_{\mathbf{w}}\mathbf{v} = \frac{\mathbf{v}\cdot\mathbf{w}}{\mathbf{w}\cdot\mathbf{w}}\mathbf{w}.$$

The component of $\mathbf{v}$ in the $\mathbf{w}$ direction is the signed length of $proj_{\mathbf{w}}\mathbf{v}$, and is given by:

$$comp_{\mathbf{w}}\mathbf{v} = \frac{\mathbf{v}\cdot\mathbf{w}}{\|\mathbf{w}\|}.$$

Remark: Some texts, in math or physics, will refer to what we've called the projection of $\mathbf{v}$ onto $\mathbf{w}$ as the component of $\mathbf{v}$ in the $\mathbf{w}$ direction. We'll reserve the word *component* for a scalar quantity in this book.

Example 2.3.5. *Projecting vectors*

Find the projection $proj_{\mathbf{w}}\mathbf{v}$ of $\mathbf{v}$ onto $\mathbf{w}$ where $\mathbf{v} = \langle 2,-3,1\rangle$ and $\mathbf{w} = \langle 1,2,1\rangle$. By the above formula,

$$proj_{\mathbf{w}}\mathbf{v} = \frac{\mathbf{v}\cdot\mathbf{w}}{\mathbf{w}\cdot\mathbf{w}}\mathbf{w} = \frac{\langle 2,-3,1\rangle\cdot\langle 1,2,1\rangle}{\langle 1,2,1\rangle\cdot\langle 1,2,1\rangle}\langle 1,2,1\rangle = -\frac{1}{2}\langle 1,2,1\rangle. \blacktriangle$$

Example 2.3.6. *Force on a pendulum*

At a particular moment, a 10kg pendulum makes an angle of $45°$ with the horizontal as in Figure 2.3.4. Let $\mathbf{F}_g = 10\langle 0,-9.8\rangle$ denote the force due to gravity. Find the projection of $\mathbf{F}_g$ in the direction perpendicular to movement.

To do so, note that the pendulum travels in a circular arc, with the rope being a radius. Thus the rope is always perpendicular to the direction of movement. Since the force $\mathbf{F}_t$ due to tension is in the direction of the rope, the question asks us to project $\mathbf{F}_g$ into $\mathbf{F}_t$.

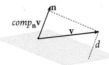

Figure 2.3.4 Projecting $\mathbf{F}_g$ into $\mathbf{F}_t$

Figure 2.3.5 Distance d to plane is $|comp_{\mathbf{n}}\mathbf{v}|$

From Figure 2.3.4 it is clear that $\mathbf{F}_t$ points in the same direction as $\mathbf{v} = \langle -1, 1 \rangle$, so we find $proj_{\mathbf{v}}\mathbf{F}_g$:

$$proj_{\mathbf{v}}\mathbf{F}_g = \frac{\langle -1,1 \rangle \cdot \langle 0,-98 \rangle}{\langle -1,1 \rangle \cdot \langle -1,1 \rangle} \langle -1,1 \rangle = -\frac{98}{2}\langle -1,1 \rangle = \langle 49,-49 \rangle. \ \blacktriangle$$

Example 2.3.7. *Distance to a plane*

Find the distance from the point $(2,4,-3)$ to the plane $3x + 3y + 4z = 1$.

To do this, pick a point in the plane and connect it to the point $(2,4,-3)$ to get a vector $\mathbf{v}$. In our case, the point $(-1,0,1)$ lies in the plane, so $\mathbf{v} = \langle 2,4,-3 \rangle - \langle -1,0,1 \rangle = \langle 3,4,-4 \rangle$. Here is the key observation: the distance d from the tip of $\mathbf{v}$ to the plane is equal to the length of the projection of $\mathbf{v}$ onto the normal direction $\mathbf{n} = \langle 3,3,4 \rangle$. This is just the absolute value of $comp_{\mathbf{n}}\mathbf{v}$, so we compute

$$\left| comp_{\mathbf{n}}\mathbf{v} \right| = \left| \frac{\langle 3,4,-4 \rangle \cdot \langle 3,3,4 \rangle}{\sqrt{\langle 3,3,4 \rangle \cdot \langle 3,3,4 \rangle}} \right| = \frac{5}{\sqrt{34}}. \ \blacktriangle$$

This example can be generalized. Let $\mathcal{P}$ be the plane through the point P and with normal vector $\mathbf{n}$, and suppose it has Cartesian equation $Ax + By + Cz = D$. Comparing Equation 2.3.1 to its vector form we see that $D = \mathbf{P} \cdot \mathbf{n}$.

Now let $Q = (f,g,h)$ be any point, and let $\mathbf{v} = \mathbf{Q} - \mathbf{P}$ be the vector from P to Q. Then the distance d from Q to the plane is the length of the projection $proj_{\mathbf{n}}\mathbf{v}$, which is $|comp_{\mathbf{n}}\mathbf{v}|$.

$$d = |comp_{\mathbf{n}}\mathbf{v}| = \left| \frac{\mathbf{v} \cdot \mathbf{n}}{\|\mathbf{n}\|} \right| = \left| \frac{(\mathbf{Q}-\mathbf{P}) \cdot \mathbf{n}}{\|\mathbf{n}\|} \right| = \left| \frac{\mathbf{Q} \cdot \mathbf{n} - \mathbf{P} \cdot \mathbf{n}}{\|\mathbf{n}\|} \right|.$$

Substituting coordinates for the vectors gives

Distance to a Plane

The distance d from the point (f, g, h) to the plane $Ax + By + Cz = D$ is the scalar

$$d = \frac{|Af + Bg + Ch - D|}{\sqrt{A^2 + B^2 + C^2}}.$$

In particular, the distance from the origin to the plane $Ax + By + Cz = D$ is seen to be

$$\frac{|D|}{\sqrt{A^2 + B^2 + C^2}}.$$

Thus we have geometric interpretations for each part of the equation

$$Ax + By + Cz = D.$$

The coefficients tell us a normal vector to the plane, while the constant can be used to find the distance to the origin. Further, if $\mathbf{n} = \langle A, B, C \rangle$ is a unit vector, the constant D *is* the distance to the origin.

Work: In physics, the work done by a force $\mathbf{F}$ on an object moving from one point to another is defined to be the component of force in the direction of movement times the distance traveled. In short, work is force times distance, as long as you consider only the component of force in the direction of motion.

This is quite reasonable, as a quick thought experiment illustrates. Suppose we have three identical cars, each traveling at 65 miles per hour along straight roads. The first along a flat road, the second uphill and the third downhill. The force of gravity is constant on the cars (well, approximately), but intuition and experience indicate that the work done by gravity is quite different on each. Gravity is doing little to no work on the car traveling horizontally, while it hinders the car traveling upward doing "negative" work on it. Gravity, on the other hand, helps the third car along, so the work done by gravity on a downhill vehicle should be positive.

This is obvious from personal experience, but can also be described considering components of gravitational force in each case. The force due to gravity is perpendicular to movement on a flat road, so for the first car gravity's component in the direction of travel is zero. Hence the work done by gravity on this car is zero. The component of gravity points opposite to the movement of the uphill car, causing negative work by gravity on it. Finally, gravity does positive work on the downhill car, since it and the velocity vector point in the same general direction. We see that changing the direction of travel while maintaining the same force affects the work done by the force. When defining work, then, one ought to consider not just the force, but the force in the direction of travel.

Work has a particularly nice description using vectors. Suppose an object moves in a straight line from point A to point B. Its displacement can be described by the vector $\mathbf{d} = \mathbf{B} - \mathbf{A}$ (here we're thinking of the points A and B as position vectors, and taking terminal minus initial to find the vector from A to B). The length $\|\mathbf{d}\|$ is the distance traveled. Recall that the component of $\mathbf{F}$ in the $\mathbf{d}$ direction is the scalar $\frac{\mathbf{F} \cdot \mathbf{d}}{\|\mathbf{d}\|}$. Since work w is the component of force in the direction of movement times the distance traveled, we have

$$w = \frac{\mathbf{F} \cdot \mathbf{d}}{\|\mathbf{d}\|} \|\mathbf{d}\| = \mathbf{F} \cdot \mathbf{d}.$$

So the work done by a constant force on an object moving in a straight line is the dot product of the force and displacement vectors!

Concept Connection: The above remarks apply to constant forces acting on objects moving linearly. We want to also consider work done by variable forces acting on objects moving along curves. This generalization is similar to calculating distance traveled and transitioning from constant to variable velocity. As in the distance scenario, it turns out that to compute the work we'll have to integrate, as we shall see in Chapter 5.

Example 2.3.8. *Work done by constant force*

A boy pulls a sled horizontally 10 meters by applying a constant force of 16 newtons at an angle of $60°$ from the horizontal. How much work did he do?

The displacement vector is $\mathbf{d} = \langle 10, 0 \rangle$ and the force vector is

$$\mathbf{F} = \langle 16 \cos 60°, 16 \sin 60° \rangle = \langle 8, 8\sqrt{3} \rangle.$$

Thus the work done is

$$w = \mathbf{F} \cdot \mathbf{d} = \langle 10, 0 \rangle \cdot \langle 8, 8\sqrt{3} \rangle = 80\text{Nm} = 80\text{J}.$$

A newton-meter is also known as a joule. ▲

Example 2.3.9. *Work by an electric force*

We briefly introduce electric fields here, for the purpose of calculating work done by a field on a point charge. An electric field is created by a system of charged objects, and it exerts a force on a test charge moving through it. The force it exerts depends on the size of the test charge, so units for an electric field are always newtons per coulomb. Given an electric field $\mathbf{E}$ and a test charge of Q coulombs, then, one finds the force by multiplying them. Please see Section 2.6 for a more detailed discussion of electric fields.

Assume that a uniform electric field $\mathbf{E} = \langle 0, 3 \rangle$ N/C has been placed on a plane, and that a point charge q of 2C is moved along a straight line from $(0, 10)$ to $(4, 3)$. Find the work that $\mathbf{E}$ does on q.

Since our charge is 2C, we find the constant force on the charge to be $\mathbf{F} = 2\langle 0,3 \rangle = \langle 0,6 \rangle$. Finally, work is the dot product of the force and displacement vectors, so

$$w = \mathbf{F} \cdot \mathbf{d} = \langle 0,6 \rangle \cdot \langle 4,-7 \rangle = -42J.$$

We follow this calculation with the observation that if the point charge travels in a horizontal line, then $\mathbf{E}$ does no work on it. Indeed, in this case the vertical component of $\mathbf{d}$ is zero, making $w = \mathbf{F} \cdot \mathbf{d}$ equal zero as well. ▲

Exercises

1. Find the angle between the following pairs of vectors.
   (a) $\mathbf{v} = \langle \sqrt{3}, 1 \rangle$ and $\mathbf{w} = \langle 7,0 \rangle$.
   (b) $\mathbf{v} = \langle 0,-1,1 \rangle$ and $\mathbf{w} = \langle 0,5,0 \rangle$.
   (c) $\mathbf{v} = \langle -3,4,2 \rangle$ and $\mathbf{w} = \langle 2,1,1 \rangle$.

2. Let $\mathbf{v}$ have initial point $(2,1,3)$ and terminal point $(3,-2,2)$, and let $\mathbf{w} = \langle 2,1,1 \rangle$. Find the angle between $\mathbf{v}$ and $\mathbf{w}$.

3. Find an equation for the line in $\mathbb{R}^2$ through the origin and orthogonal to $\mathbf{v} = \langle 3,7 \rangle$. Sketch both the line and the vector.

4. Give a geometric interpretation of $\mathbf{v} \cdot \mathbf{w}$ when both $\mathbf{v}$ and $\mathbf{w}$ are unit vectors.

5. Let $\mathbf{u} = \langle a,b,c \rangle$, $\mathbf{v} = \langle d,e,f \rangle$, and $\mathbf{w} = \langle g,h,k \rangle$ be vectors, and α a scalar. Verify the remaining parts of Theorem 2.3.1.
   (a) $(\alpha\mathbf{v}) \cdot \mathbf{w} = \alpha(\mathbf{v} \cdot \mathbf{w})$
   (b) $(\mathbf{u} + \mathbf{v}) \cdot \mathbf{w} = \mathbf{u} \cdot \mathbf{w} + \mathbf{v} \cdot \mathbf{w}$
   (c) $\|\mathbf{v}\|^2 = \mathbf{v} \cdot \mathbf{v}$

6. If $\mathbf{v} \cdot \mathbf{w} < 0$, what can you say about the angle between them?

7. Find an equation for the plane through $(4,-1,-3)$ with normal vector $\mathbf{n} = \langle 2,1,1 \rangle$.

8. Find both unit vectors orthogonal to the plane $5x - 3y + 2z = 6$.

9. Parameterize the line through the origin and orthogonal to $3x - 2y + 7z = 5$.

10. Find the projection of $\langle -2,4,3 \rangle$ onto $\langle 1,1,2 \rangle$.

11. Find the component of $\mathbf{v} = \langle 2,7,4 \rangle$ in the direction of $\mathbf{w} = \langle 3,1,-2 \rangle$.

12. Let $\mathbf{v} = \langle -3,4 \rangle$ and $\mathbf{w} = \langle 2,4 \rangle$.
    (a) Sketch the position vectors $\mathbf{v}$ and $\mathbf{w}$ on the same axes.
    (b) Find $proj_\mathbf{v}\mathbf{w}$ and include it in your sketch.
    (c) What scalar λ makes $proj_\mathbf{v}\lambda\mathbf{w}$ a unit vector?

13. Describe the set of all unit vectors $\mathbf{u}$ for which $comp_\mathbf{k}\mathbf{u} > 0$.

14. Find the distance from $(3,-1,2)$ to the plane $2x + y - 3z = 4$.

15. Find the distance to the origin from $x + y + 4z = 1$.

16. Find a formula for the set of all points 3 units from the plane $x + y + z = 3$, on the $\mathbf{n} = \langle 1, 1, 1 \rangle$ side of it.

17. Find the point on the plane $2x + y - 3z = 2$ closest to the origin.

18. Find the point on $x - 2y + z = 4$ closest to the origin.

19. Find a unit vector that makes an angle of $\pi/3$ with the y-axis and $\pi/4$ with the x-axis.

20. Find an equation for the plane parallel to $3x - 3y + 2x = 5$ and through the point $(4, 3, 5)$.

21. Find an equation for the plane orthogonal to the line $\mathbf{C}(t) = (2 + t, 3 - 2t, 1 - t)$ and through $(1, 0, -3)$.

22. The force due to gravity acting on a 5kg pendulum is $\mathbf{F}_g = \langle 0, -49 \rangle$. Find the projection of $\mathbf{F}_g$ in the direction of $\mathbf{F}_t$ when the pendulum makes an angle of $60°$ with the horizontal.

23. Find the work done by the constant force $\mathbf{F} = \langle -3, 5 \rangle$ on a particle moving a displacement $\langle d \rangle = \langle 2, 0 \rangle$.

24. A boy pulls a sled 40 meters with a force of 20N at $45°$ to the horizontal. Find the work done by the boy on the sled.

25. A test charge q of 4C is moved through the constant electric field $\mathbf{E} = \langle 0, 0, 3 \rangle$ along the line segment from $(0, 0, 2)$ to $(-1, 3, 5)$. Find the work done by $\mathbf{E}$ on q.

26. Let $\mathbf{E} = \langle 1, 2 \rangle$ N/C be a uniform electric field in the plane, and q a 2-coulomb point charge. Compute the work done by $\mathbf{E}$ as q moves from $(0, 0)$ to $(2, 1)$ along
    (a) the line from $(0, 0)$ to $(2, 1)$
    (b) the horizontal line from $(0, 0)$ to $(2, 0)$, then the vertical line from $(2, 0)$ to $(2, 1)$.

27. Compute the work done by gravity on a 1000 kilogram elevator moving 30 meters up (assume the force due to gravity is $\mathbf{F}_g = M \cdot g$ where M is mass and $g = 9.8\text{m/s}^2$.)

2.4 The Cross Product

In this section we introduce the cross product of two vectors. Whereas the dot product produces a scalar, the cross product of two vectors produces a third vector with several desirable geometric properties. Some preliminary topics must be discussed before defining the cross product. Matrices are defined, and matrix algebra introduced. Matrix addition and scalar multiplication are direct analogs of vector operations. Determinants are not treated in detail, but defined for 2×2 and 3×3 matrices, and given a geometric interpretation in each dimension. Finally matrix multiplication is defined, which will be used when discussing the multivariable chain rule in Section 3.5. After completing these preliminary topics, the algebraic definition of the cross product is given, followed by a geometric interpretation in Theorem 2.4.2. The section concludes with some applications of the cross product to geometry and physics.

Matrix Preliminaries: Matrices are an important mathematical tool that arise in a variety of contexts. We will not use them much in this course, but we do need some familiarity with them. What follows is the briefest of introductions to a subject deserving of much more. For instance, determinants are not covered in general, but one method of calculating them is introduced. We begin by defining matrices.

Definition 2.4.1. *An m-by-n matrix is a rectangular array with m rows and n columns. If A is a matrix, then a_{ij} represents the entry in the ith row and j^{th} column.*

The singular form of matrices is *matrix* NOT matrice (there is no such word).

Example 2.4.1. *Matrices*

The following are matrices. The left has numeric entries, while the first row of the middle matrix is the standard basis vectors, and the entries in the third are functions. To clarify the notation, note that in the first matrix we have $a_{23} = -2$, in the second $a_{32} = 17$, and $a_{22} = 2x^2$ in the third.

$$\begin{bmatrix} 5 & -7 & 3 \\ -14 & 3 & -2 \\ 2 & 4 & 6 \end{bmatrix}, \quad \begin{bmatrix} \mathbf{i} & \mathbf{j} & \mathbf{k} \\ 3 & 2 & 1 \\ 4 & 17 & 3 \end{bmatrix}, \quad \begin{bmatrix} 2y^2 & 2+3y^2 \\ 2+3y^2 & 2x^2 \end{bmatrix}. \quad \blacktriangle$$

The operations of matrix addition and scalar multiplication are defined in the same way as they are for vectors—they are done entry-wise. Matrices can only be added if they have the same dimension. An example illustrates this.

Example 2.4.2. *Matrix operations*

(*a*) Scalar multiplication: $3 \begin{bmatrix} 2 & -1 & 1 \\ 0 & -3 & 2 \end{bmatrix} = \begin{bmatrix} 6 & -3 & 3 \\ 0 & -9 & 6 \end{bmatrix}.$

(*b*) Matrix addition: $\begin{bmatrix} 2 & -1 & 1 \\ 0 & -3 & 2 \end{bmatrix} + \begin{bmatrix} -1 & 3 & 0 \\ 2 & -4 & 6 \end{bmatrix} = \begin{bmatrix} 1 & 2 & 1 \\ 2 & -7 & 8 \end{bmatrix}.$

(*c*) Distrubutive law:

$$2\left(\begin{bmatrix} 2 & -1 & 1 \\ 0 & -3 & 2 \end{bmatrix} + \begin{bmatrix} -1 & 3 & 0 \\ 2 & -4 & 6 \end{bmatrix}\right) = 2\begin{bmatrix} 1 & 2 & 1 \\ 2 & -7 & 8 \end{bmatrix}$$

$$= \begin{bmatrix} 2 & 4 & 2 \\ 4 & -14 & 16 \end{bmatrix}$$

$$= \begin{bmatrix} 4 & -2 & 2 \\ 0 & -6 & 4 \end{bmatrix} + \begin{bmatrix} -2 & 6 & 0 \\ 4 & -8 & 12 \end{bmatrix}$$

$$= 2\begin{bmatrix} 2 & -1 & 1 \\ 0 & -3 & 2 \end{bmatrix} + 2\begin{bmatrix} -1 & 3 & 0 \\ 2 & -4 & 6 \end{bmatrix}.$$

(*d*) General distributive law: If α is a scalar and A, B are matrices of the same dimensions, then

$$\alpha(A + B) = \alpha A + \alpha B. \ \blacktriangle \tag{2.4.1}$$

We will have occasion to calculate determinants and products of matrices, so we introduce these topics now.

Determinants: The determinant of a square matrix is a number which has significant geometric and algebraic meaning. We restrict our attention to calculating determinants of 2×2 and 3×3 matrices.

Definition 2.4.2. *The determinant of the 2×2 matrix* $\begin{bmatrix} a & b \\ c & d \end{bmatrix}$ *is*

$$det\begin{bmatrix} a & b \\ c & d \end{bmatrix} = \begin{vmatrix} a & b \\ c & d \end{vmatrix} = ad - bc.$$

In a sentence: the determinant of a 2×2 matrix is the product of the diagonal entries minus the product of the off-diagonal ones. We've introduced two notations for the determinant of a square matrix A: $detA$ and $|A|$.

For the third matrix above we calculate its determinant:

$$\begin{vmatrix} 2y^2 & 2 + 3y^2 \\ 2 + 3y^2 & 2x^2 \end{vmatrix} = (2y^2)(2x^2) - (2 + 3y^2)^2 = 4x^2y^2 - 9y^4 - 12y^2 - 4.$$

We remark that the determinant of a 2×2 matrix is the "signed" area of the parallelogram determined by the row vectors $\mathbf{v} = \langle a, b \rangle$ and $\mathbf{w} = \langle c, d \rangle$. The area is positive if the directed angle from $\mathbf{v}$ to $\mathbf{w}$ is counterclockwise, and negative if clockwise.

The justification of this remark is a nice application of polar coordinates in $\mathbb{R}^2$, so we include it. The area of a parallelogram is base times height, where height is the distance from the base to the opposite side. When the parallelogram has base $\mathbf{v}$ and adjacent side $\mathbf{w}$, the height h is the length of a perpendicular dropped from the tip of $\mathbf{w}$ to $\mathbf{v}$ (see

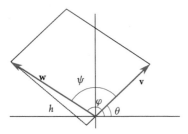

Figure 2.4.1 Area of parallelogram determined by **v** and **w**

Figure 2.4.1). If ψ is the angle between **v** and **w**, then the height is $h = \|\mathbf{w}\| \sin \psi$. In vector form, then, the area A of the parallelogram is

$$A = bh = \|\mathbf{v}\| \|\mathbf{w}\| \sin \psi. \tag{2.4.2}$$

The polar forms for the position vectors $\mathbf{v} = \langle a, b \rangle = \langle r \cos \theta, r \sin \theta \rangle$ and $\mathbf{w} = \langle c, d \rangle = \langle s \cos \varphi, s \sin \varphi \rangle$, together with the trig identity $\sin(\varphi - \theta) = \cos \theta \cdot \sin \varphi - \sin \theta \cdot \cos \varphi$, will derive the formula for us.

$$\begin{vmatrix} a & b \\ c & d \end{vmatrix} = ad - bc = r \cos \theta \cdot s \sin \varphi - r \sin \theta \cdot s \cos \varphi = rs \sin(\varphi - \theta) \tag{2.4.3}$$

Now since $r = \|\mathbf{v}\|$, $s = \|\mathbf{w}\|$, and $\psi = \varphi - \theta$, the result follows by comparing Equations 2.4.2 and 2.4.3. In Figure 2.4.1 the directed angle from **v** to **w** is counterclockwise. For the clockwise case, θ would be bigger than φ making $\sin \psi < 0$, finishing the justification that the determinant is the *signed* area of the parallelogram.

Example 2.4.3. *Area of parallelogram*

Find the area of the parallelogram determined by $\langle 3, 2 \rangle$ and $\langle -1, 5 \rangle$.
By the above remarks, the area is the absolute value of the determinant

$$\begin{vmatrix} 3 & 2 \\ -1 & 5 \end{vmatrix} = 3 \cdot 5 - (-1) \cdot 2 = 17. \ \blacktriangle$$

Example 2.4.4. *More area calculations*

Find an integer y that makes the parallelogram spanned by $\langle 2, y \rangle$ and $\langle 3, 7 \rangle$ have area 1. Can you find integers a, b that make the vectors $\langle a, b \rangle$ and $\langle 3, 7 \rangle$ determine a parallelogram of area 5?
For the first part, we need the determinant of the matrix to be ± 1. Calculating gives

$$\begin{vmatrix} 2 & y \\ 3 & 7 \end{vmatrix} = 14 - 3y = \pm 1.$$

Choosing $+1$ doesn't yield an integer for y, but solving $14 - 3y = -1$ shows that the vectors $\langle 2,5 \rangle$ and $\langle 3,7 \rangle$ span a parallelogram of area 1.

To solve the second part, think of the new parallelogram as having area 5 times that of the first part. It makes sense to multiply $\langle 2,5 \rangle$ by 5 and see what happens. We get

$$\begin{vmatrix} 10 & 25 \\ 3 & 7 \end{vmatrix} = 70 - 75 = -5,$$

so the area is 5. ▲

Determinants of 3×3 matrices can be defined using a technique called expansion by cofactors. We don't go into the general technique, but show one method for calculating it here:

$$\begin{vmatrix} a & b & c \\ d & e & f \\ g & h & i \end{vmatrix} = \begin{vmatrix} e & f \\ h & i \end{vmatrix} a - \begin{vmatrix} d & f \\ g & i \end{vmatrix} b + \begin{vmatrix} d & e \\ g & h \end{vmatrix} c$$

$$= (ei - hf)a - (di - gf)b + (dh - ge)c. \qquad (2.4.4)$$

In this instance we are expanding across the first row. Notice the entries in the first row are multiplied by the determinants of the 2×2 matrices you get by deleting the row and column they are in. For example, since b is in the first row, second column, the 2×2 matrix is the one obtained by deleting the first row and second column. Also notice that it's an alternating sum: the first and last terms are added while the middle term is subtracted. We illustrate with an example:

Example 2.4.5. *Calculating a 3×3 determinant*

$$\begin{vmatrix} i & j & k \\ 3 & 2 & 1 \\ 4 & 17 & 3 \end{vmatrix} = \begin{vmatrix} 2 & 1 \\ 17 & 3 \end{vmatrix} i - \begin{vmatrix} 3 & 1 \\ 4 & 3 \end{vmatrix} j + \begin{vmatrix} 3 & 2 \\ 4 & 17 \end{vmatrix} k$$

$$= -11i - 5j + 45k. \; ▲$$

Notice that when the entries in the first row are the standard basis vectors, the determinant of the matrix is itself a vector. This will turn out to be the cross product of the vectors in the second and third rows.

As in the case of 2×2 determinants, a 3×3 determinant has a geometric interpretation that involves parallelepipeds. What was that, you say? Parallelepiped? A parallelepiped is a box with faces that are parallelograms (usually a box has rectangular faces). In the same way that two vectors determine a parallelogram, three determine a parallelepiped by placing them with the same initial point (see Figure 2.4.2). Theorem 2.4.3 will show that the signed volume of a parallelepiped is a 3×3 determinant.

Example 2.4.6. *Parallelepiped volume*

Find the volume of the parallelepiped determined by $\langle 2,0,-1 \rangle, \langle 1,3,-1 \rangle$, and $\langle -2,1,1 \rangle$.

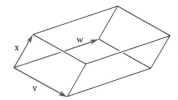

Figure 2.4.2 The parallelepiped determined by **v**, **w**, **x**

By the previous remarks the volume is the absolute value of

$$\begin{vmatrix} 2 & 0 & -4 \\ 1 & 3 & -1 \\ -2 & 1 & 1 \end{vmatrix} = \begin{vmatrix} 3 & -1 \\ 1 & 1 \end{vmatrix} \cdot 2 - \begin{vmatrix} 1 & -1 \\ -2 & 1 \end{vmatrix} \cdot 0 + \begin{vmatrix} 1 & 3 \\ -2 & 1 \end{vmatrix} \cdot (-4)$$

$$= 4 \cdot 2 - (-1) \cdot 0 + 7 \cdot (-4) = -20.$$

Thus the volume is 20. ▲

Matrix Multiplication: In addition to calculating determinants, we will also have occasion to multiply matrices. We begin by defining a matrix times a vector (thought of as a column), and illustrate how matrix multiplication can be used to define mappings. The product of two matrices is then defined using matrix-vector products, and examples given.

To begin, let A be an $m \times n$, with $\mathbf{a}_i$ denoting the ith row of A, and let $\mathbf{x}$ be a vector thought of as an $n \times 1$ column vector. The product $A\mathbf{x}$ is the $m \times 1$ vector whose i^{th} entry is the dot product of the ith row of A with $\mathbf{x}$. In symbols

$$(A\mathbf{x})_i = \mathbf{a}_i \cdot \mathbf{x}. \tag{2.4.5}$$

It should be noted that $\mathbf{x}$ has the same number of entries as A has columns; otherwise $A\mathbf{x}$ is not defined. Also, we'll jump back and forth from thinking of $\mathbf{x}$ as a column or row depending on the context.

Example 2.4.7. *Matrix-vector products*

Multiply the following, if possible:

1.

$$\begin{bmatrix} 2 & 0 & 3 \\ -3 & 1 & 1 \end{bmatrix} \begin{bmatrix} -3 \\ 2 \\ 5 \end{bmatrix} = \begin{bmatrix} \langle 2,0,3 \rangle \cdot \langle -3,2,5 \rangle \\ \langle -3,1,1 \rangle \cdot \langle -3,2,5 \rangle \end{bmatrix} = \begin{bmatrix} 9 \\ 16 \end{bmatrix}$$

2.

$$\begin{bmatrix} 1 & 0 & 0 \\ 0 & 1 & 0 \\ 0 & 0 & 1 \end{bmatrix}\begin{bmatrix} 5 \\ -2 \\ 1 \end{bmatrix} = \begin{bmatrix} \langle 1,0,0 \rangle \cdot \langle 5,-2,1 \rangle \\ \langle 0,1,0 \rangle \cdot \langle 5,-2,1 \rangle \\ \langle 0,0,1 \rangle \cdot \langle 5,-2,1 \rangle \end{bmatrix} = \begin{bmatrix} 5 \\ -2 \\ 1 \end{bmatrix}$$

Note that the product is the same as the original vector in this case. In fact, multiplying any vector $\mathbf{x}$ by the matrix with ones on the diagonal and zeros elsewhere never changes $\mathbf{x}$. For this reason the *identity matrix* I_n is the $n \times n$ matrix with ones on the diagonal and zeros elsewhere.

3.

$$\begin{bmatrix} 2 & 3 \\ -1 & 2 \end{bmatrix}\begin{bmatrix} x \\ y \end{bmatrix} = \begin{bmatrix} \langle 2,3 \rangle \cdot \langle x,y \rangle \\ \langle -1,2 \rangle \cdot \langle x,y \rangle \end{bmatrix} = \begin{bmatrix} 2x+3y \\ -x+2y \end{bmatrix}$$

4. The product

$$\begin{bmatrix} 1 & 2 & -1 \\ 4 & 0 & 3 \end{bmatrix}\begin{bmatrix} 1 \\ -3 \end{bmatrix}$$

is not defined since the required dot products aren't. ▲

Note that in Part 3 of the previous example, the matrix A times the vector $\langle x,y \rangle$ yields the vector $\langle 2x+3y, -x+2y \rangle$. If we think of these position vectors as points in the plane, then multiplication by A can be thought of as the function that maps the point (x,y) to the point $(2x+3y, -x+2y)$. In this way, matrix multiplication can be thought of as a transformation of the plane. This is indeed an important view, which we only touch on here.

Example 2.4.8. *A transformation of the plane*

Analyze the transformation $f(x,y) = A\mathbf{x}$, where

$$A = \begin{bmatrix} 1 & 2 \\ 0 & 1 \end{bmatrix}.$$

There are many ways to analyze f. First we find a formula for it:

$$f(x,y) = \begin{bmatrix} 1 & 2 \\ 0 & 1 \end{bmatrix}\begin{bmatrix} x \\ y \end{bmatrix} = \begin{bmatrix} x+2y \\ y \end{bmatrix} = (x+2y,y).$$

Now we visualize the transformation by considering where f maps the unit square. We see that f maps the square with vertices $(0,0),(1,0),(1,1),(0,1)$ to the parallelogram with vertices $f(0,0) = (0,0), f(1,0) = (1,0), f(1,1) = (3,1), f(0,1) = (2,1)$. It turns out that

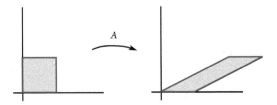

Figure 2.4.3 The transformation $\mathbf{x} \mapsto A\mathbf{x}$

f maps lines to lines, so it maps a square tessellation of the plane to one with parallelograms (see Figure 2.4.3). ▲

We pause to remark that matrix-vector multiplication distributes across linear combinations of vectors. Symbolically we have

$$A(\alpha\mathbf{v} + \beta\mathbf{w}) = \alpha A\mathbf{v} + \beta A\mathbf{w}.$$

This notation seems believable, since we've seen lots of formulas like it, but it follows directly from Definition 2.4.5 and the fact that the dot product behaves similarly (Theorem 2.3.1). When thinking of multiplication by A as a transformation, we say that it is a *linear transformation* because it distributes across linear combinations this way. The fact that matrix multiplication is a linear transformation allows us to show that Bézier curves get mapped to Bézier curves.

Example 2.4.9. *Bézier curves and matrix multiplication.*

Recall from Section 2.2 that Bézier parameterizations are linear combinations of fixed control points. A cubic Bézier curve has the parameterization

$$\mathbf{C}(t) = (1-t)^3\mathbf{P}_0 + 3(1-t)^2 t\mathbf{P}_1 + 3(1-t)t^2\mathbf{P}_2 + t^3\mathbf{P}_3,\ 0 \leq t \leq 1.$$

By the linearity of matrix multiplication we have

$$\begin{aligned} A\mathbf{C}(t) &= A\big((1-t)^3\mathbf{P}_0 + 3(1-t)^2 t\mathbf{P}_1 + 3(1-t)t^2\mathbf{P}_2 + t^3\mathbf{P}_3\big) \\ &= (1-t)^3 A\mathbf{P}_0 + 3(1-t)^2 tA\mathbf{P}_1 + 3(1-t)t^2 A\mathbf{P}_2 + t^3 A\mathbf{P}_3,\ 0 \leq t \leq 1. \end{aligned}$$

Thus the image of a Bézier curve is again a Bézier curve, just with control points $A\mathbf{P}_0, A\mathbf{P}_1, A\mathbf{P}_2, A\mathbf{P}_3$. ▲

Example 2.4.10. *Transforming Bézier curves*

Find the image $A\mathbf{C}(t)$ of the quadratic Bézier curve

$$\mathbf{C}(t) = (1-t)^2 \langle -1, 1\rangle + 2(1-t)t\langle 0, -1\rangle + t^2\langle 1, 1\rangle = (2t-1, 4t^2 - 4t + 1),\ 0 \leq t \leq 1,$$

where A is the matrix

$$A = \begin{bmatrix} 1 & 2 \\ 0 & 1 \end{bmatrix}.$$

Recall that Example 2.2.9 showed that $\mathbf{C}(t)$ parameterizes $y = x^2, -1 \le x \le 1$, so we are interested in what multiplying by A does to that curve. By linearity we need only calculate the images of the control points:

$$A\mathbf{P}_0 = \begin{bmatrix} 1 & 2 \\ 0 & 1 \end{bmatrix} \begin{bmatrix} -1 \\ 1 \end{bmatrix} = \begin{bmatrix} 1 \\ 1 \end{bmatrix}; A\mathbf{P}_1 = \begin{bmatrix} 1 & 2 \\ 0 & 1 \end{bmatrix} \begin{bmatrix} 0 \\ -1 \end{bmatrix} = \begin{bmatrix} -2 \\ -1 \end{bmatrix}; A\mathbf{P}_2 = \begin{bmatrix} 1 & 2 \\ 0 & 1 \end{bmatrix} \begin{bmatrix} 1 \\ 1 \end{bmatrix} = \begin{bmatrix} 3 \\ 1 \end{bmatrix}.$$

The image $A\mathbf{C}(t)$ of $\mathbf{C}(t)$ is the Bézier curve

$$A\mathbf{C}(t) = (1-t)^2 \langle 1,1 \rangle + 2(1-t)t \langle -2,-1 \rangle + t^2 \langle 3,1 \rangle \ 0 \le t \le 1.$$

The y-components of the control points do not change under this transformation, so the vertical scales of the pictures in Figure 2.4.4 are the same. The horizontal scales are different, however, due to shearing from the transformation.

The point is that finding the image of a curve has been reduced to keeping track of three points. This speeds up calculations required in computer graphics programs. ▲

So far we've defined how to multiply a matrix A by any vector with the same number of entries as A has columns. This can be extended to multiplying matrices together, with a similar restriction on dimensions. Let A and B be an $m \times n$ and $n \times p$ matrices, respectively, so that the number of rows in B equals the number of columns in A. Now let $\mathbf{b}_j$ represent the jth column of B. The product AB is the matrix whose j^{th} column is the product of A with the jth column of B. Since we know how to multiply A times columns, this makes sense. More formally, we have

$$AB = \begin{bmatrix} A\mathbf{b}_1 | A\mathbf{b}_2 | \cdots | A\mathbf{b}_p \end{bmatrix},$$

where the vertical lines separate columns in the product matrix.

(a)

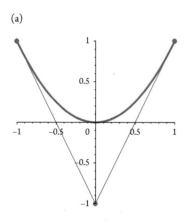

The curve $y = x^2$

(b)

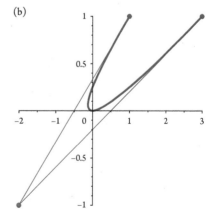

The image after multiplication by A

Figure 2.4.4 Transforming a Bézier curve

Example 2.4.11. *Multiplying matrices*

Find the following matrix products, if possible.

1.

$$\begin{bmatrix} 1 & 4 \\ 0 & -3 \\ -2 & 1 \end{bmatrix} \begin{bmatrix} 1 & 0 \\ 0 & 1 \end{bmatrix} = \begin{bmatrix} \begin{bmatrix} 1 & 4 \\ 0 & -3 \\ -2 & 1 \end{bmatrix} \begin{bmatrix} 1 \\ 0 \end{bmatrix} & \begin{bmatrix} 1 & 4 \\ 0 & -3 \\ -2 & 1 \end{bmatrix} \begin{bmatrix} 0 \\ 1 \end{bmatrix} \end{bmatrix}$$

$$= \begin{bmatrix} 1 & 4 \\ 0 & -3 \\ -2 & 1 \end{bmatrix}.$$

2. The product

$$\begin{bmatrix} 1 & 0 \\ 0 & 1 \end{bmatrix} \begin{bmatrix} 1 & 4 \\ 0 & -3 \\ -2 & 1 \end{bmatrix}$$

is not defined.

3.

$$\begin{bmatrix} 3 & 4 \\ -2 & 5 \end{bmatrix} \begin{bmatrix} 2 & 3 & 4 \\ 5 & 6 & 7 \end{bmatrix} = \begin{bmatrix} \begin{bmatrix} 3 & 4 \\ -2 & 5 \end{bmatrix} \begin{bmatrix} 2 \\ 5 \end{bmatrix} & \begin{bmatrix} 3 & 4 \\ -2 & 5 \end{bmatrix} \begin{bmatrix} 3 \\ 6 \end{bmatrix} & \begin{bmatrix} 3 & 4 \\ -2 & 5 \end{bmatrix} \begin{bmatrix} 4 \\ 7 \end{bmatrix} \end{bmatrix}$$

$$= \begin{bmatrix} 26 & 33 & 40 \\ 21 & 24 & 27 \end{bmatrix}. \blacktriangle$$

The product AB can also be defined as the $m \times p$ matrix whose ij^{th} entry is the dot product of the ith row of A with the jth column of B. With this interpretation, Part 3 of the previous example becomes

$$\begin{bmatrix} 3 & 4 \\ -2 & 5 \end{bmatrix} \begin{bmatrix} 2 & 3 & 4 \\ 5 & 6 & 7 \end{bmatrix} = \begin{bmatrix} \langle 3,4 \rangle \cdot \langle 2,5 \rangle & \langle 3,4 \rangle \cdot \langle 3,6 \rangle & \langle 3,4 \rangle \cdot \langle 4,7 \rangle \\ \langle -2,5 \rangle \cdot \langle 2,5 \rangle & \langle -2,5 \rangle \cdot \langle 3,6 \rangle & \langle -2,5 \rangle \cdot \langle 4,7 \rangle \end{bmatrix}$$

$$= \begin{bmatrix} 3 \cdot 2 + 4 \cdot 5 & 3 \cdot 3 + 4 \cdot 6 & 3 \cdot 4 + 4 \cdot 7 \\ -2 \cdot 2 + 5 \cdot 5 & -2 \cdot 3 + 5 \cdot 6 & -2 \cdot 4 + 5 \cdot 7 \end{bmatrix}$$

$$= \begin{bmatrix} 26 & 33 & 40 \\ 21 & 24 & 27 \end{bmatrix}.$$

We reiterate that the requirement that the number of columns of A equal the number of rows of B is necessary. If that doesn't hold, the product AB is not defined. The reason is that without this requirement it would be impossible to take the dot product of a row of A with a column of B.

After this, the meagerest of introductions to matrix algebra, we continue by introducing the cross product.

Algebraic Definition of v × w: With the matrix preliminaries in hand, we can define the cross product of two vectors.

Definition 2.4.3. *The cross product* $\mathbf{v} \times \mathbf{w}$ *of the vectors* $\mathbf{v} = \langle a, b, c \rangle$ *and* $\mathbf{w} = \langle d, e, f \rangle$ *is the vector*

$$\mathbf{v} \times \mathbf{w} = \begin{vmatrix} \mathbf{i} & \mathbf{j} & \mathbf{k} \\ a & b & c \\ d & e & f \end{vmatrix} = \begin{vmatrix} b & c \\ e & f \end{vmatrix} \mathbf{i} - \begin{vmatrix} a & c \\ d & f \end{vmatrix} \mathbf{j} + \begin{vmatrix} a & b \\ d & e \end{vmatrix} \mathbf{k}$$

$$= (bf - ec)\mathbf{i} - (af - dc)\mathbf{j} + (ae - db)\mathbf{k}. \tag{2.4.6}$$

Our determinant calculation in Example 2.4.5, then, is a cross product. The calculation shows that if $\mathbf{v} = \langle 3, 2, 1 \rangle$ and $\mathbf{w} = \langle 4, 17, 3 \rangle$, then $\mathbf{v} \times \mathbf{w} = \langle -11, -5, 45 \rangle$.

When we introduced the dot product, we took note of some of its algebraic properties in Theorem 2.3.1. For example, we saw that dot products distribute across vector sums, and scalars can be "factored out". Further, dot products are commutative. We take a moment to highlight some algebraic properties of the cross product.

Theorem 2.4.1. *Let* $\mathbf{v}$, $\mathbf{w}$, *and* $\mathbf{x}$ *be vectors, and let* α *be a scalar. Then*

1. $\mathbf{v} \times \mathbf{w} = -\mathbf{w} \times \mathbf{v}$
2. $(\alpha \mathbf{v}) \times \mathbf{w} = \alpha(\mathbf{v} \times \mathbf{w})$
3. $(\mathbf{v} + \mathbf{w}) \times \mathbf{x} = \mathbf{v} \times \mathbf{x} + \mathbf{w} \times \mathbf{x}$

Proof. We prove the first property. Let $\mathbf{v} = \langle a, b, c \rangle$ and $\mathbf{w} = \langle d, e, f \rangle$. Using the definition of the cross product, we see that

$$\mathbf{w} \times \mathbf{v} = \begin{vmatrix} \mathbf{i} & \mathbf{j} & \mathbf{k} \\ d & e & f \\ a & b & c \end{vmatrix} = \begin{vmatrix} e & f \\ b & c \end{vmatrix} \mathbf{i} - \begin{vmatrix} d & f \\ a & c \end{vmatrix} \mathbf{j} + \begin{vmatrix} d & e \\ a & b \end{vmatrix} \mathbf{k}$$

$$= (ec - bf)\mathbf{i} - (dc - af)\mathbf{j} + (db - ae)\mathbf{k}.$$

Comparing this with Equation 2.4.6 shows that $\mathbf{v} \times \mathbf{w} = -\mathbf{w} \times \mathbf{v}$. □

The remaining facts are left as exercises for the reader. One should observe that the cross product does not commute! Since switching the order introduces a negative sign, we say it *anti*-commutes. The cross product, however, does behave as you might expect with respect to scalar multiplication and vector addition. Scalars can be "factored" out, and the cross product "distributes" across vector addition.

Example 2.4.12. *Algebraic properties of* $\mathbf{v} \times \mathbf{w}$

Use the properties of Theorem 2.4.1 to show that $(\alpha \mathbf{v}) \times (\beta \mathbf{w}) = \alpha \beta (\mathbf{v} \times \mathbf{w})$.

Notice that property 2 of Theorem 2.4.1 only states that that you can factor scalars out of the first vector in a cross product, while this example states you can factor them from either/both vectors. This result follows from repeated applications of properties 1 and 2. The right-hand column justifies each equality.

$$
\begin{aligned}
(\alpha \mathbf{v}) \times (\beta \mathbf{w}) &= \alpha\,(\mathbf{v} \times (\beta \mathbf{w})) &&\text{Property 2} \\
&= \alpha\,(-(\beta \mathbf{w}) \times \mathbf{v}) &&\text{Property 1} \\
&= \alpha\beta\,(-\mathbf{w} \times \mathbf{v}) &&\text{Property 2} \\
&= \alpha\beta\,(\mathbf{v} \times \mathbf{w}).\, \blacktriangle &&\text{Property 1}
\end{aligned}
$$

Geometric Definition of v × w: It turns out that the cross product has geometric significance, which involves the notion of the right-hand rule for vectors. The ordered list of three vectors $\mathbf{v}, \mathbf{w}, \mathbf{x}$ satisfies the right-hand rule if when you point your right thumb along $\mathbf{x}$ your fingers curl from $\mathbf{v}$ toward $\mathbf{w}$. For example, the vectors $\mathbf{i}$, $\mathbf{j}$, and $\mathbf{k}$ satisfies the right-hand rule, but $\mathbf{j}$, $\mathbf{i}$, and $\mathbf{k}$ don't. There is an easy algebraic check to see if an ordered list of three vectors satisfy the right-hand rule. The vectors $\mathbf{v} = \langle a,b,c \rangle$, $\mathbf{w} = \langle d,e,f \rangle$ and $\mathbf{x} = \langle g,h,i \rangle$ satisfy the right hand rule if and only if the determinant of the matrix with these vectors as rows is positive. In symbols: the non-zero vectors $\mathbf{v}$, $\mathbf{w}$, and $\mathbf{x}$ satisfy the right-hand rule if and only if

$$
\begin{vmatrix} a & b & c \\ d & e & f \\ g & h & i \end{vmatrix} > 0.
$$

The proof of this fact is omitted. We are now ready to state the following theorem, which can actually be considered the definition of the cross product.

Theorem 2.4.2. *Let* $\mathbf{v}$ *and* $\mathbf{w}$ *be non-zero vectors. Then*

1. $\mathbf{v} \times \mathbf{w}$ *is orthogonal to both* $\mathbf{v}$ *and* $\mathbf{w}$*, and the vectors* $\mathbf{v}, \mathbf{w}, \mathbf{v} \times \mathbf{w}$ *satisfy the right-hand rule.*

2. $\|\mathbf{v} \times \mathbf{w}\| = \|\mathbf{v}\|\,\|\mathbf{w}\| \sin\theta$*, where* θ *is the angle between* $\mathbf{v}$ *and* $\mathbf{w}$*.*

Proof. The fact that $\mathbf{v} \times \mathbf{w}$ is orthogonal to both $\mathbf{v}$ and $\mathbf{w}$ is proven by showing that the corresponding dot products are zero. To see that $\mathbf{v}, \mathbf{w}, \mathbf{v} \times \mathbf{w}$ satisfies the right-hand rule, one can take the determinant of the matrix with these vectors as rows to see it is positive.

We'll prove part 2 letting $\mathbf{v} = \langle a,b,c \rangle$ and $\mathbf{w} = \langle d,e,f \rangle$. Let T be the area of the triangle determined by placing $\mathbf{v}$ and $\mathbf{w}$ at the same initial point. Then $2T$ is the area of the parallelogram spanned by $\mathbf{v}$ and $\mathbf{w}$. During our discussion on 2×2 determinants we derived Equation 2.4.2, which implies $2T = \|\mathbf{v}\|\,\|\mathbf{w}\| \sin\theta$. Since $\sin^2\theta = 1 - \cos^2\theta$, and $\cos\theta = \frac{\mathbf{v} \cdot \mathbf{w}}{\|\mathbf{v}\|\|\mathbf{w}\|}$, squaring the area of the parallelogram we get

$$
(2T)^2 = (\|\mathbf{v}\|\,\|\mathbf{w}\| \sin\theta)^2 = \|\mathbf{v}\|^2 \|\mathbf{w}\|^2 (1 - \cos^2\theta)
$$

$$
= \|\mathbf{v}\|^2 \|\mathbf{w}\|^2 \left(1 - \left(\frac{\mathbf{v} \cdot \mathbf{w}}{\|\mathbf{v}\|\,\|\mathbf{w}\|}\right)^2\right)
$$

$$
= \|\mathbf{v}\|^2 \|\mathbf{w}\|^2 - (\mathbf{v} \cdot \mathbf{w})^2 = (a^2 + b^2 + c^2)(d^2 + e^2 + f^2) - (ad + be + cf)^2
$$

$$
= (ac - bd)^2 + (af - dc)^2 + (bf - ce)^2 = \|\mathbf{v} \times \mathbf{w}\|^2.
$$

Therefore $\|\mathbf{v} \times \mathbf{w}\| = 2T = \|\mathbf{v}\|\,\|\mathbf{w}\| \sin\theta$, as desired. $\square$

Remark: Implicit in the proof of Theorem 2.4.2 is the fact that $\|\mathbf{v} \times \mathbf{w}\|$ is the area of the parallelogram determined by $\mathbf{v}$ and $\mathbf{w}$.

Corollary 2.4.1. *Two non-zero vectors* $\mathbf{v}$, $\mathbf{w}$ *are parallel if and only if*

$$\mathbf{v} \times \mathbf{w} = \mathbf{0}.$$

Proof. The non-zero vectors $\mathbf{v}$ and $\mathbf{w}$ are parallel if and only if $\theta = 0$ or $\theta = \pi$. By Theorem 2.4.2(2), this is true if and only if $\mathbf{v} \times \mathbf{w} = \mathbf{0}$. $\square$

Concept Connection: Since $\|\mathbf{v}\| \|\mathbf{w}\| \sin\theta$ is the area of the parallelogram determined by $\mathbf{v}$ and $\mathbf{w}$, Theorem 2.4.2 has significant geometric consequences. In particular, this will be useful when developing integral formulas for surface area and, more generally, for integrating functions on surfaces. We will approximate the surface area using parallelogram-shaped tiles, much like the tiles on a roof (see Figure 4.5.2). The cross product will allow us to calculate areas in our approximation. The area of the surface, of course, will be the integral obtained by taking the limit as our approximation gets finer.

Example 2.4.13. *Standard basis vectors*

Use the geometry definition of the cross product to compute cross products of standard basis vectors.

The product $\mathbf{i} \times -\mathbf{j}$ has to be orthogonal to both $\mathbf{i}$ and $-\mathbf{j}$, so it is either $\mathbf{k}$ or $-\mathbf{k}$ and your right hand determines which. Placing your fingers along $\mathbf{i}$ and curling them toward $-\mathbf{j}$ forces your thumb to go down, so $\mathbf{i} \times -\mathbf{j} = -\mathbf{k}$.

Applying similar reasoning computes $\mathbf{k} \times -\mathbf{j} = \mathbf{i}$. Thus the cross product of any two standard basis vectors can be computed using geometry, without resorting Equation 2.4.6. ▲

Example 2.4.14. *The cross product is* not *associative*

Separately compute $\mathbf{i} \times (\mathbf{j} \times \mathbf{j})$ and $(\mathbf{i} \times \mathbf{j}) \times \mathbf{j}$ to show they are not equal. Hence the cross product is not associative.

By Corollary 2.4.1 we see that $\mathbf{j} \times \mathbf{j} = \mathbf{0}$. It follows from Equation 2.4.6 that the cross product of any vector with the zero vector is the zero vector, hence

$$\mathbf{i} \times (\mathbf{j} \times \mathbf{j}) = \mathbf{i} \times \mathbf{0} = \mathbf{0}.$$

To compute $(\mathbf{i} \times \mathbf{j}) \times \mathbf{j}$ we reason as in Example 2.4.13. Applying the right-hand rule gives $\mathbf{i} \times \mathbf{j} = \mathbf{k}$. A second application gives

$$(\mathbf{i} \times \mathbf{j}) \times \mathbf{j} = \mathbf{k} \times \mathbf{j} = -\mathbf{i}.$$

Thus we have shown $(\mathbf{i} \times \mathbf{j}) \times \mathbf{j} \neq \mathbf{i} \times (\mathbf{j} \times \mathbf{j})$, from which it follows that the cross product is not associative. ▲

Theorem 2.4.2 is another way to see that 2×2 determinants are areas of parallelograms. The planar vectors $\mathbf{v} = \langle a, b \rangle$ and $\mathbf{w} = \langle c, d \rangle$ can be thought of as vectors in $\mathbb{R}^3$ with zero in the z-coordinate. In this case they are $\mathbf{v} = \langle a,b,0\rangle$ and $\mathbf{w} = \langle c,d,0\rangle$, and the cross product is $\mathbf{v} \times \mathbf{w} = (ad - bc)\mathbf{k}$. The second part of Theorem 2.4.2 says that $\|\mathbf{v} \times \mathbf{w}\| = \|(ad - bc)\mathbf{k}\| = |ad - bc|$ is the area of the parallelogram spanned by the vectors. This is a second proof of the fact that the (absolute value of the) 2×2-determinant is the area of the parallelogram determined by the rows. Moreover, it shows that Theorem 2.4.2 generalizes this to three dimensions.

We now consider other applications of Theorem 2.4.2. Since a plane is determined by a point and a normal direction, Theorem 2.4.2(1) can be used to find equations for planes. We can use it to parameterize lines of intersections of planes as well. Finally, we'll consider the triple scalar product, $(\mathbf{v} \times \mathbf{w}) \cdot \mathbf{x}$, and its geometric meaning.

Example 2.4.15. *Parametric to Cartesian equations of planes*

Find a Cartesian equation for the plane given by the parametric equations $S(s,t) = (2 - s + 2t, 3 + 2s - t, 4 - s + t)$.

We need a normal vector and a point on the plane. A point on the plane is $S(0,0) = (2,3,4)$. To find a normal vector, we find two vectors in the plane and take their cross product. Isolating the parameters gives

$$S(s,t) = \langle 2 - s + 2t, 3 + 2s - t, 4 - s + t\rangle = \langle 2,3,4\rangle + s\langle -1,2,-1\rangle + t\langle 2,-1,1\rangle.$$

Recall that the coefficients of the parameters give vectors in the plane. Therefore the cross product of $\langle -1,2,-1\rangle$ and $\langle 2,-1,1\rangle$ gives a normal vector. We get

$$\begin{vmatrix} \mathbf{i} & \mathbf{j} & \mathbf{k} \\ -1 & 2 & -1 \\ 2 & -1 & 1 \end{vmatrix} = \begin{vmatrix} 2 & -1 \\ -1 & 1 \end{vmatrix} \mathbf{i} - \begin{vmatrix} -1 & -1 \\ 2 & 1 \end{vmatrix} \mathbf{j} + \begin{vmatrix} -1 & 2 \\ 2 & -1 \end{vmatrix} \mathbf{k}$$

$$= \mathbf{i} - \mathbf{j} - 3\mathbf{k}.$$

Putting these together we get the equation for the plane:

$$x - y - 3z = \langle 1,-1,-3\rangle \cdot \langle 2,3,4\rangle = -13. \blacktriangle$$

Cartesian Equations of Planes Given Vectors in the Plane

To find a Cartesian equation for a plane containing the point P and the vectors $\mathbf{v}$ and $\mathbf{w}$:

1. Find the normal vector $\mathbf{n} = \mathbf{v} \times \mathbf{w}$.
2. Use the equation $\mathbf{n} \cdot \langle x,y,z\rangle = \mathbf{n} \cdot P$

Example 2.4.16. *A plane through three points*

Find a Cartesian equation for the plane through $P = (-2,3,1)$, $Q = (3,4,-2)$, and $R = (5,-3,2)$.

First we find two vectors in the plane, then take their cross product to find a normal vector to the plane. Two vectors in the plane are $\mathbf{v} = \mathbf{Q} - \mathbf{P} = \langle 3,4,-2\rangle - \langle -2,3,1\rangle = \langle 5,1,-3\rangle$ and $\mathbf{w} = \mathbf{R} - \mathbf{P} = \langle 5,-3,2\rangle - \langle -2,3,1\rangle = \langle 7,-6,1\rangle$. Their cross product is

$$\begin{vmatrix} \mathbf{i} & \mathbf{j} & \mathbf{k} \\ 5 & 1 & -3 \\ 7 & -6 & 1 \end{vmatrix} = \begin{vmatrix} 1 & -3 \\ -6 & 1 \end{vmatrix}\mathbf{i} - \begin{vmatrix} 5 & -3 \\ 7 & 1 \end{vmatrix}\mathbf{j} + \begin{vmatrix} 5 & 1 \\ 7 & -6 \end{vmatrix}\mathbf{k}$$

$$= -17\mathbf{i} - 26\mathbf{j} - 37\mathbf{k}$$

Thus $\mathbf{n} = \langle 17,26,37\rangle$ is normal to the plane (Pop quiz: what happened to the negative sign?), $(-2,3,1)$ is on the plane, and an equation for the plane is

$$17x + 26y + 37z = \langle 17,26,37\rangle \cdot \langle -2,3,1\rangle = 81.\blacktriangle$$

Example 2.4.17. *Lines of intersection of planes*

Parameterize the line of intersection of $2x - 3y + z = 5$ and $x + y - 2z = 3$.

To parameterize a line we need to know a point on the line and a vector in the line. To find a point on the line, we'll decide to find where it intersects the xz-plane. Then $y = 0$, and we've reduced the problem to finding the point of intersection of the lines $2x - z = 5$ and $x - 2z = 3$. Your favorite method of solving systems yields $x = 7/3$, $z = -1/3$, so $(7/3, 0, -1/3)$ is on the line.

To find the direction of the line, we make a key observation. If a line lies in a plane, then its direction is orthogonal to the normal vector! Since our line lies on both planes its direction is orthogonal to both normal vectors. Thus the cross product of the normal vectors gives the direction of the line. We calculate:

$$\begin{vmatrix} \mathbf{i} & \mathbf{j} & \mathbf{k} \\ 2 & -3 & 1 \\ 1 & 1 & -2 \end{vmatrix} = \begin{vmatrix} -3 & 1 \\ 1 & -2 \end{vmatrix}\mathbf{i} - \begin{vmatrix} 2 & 1 \\ 1 & -2 \end{vmatrix}\mathbf{j} + \begin{vmatrix} 2 & -3 \\ 1 & 1 \end{vmatrix}\mathbf{k}$$

$$= 5\mathbf{i} + 5\mathbf{j} + 5\mathbf{k},$$

which means that the vector $\mathbf{v} = \langle 1,1,1\rangle$ lies in the direction of the line. (Note that if $\mathbf{v}$ lies on the line, so does any scalar multiple of it.) Thus we have

$$C(t) = \left(\frac{7}{3} + t, t, -\frac{1}{3} + t\right).\blacktriangle$$

Cross Product and Rotations: Thus far we have discussed algebraic and geometric properties of the cross product. The cross product has applications to physics as well. One application is that the cross product relates angular velocity and velocity of a rotating solid; others we will investigate later.

To quantify this, we make use of the physical notion of *angular velocity*. Angular velocity, denoted $\boldsymbol{\omega}$, is a vector quantity that completely describes the rotation about an axis. The vector $\boldsymbol{\omega}$ points along the axis L of rotation, so that the rotation of the

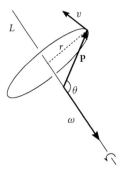

Figure 2.4.5 Rotation and the cross product

solid W satisfies the right-hand rule. More precisely, if your right thumb points along ω then your fingers curl in the direction W is rotating. The length $\|\omega\|$ of the angular velocity vector is how fast the angle is changing under the rotation, usually given in radians per second.

For example, let's compute ω in the case of "LP", or long play, vinyl records. The direction of ω is straight down since vinyl record players rotate clockwise (use your right hand to check!). An LP makes $33\frac{1}{3}$ revolutions per minute, so $\|\omega\| = 33\frac{1}{3} \cdot 2\pi/60 \approx 3.49$ rad/sec in this case. Note that every point on the LP has the same angular velocity, but not the same velocity! Points near the edge of the record move faster than those closer to the axis of rotation because they have to cover more distance to complete a single revolution in the same time.

With this intuition about angular velocity in mind, let's get back to the task at hand—relating velocity and angular velocity. First, we choose a coordinate system so that the axis of rotation L goes through the origin and place ω at the origin as well. Now let P be any point of the rotating object W, and $\mathbf{P}$ its position vector. If $\mathbf{v}$ represents the velocity vector of P as it rotates, our goal is to show $\mathbf{v} = \omega \times \mathbf{P}$.

To see that $\mathbf{v} = \omega \times \mathbf{P}$, we will show that $\mathbf{v}$ is orthogonal to both ω and $\mathbf{P}$ and that $\|\mathbf{v}\| = \|\omega\| \, \|\mathbf{P}\| \sin\theta$, where θ is the angle between ω and $\mathbf{P}$. Since the vector $\omega \times \mathbf{P}$ has these same properties, and there is only one with them, the vectors are equal.

We first check orthogonality conditions. The point P rotates in a circle around L, and $\mathbf{v}$ is tangent to that circle. Moreover, the circle is orthogonal to the half-plane with edge on L and containing the position vector $\mathbf{P}$, so the vector $\mathbf{v}$ is as well. Since both vectors ω and $\mathbf{P}$ are in the half-plane, $\mathbf{v}$ is perpendicular to them as well. Finally, since ω is chosen so that the rotation satisfies the right-hand rule, one easily checks that the ordered list ω, $\mathbf{P}$, $\mathbf{v}$ does as well. Thus $\mathbf{v}$ satisfies the same orthogonality conditions as $\omega \times \mathbf{P}$.

Now a comparison of speed and angular speed verifies the length conditions. Recall that if the angle θ cuts an arc of length s from a circle radius r, then the radian measure of θ is s/r, which implies $s = r\theta$. If r is constant, as is the case for P rotating around L, then differentiating with respect to time gives $\frac{ds}{dt} = r\frac{d\theta}{dt}$. Now in our setting we have $\|\mathbf{v}\| = \frac{ds}{dt}$, $\|\boldsymbol{\omega}\| = \frac{d\theta}{dt}$, and (using some trigonometry) $r = \|\mathbf{P}\|\sin\theta$. Substituting gives $\|\mathbf{v}\| = \|\boldsymbol{\omega}\|\,\|\mathbf{P}\|\sin\theta$, as desired.

Example 2.4.18. *Cross product and velocity*

Determine the velocity vector of a beetle standing on a playing LP 3 inches from the center.

We know $\mathbf{v} = \boldsymbol{\omega} \times \mathbf{P}$, so choose a coordinate system with the origin at the center of the record, and so that the beetle is on the point $(3\cos\theta, 3\sin\theta, 0)$. By the above remarks we have

$$\boldsymbol{\omega} = \langle 0,0,-3.49\rangle \text{ and } \mathbf{P} = \langle 3\cos\theta, 3\sin\theta, 0\rangle,$$

so that $\mathbf{v} = \langle 0,0,-3.49\rangle \times \langle 3\cos\theta, 3\sin\theta, 0\rangle = \langle 10.47\sin\theta, -10.47\cos\theta, 0\rangle.$ ▲

The Triple Scalar Product: We now combine the cross and dot products to define the triple scalar product of three vectors **v**, **w**, and **x**. It turns out that this product gives the volume of the parallelepiped determined by the vectors. This interpretation is central to the development of a formula for finding the flux of a vector field across a surface in Chapter 5.

Definition 2.4.4. *The triple scalar product of the ordered triple of vectors* **v**, **w**, *and* **x** *is* $(\mathbf{v} \times \mathbf{w}) \cdot \mathbf{x}$.

When calculating the triple scalar product, the order of operations is important. First compute the cross product $\mathbf{v} \times \mathbf{w}$, then dot the resulting vector with **x**. That's why we say an "ordered" list. There is a quick way to calculate the triple scalar product using determinants.

Lemma 2.4.1. *The triple scalar product of the vectors* $\mathbf{v} = \langle a,b,c\rangle$, $\mathbf{w} = \langle d,e,f\rangle$, *and* $\mathbf{x} = \langle g,h,i\rangle$ *is the determinant:*

$$(\mathbf{v} \times \mathbf{w}) \cdot \mathbf{x} = \begin{vmatrix} a & b & c \\ d & e & f \\ g & h & i \end{vmatrix}.$$

Proof. To compute the triple scalar product we first calculate the cross product:

$$\mathbf{v} \times \mathbf{w} = (bf - ec)\mathbf{i} - (af - dc)\mathbf{j} + (ae - db)\mathbf{k}.$$

Using this, we calculate the dot product

$$(\mathbf{v} \times \mathbf{w}) \cdot \mathbf{x} = \big((bf - ec)\mathbf{i} - (af - dc)\mathbf{j} + (ae - db)\mathbf{k}\big) \cdot \big(g\mathbf{i} + h\mathbf{j} + i\mathbf{k}\big)$$
$$= (bf - ec)g - (af - dc)h + (ae - db)i.$$

Expanding this calculation, and comparing it with Equation 2.4.4 completes the proof. □

Example 2.4.19. *Triple scalar product*

Find the triple scalar product of the vectors $\mathbf{v} = \langle -3, -2, 3 \rangle$, $\mathbf{w} = \langle 1, -3, 2 \rangle$, and $\mathbf{x} = \langle 2, 2, 3 \rangle$. By the above remarks, this amounts to calculating the determinant:

$$(\mathbf{v} \times \mathbf{w}) \cdot \mathbf{x} = \begin{vmatrix} -3 & -2 & 3 \\ 1 & -3 & 2 \\ 2 & 2 & 3 \end{vmatrix}$$

$$= -3 \cdot \begin{vmatrix} -3 & 2 \\ 2 & 3 \end{vmatrix} - (-2) \cdot \begin{vmatrix} 1 & 2 \\ 2 & 3 \end{vmatrix} + 3 \cdot \begin{vmatrix} 1 & -3 \\ 2 & 2 \end{vmatrix} = 61. \ \blacktriangle$$

Any three vectors $\mathbf{v}$, $\mathbf{w}$, and $\mathbf{x}$ determine a parallelepiped, as in Figure 2.4.6, which has a volume. When the vectors are ordered, we can define a signed volume of the parallelepiped. Take the volume to be positive if $\mathbf{v} \times \mathbf{w}$ and $\mathbf{x}$ point "in the same direction", and negative otherwise. Equivalently, the volume is positive if and only if $\mathbf{v}, \mathbf{w}, \mathbf{x}$ satisfy the right-hand rule.

Theorem 2.4.3. *The triple scalar product $(\mathbf{v} \times \mathbf{w}) \cdot \mathbf{x}$ is the signed volume of the parallelepiped determined by the ordered list of vectors $\mathbf{v}$, $\mathbf{w}$, and $\mathbf{x}$.*

Proof. The volume of a parallelepiped is the area of the base parallelogram times the height. By Theorem 2.4.2, we know that $\|\mathbf{v} \times \mathbf{w}\|$ is the area of the base parallelogram. Using Theorem 2.3.2 we see

$$(\mathbf{v} \times \mathbf{w}) \cdot \mathbf{x} = \|\mathbf{v} \times \mathbf{w}\| \cdot \|\mathbf{x}\| \cos \psi,$$

where ψ is the angle between $\mathbf{v} \times \mathbf{w}$ and $\mathbf{x}$. If we can show that $\|\mathbf{x}\| \cos \psi$ is the signed height of the parallelepiped, we are done. Since $\mathbf{v} \times \mathbf{w}$ is orthogonal to both $\mathbf{v}$ and $\mathbf{w}$, $comp_{\mathbf{v} \times \mathbf{w}} \mathbf{x} = \|\mathbf{x}\| \cos \psi$ is the signed height of the parallelepiped (see Figure 2.4.6). $\square$

At this point it may seems strange to discuss *signed* volume. In applications, however, the sign can determine whether net flow is going into or out of a solid, or whether the net charge of an electric field is positive or negative.

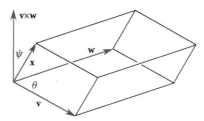

Figure 2.4.6 Triple product and volumes

Exercises

1. Compute the determinants of the following matrices

   (a) $\begin{bmatrix} 3 & -4 \\ -2 & 3 \end{bmatrix}$ (b) $\begin{bmatrix} 1 & 3 \\ -2 & 3 \end{bmatrix}$

   (c) $\begin{bmatrix} 3x & -4xy \\ 6y^2 & -8x \end{bmatrix}$ (d) $\begin{bmatrix} 1 & 3 \\ -2 & -6 \end{bmatrix}$

2. Compute the determinants of the following matrices

   (a) $\begin{bmatrix} 3 & -4 & 2 \\ -2 & 3 & 1 \\ 0 & 2 & 2 \end{bmatrix}$ (b) $\begin{bmatrix} 2 & 4 & 1 \\ -2 & 3 & 2 \\ 0 & 7 & 3 \end{bmatrix}$

3. Compare the determinants of the matrices with two rows interchanged:

   $\begin{vmatrix} a & b & c \\ d & e & f \\ g & h & i \end{vmatrix}$ $\begin{vmatrix} a & b & c \\ g & h & i \\ d & e & f \end{vmatrix}$

4. Compare the determinants of the matrices with rows interchanged:

   $\begin{vmatrix} a & b & c \\ d & e & f \\ g & h & i \end{vmatrix}$ $\begin{vmatrix} g & h & i \\ a & b & c \\ d & e & f \end{vmatrix}$

5. For what values of x and y is the determinant $\begin{vmatrix} 2y & 2x \\ 2x & 6y \end{vmatrix}$ positive?

6. For what values of x and y is the determinant $\begin{vmatrix} 6xy & 3x^2 \\ 3x^2 & 0 \end{vmatrix}$ positive?

7. Perform the following matrix multiplications:

   (a) $\begin{bmatrix} 2 & 1 \\ -2 & 3 \end{bmatrix}\begin{bmatrix} 1 & -1 \\ 2 & 3 \end{bmatrix}$ (b) $\begin{bmatrix} 1 & -1 \\ 2 & 3 \end{bmatrix}\begin{bmatrix} 2 & 1 \\ -2 & 3 \end{bmatrix}$

   Is matrix multiplication commutative?

8. Let A be the matrix

   $$A = \begin{bmatrix} 2 & 1 & -3 \\ 3 & 0 & 1 \\ -2 & 2 & 3 \end{bmatrix}.$$

(a) Form the products of A with the standard basis vectors $A\mathbf{i}, A\mathbf{j}, A\mathbf{k}$. How do the products relate to A?

(b) Let A be any $m \times 3$ matrix. What can you say about the products $A\mathbf{i}, A\mathbf{j}, A\mathbf{k}$?

9. Multiply the following matrices, if possible:

$$(a) \begin{bmatrix} -1 & 3 \\ 2 & -3 \end{bmatrix} \begin{bmatrix} 3 & 2 & 2 \\ 2 & 3 & 4 \end{bmatrix} \qquad (b) \begin{bmatrix} 4 & -3 & 2 \\ 6 & 7 & -2 \end{bmatrix} \begin{bmatrix} 1 & 2 \\ -2 & 1 \end{bmatrix}$$

10. Find the image $A\mathbf{C}(t)$ of the quadratic Bézier curve

$$\mathbf{C}(t) = (1-t)^2 \langle -1, 1 \rangle + 2(1-t)t \langle 0, -1 \rangle + t^2 \langle 1, 1 \rangle, 0 \le t \le 1,$$

where A is the matrix

$$A = \begin{bmatrix} 0 & 1 \\ 1 & 0 \end{bmatrix}.$$

Sketch $\mathbf{C}(t)$ and $A\mathbf{C}(t)$ on the same axes. Can you guess what transformation this is?

11. Find the image $A\mathbf{C}(t)$ of the quadratic Bézier curve

$$\mathbf{C}(t) = (1-t)^2 \langle -1, 1 \rangle + 2(1-t)t \langle 0, -1 \rangle + t^2 \langle 1, 1 \rangle, \ 0 \le t \le 1,$$

where A is the matrix

$$A = \begin{bmatrix} 0 & -1 \\ 1 & 0 \end{bmatrix}.$$

Sketch $\mathbf{C}(t)$ and $A\mathbf{C}(t)$ on the same axes. Can you guess what transformation this is?

12. Calculate the cross product of the following pairs of vectors:

(a) $\mathbf{v} = \langle 2, 1, 0 \rangle$, $\mathbf{w} = \langle 4, 3, 0 \rangle$ (b) $\mathbf{v} = \langle -3, 2, 0 \rangle$, $\mathbf{w} = \langle 6, -4, 0 \rangle$

(c) $\mathbf{v} = \langle 3, 0, -2 \rangle$, $\mathbf{w} = \langle 0, 4, -1 \rangle$ (d) $\mathbf{v} = \langle 1, 2, -2 \rangle$, $\mathbf{w} = \left\langle -\frac{1}{2}, -1, 1 \right\rangle$

13. Find the area of the parallelogram determined by the vectors $\mathbf{v} = \langle 3, 1 \rangle$ and $\mathbf{w} = \langle -2, 3 \rangle$, and sketch it.

14. Can you find an integer value for b that makes the parallelogram determined by $\mathbf{v} = \langle 5, b \rangle$ and $\mathbf{w} = \langle 7, 4 \rangle$ have area 1? How about the vectors $\mathbf{v} = \langle 5, b \rangle$ and $\mathbf{w} = \langle 6, 4 \rangle$?

15. For those with some number theoretic experience: What is the smallest positive area parallelogram that $\langle 36, 60 \rangle$ can make with an integer-valued vector $\langle m, n \rangle$? (Hint: the area is a linear combination of 36 and 60)

16. Show that a parallelogram in $\mathbb{R}^2$ with vertices on the integer lattice has integral area.

17. Find the area of the parallelogram determined by the vectors $\mathbf{v} = \langle 5, -2, 3 \rangle$ and $\mathbf{w} = \langle 2, -1, -3 \rangle$.

18. This problem involves scalar multiples of the vectors $\mathbf{v} = \langle 3, -2, 1 \rangle$ and $\mathbf{w} = \langle 1, 3, -2 \rangle$.

    (a) Calculate the area of the parallelogram determined by $\mathbf{v}$ and $\mathbf{w}$.

    (b) Calculate the area of the parallelogram determined by the vectors $\mathbf{v} = \Delta s \langle 3, -2, 1 \rangle$ and $\mathbf{w} = \Delta t \langle 1, 3, -2 \rangle$, where Δs and Δt are scalars. How is it related to your answer in part (a)?

19. Find a Cartesian equation for the plane through the point $(3, 3, 2)$ and containing the vectors $\mathbf{v} = \langle 2, -1, 2 \rangle$ and $\mathbf{w} = \langle 1, 3, 1 \rangle$.

20. Find a Cartesian equation for the plane through the origin and containing the vectors $\mathbf{v} = \langle 3, 1, 5 \rangle$ and $\mathbf{w} = \langle -2, 1, 1 \rangle$.

21. Find a Cartesian equation for the plane given parametrically by $\mathbf{S}(s, t) = (3 - s + 2t, 1 + 2s + t, -2 + s - 3t)$, $-\infty < s, t < \infty$.

22. Find a Cartesian equation for the plane given parametrically by $\mathbf{S}(s, t) = (4 + 3s + 2t, -3 - 2s + t, 5s + t)$, $-\infty < s, t < \infty$.

23. Parameterize the line of intersection of the planes $3y - 2z = 4$ and $6x - 2y + 3z = 2$.

24. Parameterize the line of intersection of the planes $x + y + 3z = 6$ and $2x - y + z = 4$.

25. Find a Cartesian equation for the plane containing the line $\mathbf{C}(t) = (2 - t, 3 + 2t, t)$, $-\infty < t < \infty$ and perpendicular to the plane $x - y + z = 3$.

26. Find a Cartesian equation for the plane through the points $(-2, 3, 4)$, $(1, -1, 2)$, and $(3, 5, -1)$.

27. A 45 single, instead of a 33 LP, makes 45 revolutions per minute.

    (a) Determine the angular velocity vector ω in rad/sec.

    (b) Determine the velocity vector of a point one inch from center.

    (c) Show that the speed of a point on the 45 varies linearly with its distance r from the center.

28. Find the volume of the parallelepiped spanned by the vectors $\langle 2, 1, 3 \rangle$, $\langle 4, -2, 2 \rangle$, and $\langle -1, 1, 1 \rangle$.

29. Without doing any calculations, is the triple scalar product of $(\mathbf{i} \times \mathbf{j}) \cdot \mathbf{k}$ positive or negative?

30. Without doing any calculations, find vectors $\mathbf{v}, \mathbf{w}, \mathbf{x}$ with $(\mathbf{v} \times \mathbf{w}) \cdot \mathbf{x} > 0$.

31. Verify that $(\mathbf{v} \times \mathbf{w}) \cdot \mathbf{x} = (\mathbf{x} \times \mathbf{w}) \cdot \mathbf{v}$. Compare $(\mathbf{v} \times \mathbf{w}) \cdot \mathbf{x}$ and $(\mathbf{v} \times \mathbf{x}) \cdot \mathbf{w}$.

2.5 Calculus with Parametric Curves

In this section we take a look at calculus with parametric curves, taking a decidedly geometric point of view. Despite the geometric emphasis, the analysis holds for all vector-valued functions. Sociologists or politicians, for example, may use vectors to describe current populations in a number of cities. Since populations are ever-changing, the components of the vector will be functions of time, and differentiating them gives growth rates for the populations. Despite the ubiquitous nature of vector-valued functions, we take a geometric approach and think of vector-valued functions as parametric curves in this section.

It turns out that finding the velocity vector for a curve follows from what we know about vector addition, scalar multiplication, and taking limits. Thus it is a nice application of the vector topics we've been discussing in this chapter. Arclength is also introduced, and shown to be the integral of the speed of a curve. Moreover, tangents to curves provide another context in which we encounter parameteric lines and planes, as well as applications of work. We now consider the problem of finding velocity vectors of curves.

Tangents to Curves: The parametric curve $\mathbf{C}(t) = \langle x(t), y(t), z(t) \rangle$ can be thought of as describing the position of a particle flying through space at time t. With this interpretation it makes sense to talk about the velocity of the particle. Now velocity is a vector quantity, since it has both magnitude and direction. Moreover, it should still be the rate of change of position with respect to time. In this context, the change in position is the vector $\mathbf{C}(t+h) - \mathbf{C}(t)$ (see Figure 2.5.1). So dividing by the change in time and taking a limit we get:

$$
\mathbf{C}'(t) = \lim_{h \to 0} \frac{\mathbf{C}(t+h) - \mathbf{C}(t)}{h} = \lim_{h \to 0} \frac{1}{h} \left(\mathbf{C}(t+h) - \mathbf{C}(t) \right)
$$

$$
= \lim_{h \to 0} \frac{1}{h} \left(\langle x(t+h), y(t+h), z(t+h) \rangle - \langle x(t), y(t), z(t) \rangle \right)
$$

$$
= \lim_{h \to 0} \frac{1}{h} \left(\langle x(t+h) - x(t), y(t+h) - y(t), z(t+h) - z(t) \rangle \right)
$$

$$
= \lim_{h \to 0} \left\langle \frac{x(t+h) - x(t)}{h}, \frac{y(t+h) - y(t)}{h}, \frac{z(t+h) - z(t)}{h} \right\rangle
$$

$$
= \langle x'(t), y'(t), z'(t) \rangle. \tag{2.5.1}
$$

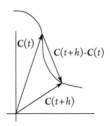

Figure 2.5.1 Differentiating parametric curves

Therefore, to find the tangent vector $\mathbf{C}'(t)$ you differentiate component-wise! The basic idea is that differentiation involves scalar multiplication (dividing by h), and vector subtraction $(\mathbf{C}(t+h) - \mathbf{C}(t))$. Since both of those operations are done component-wise, so is differentiation.

The tangent vector $\mathbf{C}'(t)$ is usually thought of as having initial point $\mathbf{C}(t)$. Notice that $\mathbf{C}'(t)$ gives the tangent vector at $\mathbf{C}(t)$ for every value of t. The unit tangent vector $\mathbf{T}(t)$ is obtained by normalizing $\mathbf{C}'(t)$:

$$\mathbf{T}(t) = \frac{\mathbf{C}'(t)}{\|\mathbf{C}'(t)\|}.$$

Example 2.5.1. *Tangent vectors to a spiral*

Let $\mathbf{C}(t) = \langle t \cos t, t \sin t \rangle$ for $t \geq 0$. The tangent vector is

$$\mathbf{C}'(t) = \langle \cos t - t \sin t, \sin t + t \cos t \rangle.$$

See Figure 2.5.2. Note that when the initial point of $\mathbf{C}'(t)$ is placed at the terminal point of the position vector $\mathbf{C}(t)$ it is tangent to the curve in the usual sense.

To find the unit tangent vector, just divide $\mathbf{C}'(t)$ by

$$\|\mathbf{C}'(t)\| = \sqrt{(\cos t - t \sin t)^2 + (\sin t + t \cos t)^2} = \sqrt{1 + t^2}.$$

Thus we have the unit tangent vector

$$\mathbf{T}(t) = \left(\frac{\cos t - t \sin t}{\sqrt{1 + t^2}}, \frac{\sin t + t \cos t}{\sqrt{1 + t^2}} \right). \blacktriangle$$

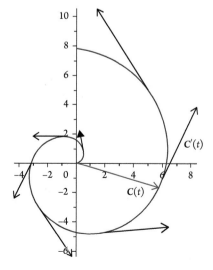

Figure 2.5.2 Tangent vectors to a spiral

Differentiation, Vector Addition and Scalar Multiplication: Differentiating vector-valued functions behaves quite nicely with the operations of vector addition and scalar multiplication. We illustrate with an example or two.

Example 2.5.2. *Differentiating vector sums*

Let $\mathbf{C}_1(t) = (t^2, t^3), \mathbf{C}_2(t) = (1 + 2t, -t), 0 \leq t \leq 1$ be two vector-valued functions. Evaluate

$$\frac{d}{dt}(\mathbf{C}_1 + \mathbf{C}_2).$$

We do so by first forming the sum, then differentiating. The sum rule from single variable calculus is seen to generalize to vector-valued functions.

$$\frac{d}{dt}(\mathbf{C}_1 + \mathbf{C}_2) = \frac{d}{dt}\left(\langle t^2, t^3 \rangle + \langle 1 + 2t, -t \rangle\right)$$
$$= \frac{d}{dt}(t^2 + 2t + 1, t^3 - t)$$
$$= \langle 2t + 2, 3t^2 - 1 \rangle = \langle 2t, 3t^2 \rangle + \langle 2, -1 \rangle$$
$$= \mathbf{C}_1'(t) + \mathbf{C}_2'(t). \ \blacktriangle$$

It is not hard to see that this generalizes, and the derivative of a sum of vector-valued functions is the sum of their derivatives. This is not surprising, since the same is true of single-variable functions, and vector differentiation is done component-wise.

Example 2.5.3. *A product rule for "scalar" multiplication*

Differentiate the spiral $\mathbf{C}(t) = (t^2 \cos t, t^2 \sin t), 0 \leq t < \infty$.
For this example, let $\mathbf{U}(t) = (\cos t, \sin t)$ be the standard parameterization of the unit circle, and note that factoring t^2 out of $\mathbf{C}(t)$ gives $\mathbf{C}(t) = t^2 \mathbf{U}(t)$. We calculate $\mathbf{C}'$ directly, then use vector algebra to derive a certain product rule.

$$\mathbf{C}'(t) = \langle 2t \cos t - t^2 \sin t, 2t \sin t + t^2 \cos t \rangle$$
$$= 2t \langle \cos t, \sin t \rangle + t^2 \langle -\sin t, \cos t \rangle$$
$$= 2t\mathbf{U}(t) + t^2 \mathbf{U}'(t).$$
$$= \frac{d}{dt}\left(t^2 \mathbf{U}(t)\right).$$

Thus the product rule for differentiating a scalar function times a vector-valued function looks much like the product rule from single variable calculus. $\blacktriangle$
We generalize the preceding example, without proof, to

$$\frac{d}{dt}\left(f(t)\mathbf{C}(t)\right) = f'(t)\mathbf{C}(t) + f(t)\mathbf{C}'(t). \tag{2.5.2}$$

This closely resembles the single-variable product rule because vector differentiation is done component-wise. Notice, in particular, if $C(t) = P$ is a constant vector, then $C'(t) = 0$ and $\frac{d}{dt}\left(f(t)C(t)\right) = f'(t)P$. This will be useful when differentiating Bézier curves.

Example 2.5.4. *Tangents to Bézier curves*

Recall from Section 2.2 that cubic Bézier curves are parameterized by a specific linear combination of control points. More precisely, given control points $P_0, \ldots, P_3$, the parameterization is

$$C(t) = (1-t)^3 P_0 + 3(1-t)^2 t P_1 + 3(1-t)t^2 P_2 + t^3 P_3$$
$$= (1-t)^3 P_0 + \left(3t^3 - 6t^2 + 3t\right) P_1 + \left(3t^2 - 3t^3\right) P_2 + t^3 P_3, \ 0 \le t \le 1.$$

By the sum rule, we differentiate $C(t)$ term by term, and applying the product rule with constant vectors means we just differentiate the scalars. Combining these observations gives

$$C'(t) = -3(1-t)^2 P_0 + \left(9t^2 - 12t + 3\right) P_1 + \left(6t - 9t^2\right) P_2 + 3t^2 P_3. \ \blacktriangle$$

Tangent Line Approximations: We pause now to remark that Equation 2.5.1 can be interpreted in terms of approximations. Instead of saying

$$C'(t) = \lim_{h \to 0} \frac{C(t+h) - C(t)}{h}, \tag{2.5.3}$$

one could say that $C'(t)$ is approximately the difference quotient

$$C'(t) \approx \frac{C(t+h) - C(t)}{h}$$

for small values of h. "Solving" this approximation for $C(t+h)$ shows that the point on the curve at time $t+h$ can be approximated by adding the position vector $C(t)$ and the scaled tangent vector $hC'(t)$. Analytically we have the approximation

$$C(t+h) \approx C(t) + hC'(t),$$

which is illustrated in Figure 2.5.3. This is completely analagous to, and follows from, the single-variable case. It will be used in Section 3.4 to approximate surface area using the area of tangent planes.

Example 2.5.5. *Approximating the helix*

Let $C(t) = (\cos t, \sin t, t)$ and use Equation 2.5.3 to approximate $C(0.1)$.

As in single variable calculus, we pick a t-value near $t = 0.1$ at which we can compute $C(t), C'(t)$, then use Equation 2.5.3. The obvious choice is to choose $t = 0$, from which we calculate

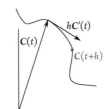

Figure 2.5.3 Tangent line approximations

$$\mathbf{C}(0) = (1,0,0), \text{ and } \mathbf{C}'(0) = (0,1,1).$$

Since $\Delta t = 0.1$, Equation 2.5.3 gives

$$\mathbf{C}(0.1) \approx \langle 1,0,0 \rangle + 0.1 \langle 0,1,1 \rangle = \langle 1,0.1,0.1 \rangle.$$

A more accurate approximation is $\mathbf{C}(0.1) = \langle 0.9950, 0.0998, 0.1 \rangle$, which shows that using tangent vectors to approximate the curve is quite reasonable in this case. ▲

Geometric Applications: The ability to compute tangent vectors to curves allows for a variety of geometric applications, including parameterizing tangent lines, finding planes normal to curves, finding tangent lines to curves of intersection, and a host of other possibilities. We explore a few of them here.

Example 2.5.6. *Parameterizing tangent lines*

Parameterize the tangent line to the curve $\mathbf{C}(t) = (\cos t, t, \sin t)$ at $t = \pi/3$.

Recall that to parameterize a line you need a point on the line and a vector in the direction of the line. In this instance $\mathbf{C}(\pi/3)$ is the point of tangency, so it is on the line, and the vector $\mathbf{C}'(\pi/3)$ is in the direction of the line. Thus the line is parameterized by

$$L(t) = \mathbf{C}(\pi/3) + t\mathbf{C}'(\pi/3) = \left\langle 1/2, \pi/3, \sqrt{3}/2 \right\rangle + t\left\langle -\sqrt{3}/2, 1, 1/2 \right\rangle$$
$$= \left\langle (1 - \sqrt{3}t)/2, \pi/3 + t, (\sqrt{3} + t)/2 \right\rangle, \ -\infty < t < \infty. ▲$$

Given a parameteric curve, then we can easily parameterize tangent lines. We summarize the strategy below.

Tangent Lines to Parametric Curves

A parameterization of the tangent line L to the parametric curve $\mathbf{C}(t)$ at $t = a$ is

$$L(t) = \mathbf{C}(a) + t\mathbf{C}'(a), \ -\infty < t < \infty.$$

Example 2.5.7. *Tangent to a quadratic Bézier curve*

Let $\mathbf{C}(t)$ be the quadratic Bézier curve with control points $(0,0), (2,1), (0,-2)$. Parameterize the tangent line to $\mathbf{C}(t)$ at $t = 1/2$.

Recall that the parameterization for $\mathbf{C}(t)$ is

$$\mathbf{C}(t) = (1-t)^2 \langle 0,0 \rangle + 2(1-t)t \langle 2,1 \rangle + t^2 \langle 0,-2 \rangle$$
$$= (4t - 4t^2, 2t - 4t^2), \ 0 \le t \le 1.$$

Thus the tangent vector is

$$\mathbf{C}'(t) = \langle 4 - 8t, 2 - 8t \rangle.$$

Evaluating these at $t = 1/2$ gives $\mathbf{C}(0.5) = (1,0)$ and $\mathbf{C}'(0.5) = \langle 0,-2 \rangle$, so the tangent line is

$$\mathbf{L}(t) = \langle 1,0 \rangle + t \langle 0,-2 \rangle = (1,-2t).$$

Note that this tangent line goes through the midpoints of the edges in the control polygon for $\mathbf{C}(t)$ (see Figure 2.5.4). A little reflection on de Casteljau's algorithm indicates that this might be expected. In de Casteljau's algorithm you recursively replace a control polygon by the one determined by the midpoints of its edges, until you get to a single point which turns out to be $\mathbf{C}(0.5)$. For a quadratic Bézier curve the first iteration of the algorithm gives a single edge (vertical in Figure 2.5.4), which is tangent to the curve at $t = 0.5$. ▲

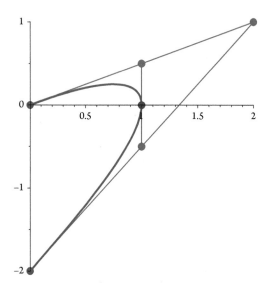

Figure 2.5.4 De Casteljau's algorithm and tangent lines

(a) (b)

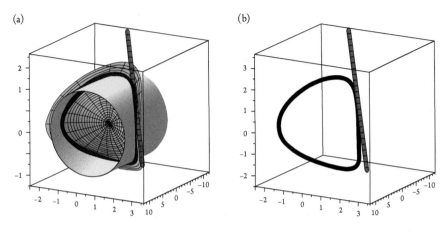

Figure 2.5.5 Tangent to curve of intersection

Example 2.5.8. *Tangent lines to curves of intersection*

Parameterize the tangent line to the curve of intersection of the generalized cylinder $y^2 + z^2 = 4$ and the surface $x = y^2 - z^2$ at the point $(0, \sqrt{2}, \sqrt{2})$ (see Figure 2.5.5).

One easily sees that the point is on both surfaces, hence on the curve of intersection. We know how to find tangent lines to parametric curves, so our strategy will be to parameterize the curve first. Recall from Section 1.4 that to parameterize the intersection of a generalized cylinder and a function, one follows:

Step 1. Parameterized the curve $y^2 + z^2 = 4$ in the yz-plane. This is a circle, so the parametric equations are

$$y = 2\cos t$$
$$z = 2\sin t, \ 0 \le t \le 2\pi.$$

Step 2. Substitute the parametric equations from step one into the function $x = y^2 - z^2$ to get the component function for x. Thus $x = (2\cos t)^2 - (2\sin t)^2 = 4\cos 2t$, using the fact that $\cos 2t = \cos^2 t - \sin^2 t$. The curve is then

$$\mathbf{C}(t) = (4\cos 2t, 2\cos t, 2\sin t), \ 0 \le t \le 2\pi.$$

Now that the curve is parameterized, we find out what value of t gives $\mathbf{C}(t) = (4\cos 2t, 2\cos t, 2\sin t) = (0, \sqrt{2}, \sqrt{2})$. Inspection verifies that $t = \pi/4$ does the trick, so the tangent line is

$$\mathbf{L}(t) = \mathbf{C}(\pi/4) + t\mathbf{C}'(\pi/4) = \left(0, \sqrt{2}, \sqrt{2}\right) + t\left(-8, -\sqrt{2}, \sqrt{2}\right)$$
$$= \left(-8t, \sqrt{2} - \sqrt{2}t, \sqrt{2} + \sqrt{2}t\right), \ -\infty < t < \infty. \ \blacktriangle$$

Example 2.5.9. *A plane normal to a curve*

Find a Cartesian equation for the plane normal to the curve $\mathbf{C}(t) = (t, t^2, t^3)$ at $t = 1$.

Recall that to find a Cartesian equation for a plane we need a normal vector and a point on the plane. In this context, $\mathbf{C}(1)$ will be a point on the plane, and $\mathbf{C}'(1)$ will be a normal vector. Calculating, we see that the point and normal direction are given by

$$\mathbf{C}(1) = (1, 1, 1), \ \mathbf{C}'(1) = (1, 2, 3).$$

Thus an equation for the plane is

$$(1, 2, 3) \cdot (x, y, z) = (1, 2, 3) \cdot (1, 1, 1)$$
$$x + 2y + 3z = 6. \ \blacktriangle$$

Speed and Arclength: Suppose we wanted to calculate how fast the bug was flying at a given instant—its speed. Speed is a scalar quantity—it has no direction, just a magnitude. Average speed is the distance traveled divided by time. The distance traveled over the time interval $(t, t + h)$ is approximately $\|\mathbf{C}(t + h) - \mathbf{C}(t)\|$, the length of the vector in Figure 2.5.1. Notice that this is not exactly the distance traveled, since the vector is the shortest distance between the points, and the curve may be longer. It is, however, a close enough estimate for small values of h. Since the instantaneous speed is the limit as $h \to 0$ of the average speed, we see

$$\text{Instantaneous speed} = \lim_{h \to 0} \frac{\|\mathbf{C}(t + h) - \mathbf{C}(t)\|}{h} = \left\| \lim_{h \to 0} \frac{\mathbf{C}(t + h) - \mathbf{C}(t)}{h} \right\|$$
$$= \|\mathbf{C}'(t)\|.$$

Because of this caluculation, we make the following definition.

Definition 2.5.1. *The* speed *of a curve* $\mathbf{C}(t)$ *is the length,* $\|\mathbf{C}'(t)\|$, *of its tangent vector.*

Example 2.5.10. *The speed of a spiral*

The speed of the spiral $\mathbf{C}(t) = (t \cos t, t \sin t)$ of the previous example is $\|\mathbf{C}'(t)\| = \sqrt{1 + t^2}$. $\blacktriangle$

Example 2.5.11. *An elliptical helix*

At what point is the curve $\mathbf{C}(t) = (2 \cos t, \sin t, t)$ traveling the fastest (see Figure 2.5.6)?

We are asked to maximize the speed. To do so, we first find the speed as a function of time. Differentiating we see $\mathbf{C}'(t) = (-2 \sin t, \cos t, 1)$. Thus the speed is $\|\mathbf{C}'(t)\| = \sqrt{4 \sin^2 t + \cos^2 t + 1}$, and we want to maximize this. To make our computations simpler, we maximize $\|\mathbf{C}'(t)\|^2$. In this case,

$$\|\mathbf{C}'(t)\|^2 = \|(-2 \sin t, \cos t, 1)\|^2$$
$$= 4 \sin^2 t + \cos^2 t + 1 = 3 \sin^2 t + 2.$$

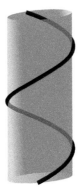

Figure 2.5.6 Elliptical helix

Figure 2.5.7 Approximating arclength using inscribed segments

This function is maximized when $|\sin t| = 1$, so when $t = \pi/2 + 2n\pi$. This is when $\mathbf{C}(t)$ is on the "flattest" portion of the ellipse. ▲

A *unit speed* curve is one whose tangent vector has unit length for all values of t. Thus the spiral $\mathbf{C} = (t\cos t, t\sin t)$ is not unit speed since $\|\mathbf{C}'(t)\| = \sqrt{1+t^2}$, which is not 1 for all values of t. However, the parameterization

$$\mathbf{C}(t) = (2\cos(t/2), 2\sin(t/2)), \ 0 \le t \le 4\pi$$

is unit speed since $\mathbf{C}'(t) = (-\sin(t/2), \cos(t/2))$, which is always a unit vector.

Recall that, in single-variable calculus, integrating speed gives distance traveled. The same is true for arbitrary parametric curves. In this context, the distance traveled is the arclength of the curve. One can think of this as approximating the length of the curve by summing the lengths of inscribed line segments. As one can see in Figure 2.5.7, increasing the number of segments used should improve the approximation. The arclength should be the limit of the approximations, as they continue to improve. The limit process turns the sum into an integral.

Definition 2.5.2. *The arclength $L(C)$ of the parametric curve $\mathbf{C}(t)$, $a \le t \le b$, is given by*

$$L(C) = \int_a^b \|\mathbf{C}'(t)\| \, dt.$$

Notation: The notation $\|\mathbf{C}'(t)\|\, dt$ is called the arclength differential form, and is common enough that writing it becomes cumbersome. For this reason we shorten it to the differential dC. When the curve C is defined parametrically, we have

$$dC = \|\mathbf{C}'(t)\|\, dt,$$

and the integral for the length $L(C)$ of $\mathbf{C}(t)$ is denoted by

$$L(C) = \int_C dC.$$

Math App 2.5.1. *Tangent vectors and arclength*

The following Math App demonstrates the tangent vector as a limit, and shows how inscribed segments approximate arclength.

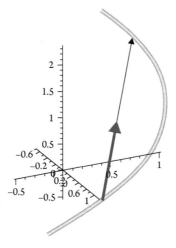

Example 2.5.12. *Arclength of a helix*

Determine the length of one revolution of the helix $\mathbf{C}(t) = (\cos t, \sin t, t)$.

Since one revolution occurs for every interval of length 2π, we choose the limits $0 \le t \le 2\pi$. Calculating the speed we get

$$\|\mathbf{C}'(t)\| = \|\langle -\sin t, \cos t, 1\rangle\| = \sqrt{\sin^2 t + \cos^2 t + 1} = \sqrt{2},$$

which implies that the arclength is

$$\int_0^{2\pi} \|\langle -\sin t, \cos t, 1\rangle\|\, dt = \int_0^{2\pi} \sqrt{2}\, dt = \sqrt{2}t \Big|_0^{2\pi} = 2\sqrt{2}\pi. \ \blacktriangle$$

Example 2.5.13. *Arclength of linear Bézier curves*

If $\mathbf{C}(t)$, $0 \leq t \leq 1$, is a linear Bézier curve from P to Q, show that the parameter value is the proportion of the distance that $\mathbf{C}(t)$ is from P to Q.

Recall from Equation 2.2.1 that the linear Bézier curve from P to Q is

$$\mathbf{C}(t) = (1-t)\mathbf{P} + t\mathbf{Q}, \ 0 \leq t \leq 1.$$

Moreover, the paragraph preceding Equation 2.2.1 uses vectors and scalar multiplication to show that the distance from P to $\mathbf{C}(t)$ is t times that from P to Q. We now use calculus to give an alternate justification of this fact.

The length of the tangent vector in this case is

$$\left\| \mathbf{C}'(t) \right\| = \| -\mathbf{P} + \mathbf{Q} \| = \| \mathbf{v} \|,$$

which is constant, and is just the length of the vector $\mathbf{v}$ from P to Q. The length of $\mathbf{C}(t)$, $0 \leq t \leq a$, then, is given by

$$\int_0^a \| \mathbf{v} \|\, dt = \| \mathbf{v} \| \int_0^a dt = a \, \| \mathbf{v} \|.$$

Since $\| \mathbf{v} \|$ is the distance from P to Q, this confirms that $\mathbf{C}(a)$ cuts segment PQ into the proportion $a : (1-a)$. ▲

Example 2.5.14. *Another curve*

Determine the arclength of $\mathbf{C}(t) = (\sqrt{2}t, \ln t, t^2/2)$ for $1 \leq t \leq 2$.

We calculate the speed to be

$$\left\| \mathbf{C}'(t) \right\| = \left\| \left(\sqrt{2}, \frac{1}{t}, t \right) \right\| = \sqrt{2 + \frac{1}{t^2} + t^2} = \sqrt{\left(\frac{1}{t} + t \right)^2} = \frac{1}{t} + t.$$

Thus the arclength is

$$\int_1^2 \frac{1}{t} + t \, dt = \ln t + \frac{t^2}{2} \Big|_1^2 = \ln 2 + 2 - \frac{1}{2} = \ln 2 + \frac{3}{2}. \ \blacktriangle$$

Work and Parametric Curves: We saw in Section 2.3 that the physical notion of work is defined to be force times distance, where one considers the component of force in the direction of movement. In the case where the force is constant and the movement linear, work turns out to be the dot product of the force and displacement vectors. Eventually, in Section 5.1, we will be able to calculate the work done by a variable force on a particle moving along a curve. For now, we content ourselves with finding the component of force in the direction of movement, when the particle moves along a curve.

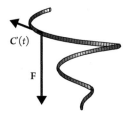

Figure 2.5.8 Gravity acting on a helicopter

Example 2.5.15. *Component of force*

A 1000kg helicopter flies along the spiral $\mathbf{C}(t) = (t\cos t, t\sin t, t)$ for $t \geq 0$, and the force due to gravity is the vector $\mathbf{F} = \langle 0, 0, -9800 \rangle$ s. Find and interpret the component of force in the direction of movement.

The key observation is that the direction of movement in this situation is the tangential direction, or $\mathbf{C}'(t)$. Differentiating coordinate-wise we get

$$\mathbf{C}'(t) = \langle \cos t - t\sin t, \sin t + t\cos t, 1 \rangle.$$

We now calculate the component of force in the tangential direction:

$$comp_{\mathbf{C}'(t)}\mathbf{F} = \frac{\mathbf{F} \cdot \mathbf{C}'(t)}{\|\mathbf{C}'(t)\|}$$

$$= \frac{\langle 0, 0, -9800 \rangle \cdot \langle \cos t - t\sin t, \sin t + t\cos t, 1 \rangle}{\sqrt{(\cos t - t\sin t)^2 + (\sin t + t\cos t)^2 + 1^2}} = \frac{-9800}{\sqrt{2 + t^2}}.$$

The fact that $comp_{\mathbf{C}'(t)}\mathbf{F}$ is negative indicates that gravity is working against the motion of the helicopter. Note also that t increases, the component of $\mathbf{F}$ tends to zero. ▲

Exercises

1. The following are all parameterizations of the circle $x^2 + y^2 = 25$ in $\mathbb{R}^2$. Find the tangent vectors for each parameterization, and determine which are unit speed curves.

   (a) $\mathbf{C}(t) = (5\cos t, 5\sin t)$, $0 \leq t \leq 2\pi$

   (b) $\mathbf{C}(t) = (5\cos(t/5), 5\sin(t/5))$, $0 \leq t \leq 10\pi$

   (c) $\mathbf{C}(t) = (5\cos(t^2), 5\sin(t^2))$, $0 \leq t \leq \sqrt{2\pi}$

2. Parameterize the tangent line to $\mathbf{C}(t) = (t^2 + 1, 2t - 3, 3 + t^3)$ at $t = -1$.

3. Parameterize the tangent line to $\mathbf{C}(t) = (\sec t, \tan t, t)$ at $t = \pi/4$.

4. Parameterize the tangent line to $\mathbf{C}(t) = (t\cos t, t\sin t, t^2)$ at $t = \pi$.

5. Parameterize the tangent line to the curve of intersection of the surfaces $x^2 + y^2 = 4$ and $x + y + z = 5$ at the point $(0, 2, 3)$.

6. Parameterize the tangent line to the curve of intersection of the generalized cylinders $z = y^2$ and $x + 2y = 4$ at the point $(-2, 3, 9)$.

7. Let $\mathbf{C}(t)$, $0 \le t \le 1$ be the Bézier curve given by control points $(0,0), (1,2)$, and $(5, -1)$. Relate $\mathbf{C}'(0)$ and $\mathbf{C}'(1)$ to the control points (the parameterization is given in Equation 2.2.3).

8. Write the tangent vectors $\mathbf{C}'(0)$ and $\mathbf{C}'(1)$ to the general cubic Bézier curve (Equation 2.2.4) in terms of the control points.

9. For the general quadratic Bézier curve (Equation 2.2.3), show that the tangent line when $t = 0.5$ is parallel to the vector from $\mathbf{P}_0$ to $\mathbf{P}_2$.

10. Recall that the degree n Bézier curve has parameterization

$$\mathbf{C}(t) = \sum_{i=0}^{n} \binom{n}{i}(1 - t)^{n-i} t^i \mathbf{P}_i, \ 0 \le t \le 1.$$

See the discussion near Equation 2.2.6 for further detail.

(a) Write $\mathbf{C}'(0)$ and $\mathbf{C}'(1)$ in terms of the control points.

(b) If $\mathbf{P}_0 = \mathbf{P}_n$ (so $\mathbf{C}(t)$ is a closed loop), what restrictions on $\mathbf{P}_1, \mathbf{P}_{n-1}$ ensure that $\mathbf{C}(t)$ has a nice tangent line at $t = 0$?

11. When is the tangent line to $\mathbf{C}(t) = (t^2 - t, 3t + 1, 2t^3 - 1)$ parallel to the line $L(t) = (2 - t, 3 + t, 5 + 2t)$?

12. When is the tangent line to $\mathbf{C}(t) = (\sqrt{t}, 2 + t, t^2 - 1)$ normal to the plane $x + 4y + 32z = 16$?

13. Show that the tangent vector to the curve $\mathbf{C}(t) = \left(\cos t, \frac{\sin t}{\sqrt{2}}, \frac{\sin t}{\sqrt{2}}\right)$ is always perpendicular to the position vector.

14. Show that the tangent vector to the curve $\mathbf{C}(t) = (\sin t \cos t, \sin^2 t, \cos t)$, $0 \le t \le \pi$ is always perpendicular to the position vector.

15. Where does the tangent vector to the helix $\mathbf{C}(t) = (\cos t, \sin t, t)$ make an angle of $\pi/3$ with the position vector?

16. Find the arclength of the curve $\mathbf{C}(t) = (e^t \cos t, e^t \sin t)$, $0 \le t \le \sqrt{2\pi}$.

17. Find the arclength of the curve $\mathbf{C}(t) = (2 \cos t, 2 \sin t, 3t)$, $0 \le t \le \sqrt{2\pi}$.

18. Find the arclength of the curve $\mathbf{C}(t) = (10t^3 + 2, 9t^2 + 5, 12t^2 - 1)$, $0 \le t \le \sqrt{2\pi}$.

19. Suppose $\mathbf{C}(t)$, $a \le t \le b$ is a unit speed curve. Show that the arclength of $\mathbf{C}(t)$ is $b - a$.

20. Show that the parameterization $\mathbf{C}(t) = (3 \cos t, 3 \sin t), 0 \le t \le 2\pi$ is a constant speed parameterization. Now find a constant a such that $\mathbf{C}(t) = (3 \cos(t/a), 3 \sin(t/a))$ is a unit speed curve. What limits on the parameter are necessary to get once around?

21. A $1kg$ particle (Large particle? Maybe just dense.) travels along the helical path $C(t) = (\cos t, \sin t, t)$. Show that the component of the force $\mathbf{F} = \langle 0, 0, -9.8 \rangle$ in the direction of travel is constant. Use that to compute the work done by gravity on the particle for $0 \le t \le 2\pi$.

22. Find the component of $\mathbf{F} = \langle 0, 0, -9.8 \rangle$ in the tangential direction of $C(t) = (\sin t \cos t, \sin^2 t, \cos t)$, $0 \le t \le \pi$.

23. Show that the helix $C(t) = (\cos t, \sin t, t)$ has constant speed.

24. Find where the tangent vector to $C(t) = \langle t, t^2, t^3 \rangle$ is parallel to the line $\mathbf{y}(s) = \langle 4 + s, -2 - 2s, 1 + 3s \rangle$.

25. Find where the tangent line to $C(t) = \langle \cos t, 2 \sin t \rangle$ has slope 2.

26. Both $C_1(t) = \langle 3 \cos t, 3 \sin t \rangle$, $0 \le t \le 2Pi$, and $C_2(t) = \langle 3 \cos(t/3), 3 \sin(t/3) \rangle$, $0 \le t \le 6\pi$ parameterize the circle centered at the origin with radius three. Compare the speeds of the two parameterizations.

27. The parametric curves $C_1(t) = (0, 2 \sin t, 2 \cos t)$, $0 \le t \le 2\pi$ and $C_2(t) = (\sqrt{3} \cos t, 2 \sin t, \cos t)$, $0 \le t \le 2\pi$ intersect at the point $(0, 2, 0)$ when $t = \pi/2$. Find an equation for the plane through $(0, 2, 0)$ that contains both tangent vectors $C_1'(t)$ and $C_2'(t)$ there.

2.6 Vector Fields—a first glance

In this section we get our first glimpse of an important, and natural, topic—vector fields. Vector fields are functions that assign vectors (rather than scalars) to points. They arise naturally, and you may have seen some during the weather forecast on your nightly news. One example of a vector field assigns to each point on a map the velocity vector of the wind at that point. Vector fields occur in other contexts as well. Electric fields, magnetic fields, gravitational fields, and fluid flow are all examples that arise in applications. Chapter 5 will investigate vector fields in greater detail, but we introduce them here briefly because the topic fits nicely with our discussion of vectors.

Definition 2.6.1. *A vector field is a function that assigns a vector to each point in its domain.*

We will use bold face letters to designate vector fields, which will frequently be given by component functions. For example, $\mathbf{F}(x,y) = \langle x + 2, -y \rangle$ denotes a vector field. When visualizing vector fields it is customary to sketch the output $\mathbf{F}(x,y)$ with its initial point at the input (x,y). This results in pictures as in Figure 2.6.1.

Planar Vector Fields: We begin with examples of vector fields in the plane, because they are easier to sketch by hand. After a few analytic examples, we also look at an example from electrostatics and introduce the circulation of a fluid along a curve.

Example 2.6.1. *Constant vector fields*

Constant vector fields are perhaps the easiest to visualize. In Figure 2.1.1 we placed the vector $\mathbf{P} = \langle 3, -3 \rangle$ in the plane with various initial points to emphasize that the vectors are the same regardless of their initial point. We can now interpret that picture differently. The vector field $\mathbf{F}(x,y) = \langle 3, -3 \rangle$ is the field that attaches to every point in the plane the vector $\mathbf{P} = \langle 3, -3 \rangle$. Figure 2.1.1 can then be seen as a picture of some representatives of the vector field $\mathbf{F}$. ▲

Example 2.6.2. *Vector fields on $\mathbb{R}^2$*

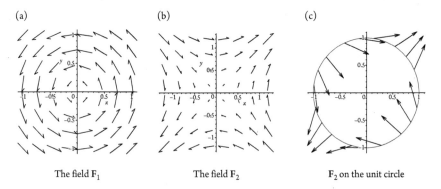

(a)	(b)	(c)
The field $\mathbf{F}_1$	The field $\mathbf{F}_2$	$\mathbf{F}_2$ on the unit circle

Figure 2.6.1 Vector fields on $\mathbb{R}^2$

A vector field in $\mathbb{R}^2$ is $\mathbf{F}_1(x,y) = \left(-\frac{y}{x^2+y^2+1}, \frac{x}{x^2+y^2+1}\right)$. The vector $\mathbf{F}_1(0,1) = \langle -0.5, 0 \rangle$ should be drawn with initial point at $(0,1)$. Notice that the vector $\mathbf{F}_1(x,y)$ is a scalar multiple of $\langle -y, x \rangle$, which is orthogonal to the position vector $\langle x,y \rangle$. This relationship is clear (is it?) from Figure 2.6.1(a).

A second vector field is $\mathbf{F}_2(x,y) = \langle y, x \rangle$. You see that the vectors on the line $y = x$ run parallel to the line, while those on either axis are perpendicular to the axis (why?).

We can also evaluate a vector field on a curve simply by restricting the domain. For example, the domain of $\mathbf{F}_2$ is the whole plane, but we can evaluate it on the unit circle. Analytically, this just amounts to substituting the parametric equations $\mathbf{C}(t) = (\cos t, \sin t)$ for x and y in the formula for $\mathbf{F}_2$. We have

$$\mathbf{F}_2(\mathbf{C}(t)) = \mathbf{F}_2(\cos t, \sin t) = \langle \sin t, \cos t \rangle.$$

Please understand the notation $\mathbf{F}_2(\mathbf{C}(t))$. It is cumbersome, but not incomprehensible. In English it means "Evaluate the vector field $\mathbf{F}_2$ on the curve $\mathbf{C}(t)$". Geometrically this corresponds to drawing only those vectors of $\mathbf{F}_2$ (seen in Figure 2.6.1(b)) that originate on the unit circle (see Figure 2.6.1(c)). ▲

Electric Fields: We begin by using Coulomb's law, together with the law of superposition, to describe the electric field generated by fixed point charges $Q_1, \ldots, Q_n$. Recall that Coulomb's law says that a charge Q exerts a force on another q given by

$$\mathbf{F} = k\frac{Q \cdot q}{\rho^2}\hat{\rho},$$

where ρ is the distance between them, $\hat{\rho}$ is the unit vector pointing from Q to q, and k is Coulomb's constant. The principle of superposition implies that the net force a system $Q_1, \ldots, Q_n$ of charges exerts on a test charge q is the sum of the forces of the individual charges:

$$\mathbf{F} = \sum k\frac{Q_i \cdot q}{\rho_i^2}\hat{\rho}_i = q\sum k\frac{Q_i}{\rho_i^2}\hat{\rho}_i.$$

Notice that the size q of the test charge can be factored out of $\mathbf{F}$, and what remains in the sum depends only on the position of q. Indeed, ρ_i is the distance between charges q and Q_i while $\hat{\rho}_i$ is the unit vector from Q_i to q. The electric field $\mathbf{E}$ generated by fixed point charges $Q_1, \ldots, Q_n$ is defined to be that portion of $\mathbf{F}$ that depends only on the position of q and not its magnitude, so that

$$\mathbf{E} = \sum k\frac{Q_i}{\rho_i^2}\hat{\rho}_i. \qquad (2.6.1)$$

Clearly $\mathbf{F} = q\mathbf{E}$, so we interpret the electric field $\mathbf{E}$ as the *force per unit charge* of the system $Q_1, \ldots, Q_n$. In other words, multiplying $\mathbf{E}$ by the size of a test charge q gives the force of the system on q. Take a moment to convince yourself that "force per unit charge" makes sense...don't just let it roll off your tongue!

Example 2.6.3. *An electric dipole*

A positive point charge placed at $(-1,0)$ and a negative one at $(1,0)$ creates an electric field which exerts a force on a stationary test charge anywhere in the plane. Find an explicit formula for **E**.

We remark that the electric field is really three-dimensional, but by symmetry we consider only that portion in the plane containing the charges. This could be revolved around the line between the charges to obtain the entire field.

We begin by stating the formula and then justify it. In this case, the force experienced by the point charges at $(\pm 1,0)$ is given by

$$\mathbf{E}(x,y) = \frac{k}{((x+1)^2 + y^2)^{3/2}}\langle x+1, y\rangle - \frac{k}{((x-1)^2 + y^2)^{3/2}}\langle x-1, y\rangle. \qquad (2.6.2)$$

Take a moment to understand this formula. The vectors $\langle x+1, y\rangle$ and $\langle x-1, y\rangle$ are from the charges at $(-1,0)$ and $(1,0)$, respectively, to the point (x,y). Normalizing these vectors results in dividing by $\rho_\pm = \sqrt{(1\pm x)^2 + y^2}$. The additional $\rho_\pm^2 = (1\pm x)^2 + y^2$ in Equation 2.6.1, results in the $3/2$ exponent in the denominator of Equation 2.6.2. The field **E** is depicted in Figure 2.6.2(a).

Physicists usually use *field lines* to represent electric fields, obtained by drawing curves tangent to the vectors as in Figure 2.6.2(b) (for a nice description of how to parameterize these field lines see [7]). These are the paths that a positively charged particle would follow if released in the field. We will typically use the vectors of Figure 2.6.2(a) to emphasize that electric fields are physical examples of vector fields. ▲

Velocity Fields: An ideal fluid flow can be modeled using the velocity potential and stream functions, as in Example 1.2.10 and the discussion preceding it. Rather than using the velocity potential and stream functions, a velocity vector field can be used to specify a

(a)   (b)   (c)

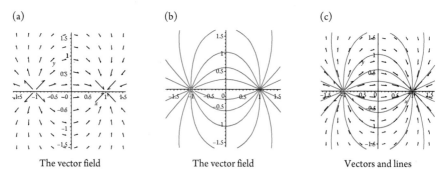

The vector field   The vector field   Vectors and lines

Figure 2.6.2 Electric field from two point charges

fluid flow. This is the field that attaches to each point the velocity vector of a particle in the flow at that point. We begin with examples of velocity fields for some standard models, then move on to introduce the notions of flux and circulation.

Example 2.6.4. *Some standard velocity fields*

A uniform flow is given by a constant velocity vector, which corresponds to fluid flowing always in the same direction at a constant speed. Figure 2.6.3(a) illustrates the field $\mathbf{V}_1(x,y) = \langle 1,2 \rangle$.

An ideal flow around a right angle was introduced in Example 1.2.10 as having velocity potential $\varphi = x^2 - y^2$ and stream function $\psi = 2xy$. This flow can also be defined by the velocity field $\mathbf{V}_2 = \langle -2x, 2y \rangle$ pictured in Figure 2.6.3(b). The flow begins on the right, flows toward the y-axis while turning vertical. The velocity vectors of $\mathbf{V}_2$ are tangent to the streamlines of Example 1.2.10.

The flow from a point vortex was introduced in Section 1.2, Exercise 22, as having the polar forms of velocity potential and stream functions

$$\varphi = K\theta, \ \psi = -K\ln r.$$

The velocity field, in Cartesian coordinates, for this flow is

$$\mathbf{V}_3 = K\left\langle \frac{-y}{x^2+y^2}, \frac{x}{x^2+y^2} \right\rangle = \frac{K}{x^2+y^2}\langle -y,x \rangle,$$

which is seen in Figure 2.6.3(c). Note that the flow is counterclockwise about the origin, moving more swiftly the closer you are to it.

Disclaimer: All vector fields rendered are scalar multiples of the fields themselves. Vectors have been rescaled for aesthetic reasons. ▲

Math App 2.6.1. *Flow lines in fluid flow*

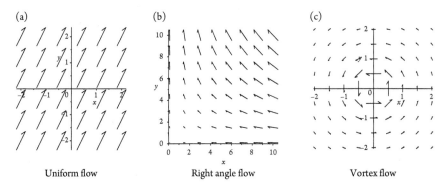

(a) Uniform flow   (b) Right angle flow   (c) Vortex flow

Figure 2.6.3 Velocity fields of ideal flows

The following Math App allows you to visualize an ideal fluid flow around a π/n-corner. Click the hyperlink, or print users open manually, to take a quick peak.

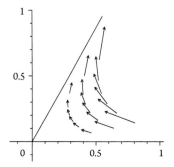

We now introduce the flux of a flow *across* and circulation of a flow *along* a curve. Circulation and flux are key ideas in Chapter 5, playing roles in the line integrals of Section 5.1, and the Theorems of Green, Stokes and Gauss (Sections 5.3, 5.6, and 5.7). *Flux* is the amount of fluid that crosses a fixed curve per unit time, or the rate at which the fluid crosses the curve. The *circulation* of a flow around a loop is the rate at which it flows along the curve times the length of the loop. The rates needed to calculate circulation and flux can be found by projecting the velocity field onto tangential and normal directions to the curve, respectively. More precisely, the rate at which $\mathbf{V}$ crosses $\mathbf{C}(t)$ is the component of $\mathbf{V}$ normal to the curve, while $comp_{\mathbf{C}'}\mathbf{V}$ is the rate at which fluid flows along the curve (see Figure 2.6.4). We remark that Chapter 11 of [5] has quite a nice description of flux, circulation, and related topics, not to mention that it's a wonderful book to read at some point in your career!

Since there are two normal directions at each point on a curve, we make the convention that $\mathbf{n}$ will be $\mathbf{C}'(t)$ rotated clockwise $90°$, as in Figure 2.6.4. With this convention, positive flux flows from left-to-right across $\mathbf{C}(t)$ while right-to-left flow results in negative flux. We remark that it is easy to rotate vectors $90°$ clockwise in the plane. Indeed, note that $\langle b, -a \rangle$ is orthogonal to $\langle a, b \rangle$ and a quick sketch will convince you that it is rotated clockwise from $\langle a, b \rangle$.

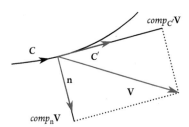

Figure 2.6.4 Tangential and normal components

Example 2.6.5. *Vortex flow*

(a) Find the tangential and normal components of the velocity of a vortex $\mathbf{V} = \frac{K}{x^2+y^2}\langle -y, x\rangle$ along the unit circle $\mathbf{C}(t) = (\cos t, \sin t)\ 0 \le t \le 2\pi$.

We only consider the velocity of the fluid along the curve, and must evaluate it there. This amounts to replacing x, y in the formula for $\mathbf{V}$ with their parametric equations. We have

$$\mathbf{V}(\mathbf{C}(t)) = \frac{K}{\cos^2 t + \sin^2 t}\langle -\sin t, \cos t\rangle = K\langle -\sin t, \cos t\rangle.$$

The tangential direction is $\mathbf{C}'(t) = \langle -\sin t, \cos t\rangle$, so we compute

$$comp_{\mathbf{C}'}\mathbf{V} = \mathbf{V} \cdot \frac{\mathbf{C}'}{\|\mathbf{C}'\|} = K\langle -\sin t, \cos t\rangle \cdot \langle -\sin t, \cos t\rangle = K.$$

To find the component of $\mathbf{V}$ in the normal direction, we first rotate the tangent vector $\mathbf{C}'(t)$ clockwise 90° by swapping coordinates and negating the second. The normal direction is then $\mathbf{n} = \langle \cos t, \sin t\rangle$, which makes sense because radii are normal to circles. Computation yields

$$comp_{\mathbf{n}}\mathbf{V} = \mathbf{V} \cdot \frac{\mathbf{n}}{\|\mathbf{n}\|} = K\langle -\sin t, \cos t\rangle \cdot \langle \cos t, \sin t\rangle = 0.$$

This shows that the flow of a vortex is tangent to the unit circle—or that the unit circle is a streamline for the flow. Hopefully this is what you found in Section 1.2, Exercise 22!

(b) Find the tangential and normal components of the velocity of a vortex $\mathbf{V} = \frac{K}{x^2+y^2}\langle -y, x\rangle$ along the line segment $\mathbf{C}(t) = (t, 2t)\ 1 \le t \le 2$.

As in part (a), the first step is to evaluate the vector field on the curve, giving

$$\mathbf{V}(\mathbf{C}(t)) = \frac{K}{t^2 + 4t^2}\langle -2t, t\rangle = \frac{K}{5t}\langle -2, 1\rangle.$$

The tangential direction is $\mathbf{C}'(t) = \langle 1, 2\rangle$, so the tangential component of $\mathbf{V}$ is

$$comp_{\mathbf{C}'}\mathbf{V} = \mathbf{V} \cdot \frac{\mathbf{C}'}{\|\mathbf{C}'\|} = \frac{K}{5t}\langle -2, 1\rangle \cdot \left(1/\sqrt{5}, 2/\sqrt{5}\right) = 0.$$

This means the fluid is not flowing along $\mathbf{C}$ at all, and must be normal to the curve! This is because our curve is on a ray emanating from the origin. Convince yourself that if circles centered at the origin are streamlines, then rays from the origin are perpendicular to the field. The whole switch-and-negate trick makes finding $comp_{\mathbf{n}}\mathbf{V}$ straightforward:

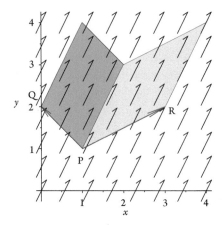

Figure 2.6.5 Flux of uniform flow across vectors

$$comp_{\mathbf{n}}\mathbf{V} = \mathbf{V} \cdot \frac{\mathbf{n}}{\|\mathbf{n}\|} = \frac{K}{5t}\langle -2, 1\rangle \cdot \left\langle 2/\sqrt{5}, -1/\sqrt{5}\right\rangle = -\frac{K}{\sqrt{5}t}.$$

The interested reader should verify that $comp_{\mathbf{n}}\mathbf{V} = -\|\mathbf{V}\|$. In fact, draw a few pictures and see why this makes sense. ▲

We now consider some examples of flux computation, starting with a uniform flow (i.e. one with a constant velocity field). The computation of the flux of a uniform flow across a line segment is intuitive and straightforward. Let $\mathbf{V}$ be the velocity field of a uniform flow and PQ the segment from P to Q. All particles crossing PQ in one unit of time form a parallelogram with $\overrightarrow{PQ}$ as one side and $\mathbf{V}$ the other (see Figure 2.6.5). This is because particles initially on PQ will travel $\mathbf{V}$ in unit time, forming another copy of $\overrightarrow{PQ}$ at the end of $\mathbf{V}$ after one unit of time. The region between will be filled in by the rest of the flow, so the "volume" per unit time that crosses $\overrightarrow{PQ}$ is the area of the parallelogram (see Figure 2.6.5).

Letting $\mathbf{n}$ be the rightward normal to $\overrightarrow{PQ}$, the signed area is

$$A = \left\|\overrightarrow{PQ}\right\| comp_{\mathbf{n}}\mathbf{V};$$

however, there is an alternative way to compute it. Recall that Equation 2.4.3 shows that the signed area of the parallelogram is the determinant of the matrix with rows $\mathbf{V}$ and $\overrightarrow{PQ}$. We use this approach in the following example.

Example 2.6.6. *Uniform flux*

We calculate the flux of a vector field across two directed line segments.

(a) Find the flux of the uniform flow $\mathbf{V} = \langle 1, 2\rangle$ across the directed line segment from $P = (1, 1)$ to $Q = (0, 2)$.

In this case, $\overrightarrow{PQ} = \langle -1, 1 \rangle$, and the flux is

$$\begin{vmatrix} 1 & 2 \\ -1 & 1 \end{vmatrix} = 3.$$

This is positive because the flow is from left to right across $\overrightarrow{PQ}$, and it is the area of the parallelogram above it (see Figure 2.6.5).

(b) Find the flux of the uniform flow $\mathbf{V} = \langle 1, 2 \rangle$ across the directed line segment from $P = (1, 1)$ to $R = (3, 2)$.

In this case, $\overrightarrow{PR} = \langle 2, 1 \rangle$, and the flux is

$$\begin{vmatrix} 1 & 2 \\ 2 & 1 \end{vmatrix} = -3.$$

This is negative of the area above $\overrightarrow{PR}$, since the flow across $\overrightarrow{PR}$ is left to right (see Figure 2.6.5). ▲

Now suppose the curve is not a line segment, and the flow is not uniform. If the curve was short enough, we could reasonably approximate the flux by making some simplifying assumptions. For a short curve the flow would be close to uniform, so we assume it's the same as the velocity at the beginning of the curve. Moreover, the tangent line will make a reasonable approximation of the curve as discussed in Section 2.5 (see Figure 2.5.3 in particular). We apply these principles in the following example.

Example 2.6.7. *Approximating flux*

Approximate the flux of $\mathbf{V}(x, y) = \langle -2x, 2y \rangle$ across the curve $\mathbf{C}(t) = (t^2, t)$, $1 \le t \le 1.25$.

The curve begins at $\mathbf{C}(1) = (1, 1)$, so we assume that the flow has constant velocity $\mathbf{V}(1, 1) = \langle -2, 2 \rangle$. Moreover, we replace the curve with its tangent line at $\mathbf{C}(1)$, which is

$$\mathbf{L}(t) = \mathbf{C}(1) + t\mathbf{C}'(1) = \langle 1, 1 \rangle + t \langle 2, 1 \rangle = (1 + 2t, 1 + t), \ 0 \le t \le .25.$$

A zoomed in view of this situation is depicted in Figure 2.6.6(a). Note that the curve is starting to pull away from the tangent line, but not dramatically so. Also note that the vector field in this view is fairly uniform (admittedly, the depiction is a scalar multiple of the actual field, but it gives the correct intuition).

Thus we want the flux of $\langle -2, 2 \rangle$ across the vector from $(1, 1)$ to $\mathbf{L}(.25) = (1.5, 1.25)$, which is $\langle .5, .25 \rangle$. Using the determinant we compute the flux to be

$$\begin{vmatrix} -2 & 2 \\ .5 & .25 \end{vmatrix} = -1.5.$$

The parallelogram in Figure 2.6.6(b) represents what the flow would have done had it been uniform. Clearly this doesn't occur in the flow around a corner (again, recall that the

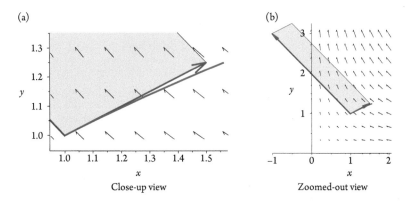

(a)   (b)

Close-up view   Zoomed-out view

Figure 2.6.6 Approximating flux

vectors in the velocity field should be comparable in length to $\langle -2, 2 \rangle$, but they have been rescaled to give the idea of the flow). ▲

In Chapter 5, local approximations of flux will be added to approximate the total flux across a curve. Taking a limit as the local approximations become *more* local leads to an integral that computes the actual flux.

Math App 2.6.2. *Approximating flux*

The following Math App allows you to engage in the process of approximating the flux of flows across the unit circle. Experiment and gain deeper intuition.

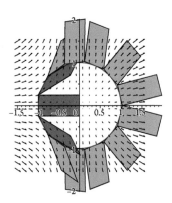

Vector Fields in $\mathbb{R}^3$: After having investigated some vector fields in the plane, we turn our attention to three-dimensional vector fields. In Chapter 5 we will regularly evaluate vector fields on surfaces, so some introductory examples are given here.

Example 2.6.8. *Vector fields in $\mathbb{R}^3$*

As in $\mathbb{R}^2$, vector fields in space can be specified by their component functions. The fields $\mathbf{F}_1(x,y,z) = \langle 2x, 2y, 1 \rangle$ and $\mathbf{F}_2(x,y,z) = \langle -x, -y, -z \rangle$ are pictured below. The vectors in $\mathbf{F}_1$ all change one unit vertically. As one moves horizontally away from the

(a) (b) (c)

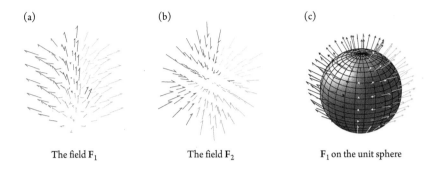

The field $\mathbf{F}_1$ The field $\mathbf{F}_2$ $\mathbf{F}_1$ on the unit sphere

Figure 2.6.7 Vector fields on $\mathbb{R}^3$

origin, the x and y coordinates increase and the vertical component of $\mathbf{F}_1$ is smaller relative to the horizontal components. On the z-axis, $\mathbf{F}_1$ yields the vector $\langle 0,0,1 \rangle$. If we restrict the field $\mathbf{F}_1$ to the unit sphere, we get a "hairy" sphere, with lots of mousse to make it straight (see Figure 2.6.7(c)). The field $\mathbf{F}_2$ attaches the vector $\langle -x, -y, -z \rangle$ to the point (x,y,z). Thus the tip of the vector should touch the origin. The vectors in Figure 2.6.7(b) are not drawn to scale, to give a better geometric understanding of $\mathbf{F}_2$: all vectors point toward the origin. ▲

Example 2.6.9. *Normal to a plane*

Find the component of $\mathbf{F}(x,y,z) = \langle x+y,z,-2y \rangle$ in the upward normal direction to the plane $x - 2y + z = 7$. Implicit in this problem is the assumption that we are evaluating $\mathbf{F}$ on the plane. We will see how we use this analytically in the last step. Since an upward-pointing normal to the plane is $\mathbf{n} = \langle 1,-2,1 \rangle$ (its negative is a downward-pointing normal since the z-component would be negative), we get the component of $\mathbf{F}$ in the $\mathbf{n}$ direction is

$$comp_{\mathbf{n}}\mathbf{F} = \frac{\mathbf{F} \cdot \mathbf{n}}{\|\mathbf{n}\|} = \frac{1}{\sqrt{6}} \langle x+y,z,-2y \rangle \cdot \langle 1,-2,1 \rangle = \frac{1}{\sqrt{6}}(x-y-2z).$$

At no point in this calculation did we use the fact that we evaluated $\mathbf{F}$ on the plane. We can use this now to reduce our calculation to a function of two variables. The coordinates of any point on the plane satisfy $x - 2y + z = 7$, so we can replace z with $7 - x + 2y$. Making this simplification, we get

$$comp_{\mathbf{n}}\mathbf{F} = \frac{1}{\sqrt{6}}(x-y-2z) = \frac{1}{\sqrt{6}}(3x - 5y - 14).$$

Evaluating $\mathbf{F}$ on the surface will be required later when performing certain integrals. ▲

Example 2.6.10. *Flux in three dimensions*

Suppose the field $\mathbf{F}_1(x,y) = \langle 2x, 2y, 1 \rangle$ pictured above represents the velocity field of a fluid flowing through space. Perhaps hot water in a cup of tea. Now suppose the unit

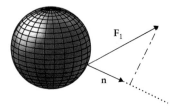

Figure 2.6.8 Component of $\mathbf{F}_1$ in normal direction — to calculate flux

sphere represents a tea ball in which tea is steeping. The water flows through the ball, making a delicious cup of tea. The component of the velocity $\mathbf{F}_1$ in the normal direction to the sphere measures how fast tea is flowing across it.

We want to calculate the component of $\mathbf{F}_1$ orthogonal to the sphere. We know from geometry that radii are orthogonal to spheres, so if (x,y,z) is a point on the sphere, then the position vector $\langle x,y,z\rangle$ is an outward-pointing normal to the sphere. Moreover, since (x,y,z) is on the unit sphere, $\langle x,y,z\rangle$ is a unit vector, and the component of $\mathbf{F}_1$ in the normal direction is

$$\mathbf{F}_1(x,y,z) \cdot \langle x,y,z\rangle = \langle 2x, 2y, 1\rangle \cdot \langle x,y,z\rangle = 2x^2 + 2y^2 + z.$$

Since the sphere is not the graph of a function, evaluating $\mathbf{F}$ on it is ambiguous, and the above calculation suffices for our present purposes. ▲

Example 2.6.11. *Electric field from point charge*

A point charge Q at the origin generates the electric field

$$\mathbf{E} = \frac{kQ}{\rho^2}\hat{\rho}. \tag{2.6.3}$$

Find the component of $\mathbf{E}$ in the upward-pointing normal direction to the plane $z = 1$.

The problem can be completed using brute force, converting to Cartesian coordinates. One can also take advantage of symmetry and trigonometry to find this component. We'll complete both approaches, tackling the brute force method first.

Every point on the plane $z = 1$ is of the form $(x,y,1)$, and we compute $\mathbf{E}(x,y,1)$. Since $\hat{\rho}$ is the unit vector pointing from the origin to $(x,y,1)$, we compute $\hat{\rho} = \frac{1}{\sqrt{1+x^2+y^2}}\langle x,y,1\rangle$, and $\rho = \sqrt{1 + x^2 + y^2}$. With these calculations we find

$$\mathbf{E}(x,y,1) = \frac{kQ}{(1 + x^2 + y^2)^{3/2}}\langle x,y,1\rangle.$$

Since $\mathbf{k}$ is the upward-pointing normal to $z = 1$, we compute

$$comp_{\mathbf{k}}\mathbf{E} = \frac{kQ}{(1 + x^2 + y^2)^{3/2}}\langle x,y,1\rangle \cdot \langle 0,0,1\rangle = \frac{kQ}{(1 + x^2 + y^2)^{3/2}}.$$

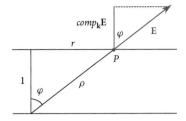

Figure 2.6.9 Vertical component of electric field

Now let's use symmetry and other coordinates. Let P be a point on $z = 1$. The "magic triangle" Figure 1.1.6(a), has vertical edge length 1 and hypotenuse ρ (see Figure 2.6.9). Placing $\mathbf{E}(P)$ with its initial point at P, the projection in the vertical direction makes a right triangle similar to the magic triangle. By ratios of corresponding sides, we have

$$comp_{\mathbf{k}}\mathbf{E} = \frac{\|\mathbf{E}\|}{\rho} = \frac{kQ}{\rho^3}.$$

The last equality follows because $\hat{\rho}$ is a unit vector, so Equation 2.6.3 yields $\|\mathbf{E}\| = \frac{k|Q|}{\rho^2}$. Note that both computations yield the same result. ▲

Exercises

1. Sketch the vectors $\mathbf{F}(\pm1,\pm1)$, $\mathbf{F}(0,\pm1)$, and $\mathbf{F}(\pm1,0)$, for the vector field $\mathbf{F}(x,y) = \langle x - y, x + y \rangle$.

2. Describe the vectors of $\mathbf{F}(x,y) = \langle y + 1, x - 2 \rangle$ on the line $x = 2$ in the plane. Sketch a few to illustrate your description.

3. The vector fields pictured are (scalar multiples of) one of the following: $\mathbf{F}_1(x,y) = \langle y,x \rangle$, $\mathbf{F}_2(x,y) = \langle x,y \rangle$, $\mathbf{F}_3(x,y) = \langle -y,x \rangle$, $\mathbf{F}_4(x,y) = \langle -x,y \rangle$. Label each picture with its formula, and justify your choice.

(a)

(b)

4. Is there a point on the unit sphere where the vector field $\mathbf{F}(x,y,z) = \langle x+y, y-2z, x-z\rangle$ vanishes (i.e. is the zero vector)?

5. Let $\mathbf{C}(t) = (\cos t, \sin t), 0 \le t \le \pi/2$ be the quarter unit circle in the first quadrant, and consider the vector field $\mathbf{F}(x,y) = \langle x+y, x-y\rangle$.
   (a) Sketch $\mathbf{C}(t)$, and include $\mathbf{C}'(\pi/4)$ in your sketch.
   (b) Include $\mathbf{F}(\sqrt{2}/2, \sqrt{2}/2)$, together with its projection onto $\mathbf{C}'(\pi/4)$, in the sketch.
   (c) Compute $proj_{\mathbf{C}'(\pi/4)}\mathbf{F}(\sqrt{2}/2, \sqrt{2}/2)$.

6. Let $\mathbf{F}(x,y) = \langle x+y, x\rangle$, and consider the restriction of $\mathbf{F}$ to the unit circle $\mathbf{C}(t) = (\cos t, \sin t)$. Find the component of $\mathbf{F}$ in the direction of the tangent vector to the unit circle. How would this change, using the circle radius 2, given by $\mathbf{C}(t) = (2\cos t, 2\sin t)$?

7. This problem concerns the uniform flow with velocity $\mathbf{V} = \langle -1, 1\rangle$.
   (a) Sketch some representative vectors of the flow in the plane, and describe the streamlines of the flow.
   (b) Sketch the directed line segment from $P = (2,2)$ to $Q = (3,2)$ and compute the flux across it.
   (c) Which unit position vector $\hat{\mathbf{v}}$ will have maximum flux across it? minimum flux?
   (d) If the flux across $\vec{PQ}$ is zero, what can you say about $\mathbf{V}$ and $\vec{PQ}$?

8. Approximate the flux of $\mathbf{V}(x,y) = \langle -2x, 2y\rangle$ across the curve $\mathbf{C}(t) = (t+1, t^2 + 1), 1 \le t \le 1.1$ by assuming that the vector field is constant $\mathbf{V}(\mathbf{C}(1))$ and the curve is approximated by $0.1\mathbf{C}'(1)$.

9. The velocity field, in Cartesian coordinates, for the flow near a point source is

$$\mathbf{V} = K\left(\frac{x}{x^2+y^2}, \frac{y}{x^2+y^2}\right) = \frac{K}{x^2+y^2}\langle x,y\rangle.$$

   (a) Sketch enough vectors to get a feel for the flow. (Notice that they are scalar multiples of position vectors. What happens as points get further from the origin? closer to it?)
   (b) Let $\mathbf{C}(t) = (3\cos t, 3\sin t)$. Find the tangential and normal components of $\mathbf{V}$ for this curve.
   (c) Let $\mathbf{C}(t) = (3t, 4t), t > 0$. Find the tangential and normal components of $\mathbf{V}$.
   (d) Are either of these curves streamlines? Justify your answer.

10. Evaluate the vector field $\mathbf{F}(x,y,z) = \langle x^2 + y^2, x+y, x-y\rangle$ on the helix $\mathbf{C}(t) = (\cos t, \sin t, t)$, then find the component of $\mathbf{F}$ in the $\mathbf{C}'(t)$ direction.

11. Find the component of $\mathbf{F}(x,y,z) = \langle z, 2y, x\rangle$ in the direction of the upward pointing normal to the plane $3x + 2y - 4z = 12$ (be sure to eliminate z from the component).

12. Find the component of $\mathbf{F}(x,y,z) = \langle y, -x, z\rangle$ in the direction of the outward pointing normal to the unit sphere. Write your answer in terms of x and y (i.e. use the equation of the sphere to eliminate z).

13. A point charge Q is placed at the origin, generating the electric field

$$\mathbf{E} = \frac{kQ}{\rho^2}\hat{\boldsymbol{\rho}}.$$

(a) Find the component of $\mathbf{E}$ normal to the unit sphere.

(b) Find the component of $\mathbf{E}$ normal to the plane $x+y+z=0$.

14. A point charge Q is placed at the origin, generating the electric field

$$\mathbf{E} = \frac{kQ}{\rho^2}\hat{\boldsymbol{\rho}}.$$

(a) Find the component of $\mathbf{E}$ tangent to the curve $\mathbf{C}(t) = (t, 3t, -2t)$, $-\infty < t < \infty$.

(b) Find where $\mathbf{E}$ is orthogonal to $\mathbf{C}(t) = (\cos t, \sin t, t)$, $-\infty < t < \infty$.

3 Differentiation

This chapter introduces differentiation of multivariable functions. As in the single-variable setting derivatives tell a lot about functions, from geometric properties to rates of change to optimization. These topics are all addressed in turn.

We have already encountered functions of several variables as well as vector-valued functions. Nonetheless, Section 3.1 takes a closer look at functions and some of their important properties. Limits and continuity are discussed in Section 3.2, paving the way for an introduction to partial differentiation in Section 3.3. Partial derivatives are used to find tangent planes to surfaces in Section 3.4, as well as to approximate function values. One of the most important features of differentiation, the multivariable chain rule, is discussed in Section 3.5, and some of its applications presented in Section 3.6. The subject of optimization is treated in Sections 3.7 and 3.8, rounding out our discussion of differentiation.

3.1 Functions

In single-variable calculus you studied functions whose domain and range were subsets of $\mathbb{R}^1$, like $f(x) = xe^x$. When thinking of these as maps from the real numbers to itself, we use the notation $f : \mathbb{R}^1 \to \mathbb{R}^1$. When introducing surfaces we considered graphs of functions of two variables, which are maps $f : \mathbb{R}^2 \to \mathbb{R}^1$. A map $f : \mathbb{R}^n \to \mathbb{R}$ will be called a function of several variables because it has n independent variables (the "several variables" part), and one dependent variable (the "function" part). For example $f(x, y, z) = xz^2 \sin y$ is a function of several variables since it requires three inputs to produce a single number output.

We also considered parametric curves $\boldsymbol{C} : \mathbb{R} \to \mathbb{R}^n$, which are examples of *vector-valued functions*, and parametric surfaces $\boldsymbol{S} : \mathbb{R}^2 \to \mathbb{R}^3$, which are examples of vector-valued functions *of several variables*. The "several variables" part refers to the domain having dimension greater than one, and the "vector-valued" part to the range being multidimensional. To study parametric surfaces $S(s,t) = (x(s,t), y(s,t), z(s,t))$ we focused on their component functions $x(s,t)$, $y(s,t)$ and $z(s,t)$. Notice that the component functions of a vector-valued function are functions of several variables since their output is a single number. For example, the vector-valued function of two variables $S(r,\theta) = (r\cos\theta, r\sin\theta, r^2)$ is made up of three functions of two variables, $x(r,\theta) = r\cos\theta, y(r,\theta) = r\sin\theta$, and $z(r,\theta) = r^2$. Note that the vector fields introduced in Section 2.6 are also examples of vector-valued functions of several variables.

We now introduce some terminology with a familiar example. Thus the most general functions we consider are mappings from $\mathbb{R}^n$ to $\mathbb{R}^m$ (in symbols $f : \mathbb{R}^n \to \mathbb{R}^m$), where $\mathbb{R}^n$ is the *domain* and $\mathbb{R}^m$ the *codomain* of f. The domain of the parametric plane $S(s,t) = (2 - s + 2t, 3 + s - t, 2s + 3t)$ is the *st*-plane, or $\mathbb{R}^2$, while the codomain is $\mathbb{R}^3$ since the surface lives in $\mathbb{R}^3$. The points on the surface form the *image* of f in $\mathbb{R}^3$. Using techniques of Chapter 2 one can show that the image of S is the plane with Cartesian equation $5x + y - 3z = 13$. For this parametric surface, the domain is all of $\mathbb{R}^2$ while the image is a subset of the codomain $\mathbb{R}^3$ (the plane is not all of $\mathbb{R}^3$). There are many important examples of functions whose domain is not all of $\mathbb{R}^n$, but we will not typically specify restrictions on domains in this text. We now give two examples of maps $f : \mathbb{R}^2 \to \mathbb{R}^2$, and illustrate how to analyze them geometrically.

Example 3.1.1. *A linear map of the plane to itself*

The goal of this example is to understand the function $f : \mathbb{R}^2 \to \mathbb{R}^2$ given by $f(x,y) = (x - 3y, x + 3y)$.

We can think of f as being defined by component functions

$$f(x,y) = (f_1(x,y), f_2(x,y)),$$

where $f_1(x,y) = x - 3y$ and $f_2(x,y) = x + 3y$. Studying general functions through their component functions is a useful technique. In particular the most general form of the multivariable chain rule for differentiation, derived in Section 3.5, thinks of a general function as being comprised of component functions.

We can also think of this particular function using matrix multiplication, as in Example 2.4.8. The matrix corresponding to f is the matrix of coefficients of the component functions, each row being the coefficients of the corresponding component function. As in Section 2.4, the image of the square $\square ABCD$ with vertices $(0,0), (1,0), (1,1), (0,1)$ gives an idea of how f transforms the plane. Algebraically, we have

$$f(x,y) = \begin{bmatrix} 1 & -3 \\ 1 & 3 \end{bmatrix} \begin{bmatrix} x \\ y \end{bmatrix},$$

and the image of $\square ABCD$ is pictured in Figure 3.1.1.

Both of its component functions are linear in x and y, so f maps lines to lines. One way to see this is to consider the image under f of some parametric curves. In particular, we'll study what f does to horizontal and vertical lines, as well as to the unit circle.

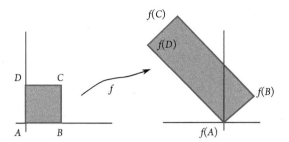

Figure 3.1.1 The image of □ABCD

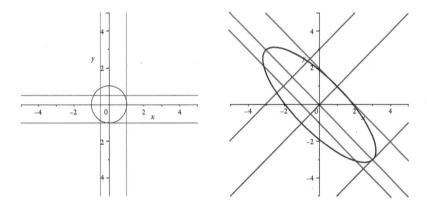

Figure 3.1.2 A vector-valued function of two variables

Horizontal lines in $\mathbb{R}^2$ are given parametrically by $\mathbf{C}(t) = (t, c)$, where c is a constant. Applying f we see that $f(\mathbf{C}(t)) = f(t, c) = (t - 3c, t + 3c)$. Recall that, given a parametric line, the constants give a point on the line and the coefficients of t give a vector $\mathbf{v}$ in the line. Thus $f(\mathbf{C}(t))$ is a line through the point $(-3c, 3c)$, and in direction $\mathbf{v} = \langle 1, 1 \rangle$. In other words, horizontal lines get mapped to lines with slope one by f.

Similarly, vertical lines are $\mathbf{C}(t) = (c, t)$, giving $f(\mathbf{C}(t)) = f(c, t) = (c - 3t, c + 3t)$. These are lines through the point (c, c) in the direction of $\mathbf{v} = \langle -3, 3 \rangle$, and f maps vertical lines to lines with slope -1. See Figure 3.1.2 for the images of several lines.

Finally, we can see that the image of the unit circle $\mathbf{C}(t) = (\cos t, \sin t)$ is the ellipse given parametrically by $f(\mathbf{C}(t)) = (\cos t - 3 \sin t, \cos t + 3 \sin t)$ pictured on the right. ▲

Note that the previous examples involved composing functions. Indeed, one way to find the image of a curve C in the domain of f is to parameterize it by $\mathbf{C}(t)$, then apply $f(x, y)$ to the parameterization. Notationally, this looks like $f(\mathbf{C}(t))$, and evaluating general functions on curves plays an important role in the chain rule in this chapter, as well as in integration of functions and vector fields along curves in Chapters 4 and 5.

Math App 3.1.1. *Visualizing linear transformations*

Linear transformations can be considered motions of the plane. This Math App allows you to investigate just how this is done.

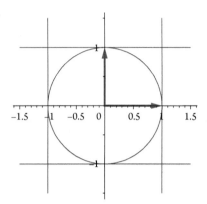

Example 3.1.2. *The polar change of coordinates*

A second map $f : \mathbb{R}^2 \to \mathbb{R}^2$ is the traditional change of coordinates from the $r\theta$-plane to the xy-plane. In this case $f(r,\theta) = (r\cos\theta, r\sin\theta)$. The mapping takes the rectangle $0 \le r \le 1, 0 \le \theta \le 2\pi$ to the unit disk, mapping horizontal segments like $\theta = \pi/3$ to rays and vertical segments like $r = 0.4$ to circles.

If you extend the domain of f to the entire $r\theta$-plane, observe that infinitely many points have the same image under f. An extreme case happens when $r = 0$. Note that $f(0,\theta) = (0,0)$, so the θ-axis in the $r\theta$-plane is mapped to the origin. Further, $f(r,\theta + 2n\pi) = f(r,\theta)$ for any integer n, and $f(-r,\theta) = f(r,\theta + \pi)$. ▲

To convince you that you already know a lot of vector-valued functions, we recall examples of vector-valued functions that we've already encountered.

Example 3.1.3. *Familiar vector-valued functions*

The following are vector-valued functions you know a lot about.

(a) The helix $\mathbf{C}(t) = (\cos t, \sin t, t)$, $-\infty < t < \infty$, is a map $\mathbf{C} : \mathbb{R} \to \mathbb{R}^3$.

(b) The paraboloid $\mathbf{S}(s,t) = (s\cos t, s\sin t, s^2)$, $s \ge 0, 0 \le t \le 2\pi$, is a function whose domain is a subset of $\mathbb{R}^2$, whose codomain is $\mathbb{R}^3$, and whose image is the graph of $z = x^2 + y^2$. In functional notation we have $\mathbf{S} : \mathbb{R}^2 \to \mathbb{R}^3$.

(c) The electric field generated by a point charge Q at the origin in space is

$$\mathbf{E} = \frac{kQ}{\rho^2}\widehat{\rho}.$$

This takes points in space as input, and spits out vectors in space, so can be thought of as $\mathbf{E} : \mathbb{R}^3 \to \mathbb{R}^3$. Technically $\mathbf{E}$ is not defined at the origin because of the ρ^2

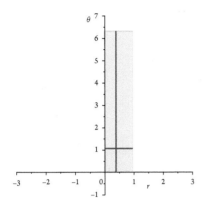
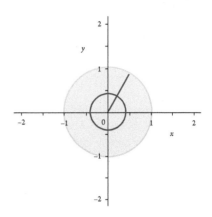

Figure 3.1.3 Change of coordinates

in the denominator, so its domain is $\mathbb{R}^3$ minus the origin. This is an important example where the domain must be restricted. ▲

Example 3.1.4. *Two functions from $\mathbb{R}^3$ to $\mathbb{R}^2$*

In this example we consider two maps that can be defined using matrix multiplication of Section 2.4.

(a) The function $f : \mathbb{R}^3 \to \mathbb{R}^2$ defined by $f(x,y,z) = (x,y)$ can be thought of as projection onto the xy-plane. The function f can be defined using matrix multiplication as follows:

$$f(x,y,z) = \begin{bmatrix} 1 & 0 & 0 \\ 0 & 1 & 0 \end{bmatrix} \begin{bmatrix} x \\ y \\ z \end{bmatrix} = \begin{bmatrix} x \\ y \end{bmatrix} = (x,y).$$

Here we are jumping back and forth between thinking of (x,y) as a point, and thinking of it as a position vector $\begin{bmatrix} x \\ y \end{bmatrix}$.

Notice that the domain of f is all of $\mathbb{R}^3$, and the image of f is all of the codomain $\mathbb{R}^2$. Further note that the z-axis gets mapped to the origin by f, and more generally f maps any vertical line to its point of intersection with the xy-plane. In this sense, f can be thought of as projecting points in $\mathbb{R}^3$ to the xy-plane.

(b) The function $f : \mathbb{R}^3 \to \mathbb{R}^2$ defined by $f(x,y,z) = (x+y+z, 2x+2y+2z)$ can also be defined using matrix multiplication by

$$f(x,y,z) = \begin{bmatrix} 1 & 1 & 1 \\ 2 & 2 & 2 \end{bmatrix} \begin{bmatrix} x \\ y \\ z \end{bmatrix} = \begin{bmatrix} x+y+z \\ 2x+2y+2z \end{bmatrix} = (x+y+z, 2x+2y+2z).$$

Since the second coordinate of $f(x,y,z)$ is always twice the first, the image of f is the line $y = 2x$ in the xy-plane. Note that the plane $x+y+z=0$ in $\mathbb{R}^3$

gets mapped by f to the origin, and more generally the plane $x + y + z = c$ gets mapped to the point $(c, 2c)$. ▲

Before moving on we introduce the notions of one-to-one and onto for functions. These ideas are important in mathematics, but not emphasized in the remainder of the text.

One property of functions in general is that of being *one-to-one*. A function is one-to-one if different inputs yield different outputs. An alternative way to say it is: if two outputs are the same, they came from the same input. It's probably a good idea to remember what one-to-one means in English (distinct inputs yield distinct outputs), but occasionally it's helpful to translate the English into symbols. A function of two variables is one-to-one if

$$(a, b) \neq (c, d) \implies f(a, b) \neq f(c, d).$$

Equivalently, $f(x, y)$ is one-to-one if

$$f(a, b) = f(c, d) \implies (a, b) = (c, d). \tag{3.1.1}$$

These symbolic forms are useful when proving that a given function is one-to-one. Let's look at some examples.

Example 3.1.5. *Checking one-to-one*

We check whether several functions are one-to-one or not.

(a) Let $f : \mathbb{R} \to \mathbb{R}$ be given by $f(x) = e^x$. Determine if f is one-to-one.
    We'll use single-variable calculus to solve this. Since $f'(x) > 0$ for all x, the function $f(x)$ is always increasing. Thus if $a < b$, then $f(a) < f(b)$. This implies that different inputs yield different outputs, and f is one-to-one (pause...now confirm this!).

(b) The change of coordinates of Example 3.1.2 given by $f(r, \theta) = (r \cos \theta, r \sin \theta)$ is *not* one-to-one. In fact, the discussion of that example shows that it's infinite-to-one (meaning infinitely many points in the domain go to the same point in the codomain).

(c) Determine if $f(x, y) = (x - 3y, x + 3y)$ is one-to-one.
    For this example we demonstrate how to use the implication of Equation 3.1.1 to prove that a function is one-to-one. The strategy is as follows. Start by assuming there are points $(a, b), (c, d)$ in the domain with the same images under f (i.e. such that $f(a, b) = f(c, d)$). Using the formula for f, show that this implies $(a, b) = (c, d)$. This proves the implication of Equation 3.1.1, so f is one-to-one. We now apply this strategy.
    Supposing $f(a, b) = f(c, d)$, using the formula for f gives

$$f(a,b) = f(c,d)$$
$$(a - 3b, a + 3b) = (c - 3d, c + 3d),$$

and this single vector equation leads to the system of scalar equations

$$a - 3b = c - 3d$$
$$a + 3b = c + 3d.$$

Adding the equations gives $2a = 2c$ while subtracting the equations gives $6b = 6d$. Together these imply that $(a, b) = (c, d)$, showing that f is one-to-one. ▲

A second important property of functions is that of being *onto*. A function is onto if the codomain is its image. On a more detailed level: a function is onto if every point in the codomain is the image of some point in the domain. Let's consider some examples.

Example 3.1.6. *Checking onto*

We check whether several functions are onto or not.

(a) Let $f : \mathbb{R} \to \mathbb{R}$ be given by $f(x) = e^x$. Determine if f is onto.
This function is onto if for every real number y, there is a number x such that $e^x = y$. Since e^x is always positive, the function f is not onto. For example, there is no number x with $e^x = -17$.

(b) The change of coordinates of Example 3.1.2 given by $f(r, \theta) = (r\cos\theta, r\sin\theta)$ is onto. In fact, for every point (x, y) in the codomain, there are infinitely many points (r, θ) with $f(r, \theta) = (x, y)$.

(c) Determine if $f(x, y) = (x - 3y, x + 3y)$ is onto.
For this example we demonstrate a more analytic approach. The function f is onto if for every point (a, b) in the codomain there is a point (x, y) in the domain with $f(x, y) = (a, b)$. Thus, to demonstrate onto, one need only solve $f(x, y) = (a, b)$ for x and y in terms of a, b.

Let (a, b) be an arbitrary point in the codomain. The one vector equation $(x - 3y, x + 3y) = (a, b)$ is equivalent to the system of scalar equations

$$x - 3y = a$$
$$x + 3y = b.$$

Adding the equations gives $x = (a + b)/2$ while subtracting the equations gives $y = (b - a)/6$. Thus any point (a, b) is the image of $((a + b)/2, (b - a)/6)$, and f is onto.

It should be noted that in this case we actually found the inverse function of our original map f. This technique is analogous to the familiar "switch-and-solve" technique for finding inverse functions. ▲

Exercises

1. Consider the parametric plane $S(s,t) = (s+t, 2-3t, 1+3s-t)$, $-\infty < s,t < \infty$. What are the domain and codomain of S? Find a Cartesian equation for the image of S.

2. Consider the parametric surface $S(s,t) = (\cos t, \sin t, s)$, $-\infty < s,t < \infty$. What are the domain and codomain of S? Find a cylindrical equation for the image of S.

3. Consider the parametric curve $C(t) = (\cos t, \sin t)$, $-\infty < t < \infty$ (note the unusual limits on t!). What are the domain, codomain and image of C?

4. Consider the parametric curve $C(t) = (2-t, 3+t)$, $-\infty < t < \infty$. What are the domain, codomain of C? Find a Cartesian equation for the image of C.

5. The line $x - 2y = 1$ is parameterized by $C(t) = (1+2t, t)$. Find the image of $x - 2y = 1$ under the function $f(x,y) = (x-y, x+y)$ by evaluating $f(C(t))$. Now find a Cartesian equation for it.

6. Let $f(x,y) = (2x, 3y)$ and find the image of the unit circle under f (hint: parameterize first!).

7. Sketch the images of several horizontal and vertical lines to analyze the map $f(x,y) = (x, 2y)$. What is the image of the unit circle?

8. Sketch the images of several horizontal and vertical lines to analyze the map $f(x,y) = (-y, x)$. What is the image of the unit circle?

9. Let $f : \mathbb{R}^3 \to \mathbb{R}^2$ be defined by

$$f(x,y,z) = \begin{bmatrix} 0 & 1 & 0 \\ 0 & 0 & 1 \end{bmatrix} \begin{bmatrix} x \\ y \\ z \end{bmatrix}.$$

Describe all points in $\mathbb{R}^3$ that f maps to $(0,0)$. Describe the map f.

10. Find matrix defining

$$f(x,y,z) = \begin{bmatrix} a & b & c \\ d & e & f \end{bmatrix} \begin{bmatrix} x \\ y \\ z \end{bmatrix},$$

so that the image of f is the line $y = -x$ in $\mathbb{R}^2$. What points does your f map to the origin?

11. Let $f : \mathbb{R}^2 \to \mathbb{R}^2$ be the function $f(r, \theta) = (r\cos\theta, r\sin\theta)$. Find all points in the $r\theta$-plane (the domain) that map to the unit circle in the codomain.

12. Let $f : \mathbb{R}^3 \to \mathbb{R}^3$ be the coordinate transformation

$$f(r, \theta, z) = (r\cos\theta, r\sin\theta, z).$$

Axes in the domain are labeled r, θ, z, while axes in the codomain are labeled x, y, z. In the domain, $r = 2$ is a plane parallel to the θz-plane. Describe its image under f.

13. Let $f : \mathbb{R}^3 \to \mathbb{R}^3$ be the coordinate transformation

$$f(\rho, \theta, \phi) = (\rho \sin\phi \cos\theta, \rho \sin\phi \sin\theta, \rho \cos\phi).$$

Find all points in the domain that map to the unit sphere in the codomain.

14. Determine if $f(x, y) = (x + 2y, -2x - 4y)$ is one-to-one. Determine if it is onto.

15. Is the function $f(x, y) = (x, e^y)$ one-to-one? Onto?

16. Is the function $f(x, y) = (|x|, y)$ one-to-one? Onto?

17. Let $f : \mathbb{R}^3 \to \mathbb{R}^3$ be the change of coordinates

$$f(\rho, \theta, \phi) = (\rho \sin\phi \cos\theta, \rho \sin\phi \sin\theta, \rho \cos\phi)$$

(so the domain in this case is all of $\mathbb{R}^3$). Is f one-to-one? Onto?

18. What is your favorite function that is both one-to-one and onto? Neither one-to-one nor onto? Justify these properties.

19. What is your favorite function that is one-to-one and not onto? Onto and not one-to-one? Justify these properties.

3.2 Limits

The last section was a brief look at (vector-valued) functions of several variables. It is time to discuss the notion of limits in the several variable setting. There are many similarities between single and several variable limits, as well as some subtleties not found in the single-variable case. In this section we first illustrate some multivariable limit techniques that are reminiscent of single-variable limits. After becoming more comfortable with them, we provide a technical definition of multivariable limits and investigate them more carefully.

The intuitive definition of the limit of $f(x)$ as $x \to c$ is that it is the *expected* function value. The key term there is *expected*. That means that $\lim_{x \to c} f(x)$ is the value of $f(c)$ you'd expect to get by looking at the function for x near c. Let's say that again, slightly differently, because it is worth having some intuition about limits. To calculate $\lim_{x \to c} f(x)$ you look see what $f(x)$-values are trending toward as x gets closer to c. Thus $\lim_{x \to c} f(x)$ doesn't care what you get when you plug c into $f(x)$; it just cares what happens for x near c. These intuitive descriptions are worth pondering over a cup of coffee, together with some examples of limits from single-variable calculus. We begin with a sample limit that should jog your memory.

Example 3.2.1. $\lim_{x \to 3} \dfrac{\sqrt{x+1} - 2}{x - 3}$

When calculating limits, you first evaluate $f(c)$ to see if it works. Letting $x = 3$ in $f(x)$ you get $\frac{0}{0}$, so you have to do tricks, and we rationalize the numerator. Here goes:

$$\lim_{x \to 3} \frac{\sqrt{x+1} - 2}{x - 3} = \lim_{x \to 3} \frac{\sqrt{x+1} - 2}{x - 3} \cdot \frac{\sqrt{x+1} + 2}{\sqrt{x+1} + 2}$$
$$= \lim_{x \to 3} \frac{x - 3}{(x - 3)(\sqrt{x+1} + 2)}$$
$$= \lim_{x \to 3} \frac{1}{\sqrt{x+1} + 2} = \frac{1}{4}.$$

This example underscores that limits don't care what happens at $x = 3$, since the function is undefined there. Limits only care what happens near $x = 3$. The above calculations show that for $x \neq 3$ we know $f(x) = \dfrac{1}{\sqrt{x+1} + 2}$, so we can calculate the limit. ▲

This is an example of one trick from single-variable calculus for calculating limits. You should be familiar with many others, like factoring and cancelling, L'Hôpital's rule, and logarithmic limits. Here's another single-variable example.

Example 3.2.2. $\lim_{x \to \infty} \left(1 - \dfrac{2}{x} \right)^x$

In this example the exponent contains a variable. Recall that the strategy in this case is to take the limit of the natural logarithm of the function, then exponentiate it. Formally we have

$$\lim_{x \to c} f(x) = \lim_{x \to c} e^{\ln f(x)} = e^{\lim_{x \to c} \ln f(x)},$$

where the last equality is justified by a theorem on limits of compositions of functions. Thus we begin by taking the limit of the natural logarithm of our function.

$$\lim_{x \to \infty} \ln \left(1 - \frac{2}{x} \right)^x = \lim_{x \to \infty} x \ln \left(1 - \frac{2}{x} \right)$$

$$= \lim_{x \to \infty} \frac{\ln \left(1 - \frac{2}{x} \right)}{1/x} \qquad \text{(property of logarithms)}$$

$$= \lim_{x \to \infty} \frac{\frac{1}{1 - \frac{2}{x}} \cdot \frac{2}{x^2}}{\frac{-1}{x^2}} \qquad \text{(L'Hôspital's Rule)}$$

$$= \lim_{x \to \infty} \frac{-2}{1 - \frac{2}{x}} = -2.$$

Exponentiating we have

$$\lim_{x \to \infty} \left(1 - \frac{2}{x} \right)^x = e^{\lim_{x \to \infty} \ln \left(1 - \frac{2}{x} \right)^x} = e^{-2}. \ \blacktriangle$$

Limits of functions of several variables have the same intuitive idea as in single-variable calculus. The limit of $f(x,y)$ as $(x,y) \to (a,b)$ is the expected function value. Some of the same tricks also work, and we illustrate with some examples. In particular, you can evaluate limits of polynomials, rational functions, trigonometric functions, and their compositions just by evaluating the function at the desired point (as long as it is defined there). This follows from some of the same theorems as single variable calculus, ramped up to the several-variable case. The theorems are long lists of properties of limits, all akin to "the limit of a sum is the sum of the limits" (provided the individual limits exist). In limit notation this looks like

$$\lim_{(x,y) \to (a,b)} f(x,y) + g(x,y) = \lim_{(x,y) \to (a,b)} f(x,y) + \lim_{(x,y) \to (a,b)} g(x,y).$$

Rather than listing all properties, we assume familiarity with the single-variable results and use them at will. We illustrate with some examples.

Example 3.2.3. $\displaystyle\lim_{(x,y) \to (2,-1)} x^2 + y^2$

Simply by evaluating we get

$$\lim_{(x,y)\to(2,-1)} x^2 + y^2 = 2^2 + (-1)^2 = 5. \blacktriangle$$

Example 3.2.4. $\displaystyle\lim_{(x,y)\to(\sqrt{\pi},0)} \sin\left(\frac{x^2}{2} - 2xy\right)$

Since this is a composition of a trigonometric function and polynomial, we simply substitute to evaluate the limit.

$$\lim_{(x,y)\to(\sqrt{\pi},0)} \sin\left(\frac{x^2}{2} - 2xy\right) = \sin(\pi/2) = 1. \blacktriangle$$

Thus some several-variable versions of nice functions still behave nicely. Substitution is another trick you can use to relate multivariable limits to familiar single variable ones. We demonstrate with some examples.

Example 3.2.5. $\displaystyle\lim_{(x,y)\to(0,0)} \frac{\sin\left(x^2+y^2\right)}{x^2+y^2}$

When substituting $(0,0)$ into the function, you get $\frac{0}{0}$, which is undefined. However, substituting θ for $x^2 + y^2$ in the function gives a more familiar $\frac{\sin\theta}{\theta}$. Moreover, notice that $(x,y) \to (0,0)$ if and only if $\theta = x^2 + y^2 \to 0$. Therefore we have

$$\lim_{(x,y)\to(0,0)} \frac{\sin\left(x^2+y^2\right)}{x^2+y^2} = \lim_{\theta\to0} \frac{\sin\theta}{\theta} = 1. \blacktriangle$$

In the last example it is imperative that $(x,y) \to (0,0)$ be equivalent to $\theta \to 0$. This means that no matter what manner (x,y) tends to $(0,0)$ the corresponding θ-values will approach 0, AND if $\theta \to 0$, then x and y both must approach 0. This will be made precise after the $\epsilon - \delta$ definition of a limit is given. In the meantime, we consider another substitution example.

Example 3.2.6. $\displaystyle\lim_{(x,y)\to(1,1)} \frac{\sqrt{x^2+y^2-2x-2y+3}-1}{x^2+y^2-2x-2y+2}$

First notice that completing the square on both x and y gives the limit

$$\lim_{(x,y)\to(1,1)} \frac{\sqrt{(x-1)^2+(y-1)^2+1}-1}{(x-1)^2+(y-1)^2}.$$

When substituting $(1,1)$ into the function, you get $\frac{0}{0}$, which is undefined. However, letting $z = (x-1)^2 + (y-1)^2$ in the function gives $\frac{\sqrt{z+1}-1}{z}$. We see that $(x,y) \to (1,1)$ if and only if $z \to 0$. Therefore we have

$$\lim_{(x,y)\to(1,1)} \frac{\sqrt{x^2+y^2-2x-2y+3}-1}{x^2+y^2-2x-2y+2} = \lim_{z\to0} \frac{\sqrt{z+1}-1}{z}$$

$$= \lim_{z\to0} \frac{\sqrt{z+1}-1}{z} \cdot \frac{\sqrt{z+1}+1}{\sqrt{z+1}+1}$$

$$= \lim_{z\to0} \frac{z}{z(\sqrt{z+1}+1)} = \frac{1}{2}. \blacktriangle$$

Given the above examples, one may be tempted to just use algebra when calculating several variable limits. There are technicalities that we've glossed over in the previous two substitution examples. In order to discuss this further we need the technical definition of a limit, which will take some doing, so grab some coffee and settle in.

First we recall the single-variable $\epsilon - \delta$ definition of a limit.

Definition 3.2.1. *The limit of $f(x)$ as x approaches c is L if for every $\epsilon > 0$ there is a $\delta > 0$ such that the inequality $0 < |x - c| < \delta$ implies that $|f(x) - L| < \epsilon$ is true as well.*

This was one of those daunting definitions you hurried to forget, so let's spend a minute going back over what it says. This definition tells you when a number L is the limit of a function $f(x)$. Intuitively (again) it is the technical way of saying that as x approaches c the $f(x)$-values approach L. To interpret the definition, recall that the inequality $|f(x) - L| < \epsilon$ means that the function value $f(x)$ is within ϵ of the number L. Equivalently, the value $f(x)$ is greater than $L - \epsilon$ and less than $L + \epsilon$, or $f(x)$ is in the interval $(L - \epsilon, L + \epsilon)$. I like to say:

The inequality $|f(x) - L| < \epsilon$ says $f(x)$ is in an open ball centered at L with radius ϵ.

Since Definition 3.2.1 says "for every $\epsilon > 0$", we are allowed to specify the radius of the open ball to be as small as we like. So we can specify how close we want $f(x)$ to be to L.

Similarly, the inequality $|x - c| < \delta$ says that x is in an open ball centered at c with radius δ.

Thus Definition 3.2.1 says that no matter how close we want $f(x)$ to be to L ("for every $\epsilon > 0$"), we can find a radius $\delta > 0$, so that whenever x is within δ of c we know the function value $f(x)$ is close enough to L (see Figure 3.2.1).

To generalize Definition 3.2.1 to functions of several variables, the ϵ portion of the definition translates directly but the δ part needs some work. To say that x is within δ of c

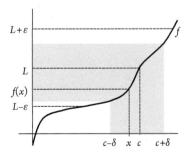

Figure 3.2.1 A single-variable limit

in the one-variable case means x is in the interval $(c - \delta, c + \delta)$ of real numbers (which we think of as an open ball in one dimension). For two variables, to say that the point (x, y) is within δ of the point (a, b), means that (x, y) is inside the open ball centered at (a, b) with radius δ, which is a disk that we denote by $B_\delta(a, b)$. The easiest way to formalize this in higher dimensions is to use vector notation. The point $\mathbf{x} \in \mathbb{R}^n$ is within δ of the point $\mathbf{c}$ if the length of the vector $\mathbf{x} - \mathbf{c}$ is less than δ, or formally $\|\mathbf{x} - \mathbf{c}\| < \delta$. Open balls will be important in the following.

In many contexts, one needs to calculate a limit where a function is not defined. Derivatives are a classic example of this, as are $\lim_{\theta \to 0} \frac{\sin \theta}{\theta}$ and limits involving L'Hôpital's rule. For this reason it is important to know where a function is defined and where it isn't, which makes domains of functions of several variables important. Frequently domains are so-called "open" subsets of $\mathbb{R}^n$, which we now introduce.

An *open* subset of $\mathbb{R}^n$ is one that contains an open ball around each of its points. For example, all points (x, y) in $\mathbb{R}^2$ satisfying $x^2 + y^2 < 1$ is an open set (called the open unit disk D), as is the plane minus the line $y = x$ (see Figure 3.2.2 (a) and (c)). The reason is if you start with any point in the set, you can find an open ball centered at that point and completely contained in the set. Basically a set is open if there's a little wiggle room within the set around each of its points.

The closed unit disk $\overline{D}$, points satisfying $x^2 + y^2 \leq 1$, is not open. The reason is that there isn't wiggle room *within the set* around points on the unit circle. More formally, if $\mathbf{c}$ is a point on the unit circle, then there is no open ball centered at $\mathbf{c}$ that is contained entirely in $\overline{D}$ (see Figure 3.2.2(b)).

To put it another way, if $\mathbf{c}$ is on the unit circle then every open ball $B_\delta(\mathbf{c})$ contains both points within $\overline{D}$ and points outside it. The *boundary* of a set is all points with this property. Thus the boundary of $\overline{D}$ is the unit circle, which is also the boundary of the open unit disk D! The difference is that $\overline{D}$ contains its boundary while D does not. A set that contains its boundary points is called *closed*. Thus $\overline{D}$ is closed while D is open.

As another example of the boundary of a set, let $U = \mathbb{R}^2 \setminus \{y = x\}$ be the plane minus the line $y = x$. Every point $\mathbf{c}$ on $y = x$ is in the boundary of U since open balls centered at these points contain both points within U and outside it (see Figure 3.2.2(c)).

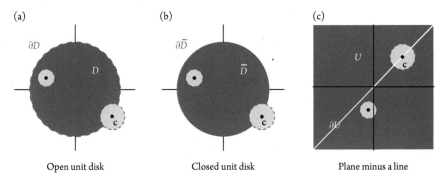

(a) Open unit disk
(b) Closed unit disk
(c) Plane minus a line

Figure 3.2.2 Open and closed sets

The notions of open sets and boundary points are illustrated in the following examples of function domains.

Example 3.2.7. *Domains of functions*

Find the domains of the following functions, and their boundaries.

(a) Let $f(x,y) = \dfrac{\sin(x^2 + y^2)}{x^2 + y^2}$.

The origin is the only point where f is undefined, so its domain is the punctured plane $U = \mathbb{R}^2 \setminus \{(0,0)\}$. This set is open. If $\mathbf{x}$ is not the origin, then the open ball of radius $\|\mathbf{x}\| / 2$ does not contain the origin (since the origin is $\|\mathbf{x}\|$ units away from $\mathbf{x}$). Thus U contains an open ball around each of its points, which is the definition of an open set.

The boundary of the punctured plane is the origin. Indeed, every open ball centered at the origin contains a point not in U, namely the origin itself. Such an open ball also contains many points in U (all but the origin!). This, of course, is the definition of a boundary point (every open ball contains points inside *and* outside the set).

(b) Let $f(x,y) = \dfrac{x^2 - y^2}{x - y}$.

The domain of f is all points with $y \neq x$, so the set $U = \mathbb{R}^2 \setminus \{y = x\}$. We've already seen that this is an open set, and that its boundary is the line $y = x$.

(c) Let $f(x,y) = \ln\left(1 - \sqrt{x^2 + y^2}\right)$.

The natural logarithm is defined only for positive numbers, therefore the domain of f is all points satisfying $1 - \sqrt{x^2 + y^2} > 0$. Since $\sqrt{x^2 + y^2}$ is the distance to the origin, this is all points less than one unit from the origin, or the open unit disk. ▲

With this set terminology in hand, we make the following definition.

Definition 3.2.2. *Let A be an open subset of $\mathbb{R}^n$, $f : A \to \mathbb{R}$ a function of n variables, and $\mathbf{c} \in \mathbb{R}^n$ a point in A or its boundary. The* limit *of $f(\mathbf{x})$ as $\mathbf{x} \to \mathbf{c}$ is the number L if for every $\epsilon > 0$ there is a $\delta > 0$ such that for all $\mathbf{x} \in A$ with $0 < \|\mathbf{x} - \mathbf{c}\| < \delta$ we have $\left|f(\mathbf{x}) - L\right| < \epsilon$.*

Thus Definition 3.2.2 says that no matter how close we want $f(\mathbf{x})$ to be to L ("for every $\epsilon > 0$"), we can find a radius $\delta > 0$, so that whenever $\mathbf{x}$ is in A and within δ of $\mathbf{c}$ we know the function value $f(\mathbf{x})$ is close enough to L.

Notice that the previous paragraph is identical—literally—to the one we used to describe single-variable limits, except the real numbers x and c from one variable are replaced by vectors $\mathbf{x}$ and $\mathbf{c}$. I guess we also replace the absolute value in $0 < |x - c| < \delta$ with the norm of a vector $0 < \|\mathbf{x} - \mathbf{c}\| < \delta$. Oh, and we add the restriction that $\mathbf{x}$ be in the domain A of f.

Example 3.2.8. *A simple start*

Show that $\lim\limits_{(x,y)\to(a,b)} 5x = 5a.$

Note that the domain of $f(x,y) = 5x$ is $\mathbb{R}^2$, so we can consider the limit. For every $\epsilon > 0$ we must find a $\delta > 0$ such that

$$\text{If } 0 < \sqrt{(x-a)^2 + (y-b)^2} < \delta, \text{ then } |5x - 5a| < \epsilon. \tag{3.2.1}$$

Note that $|5x - 5a| < \epsilon$ is equivalent to $|x - a| < \epsilon/5$. Thus we must find a $\delta > 0$ so that if (x,y) is in $B_\delta(a,b)$, we know $|x - a| < \epsilon/5$.

Since $(y-b)^2 \geq 0$, we have the following implication:

$$\text{If } 0 < \sqrt{(x-a)^2 + (y-b)^2} < \delta, \text{ then } \sqrt{(x-a)^2} = |x - a| < \delta. \tag{3.2.2}$$

Thus given $\epsilon > 0$, choosing $\delta = \epsilon/5$ gives the desired implication. Indeed, we have

$$\text{If } 0 < \sqrt{(x-a)^2 + (y-b)^2} < \epsilon/5, \text{ then } \sqrt{(x-a)^2} = |x - a| < \epsilon/5.$$

Woah, what just happened?! Take a moment to read the definition again. Now, using the notation of the definition, verify that the above reasoning shows that if $0 < \|\mathbf{x} - \mathbf{c}\| < \epsilon/5$ then $|f(\mathbf{x}) - L| < \epsilon$ in this case (i.e. what are $\mathbf{c}$, L, $f(\mathbf{x})$, etc.). ▲

The main properties of limits from single-variable calculus can be shown to hold in this more general setting as well. With all the necessary assumptions, it can be shown that

$$\lim_{\mathbf{x}\to\mathbf{c}} f(\mathbf{x}) + g(\mathbf{x}) = \lim_{\mathbf{x}\to\mathbf{c}} f(\mathbf{x}) + \lim_{\mathbf{x}\to\mathbf{c}} g(\mathbf{x})$$

$$\lim_{\mathbf{x}\to\mathbf{c}} f(\mathbf{x})g(\mathbf{x}) = \lim_{\mathbf{x}\to\mathbf{c}} f(\mathbf{x}) \cdot \lim_{\mathbf{x}\to\mathbf{c}} g(\mathbf{x}),$$

constants can be pulled out of limits, limits of quotients are the quotient of the limits, etc.

Example 3.2.9. *Using limit properties*

Using the fact that $\lim\limits_{\mathbf{x}\to(a,b)} x = a$ and $\lim\limits_{\mathbf{x}\to(a,b)} y = b$, use properties of limits to show

$$\lim_{\mathbf{x}\to(a,b)} 5x - 4y = 5a - 4b.$$

We proceed as follows, justifying each step taken.

$$\lim_{\mathbf{x}\to(a,b)} 5x - 4y = \lim_{\mathbf{x}\to(a,b)} 5x - \lim_{\mathbf{x}\to(a,b)} 4y \qquad \text{(limit sum rule)}$$

$$= 5 \cdot \lim_{\mathbf{x}\to(a,b)} x - 4 \cdot \lim_{\mathbf{x}\to(a,b)} y \qquad \text{(constant multiple rule)}$$

$$= 5a - 4b. \qquad \text{(assumed limits)}$$

We see that two basic limits, together with limit properties, allow us to avoid the epsilon-delta method in many situations. ▲

As in the previous example, multivariable limits of many functions we encounter can be obtained by evaluating $f(\mathbf{c})$ when it's defined (e.g. polynomials, rational functions, exponential functions, logarithmic functions, etc.).

Example 3.2.10. *Domains and limits*

Notice that to evaluate $\lim_{\mathbf{x}\to\mathbf{c}} f(x,y)$ it is assumed that $\mathbf{c}$ is either in the domain of f, or a boundary point of the domain. Use this to determine which limits cannot be considered.

(a) $\displaystyle\lim_{(x,y)\to(0,0)} \frac{\sin\left(x^2+y^2\right)}{x^2+y^2}$

The domain of $f(x,y)$ is the punctured plane, and the origin is a boundary point, so the limit can be considered.

(b) $\displaystyle\lim_{(x,y)\to(-1,1)} \sqrt{xy}$

The domain of $f(x,y) = \sqrt{xy}$ is all points with $xy \geq 0$, which is the first and third quadrant (including the axes). The point $(-1,1)$ is not in the domain, nor is it a boundary point because any open ball centered at $(-1,1)$ with radius at most 1 misses the domain of f entirely. Thus this limit cannot be considered.

(c) $\displaystyle\lim_{(x,y)\to(0,0)} \ln\left(1 - \sqrt{x^2+y^2}\right)$

This limit can be taken since the origin is in the domain of $f(x,y)$. ▲

Example 3.2.11. $\displaystyle\lim_{(x,y)\to(3,3)} \frac{x^2-y^2}{x-y}$

We've already seen that the domain of f is all points $y \neq x$, and that the line $y = x$ is the boundary of this domain. Thus $(3,3)$ is a boundary point of the domain A, and we can consider the above limit.

Now when taking the limit we consider those points *in* A and within δ of $(3,3)$. Since we restrict our attention to points within A, we know $y \neq x$, and the calculation proceeds as follows:

$$\lim_{(x,y)\to(3,3)} \frac{x^2-y^2}{x-y} = \lim_{(x,y)\to(3,3)} \frac{(x-y)(x+y)}{x-y}$$
$$= \lim_{(x,y)\to(3,3)} x+y = 6.$$

The last limit is calculated using by substituting (technically using techniques similar to Example 3.2.9). ▲

Some limits can be calculated using basic properties, or techniques from single-variable calculus. For these limits we can avoid the epsilon-delta approach, as in previous examples. Other limits require more ingenuity, and we proceed with some examples that use the epsilon-delta approach directly.

Example 3.2.12. $\displaystyle\lim_{(x,y)\to(0,0)} \frac{5x^2y}{x^2+y^2} = 0$

The domain of f is the punctured plane, and $(0,0)$ is its boundary, so we can compute the limit. For an epsilon-delta proof we must find a δ for every $\epsilon > 0$ such that

$$\left| \frac{5x^2 y}{x^2 + y^2} \right| < \epsilon \text{ whenever } 0 < \sqrt{x^2 + y^2} < \delta.$$

The trick in this case is the following string

$$\left| \frac{5x^2 y}{x^2 + y^2} \right| = \left| \frac{x^2}{x^2 + y^2} \right| |5y| \le 5 |y|,$$

where the last inequality follows from the fact that $x^2 + y^2 \ge x^2$. Similarly,

$$|y| = \sqrt{y^2} \le \sqrt{x^2 + y^2},$$

and putting these together we get

$$\left| \frac{5x^2 y}{x^2 + y^2} \right| \le 5 \sqrt{x^2 + y^2}. \tag{3.2.3}$$

This last inequality shows that

$$\left| \frac{5x^2 y}{x^2 + y^2} \right| < \epsilon \text{ whenever } 0 < \sqrt{x^2 + y^2} < \epsilon/5$$

(just replace $\sqrt{x^2 + y^2}$ with $\epsilon/5$ in the inequality of Equation 3.2.3).

Run through that reasoning one more time, then consider what it means. Give me any $\epsilon > 0$, I've just demonstrated that if $0 < \sqrt{x^2 + y^2} < \epsilon/5$, then $f(x,y)$ is within ϵ of 0. This proves the desired limit. ▲

Example 3.2.13. $\displaystyle \lim_{(x,y) \to (0,0)} \frac{\sin \left(x^2 + y^2 \right)}{x^2 + y^2} = 1$

In Example 3.2.5 we showed this limit is one by relating it to a single-variable limit. Now lets use the epsilon-delta approach.

The domain of f is the punctured plane, and the origin is a boundary point of the domain, so we can consider the limit.

To satisfy Definition 3.2.2, we must find a $\delta > 0$ for each $\epsilon > 0$ such that

$$\left| \frac{\sin \left(x^2 + y^2 \right)}{x^2 + y^2} - 1 \right| < \epsilon \text{ whenever } 0 < \sqrt{x^2 + y^2} < \delta. \tag{3.2.4}$$

The restriction that $0 < \sqrt{x^2 + y^2}$ ensures that the points we consider are in the domain of f, another requirement of Definition 3.2.2 (i.e. that $\mathbf{x} \in A$).

We will use the substitution $\theta = x^2 + y^2$ together with the fact from single variable calculus that

$$\lim_{\theta \to 0} \frac{\sin \theta}{\theta} = 1.$$

By Definition 3.2.1, the single-variable limit implies that for any $\epsilon > 0$ there is a $\delta > 0$ such that

$$\left| \frac{\sin \theta}{\theta} - 1 \right| < \epsilon \text{ whenever } 0 < \theta < \delta. \tag{3.2.5}$$

Note what we just did! Rather than using algebra to find δ we used the definition of the single-variable limit. There's still some algebraic trickery needed, but that is the big point of this example—the existence of the single-variable limit provides us with a δ.

The reason more trickery is needed is that the δ we found bounds $\theta = x^2 + y^2$, but we need a bound on $\sqrt{\theta} = \sqrt{x^2 + y^2}$. Now if $0 < \delta < 1$, then $\delta^2 < \delta$, so the above conditional implies

$$\text{If } 0 < \theta < \delta^2, \text{ then } \left| \frac{\sin \theta}{\theta} - 1 \right| < \epsilon. \tag{3.2.6}$$

Now, since $0 < \delta < 1$, if $0 < \theta < \delta^2$ then $0 < \theta < \delta$, and the implication of Equation 3.2.5 yields that of Equation 3.2.6.

Now use the fact that if $0 < \theta < \delta^2$ then $0 < \sqrt{\theta} < \delta$ to change the hypothesis of Equation 3.2.6. Writing the result in terms of x and y gives

$$\text{If } 0 < \sqrt{x^2 + y^2} < \delta, \text{ then } \left| \frac{\sin (x^2 + y^2)}{x^2 + y^2} - 1 \right| < \epsilon,$$

which is equivalent to the desired implication of Equation 3.2.4. This makes the calculations of Example 3.2.5 precise, and a similar trick could be used to give an epsilon-delta proof of the limit in Example 3.2.6. ▲

The section concludes with some examples of when multivariable limits do *not* exist. Recall that in single-variable calculus a limit didn't exist if the right-hand and left-hand limits were different. The two-sided limit didn't exist because coming at c from different directions gave different expected values. The classic example of this is $\lim_{x \to 0} \frac{|x|}{x}$. Since

$$\frac{|x|}{x} = \begin{cases} 1 & x > 0 \\ -1 & x < 0 \end{cases},$$

the right-hand limit is 1 while the left-hand limit is -1. Thus the two-sided limit $\lim_{x \to 0} \frac{|x|}{x}$ doesn't exist. Basically, if the limit depended on which path you took to get to c, it didn't exist. Analogously, one way to show multivariable limits don't exist is to show the limit is path dependent. In other words, different paths to **c** lead to different limits. We illustrate with examples.

Example 3.2.14. $\displaystyle\lim_{(x,y)\to(0,0)} \frac{x^2}{x^2+y^2}$

To show this limit doesn't exist, take the limit as $(x, y) \to (0,0)$ along two different lines and show you get two different numbers. The x-axis has equation $y = 0$, so substituting 0 for y then taking the limit gives the expected function values as $(x, y) \to (0,0)$ along the x-axis. We calculate along the x-axis

$$\lim_{(x,y)\to(0,0)} \frac{x^2}{x^2+y^2} = \lim_{(x,0)\to(0,0)} \frac{x^2}{x^2} = 1.$$

Similarly, letting $x = 0$, then taking the limit gives the limit along the y-axis. We have

$$\lim_{(x,y)\to(0,0)} \frac{x^2}{x^2+y^2} = \lim_{(0,y)\to(0,0)} \frac{0}{y^2} = 0.$$

Since the limits as **x** approaches **0** along different directions are different, the limit doesn't exist (see Figure 3.2.3).

We can be more general than we have been so far. If we want $\mathbf{x} \to \mathbf{0}$ along the line $y = mx$, we can substitute mx for y and take the limit as follows:

$$\lim_{(x,y)\to(0,0)} \frac{x^2}{x^2+y^2} = \lim_{(x,mx)\to(0,0)} \frac{x^2}{x^2+m^2x^2} = \lim_{(x,mx)\to(0,0)} \frac{1}{1+m^2} = \frac{1}{1+m^2}.$$

This calculation shows that the limit changes depending on the slope of the line you approach **0** on. This phenomenon is pictured in Figure 3.2.3. ▲

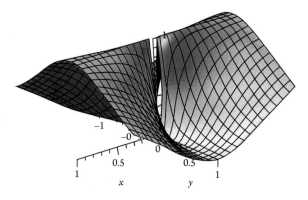

Figure 3.2.3 The surface $z = x^2/(x^2+y^2)$

Math App 3.2.1. *Two-variable limits*

This Math App takes a more geometric look at the previous example. Click the hyperlink, or print users open manually, to investigate animations which give more insight into our previous calculations.

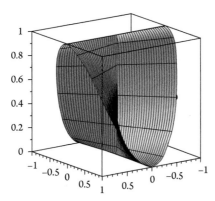

In the previous example we approached the origin along different lines to get different limits. In the next example we approach along different cubics.

Example 3.2.15. $\displaystyle\lim_{(x,y)\to(0,0)} \frac{x^6}{x^6 + y^2}$

Note that approaching the origin along the cubic $y = mx^3$ yields the following limit:

$$\lim_{(x,y)\to(0,0)} \frac{x^6}{x^6 + y^2} = \lim_{(x,mx)\to(0,0)} \frac{x^6}{x^6 + m^2 x^6} = \lim_{(x,mx)\to(0,0)} \frac{1}{1 + m^2} = \frac{1}{1 + m^2}.$$

Thus different paths yield different limits, and the limit does not exist. ▲

Example 3.2.16. $\displaystyle\lim_{(x,y)\to(0,0)} \frac{x^2 + y^2}{x + y}$

This example is a little trickier than the previous ones. Substituting $y = mx$ (with $m \neq -1$) yields

$$\lim_{(x,y)\to(0,0)} \frac{x^2 + y^2}{x + y} = \lim_{(x,mx)\to(0,0)} \frac{(1 + m^2)x^2}{(1 + m)x} = \lim_{x\to0} \frac{(1 + m^2)}{(1 + m)} x = 0,$$

which makes it seem like the limit should be zero. The problem with this substitution is that the powers of x are different in top and bottom, so they don't cancel as in the previous examples.

In this instance, we can determine other curves to travel on. The key is to consider level curves of the function $f(x,y) = (x^2 + y^2)/(x + y)$. For example, the level curve at level 2 has an equation which simplifies as follows:

$$\frac{x^2 + y^2}{x + y} = 2$$

$$x^2 + y^2 = 2x + 2y$$

$$(x - 1)^2 + (y - 1)^2 = 2.$$

The level curve is a circle centered at $(1,1)$ and through the origin! This is valid for all points on the circle, *except* the origin, of course, since our original function is not defined there. Thus if $(x,y) \to (0,0)$ along this circle, the limit is 2. Since this is not the same value as approaching the origin along lines, the limit does not exist. This is shown in Figure 3.2.4.

More generally, for any $c \neq 0$ you can verify that the level curve is

$$(x - c/2)^2 + (y - c/2)^2 = c^2/2,$$

for every point except the origin. Approaching the origin along these curves gives a limit of c, further indicating the limit doesn't exist. ▲

Remark:

- So far we have defined limits of functions $f : \mathbb{R}^n \to \mathbb{R}$. Limits of vector-valued functions $f : \mathbb{R}^n \to \mathbb{R}^m$ follow from Definition 3.2.2, and the fact that $\lim_{\mathbf{x} \to \mathbf{c}} f(\mathbf{x}) = \mathbf{L}$ if and only if each component function approaches the corresponding component of $\mathbf{L}$.

- Continuity at a point is defined analogously to the single-variable case. Namely, f is continuous at $\mathbf{c}$ if

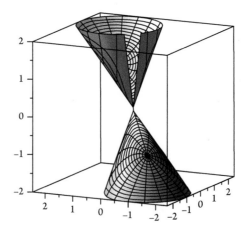

Figure 3.2.4 Using level curves to show limits don't exist

$$\lim_{\mathbf{x}\to\mathbf{c}} f(\mathbf{x}) = f(\mathbf{c}).$$

This means f is continuous if what you get (the $f(\mathbf{c})$ part) is what you expected (the $\lim_{\mathbf{x}\to\mathbf{c}} f(\mathbf{x})$ part). Of course, the same terminology from single-variable carries over. The function f is *continuous* if it is continuous at each point in its domain, etc. One can prove that sums, differences, products and compositions of continuous functions are continuous (See, for example, Shifrin's *Multivariable Mathematics*, [8]).

Example 3.2.17. *Continuity at the origin*

Show that the function

$$f(x,y) = \frac{|x|\,y}{\sqrt{x^2+y^2}}, \, f(0,0) = 0,$$

is continuous (see Figure 3.2.5).

Since $f(0,0)$ is defined to be zero, the domain of f is the entire plane. At every point other than the origin, the limit and function value agree. To show that f is continuous on its domain, we need only show

$$\lim_{\mathbf{x}\to 0} \frac{|x|\,y}{\sqrt{x^2+y^2}} = 0,$$

and we'll use the epsilon-delta approach. Given any $\epsilon > 0$, we must find a $\delta > 0$ such that

$$\text{If } 0 < \sqrt{x^2+y^2} < \delta, \text{ then } \left| \frac{|x|\,y}{\sqrt{x^2+y^2}} - 0 \right| < \epsilon.$$

To find the δ we'll bound $f(x,y)$ as follows. Since $|x| = \sqrt{x^2}$ and $\sqrt{x^2} \le \sqrt{x^2+y^2}$, dividing the second inequality by $\sqrt{x^2+y^2}$ gives

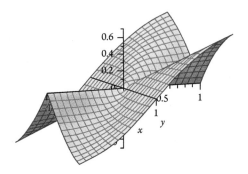

Figure 3.2.5 A continuous function

$$\frac{|x|}{\sqrt{x^2+y^2}} \leq 1.$$

This implies

$$\left|\frac{|x|y}{\sqrt{x^2+y^2}} - 0\right| = \frac{|x|}{\sqrt{x^2+y^2}}|y| \leq |y|.$$

Since $|y| = \sqrt{y^2} \leq \sqrt{x^2+y^2}$, this implies that

$$\left|\frac{|x|y}{\sqrt{x^2+y^2}} - 0\right| \leq |y| \leq \sqrt{x^2+y^2}. \tag{3.2.7}$$

If $0 < \sqrt{x^2+y^2} < \epsilon$, Equation 3.2.7 implies that

$$\left|\frac{|x|y}{\sqrt{x^2+y^2}} - 0\right| < \epsilon$$

as well. Letting $\delta = \epsilon$ gives the implication of Definition 3.2.2, proving the limit. Thus f is a continuous function. ▲

Exercises

1. Find the domain of each of the following functions. Determine if the domain is an open set, and the boundary of each domain.

   (a) $f(x,y) = \dfrac{x^2-y^2}{x+y}$

   (b) $f(x,y) = \dfrac{x^2+y^2}{x^2-y^2}$

   (c) $f(x,y) = \dfrac{x^2-y^2}{x^2+y^2}$

   (d) $f(x,y) = e^{-\frac{1}{x^2+y^2}}$

2. Give an epsilon-delta proof that $\lim\limits_{x\to(a,b)} 4y = 4b$.

3. Give an epsilon-delta proof that $\lim\limits_{x\to(a,b)} x = a$.

4. Give an epsilon-delta proof that $\lim\limits_{x\to(a,b)} y = b$.

5. Give an epsilon-delta proof that $\displaystyle\lim_{(x,y)\to(0,0)} \frac{4x^3}{x^2+y^2} = 0$.

6. Assuming $\displaystyle\lim_{\mathbf{x}\to(a,b)} x = a$ and $\displaystyle\lim_{\mathbf{x}\to(a,b)} y = b$, use properties of limits to show

   (a) $\displaystyle\lim_{\mathbf{x}\to(a,b)} 3x^2 = 3a^2$.

   (b) $\displaystyle\lim_{\mathbf{x}\to(a,b)} 5xy = 5ab$.

7. Use properties of limits (and algebra) to evaluate

   (a) $\displaystyle\lim_{(x,y)\to(-1,-1)} \frac{x-y}{x^2-y^2}$

   (b) $\displaystyle\lim_{(x,y)\to(2,2)} \frac{\sqrt{x-y+1}-1}{x-y}$

8. Which of the following limits tend toward points that violate the domain restriction of Definition 3.2.2?

   (a) $\displaystyle\lim_{(x,y)\to(1,0)} \ln\left(1 - \sqrt{x^2+y^2}\right)$

   (b) $\displaystyle\lim_{(x,y)\to(2,0)} \ln\left(1 - \sqrt{x^2+y^2}\right)$

   (c) $\displaystyle\lim_{(x,y)\to(0,1)} \sqrt{x^2-y^2}$

   (d) $\displaystyle\lim_{(x,y)\to(1,0)} \sqrt{x^2-y^2}$

9. Calculate the following limits directly.

   (a) $\displaystyle\lim_{\mathbf{x}\to(1,2)} \frac{x^2-3x}{y^3+2y}$

   (b) $\displaystyle\lim_{\mathbf{x}\to(3,-1)} \sin(\pi(x^2+y^2)/4)$

   (c) $\displaystyle\lim_{\mathbf{x}\to(e,e^3)} \ln(xy)$

   (d) $\displaystyle\lim_{\mathbf{x}\to(\ln 3,\ln 2)} e^{x-3y}$

10. Calculate the following limits using a substitution and single-variable techniques.

    (a) $\displaystyle\lim_{\mathbf{x}\to(0,0)} \arctan\left(\frac{5}{x^2+y^2}\right)$

    (b) $\displaystyle\lim_{\mathbf{x}\to(0,0)} \left(1 + \frac{1}{x^2+y^2}\right)^{x^2+y^2}$

    (c) $\displaystyle\lim_{\mathbf{x}\to(0,0)} \left(1 + 3x^2 + 3y^2\right)^{\frac{1}{x^2+y^2}}$

(d) $\displaystyle\lim_{\mathbf{x}\to(0,0)} \frac{1-\cos\sqrt{x^2+y^2}}{\sqrt{x^2+y^2}}$

(e) $\displaystyle\lim_{\mathbf{x}\to(0,0)} e^{-\frac{1}{x^2+y^2}}$

11. Use the epsilon-delta approach of Example 3.2.13 to prove the limits of Exercise 10.

12. Show that the following limits don't exist.

(a) $\displaystyle\lim_{\mathbf{x}\to(0,0)} \frac{x^3}{x^3+y^3}$

(b) $\displaystyle\lim_{\mathbf{x}\to(0,0)} \frac{y^2}{x^2+y^2}$

(c) $\displaystyle\lim_{\mathbf{x}\to(0,0)} \frac{x^2}{x^2+y}$

(d) $\displaystyle\lim_{\mathbf{x}\to(0,0)} \frac{x^2+y^2}{x}$

(e) $\displaystyle\lim_{\mathbf{x}\to(0,0)} \frac{x^2}{y}$

3.3 Partial Derivatives

In single-variable calculus we saw that derivatives were instantaneous rates of change, with respect to the independent variable. Analytically, recall that the average rate of change of $f(x)$ on the interval $(x, x+h)$ is given by the difference quotient $\left(f(x+h) - f(x)\right)/h$. This approximates the instantaneous rate of change of f at x for small h, and the approximation becomes better as h gets smaller. Thus we define the instantaneous rate of change $f'(x)$ to be

$$f'(x) = \lim_{h \to 0} \frac{f(x+h) - f(x)}{h}, \tag{3.3.1}$$

if the limit exists.

It doesn't hurt to recall the geometric interpretation of the pieces of the definition of the derivative, and we refer to Figure 3.3.1 for the following notation. Let P and Q be the points $(x, f(x))$ and $(x+h, f(x+h))$, respectively, on the graph of $y = f(x)$. The difference quotient $\left(f(x+h) - f(x)\right)/h$ gives the slope m_{sec} of the secant line through P and Q, while the derivative $f'(x)$ is the slope m_{tan} of the tangent line at P. Taking the limit as h tends to 0 in Equation 3.3.1 is equivalent to letting the point Q get closer to P. Equation 3.3.1 simply states that the slopes m_{sec} approach m_{tan} as $Q \to P$. Symbolically we have

$$m_{tan} = \lim_{Q \to P} m_{sec},$$

and in English this just means the slopes of the secant lines approach the slope of the tangent line to f at P as Q gets closer to P.

To begin our study of differentiation we focus on functions $f : \mathbb{R}^n \to \mathbb{R}$ of several variables. We now have several independent variables, and could ask for the instantaneous

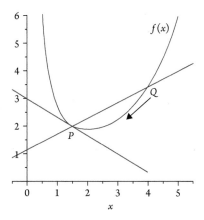

Figure 3.3.1 The derivative as slope of tangent line

rate of change of f with respect to any of them. This will be called the partial derivative of f, because it focuses on the rate of change with respect to just one of many variables. We treat partial derivatives analytically first, then discuss their geometric interpretation.

For simplicity, we restrict our attention to functions $f(x, y, z)$ of three variables, and consider the instantaneous rate of change of f with respect to the independent variable x—the partial derivative of f with respect to x. The first observation is that if we want to know how fast f is changing in the x-direction, we think of y and z as fixed. In other words: to define the partial derivative of f with repsect to x, don't change y and z, just change x. We now take our cue from single-variable calculus. To find the instantaneous rate of change, we approximate it with the average rate $\left(f(x+h, y, z) - f(x, y, z)\right)/h$ over the interval $(x, x+h)$. Note that y and z stay fixed in this difference quotient because we want the change of f in the x-direction. Now take a limit as $h \to 0$ to find the instantaneous rate. Formally, we define

Definition 3.3.1. *The partial derivative $\frac{\partial f}{\partial x}$ of $f(x, y, z)$ with respect to x is*

$$\frac{\partial f}{\partial x} = \lim_{h \to 0} \frac{f(x+h, y, z) - f(x, y, z)}{h}$$

when the limit exists.

Thus partial derivatives can be interpreted as instantaneous rates of change. The difference quotient in the definition is the average rate of change of f on the x-interval $(x, x+h)$ (fixing y and z). In taking a limit, the average rate of change approaches the instantaneous rate of change, so $\frac{\partial f}{\partial x}(x, y, z)$ is the instantaneous rate of change of f in the x-direction. Of course, $\frac{\partial f}{\partial y}(x, y, z)$ is the instantaneous rate of change in the y-direction, etc. This highlights one difference between single- and multivariable differentiation. In single-variable calculus there's essentially one direction to differentiate with respect to. In multivariable calculus, there's a derivative in every direction of the domain! We'll make this more precise when we investigate directional derivatives.

To calculate the partial derivative of $f(x, y, z)$ with respect to x, think of y and z as constants and differentiate x normally. WARNING: This is **not** implicit differentiation from single-variable calculus! In implicit differentiation we thought of y as being a *function* of x defined implicitly by an equation. In partial differentiation with respect to x we think of y as a *constant*.

Example 3.3.1. *Calculating partial derivatives.*

Find all first-order partial derivatives of $f(x, y, z) = x^2 \sin(xy^2 z^3)$.

$$\frac{\partial f}{\partial x}(x, y, z) = 2x \sin(xy^2 z^3) + x^2 y^2 z^3 \cos(xy^2 z^3),$$

$$\frac{\partial f}{\partial y}(x, y, z) = x^2(2xyz^3 \cos(xy^2 z^3)) = 2x^3 yz^3 \cos(xy^2 z^3),$$

$$\frac{\partial f}{\partial z}(x, y, z) = x^2(3xy^2 z^2 \cos(xy^2 z^3)) = 3x^3 y^2 z^2 \cos(xy^2 z^3). \ \blacktriangle$$

When changing between different coordinate systems (from cylindrical to Cartesian, for example), it is sometimes useful to relate their partial derivatives using the change-of-variables conversions. We illustrate with an example.

Example 3.3.2. *Partial derivatives and coordinate transformations*

Recall that changing from Cartesian to polar coordinates in the plane is achieved by the equations

$$r = \sqrt{x^2 + y^2}$$

$$\theta = \tan^{-1} \frac{y}{x}.$$

Differentiating these equations with respect to x and y gives the following:

$$\frac{\partial r}{\partial x} = \frac{x}{\sqrt{x^2 + y^2}} \qquad \frac{\partial r}{\partial y} = \frac{y}{\sqrt{x^2 + y^2}}$$

$$\frac{\partial \theta}{\partial x} = \frac{1}{1 + (y/x)^2} \frac{-y}{x^2} = \frac{-y}{x^2 + y^2} \qquad \frac{\partial \theta}{\partial y} = \frac{1}{1 + (y/x)^2} \frac{1}{x} = \frac{x}{x^2 + y^2}. \blacktriangle$$

Just as we have second derivatives in single variable calculus, there are second-order partial derivatives. Since there are several variables, there are a fair number of second-order partial derivatives. The notation for first taking the partial derivative with respect to x, then z is $\frac{\partial^2 f}{\partial z \partial x}$. So, as with composition of functions, the order you differentiate in is read from right to left in this notation. There is another common notation for partial derivatives that uses subscripts. We let $f_x = \frac{\partial f}{\partial x}$. Higher-order partial derivatives in subscript notation are read left-to-right, so f_{xyz} means differentiate with respect to x, then y, and finally z.

Example 3.3.3. *Higher-order partial derivatives*

Recall from Example 1.2.11 that the stream function for an ideal fluid flow around a $\pi/3$-corner is

$$\psi(x,y) = 3x^2 y - y^3,$$

where we saw that level curves of ψ correspond to streamlines of the flow. Find all second-order partial derivatives of ψ.

$$\psi_x(x,y) = 6xy \quad \psi_y(x,y) = 3x^2 - 3y^2$$

Now we calculate the second-order partial derivatives

$$\psi_{xx}(x,y) = 6y \qquad \psi_{xy}(x,y) = 6x$$
$$\psi_{yx}(x,y) = 6x \qquad \psi_{yy}(x,y) = -6y. \blacktriangle$$

Notice that the mixed partials ψ_{xy} and ψ_{yx} are equal. This is no coincidence! In fact, as long as f is reasonably nice, mixed partials will always be equal.

Theorem 3.3.1. *If $f(x,y)$ has continuous second-order partial derivatives, then mixed partial derivatives are equal. Symbolically*

$$\frac{\partial^2 f}{\partial y \partial x} = \frac{\partial^2 f}{\partial x \partial y}, \ or \ f_{xy} = f_{yx}.$$

This theorem generalizes to functions of three or more variables, with the analogous conclusion.

Example 3.3.4. *Verifying the equality of mixed partials*

Verify that the mixed partial derivatives of $f(x,y) = x^3 - 3x^2y + xy^3 - 5x + 7$ are equal.

Calculating first partial derivatives gives $f_x(x,y) = 3x^2 - 6xy + y^3 - 5$ and $f_y(x,y) = -3x^2 + 3xy^2$. From these we compute

$$f_{xy}(x,y) = -6x + 3y^2 = f_{yx}(x,y). \ \blacktriangle$$

After learning about differentiation in single-variable calculus, the inverse problem was considered. Rather than being given a function to differentiate, you were given the derivative and asked to find the original function. An analogous question can be asked in the multivariable setting, and we demonstrate with some examples.

Example 3.3.5. *Partial differentiation backwards*

Find a function $f(x,y)$ whose partial derivative with respect to x is $f_x(x,y) = 2xe^{x^2+y^2} - 3y$.

To find $f(x,y)$ we do "partial integration". In other words, we integrate f_x with respect to x, thinking of any function of y as a constant. We see

$$\int 2xe^{x^2+y^2} - 3y \ dx = e^{x^2+y^2} - 3xy + g(y). \ \blacktriangle$$

This illustrates one interesting feature of "partial" integration—the arbitrary constant is really an arbitrary function of y! This makes sense since $\frac{\partial}{\partial x}g(y) = 0$, so adding the function of y does not change the partial derivative with respect to x.

Example 3.3.6. *Finding a function given both partial derivatives*

Find a function $f(x,y)$ given

$$f_x(x,y) = x^2y - 2x + 5, \ f_y(x,y) = \frac{1}{3}x^3 + 2y.$$

We outline the steps of the process:

Step 1: Do partial integration $\int f_y(x,y)\,dy$, remembering to add an arbitrary function of x. This determines the portion of $f(x,y)$ that depends on y.

$$f(x,y) = \int \frac{1}{3}x^3 + 2y\,dy = \frac{1}{3}x^3y + y^2 + g(x)$$

Step 2: Set the partial derivative of your answer from step 1 equal to the given $f_x(x,y) = x^2y - 2x + 5$, and solve for $g'(x)$. In our case we get

$$f_x(x,y) = x^2y + g'(x) = x^2y - 2x + 5,$$

which gives $g'(x) = 2x - 5$. The result must be void of y's. If there are any y's in the equation, something went wrong in Step 1, or the problem is impossible to solve.

Step 3: Integrate to find $g(x)$, and write the concluding sentence. In our case,

$$g(x) = \int 2x - 5\,dx = x^2 - 5x + C.$$

where C is the friendly arbitrary constant from single-variable calculus. We conclude that

$$f(x,y) = \frac{1}{3}x^3y + y^2 + x^2 - 5x + C. \blacktriangle$$

Example 3.3.7. *Inconsistent partial derivatives*

Is there a function whose partial derivatives are

$$f_x(x,y) = 3y - 2x, \ f_y(x,y) = 4x + 2y?$$

Naively, we apply the steps of the previous example:

$$f(x,y) = \int 3y - 2x\,dx = 3xy - x^2 + g(y).$$

Setting the partial of the $f(x,y)$ just found equal to the given one, and solving for $g(y)$ we get

$$3x + g'(y) = 4x + 2y$$
$$g'(y) = x + 2y.$$

Of course, this is a contradiction because $g(y)$ is a function of y alone. This means its derivative can only have y's as well—no x's! For this reason, there is no function $f(x,y)$ with the given partial derivatives.

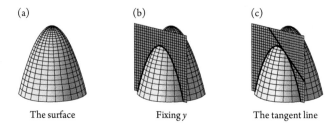

(a)   (b)   (c)

The surface   Fixing y The tangent line

Figure 3.3.2 The geometry of partial derivatives

An alternative approach is to use equality of mixed partials (Theorem 3.3.1) to show no such $f(x,y)$ exists. If such an f existed, then $f_{xy} = f_{yx}$. However, calculating mixed partials directly from the given partial derivatives gives

$$f_{xy} = \frac{\partial}{\partial y}\left(3y - 2x\right) = 3 \neq 4 = \frac{\partial}{\partial x}\left(4x + 2y\right) = f_{yx}.$$

Since the given partial derivatives imply $f_{xy} \neq f_{yx}$, a function with those partial derivatives does not exist. This test will be used in Chapter 5 to determine whether or not a vector field is conservative. ▲

The Geometry of Partial Derivatives: Now that we know how to calculate partial derivatives, let's take a look at what they tell us geometrically. We'll focus on the case of a function of two variables $f(x,y)$. We know that the graph of $z = f(x,y)$ is a surface in $\mathbb{R}^3$, and that fixing y to be a constant is equivalent to slicing the surface with a plane parallel to the xz-plane. For example, the surface $z = 4 - x^2 - y^2$ is pictured in Figure 3.3.2(a), and the intersection with the plane $y = 1$ in Figure 3.3.2(b). Note that the intersection of $z = f(x,y)$ with the plane is a curve in the plane. It turns out that $\frac{\partial f}{\partial x}(x,y)$ is the slope of the tangent line to this curve of intersection.

One can see this analytically as well. An equation for the curve of intersection of $z = 4 - x^2 - y^2$ and the plane $y = 1$ can be obtained by substituting 1 for y. Thus the curve is given by the equation $z(x) = f(x,1) = 3 - x^2$ in the plane $y = 1$. The derivative of z, therefore, is the same limit as in Definition 3.3.1 where 1 is substituted for y (and there is no third variable in f, of course). Symbolically we have

$$\frac{dz}{dx} = \lim_{h \to 0} \frac{z(x+h) - z(x)}{h} = \lim_{h \to 0} \frac{f(x+h,1) - f(x,1)}{h} = \frac{\partial f}{\partial x}(x,1).$$

Math App 3.3.1. *Partial differentiation*

In this Math App we illustrate the previous discussion with an animation showing the limiting process in partial differentiation. Interactive plots also demonstrate how partial derivatives of parametric surfaces (see below) are always tangent to the surface.

Extending Partial Differentiation to Vector-Valued Functions: Recall that a vector-valued function of several variables $f : \mathbb{R}^n \rightarrow \mathbb{R}^m$ can be thought of as being composed of its component functions. For example, if $f(x,y) = (xy^2, xe^y, x^2 \ln y)$ its component functions are $f_1(x,y) = xy^2$, $f_2(x,y) = xe^y$, and $f_3(x,y) = x^2 \ln y$. To differentiate f with respect to x, just differentiate component-wise. This is analogous to how we differentiated parametric curves, the only difference being that now we're taking partial derivatives of component functions rather than single-variable derivatives. Sometimes we combine like partial derivatives of component functions into a single vector, as with parametric curves and as in the following examples.

Example 3.3.8. *Partial differentiation with vector-valued functions*

Find the partial derivatives of $f(x,y) = (xy^2, xe^y, x^2 \ln y)$. Differentiating component-wise gives

$$f_x(x,y) = \langle y^2, e^y, 2x \ln y \rangle$$
$$f_y(x,y) = \langle 2xy, xe^y, x^2/y \rangle \quad \blacktriangle$$

Recall that in function terminology a parametric surface is a vector-valued function of two variables whose range is in three dimensions. We can now differentiate parametric surfaces! We close this section with an example illustrating the calculation, but in the next section we determine the geometric significance of these derivatives.

Example 3.3.9. *Partial derivatives of parametric surfaces*

Parameterize the sphere $\rho = 3$ and find its partial derivatives.

To parameterize a constant-coordinate surface, substitute the value into the coordinate transformations, giving

$$S(\theta, \phi) = (3 \sin \phi \cos \theta, 3 \sin \phi \sin \theta, 3 \cos \phi), \ 0 \leq \theta \leq 2\pi, \ 0 \leq \phi \leq \pi.$$

One immediately computes the vectors of partial derivatives

$$S_\phi(\theta, \phi) = \langle 3 \cos \phi \cos \theta, 3 \cos \phi \sin \theta, -3 \sin \phi \rangle,$$
$$S_\theta(\theta, \phi) = \langle -3 \sin \phi \sin \theta, 3 \sin \phi \cos \theta, 0 \rangle.$$

Observe that when $\phi = 0$ we have that $S_\theta(\theta, 0)$ is the zero vector. This is because the image of the line segment from $(0,0)$ to $(2\pi, 0)$ in the $\theta\phi$-plane on our parametric

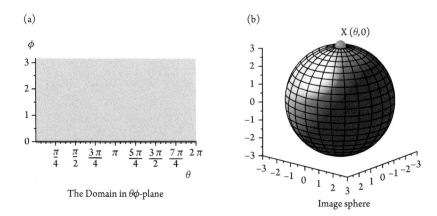

Figure 3.3.3 Parametric sphere and singular point

surface is the single point $(0,0,3)$. Since a whole line gets mapped to a single point, the parameterization is singular there, as indicated by a zero derivative. ▲

Exercises

1. Find all first-order partial derivatives for the following functions:
   (a) $f(x,y) = \frac{x^3 y}{1+x^2+y^2}$
   (b) $f(x,y,z) = x\tan^{-1}(yz)$
   (c) $f(x,y) = \sec(x^2 + 2xy + y^2)$
   (d) $f(x,y,z) = \ln(xy) - e^{yz}$

2. Find all first-order partial derivatives for the following functions:
   (a) $f(x,y) = \sec(xy)\tan(x^2)$
   (b) $f(x,y,z) = xyze^{-x^2-y^2}$
   (c) $f(x,y) = \sin^{-1}\left(\frac{x}{x^2+y^2}\right)$
   (d) $f(x,y,z) = 4x^2 + 3y^2 - 2z^2 - 3xy + 2yz - 6xz + 5x - 2y + 4z + 24$

3. Using the change of coordinate formulas, find the partial derivatives of x,y, and z with respect to the cylindrical coordinates r,θ,z.

4. Using change of coordinate formulas, find the partial derivatives of x,y, and z with respect to the spherical coordinates ρ,θ,ϕ.

5. Find the change of coordinate formulas to translate from Cartesian to spherical (two are $\rho = \sqrt{x^2+y^2+z^2}$, and we know $\theta = \tan^{-1}(y/x)$, so what's ϕ?). Then find the partial derivatives of ρ,θ,ϕ with respect to x,y,z.

6. A critical point of $f(x,y)$ is any point where both partial derivatives vanish. Find all the critical points of
   (a) $f(x,y) = x^2 - 3xy + y^2 + 2x + 5$
   (b) $f(x,y) = x^2 - 2xy + y^2$
   (c) $f(x,y) = Ax^2 + Bxy + Cy^2 + Dx + Ey + F$

7. Find the second-order partial derivatives of the general quadratic in two variables $f(x,y) = Ax^2 + Bxy + Cy^2 + Dx + Ey + F$.

8. Find all second-order partial derivatives of $f(x,y) = \cos x \sin y - xy^3$.

9. Find a function $f(x,y)$ whose partial derivatives are $f_x(x,y) = 3x^2 - 2x + 1$ and $f_y(x,y) = 2y$.

10. Find a function $f(x,y)$ whose partial derivatives are $f_x(x,y) = ye^{xy} + 2xy + 3$ and $f_y(x,y) = xe^{xy} + x^2 - 3y$.

11. Which of the following are the partial derivatives of a function of two variables? If they are, find the function.
    (a) $f_x(x,y) = 2xy^3 - \sin x$ and $f_y(x,y) = 3x^2y^2 + \sec y \tan y$.
    (b) $f_x(x,y) = 6xy + 6x$ and $f_y(x,y) = 6xy + 6y$.
    (c) $f_x(x,y) = \left(\frac{y}{x}\right)^{2/3}$ and $f_y(x,y) = \left(\frac{x}{y}\right)^{1/3}$.
    (d) $f_x(x,y) = x^2y^3 + y$ and $f_y(x,y) = y^2x^3 - x$.

12. The Cobb–Douglas production function is sometimes used in economics to model production based on labor and capital. One form is $Y = L^{2/3}K^{1/3}$, where Y is the value of all goods produced in a year, L is the number of hours worked, and K is the capital input. Find the first-order partial derivatives of Y.

13. Find the vectors of partial derivatives of the parametric surface
$$S(r,\theta) = (r\cos\theta, r\sin\theta, r), \ r \geq 0, 0 \leq \theta \leq 2\pi.$$

14. Obtain a parameterization $S(x,y)$ for the graph of $f(x,y) = x^2 + y^2 - 2y$, then find the partial derivative vectors S_x and S_y.

15. Parameterize the graph of $f(x,y) = x^2y^3 - 3xy + 2y^2$, then find S_x and S_y.

16. Let $S(x,y) = (x,y,f(x,y))$ be the parameterization of the graph of $z = f(x,y)$. Describe the relationship between S_x and S_y to the partial derivatives of $f(x,y)$.

17. Parameterize the surface $r = 7$, and find the vectors of partial derivatives.

18. Parameterize the surface $\phi = \pi/4$ and find the vectors of partial derivatives.

19. Find the vectors of partial derivatives $f_x, f_y,$ and f_z of the vector-valued function
$$f(x,y,z) = \left(\sin(2x - y + 3z), \frac{z}{x^2 + y^2 + 1}\right).$$

20. Find the vectors of partial derivatives of $f(x,y,z) = (x^2z, e^{x^2+y^2+z^2}, xyz^3)$.

3.4 Tangent Planes to Surfaces

In single-variable calculus, differentiation can be introduced as a method for finding the equation of a tangent line to a curve. In multivariable calculus, we can use partial differentiation to find equations for tangent planes to surfaces. Recall that surfaces can be described parametrically, as a graph, or as a level surface. In each case we can use derivatives to find equations for tangent planes, although the level surface techniques will have to wait until we introduce the multivariable chain rule. We begin by defining the tangent plane to a surface.

Definition 3.4.1. *Let p be a point on the surface S. The tangent plane to S at p is the plane, if it exists, containing the tangent vectors to all curves on S that pass through p.*

We first mention that not all points on surfaces have tangent planes. The function

$$f(x,y) = \frac{|x|\,y}{\sqrt{x^2+y^2}}, f(0,0) = 0,$$

of Example 3.2.17, has no tangent plane at any point on the y-axis. Indeed, Figure 3.4.1 shows a crease in the surface along the y-axis, indicating there that is no well-defined tangent plane there. To avoid such situations, it is enough to assume that all component functions have continuous partial derivatives on their domains (see [8]). We make this assumption in this section.

Let $S(s,t) = (x(s,t),y(s,t),z(s,t))$ be a parametric surface (whose component functions have continuous partial derivatives). We have already seen that fixing s and letting t run yields a curve on the surface (they arise as grid lines in the diagrams of Section 1.5). The derivative of a parametric curve is a tangent vector to it. Since partial differentiation treats other variables as constant, and fixing a parameter to be constant yields a curve, the partial derivative of a parametric surface is a vector tangent to the surface. Symbolically, we see that $S_s(s,t) = \langle x_s(s,t),y_s(s,t),z_s(s,t)\rangle$ and $S_t(s,t) = \langle x_t(s,t),y_t(s,t),z_t(s,t)\rangle$ are tangent vectors to the surface at $S(s,t)$. Since both are tangent to the surface, their cross product is a normal vector to the surface. This observation leads to

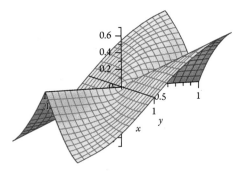

Figure 3.4.1 No well-defined tangent plane

> **Tangent Planes to Parametric Surfaces**
>
> The cross product $S_s(s_0, t_0) \times S_t(s_0, t_0)$ is normal to the parametric surface $S(s, t)$ at the point $S(s_0, t_0)$.

Example 3.4.1. *A paraboloid*

Find an equation for the tangent plane to the paraboloid

$$S(r, \theta) = (r\cos\theta, r\sin\theta, 4 - r^2)$$

at the point $S(1, \pi/4) = (\sqrt{2}/2, \sqrt{2}/2, 3)$.

Rather than merely doing the calculation, we parallel the justification given in the preceeding paragraph. The curve $S(1, \theta) = (\cos\theta, \sin\theta, 3)$ has tangent vector $S_\theta(1, \theta) = \langle -\sin\theta, \cos\theta, 0 \rangle$. Evaluating this at $\theta = \pi/4$ gives a tangent vector to the curve $S_\theta(1, \pi/4) = \langle -\sqrt{2}/2, \sqrt{2}/2, 0 \rangle$ as in Figure 3.4.2(a). Since it is tangent to a curve on the surface, it is a vector in the tangent plane by Definition 3.4.1.

Similarly, the curve on the surface obtained by fixing $\theta = \pi/4$ is given parameter-ically by $S(r, \pi/4) = (\sqrt{2}/2\,r, \sqrt{2}/2\,r, 4 - r^2)$. A tangent to the curve is $S_r(r, \pi/4) = \langle \sqrt{2}/2, \sqrt{2}/2, -2r \rangle$. Thus when $r = 1$, we see that $S_r(1, \pi/4) = \langle \sqrt{2}/2, \sqrt{2}/2, -2 \rangle$ is tangent to a curve on the surface, hence in the tangent plane (see Figure 3.4.2(b)).

Thus we know two vectors in the tangent plane, and their cross product is normal to the plane. We calculate

$$n = S_s \times S_t = \begin{vmatrix} i & j & k \\ -\sqrt{2}/2 & \sqrt{2}/2 & 0 \\ \sqrt{2}/2 & \sqrt{2}/2 & -2 \end{vmatrix}$$

$$= -\sqrt{2}i - \sqrt{2}j - k.$$

Any scalar multiple of n is normal to the tangent plane, so we choose $\langle \sqrt{2}, \sqrt{2}, 1 \rangle$ and get the equation

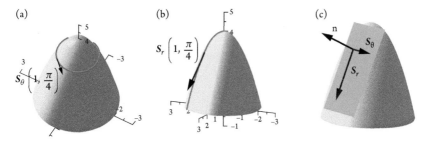

(a)   (b)   (c)

Figure 3.4.2 Parametric tangent plane

Figure 3.4.3 Tangent plane to a parametric torus

$$\sqrt{2}x + \sqrt{2}y + z = \left(\sqrt{2}, \sqrt{2}, 1\right) \cdot \left(\sqrt{2}/2, \sqrt{2}/2, 3\right) = 5. \ \blacktriangle$$

Example 3.4.2. *The tangent plane to a torus*

Recall that $S(s,t) = ((\cos t + 2)\cos s, (\cos t + 2)\sin s, \sin t)$, for $0 \le s, t \le 2\pi$, parameterizes a torus. Calculating partial derivatives gives us tangent vectors to the surface, and we calculate them coordinate-wise:

$$S_s(s,t) = \langle -(\cos t + 2)\sin s, (\cos t + 2)\cos s, 0 \rangle,$$
$$S_t(s,t) = \langle -\sin t \cos s, -\sin t \sin s, \cos t \rangle.$$

To get the tangent plane to $S(s,t)$ we need a point and a normal direction. The parametric equations give the point, while the cross product of the partial derivatives gives the normal direction. Thus a normal direction at $S(s,t)$ is

$$S_s \times S_t = \begin{vmatrix} \mathbf{i} & \mathbf{j} & \mathbf{k} \\ -(\cos t + 2)\sin s & (\cos t + 2)\cos s & 0 \\ -\sin t \cos s & -\sin t \sin s & \cos t \end{vmatrix}$$
$$= ((\cos t + 2)\cos s \cos t)\mathbf{i} + ((\cos t + 2)\sin s \cos t)\mathbf{j} + ((\cos t + 2)\sin t)\mathbf{k}$$

We use these calculations to find an equation for the tangent plane at $(s,t) = (\pi/4, \pi/4)$. The point on the tangent plane is $S(\pi/4, \pi/4) = (\sqrt{2} + 1/2, \sqrt{2} + 1/2, \sqrt{2}/2)$ and a normal vector is

$$S_s(\pi/4, \pi/4) \times S_t(\pi/4, \pi/4) = (\sqrt{2}/2 + 2)\left(1/2, 1/2, \sqrt{2}/2\right).$$

The scalar multiple $\mathbf{n} = \left(1, 1, \sqrt{2}\right)$ is also normal, so we get the equation

$$x + y + \sqrt{2}z = \left(1, 1, \sqrt{2}\right) \cdot \left(\sqrt{2} + 1/2, \sqrt{2} + 1/2, \sqrt{2}/2\right) = 2\sqrt{2} + 2. \ \blacktriangle$$

Tangent Planes to Graphs: We now use the above remarks to find tangent planes to graphs of functions of two variables. Recall that $S(x,y) = (x, y, f(x,y))$ parameterizes the graph of $f(x,y)$. One calculates the normal vector

$$S_x \times S_y = \begin{vmatrix} \mathbf{i} & \mathbf{j} & \mathbf{k} \\ 1 & 0 & f_x \\ 0 & 1 & f_y \end{vmatrix} = -f_x\mathbf{i} - f_y\mathbf{j} + \mathbf{k}$$

Using this we get:

Tangent Planes to Graphs of Functions

The vector $\mathbf{n} = \langle -f_x(x_0,y_0), -f_y(x_0,y_0), 1 \rangle$ is normal to the surface $z = f(x,y)$ at the point $(x_0, y_0, f(x_0, y_0))$.

Math App 3.4.1. *Tangent planes*

Click the hyperlink, or print users open manually, to investigate tangent planes of two surfaces.

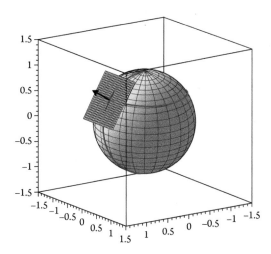

Example 3.4.3. *Horizontal tangent planes*

Find points on the graph of $f(x,y) = x^3 - x + y^2$ where the tangent planes are horizontal. This is equivalent to finding where the normal vector is parallel to the standard basis vector $\mathbf{k}$, so we solve the vector equation $\mathbf{n} = \lambda \mathbf{k}$. By the previous observations we have

$$\mathbf{n} = \langle -f_x(x_0,y_0), -f_y(x_0,y_0), 1 \rangle = \langle 3x^2 - 1, 2y, 1 \rangle = \langle 0, 0, \lambda \rangle.$$

Solving the one vector equation is equivalent to solving the system of three coordinate equations:

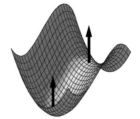

Figure 3.4.4 Horizontal tangents to a graph

$$3x^2 - 1 = 0$$
$$2y = 0$$
$$1 = \lambda$$

Whose solution yields $(x,y) = (\pm 1/\sqrt{3}, 0)$. Substituting into $f(x,y)$ to find the z-coordinate, we find the points on the surface that have horizontal tangent planes are $\left(\pm \frac{1}{\sqrt{3}}, 0, \mp \frac{2}{3\sqrt{3}} \right)$. ▲

Note that one of the points is a saddle similar to Figure 1.5.1, while the other is a local minimum. In Section 3.7 we'll see how to determine the nature of these critical points using tests similar to the second-derivative test in single-variable calculus.

Tangent Plane Approximations: The tangent line in single-variable calculus was used to approximate function values or changes in function values. Suppose $L(x)$ is an equation for the tangent line to $f(x)$ at $x = a$. Then $L(x)$ goes through the point $(a, f(a))$ and has slope $f'(a)$, and solving the point-slope formula for $L(x)$ gives the equation

$$L(x) = f(a) + f'(a)(x - a).$$

Since the tangent line $y = L(x)$ is close to the curve $y = f(x)$ for x-values near a, we said $f(x) \approx L(x)$ for x near a (the wavy equals sign means "approximately"). This approximation took on two forms:

$$f(x) \approx f(a) + f'(a)(x - a),$$
$$\Delta f \approx f'(a)\Delta x, \tag{3.4.1}$$

where $\Delta f = f(x) - f(a)$ is the change in function values corresponding to a change $\Delta x = x - a$ in x-values. For example, to approximate $\sqrt{3.9}$ use $f(x) = \sqrt{x}$, $a = 4$, and $x = 3.9$ in the approximation of Equation 3.4.1. Straightforward calculation (try this mentally—especially the derivative part) gives

$$\sqrt{3.9} \approx 2 + \frac{1}{4}(3.9 - 4) = 2 - .025 = 1.975,$$

and a calculator gives $\sqrt{3.9} \approx 1.97484$, accurate to a few more decimal places.

The second approximation of Equation 3.4.1 can be used to approximate changes in function values. For example, if the radius of a circle is measured to be 2 cm with an error of at most 0.1 cm, estimate the error in the area calculation. In this case, error is a change in function value. The error in the area is the difference between the calculated area and the actual area (which is unknown). We are told that the error in the radius measurement, or Δr, is at most 0.1 cm. Using $A = \pi r^2$, $r = 2$, and $\Delta r = 0.1$ in Equation 3.4.1 gives

$$\Delta A \approx 2\pi r \Delta r = 4\pi \cdot 0.1 = 0.4\pi.$$

In an analogous way, the tangent plane to $f(x,y)$ can be used to approximate function values or changes in function values. The tangent plane $L(x,y)$ to $z = f(x,y)$ at the point (a,b) can be shown to have equation (try it!)

$$L(x,y) = f(a,b) + f_x(a,b)(x-a) + f_y(a,b)(y-b).$$

This yields the approximations

$$f(x,y) \approx f(a,b) + f_x(a,b)(x-a) + f_y(a,b)(y-b), \text{ and}$$
$$\Delta f \approx f_x(a,b)\Delta x + f_y(a,b)\Delta y. \tag{3.4.2}$$

Math App 3.4.2. *Approximations*

The following Math App gives a geometric interpretation of the approximations just discussed.

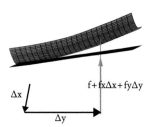

Δx

Δy

$f + f_x\Delta x + f_y\Delta y$

We illustrate how to use these approximations with some examples.

Example 3.4.4. *Approximating a function value*

Approximate $\sqrt{4.9^2 + 12.1^2}$.

The value to be approximated is $f(4.9, 12.1)$ where $f(x,y) = \sqrt{x^2 + y^2}$. To use the approximation of Equation 3.4.2, we have to find a point (a,b) near the point $(4.9, 12.1)$ *and* where we can calculate $f(a,b)$. Since $f(5, 12) = \sqrt{5^2 + 12^2} = 13$, we choose to linearize f at $(5, 12)$. Evaluating $f_x(x,y) = x/\sqrt{x^2 + y^2}$ and $f_y(x,y) = y/\sqrt{x^2 + y^2}$ at the point $(5, 12)$ gives the linearization

$$f(x,y) \approx 13 + \frac{5}{13}(x-5) + \frac{12}{13}(y-12),$$

and evaluating at $(4.9, 12.1)$ yields

$$\sqrt{4.9^2 + 12.1^2} \approx 13 + \frac{5}{13}(-0.1) + \frac{12}{13}(0.1) = 13 + \frac{7}{130}. \; \blacktriangle$$

Example 3.4.5. *Error estimate*

The radius and height of a cone are measured to be 3 inches and 4 inches, respectively. If the error in the radius measurement is at most 0.04 in, and that of the height at most 0.01 in, estimate the error in the calculation of volume using these measurements.

The idea is that any error in your measurements will result in an error in the corresponding volume calculation. To get an upper estimate of the error, use the (absolute value of the) maximal possible error. In this case the maximal errors in measurement lead to $\Delta r = 0.04$ and $\Delta h = 0.01$. Since the volume of a cone is $V(r,h) = \frac{1}{3}\pi r^2 h$, the error in the volume calculation is approximated by

$$\Delta V \approx V_r(3,4)\Delta r + V_h(3,4)\Delta h = \frac{2}{3}\pi \cdot 3 \cdot 4 \cdot 0.04 + \frac{1}{3}\pi \cdot 3^2 \cdot 0.01 = 0.35\pi. \; \blacktriangle$$

In addition to approximating function values, tangent planes can be used to approximate surface area. This section concludes with two such examples.

Example 3.4.6. *Approximating surface area*

Use a parallelogram in the tangent plane to approximate the area of the patch of the unit sphere parameterized by

$$S(\theta, \phi) = (\sin\phi \cos\theta, \sin\phi \sin\theta, \cos\phi), \; \pi/2 \le \phi \le 2\pi/3, \; 0 \le \theta \le \pi/6.$$

The portion of the sphere is highlighted in Figure 3.4.5(*a*), and the question is, what parallelogram should we use? The vectors of partial derivatives $S_\theta(0, \pi/2)$ and $S_\phi(0, \pi/2)$ lie in the tangent plane to the patch at the corner $S(0, \pi/2) = (1,0,0)$. It stands to reason that some scalar multiple of these should be used to approximate the area, and the limits on θ and ϕ determine which. In general, you multiply each vector by the change in the parameter over the limits of the parameterization: $(\Delta\theta)S_\theta$ and $(\Delta\phi)S_\phi$ (the choice of scalars $\Delta\phi$, $\Delta\theta$ is justified using the multivariable notion of differentiability given in Definition 3.5.3 of Section 3.5). The change $\Delta\theta$ and $\Delta\phi$ over the region we are parameterizing is $\pi/6$. Thus the area of the spherical patch is approximately the area of the parallelogram determined by the vectors $(\pi/6)S_\theta$ and $(\pi/6)S_\phi$. This area, you recall, is the length of the cross product of the two vectors. We proceed with the calculations.

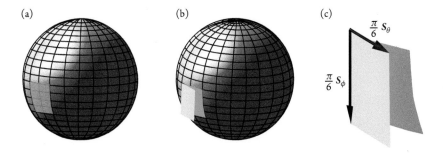

Figure 3.4.5 Approximating surface area with parallelograms

$$S_\theta(\theta,\phi) = \langle -\sin\phi\sin\theta, \sin\phi\cos\theta, 0 \rangle$$
$$S_\phi(\theta,\phi) = \langle \cos\phi\cos\theta, \cos\phi\sin\theta, -\sin\phi \rangle$$

We are interested in these vectors at the point $(\theta,\phi) = (0,\pi/2)$, which are $S_\theta(0,\pi/2) = \langle 0,1,0 \rangle$ and $S_\phi(0,\pi/2) = \langle 0,0,-1 \rangle$. Further, to approximate the given patch, we multiply each by the scalar $\pi/6$ and find the length of the cross product. The cross product is:

$$(\pi/6)S_\theta \times (\pi/6)S_\phi = \begin{vmatrix} \mathbf{i} & \mathbf{j} & \mathbf{k} \\ 0 & \pi/6 & 0 \\ 0 & 0 & -\pi/6 \end{vmatrix}$$

$$= -\frac{\pi^2}{36}\mathbf{i}.$$

The length is evidently $\frac{\pi^2}{36} \approx 0.2746$. One can compute the actual area of the patch is $\frac{\pi}{12} \approx 0.2618$. ▲

We wish to generalize the problem slightly. If we wish to approximate the patch of the sphere given by the limits $\pi/2 \le \phi \le \pi/2 + \Delta\phi$, $0 \le \theta \le \Delta\theta$, we would use the parallelogram determined by $(\Delta\theta)S_\theta$ and $(\Delta\phi)S_\phi$. The cross product calculation becomes

$$(\Delta\theta)S_\theta \times (\Delta\phi)S_\phi = \begin{vmatrix} \mathbf{i} & \mathbf{j} & \mathbf{k} \\ 0 & \Delta\theta & 0 \\ 0 & 0 & -\Delta\phi \end{vmatrix}$$

$$= -\Delta\theta\,\Delta\phi\mathbf{i}.$$

Thus the approximate area of the patch is $\Delta\theta\,\Delta\phi$. One can show that the actual area of the patch is $\Delta\theta\sin(\Delta\phi)$. Recall that for small $\Delta\phi$ we know $\sin(\Delta\phi) \approx \Delta\phi$, so this approximation gets better as our patch gets smaller. This idea is summarized below.

Approximating Parametric Surface Area with Parallelograms

The approximate area of the surface

$$S(s,t), \; a \le s \le a + \Delta s, \; b \le t \le b + \Delta t,$$

is

$$Area \approx \Delta s \Delta t \, \| S_s(a,b) \times S_t(a,b) \|.$$

This is the area of the parallelogram determined by $\Delta s S_s(a,b)$ and $\Delta t S_t(a,b)$ (we are assuming that both Δs and Δt are positive, and neither $S_s(a,b)$ nor $S_t(a,b)$ is the zero vector).

Example 3.4.7. *Cylindrical surface area*

Approximate the surface area of the cylindrical patch

$$S(s,t) = (3\cos t, 3\sin t, s), \; 0 \le s \le 2, \; 0 \le t \le \pi/6.$$

Using the above strategy, we need to calculate the length

$$\| 2 S_s(0,0) \times (\pi/6) S_t(0,0) \|.$$

Since $S_s(s,t) = \langle 0,0,1 \rangle$ and $S_t(s,t) = \langle -3\sin t, 3\cos t, 0 \rangle$, we have

$$2 S_s(0,0) \times (\pi/6) S_t(0,0) = \begin{vmatrix} \mathbf{i} & \mathbf{j} & \mathbf{k} \\ 0 & 0 & 2 \\ 0 & \pi/2 & 0 \end{vmatrix} = -\pi \mathbf{i}.$$

Thus the area is approximately π square units. In fact, for cylinders, we know the area is the height times the circumference. In this case the height of the patch is 2, and the circumference is $3 \cdot (\pi/6) = \pi/2$. Therefore the actual area is $2(\pi/2) = \pi$. So our approximation gives the actual area! This is not common, but follows from the fact that cylinders can be flattened without stretching. ▲

Concept Connection: The idea of approximating surface area with parallelogram tiles will play an important role in deriving an integral formula for the area of a parametric surface. We will approximate the surface with parallelogram shingles, and take a limit as the approximation gets finer. In this example, we saw that as the patch got smaller, the approximation got better. This is evidence that the integral we derive will give the desired area of the surface.

We have begun our analysis of differentiation by defining partial derivatives. Then we saw how partial differentiation can be used to find tangent planes to surfaces given parametrically and as the graph of $f(x,y)$. These are two ways we defined surfaces in Chapter 1. We also encountered level surfaces of a function of three variables. After introducing gradients and the multivariable chain rule we will be able to find tangent planes to level surfaces as well.

Exercises

1. Find an equation for the tangent plane to the surface

$$S(s,t) = (\sin s \cos t, \sin s \sin t, \cos s)$$

   at the point $(s,t) = (\pi/4, \pi/2)$. Sketch the surface and its tangent plane.

2. Parameterize the surface $\phi = \pi/3$ and find an equation for the tangent plane to it at the point $(\sqrt{3}/2, 0, 1/2)$.

3. Find where the tangent plane to $S(s,t) = (s+t, s-t, 4st+4t-2s)$, $-\infty < s,t < \infty$, is horizontal.

4. Find an equation for the tangent plane to the paraboloid

$$f(x,y) = 6 - x^2 - y^2 + 2x - 4y$$

   at the point $(0,0,6)$.

5. Find an equation for the tangent plane to the graph of $f(x,y) = 2x^2 - 3xy^3$ at $(x,y) = (2,1)$.

6. Find where the tangent plane to $z = x^2 - 2xy - 3y^2 + 4x$ is horizontal.

7. Find where the tangent plane to $f(x,y) = (x^2 + y^2)^2 - (x^2 + y^2)$ is horizontal. Sketch the intersection of the surface with the yz-plane, then rotate it about the z-axis to see what it looks like.

8. Show that the equation for the tangent plane to $z = f(x,y)$ at the point (a,b) is given by

$$z = f(a,b) + f_x(a,b)(x-a) + f_y(a,b)(y-b).$$

9. Approximate $\sqrt{2.99^2 + 4.02^2}$.

10. Approximate $1.02e^{-0.01}$.

11. The radius and height of a cylinder are measured to be 2 inches and 4 inches, respectively. If the error in the radius measurement is at most 0.2 in, and that of the height at most 0.1 in, estimate the error in the volume calculation using these measurements.

12. The production of a product is given by $Q = K^{1/3}L^{2/3}$, where K is capital and L is labor. If $(K,L) = (1,8)$, estimate the change in output corresponding to a change of $\Delta K = 0.1$ and $\Delta L = -0.2$.

13. Approximate the area of the paraboloid patch

$$\mathbf{S}(s,t) = (s\cos t, s\sin t, s^2), \ 1 \leq s \leq 2, \pi/6 \leq t \leq \pi/3.$$

14. Approximate the area of the conical patch

$$\mathbf{S}(s,t) = (s\cos t, s\sin t, s), \ 1 \leq s \leq 2, 0 \leq t \leq \pi/6.$$

15. Approximate the area of the conical patch

$$\mathbf{S}(s,t) = (s\cos t, s\sin t, s), \ 1 \leq s \leq 1 + \Delta s, 0 \leq t \leq \Delta t.$$

Use geometry to determine the actual area, and compare the two.

3.5 The Chain Rule

In this section we discuss the multivariable chain rule, which generalizes the single-variable one. Recall that the single-variable chain rule says that the derivative of a composition of functions is the product of the derivatives. Let's say that again:

To differentiate a composition of functions, multiply the derivatives.

It turns out that this idea generalizes beautifully to the multivariable setting, with the right interpretation of the terms "derivative" and "multiply". We open by introducing gradients and derivative matrices, followed by a quick note on differentiability. After these preliminaries we discuss the chain rule.

Gradients and the Derivative Matrix: We begin this section by making sense of the derivative Df of a vector-valued function $f : \mathbb{R}^n \to \mathbb{R}^m$ of several variables. The first step is to consider functions of several variables $f : \mathbb{R}^n \to \mathbb{R}$, where the gradient is a useful tool.

Definition 3.5.1. *The gradient $\nabla f(x,y,z)$ of a function $f(x,y,z)$ of several variables is the vector of first partial derivatives. In symbols,*

$$\nabla f(x,y,z) = \langle f_x(x,y,z), f_y(x,y,z), f_z(x,y,z) \rangle.$$

The gradient of a function encodes significant information about its rate of change, as well as about the geometry of it. We will investigate this further in Section 3.6. For now notice that ∇f is another example of a vector field, since it assigns a vector to every point in the domain of f.

Example 3.5.1. *Gradients of functions*

Find the gradients of $g(x,y) = 2x^3y - xy^2$ and $f(x,y,z) = x\tan^{-1}y - y\sin z$. Taking partial derivatives we see

$$\nabla g(x,y) = \langle 6x^2y - y^2, 2x^3 - 2xy \rangle$$

$$\nabla f(x,y,z) = \left\langle \tan^{-1}y, \frac{x}{1+y^2} - \sin z, -y\cos z \right\rangle. \ \blacktriangle$$

Notice that the gradient of a function of several variables is a vector field on the domain of the function. In the above example, the domain of $g(x,y)$ is $\mathbb{R}^2$. The gradient $\nabla g(x,y) = \langle 6x^2y - y^2, 2x^3 - 2xy \rangle$ is a vector field on the domain $\mathbb{R}^2$ (see Figure 3.5.1). More generally, if $f : \mathbb{R}^n \to \mathbb{R}$, then there are n independent variables, so n partial derivatives. Therefore ∇f can be thought of as a vector field on $\mathbb{R}^n$.

Example 3.5.2. *Electric Potential*

The electric potential due to a point charge Q placed at the origin was introduced in Section 1.3, Exercise 24, to be

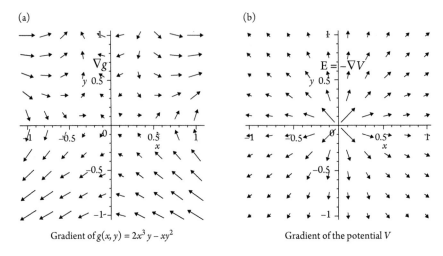

(a)

∇g

Gradient of $g(x, y) = 2x^3 y - xy^2$

(b)

$\mathbf{E} = -\nabla V$

Gradient of the potential V

Figure 3.5.1 Gradient vector fields

$$V = \frac{kQ}{\rho} = \frac{kQ}{\sqrt{x^2 + y^2 + z^2}},$$

where k is Coulomb's constant. Level surfaces of V have equations of the form $c = \frac{kQ}{\rho}$, and solving for ρ gives the constant coordinate surfaces $\rho = kQ/c$ (what types of surfaces are these?).

It turns out that electric potential is related to the electric fields derived using Coulomb's law and the law of superposition in Section 2.6. In fact, any electric field is the (negative of the) gradient of a potential function. In symbols, we have $\mathbf{E} = -\nabla V$, where $\mathbf{E}$ is the force per unit charge (a vector) and V is the potential (a scalar). Let's calculate the electric field due to a point charge as a gradient.

Taking partial derivatives gives the gradient

$$\nabla V(x,y,z) = \left\langle \frac{-kQx}{(x^2 + y^2 + z^2)^{3/2}}, \frac{-kQy}{(x^2 + y^2 + z^2)^{3/2}}, \frac{-kQz}{(x^2 + y^2 + z^2)^{3/2}} \right\rangle$$

$$= \frac{-kQ}{(x^2 + y^2 + z^2)^{3/2}} \langle x, y, z \rangle.$$

The electric field generated by the point charge is then

$$\mathbf{E} = -\nabla V = \frac{kQ}{(x^2 + y^2 + z^2)^{3/2}} \langle x, y, z \rangle.$$

A two-dimensional slice of this gradient field is illustrated in Figure 3.5.1(b). The reader should confirm that this is the same field of Equation 2.6.3 which states

$$\mathbf{E} = \frac{kQ}{\rho^2}\widehat{\rho},$$

and followed from Coulomb's law (recall that $\widehat{\rho}$ is a *unit* vector from the origin to the point, and $\rho = \sqrt{x^2 + y^2 + z^2}$). ▲

Example 3.5.3. *Ideal fluid flow revisited*

Recall from Example 1.2.10 that the velocity potential and stream function of an ideal flow around a right-angled corner are

$$\varphi = x^2 - y^2, \text{ and}$$
$$\psi = 2xy,$$

respectively. We now demonstrate that the gradients of φ and ψ are perpendicular at every point.

$$\nabla\varphi \cdot \nabla\psi = (2x, -2y) \cdot (2y, 2x) = 4xy - 4xy = 0.$$

Since the dot product is always zero, for *every* point (x, y), the vectors $\nabla\varphi$ and $\nabla\psi$ are always orthogonal (see Figure 3.5.2). It turns out that this is not a coincidence, but follows from the assumptions that an ideal fluid flow is both incompressible and irrotational. ▲

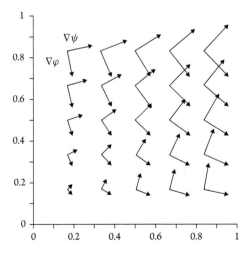

Figure 3.5.2 Orthogonal velocity and stream gradients

The gradient, then, associates to every function of several variables a "derivative vector". We wish to generalize this to vector-valued functions. Parametric surfaces $S(s,t)$ are examples of vector-valued functions of several variables. The component functions $x(s,t)$, $y(s,t)$, and $z(s,t)$ are functions of several variables, and so have gradients. We define the derivative matrix $DS(s,t)$ of $S(s,t)$ to be

$$DS(s,t) = \begin{bmatrix} x_s & x_t \\ y_s & y_t \\ z_s & z_t \end{bmatrix}.$$

Notice that there is a row for each component function, and a column for each parameter. The columns are what we called S_s and S_t in the previous section. The rows are simply the gradients of the component functions. We can define the derivative matrix for any vector-valued function of several variables, and it is called the derivative matrix.

Definition 3.5.2. *Let* $f : \mathbb{R}^n \to \mathbb{R}^m$ *have component functions* $f_i(x_1,\ldots,x_n)$ *for* $i = 1,\ldots,m$. *The derivative matrix* Df *of* f *at a point* $\mathbf{x} \in \mathbb{R}^n$ *is*

$$Df(\mathbf{x}) = \left[\frac{\partial f_i}{\partial x_j}(\mathbf{x}) \right].$$

That is, the ij^{th} *entry of the derivative matrix* Df *is the partial derivative of the* i^{th} *component function with respect to the* j^{th} *variable.*

First observe that if $f : \mathbb{R}^n \to \mathbb{R}$ is simply a function of several variables (not vector-valued), then the derivative matrix is simply the gradient defined above. More generally, the ith row of the derivative matrix is ∇f_i, the gradient of the ith component function. So Df is the matrix whose rows are the gradients of the component functions of f.

Example 3.5.4. *Derivative matrices*

Find the derivative matrices of the following functions.

(a) $f(x,y) = (2x + 3y, x - 2y)$.
   In this case the component functions are

$$f_1(x,y) = 2x + 3y \text{ and } f_2(x,y) = x - 2y.$$

Taking partial derivatives we have

$$Df(x,y) = \begin{bmatrix} \frac{\partial f_1}{\partial x} & \frac{\partial f_1}{\partial y} \\ \frac{\partial f_2}{\partial x} & \frac{\partial f_2}{\partial y} \end{bmatrix} = \begin{bmatrix} 2 & 3 \\ 1 & -2 \end{bmatrix}.$$

Notice that the function f can be defined using matrix multiplication, as described in Section 2.4, by

$$f(x,y) = \begin{bmatrix} 2 & 3 \\ 1 & -2 \end{bmatrix} \begin{bmatrix} x \\ y \end{bmatrix} = (2x + 3y, x - 2y).$$

The fact that the derivative matrix for a linear transformation is the same as the coefficient matrix is analogous to the fact that the derivative of a line $y = mx + b$ is its slope (the coefficient of x). In the single-variable case the coefficient matrix is a 1×1 matrix—the scalar m.

(b) The derivative matrix for the parametric torus

$$S(s,t) = ((\cos t + 2)\cos s, (\cos t + 2)\sin s, \sin t)$$

is the 3×2 matrix

$$DS(s,t) = \begin{bmatrix} -(\cos t + 2)\sin s & -\sin t \cos s \\ (\cos t + 2)\cos s & -\sin t \sin s \\ 0 & \cos t \end{bmatrix}. \; \blacktriangle$$

Differentiability: This is a good time to mention the multivariable definition of differentiability. It turns out that for a function to be differentiable it needs more than just the existence of first partial derivatives, as we will see. One important consequence of differentiability is that it allows us to justify approximating functions with vectors. In particular, it will provide the basis for approximating surface area with parallelograms. We won't emphasize the concept of differentiability in this text, for more details consult [3] or [8].

Definition 3.5.3. *The function $f : \mathbb{R}^n \to \mathbb{R}^m$ is differentiable at $\mathbf{a} \in \mathbb{R}^n$ if the derivative matrix $Df(\mathbf{a})$ exists and*

$$\lim_{\mathbf{h} \to 0} \frac{f(\mathbf{a} + \mathbf{h}) - f(\mathbf{a}) - Df(\mathbf{a}) \cdot \mathbf{h}}{\|\mathbf{h}\|} = 0.$$

If f is differentiable at $\mathbf{a}$ then its derivative matrix exists, which implies that all first partial derivatives exist. Differentiability, however, requires more than just the existence of all first partial derivatives! It requires that this weird limit vanish as well. To see why differentiability should require more than the existence of first partials, we consider the function of Example 3.2.17.

Example 3.5.5. *Differentiability is more than the existence of first partials*

Show that both first partial derivatives of

$$f(x,y) = \frac{|x| y}{\sqrt{x^2 + y^2}}, \; f(0,0) = 0,$$

exist at the origin, but it is not differentiable there.

We compute the partial derivate $f_x(0,0)$ from the definition:

$$f_x(0,0) = \lim_{h \to 0} \frac{f(h,0) - f(0,0)}{h} = \lim_{h \to 0} \frac{0-0}{h} = 0.$$

Similarly, $f_y(0,0) = 0$. These calculations are clear because the function $f(x,y)$ vanishes on the x- and y-axes, and so is constant there (see Figure 3.2.5). Since the partials are the rates of change in the horizontal and vertical directions, and the function is constant in those directions, the partials are zero.

Now consider the limit of Definition 3.5.3. Since $f(x,y)$ is a function of several variables, its derivative matrix is its gradient, which we just calculated to be the zero vector. Letting $\mathbf{a} = (0,0)$ and $\mathbf{h} = (x,y)$, the limit becomes

$$\lim_{(x,y) \to (0,0)} \frac{\frac{|x|y}{\sqrt{x^2+y^2}} - 0 - \langle 0,0 \rangle \cdot \langle x,y \rangle}{\sqrt{x^2+y^2}} = \lim_{(x,y) \to (0,0)} \frac{\frac{|x|y}{\sqrt{x^2+y^2}}}{\sqrt{x^2+y^2}}$$

$$= \lim_{(x,y) \to (0,0)} \frac{|x| \, y}{x^2 + y^2},$$

which does not exist. Check, for example, that the limit as $(x,y) \to (0,0)$ along the line $y = mx$ depends on m (and on whether x approaches zero from the right or left!).

Since the limit is not zero, the function is not differentiable at the origin even though both partial derivatives exist there.

Finally, this makes intuitive sense. If $f(x,y)$ is differentiable at a point, it should have a nice tangent plane there. A quick look at Figure 3.2.5 will convince you there isn't a nice tangent plane to $f(x,y)$ at the origin. ▲

The definition of differentiability also justifies approximating small patches of surface with parallelograms, as we see in the next example.

Example 3.5.6. *Parallelograms approximate surface area*

In Example 3.4.6 we approximated the spherical patch

$$\mathbf{S}(\theta,\phi) = (\sin\phi\cos\theta, \sin\phi\sin\theta, \cos\phi), \ \pi/2 \leq \phi \leq 2\pi/3, \ 0 \leq \theta \leq \pi/6.$$

using the parallelogram determined by the vectors $\frac{\pi}{6}\mathbf{S}_\phi(0,\pi/2)$, $\frac{\pi}{6}\mathbf{S}_\theta(0,\pi/2)$ (see Figure 3.4.5). The scalars $\pi/6$ were determined by the changes in parameters $\Delta\phi$, $\Delta\theta$ for the surface $\mathbf{S}$. Use Definition 3.5.3 to justify this choice.

Before embarking on a notational nightmare, we outline the argument: Differentiability implies a certain limit is zero, which means the numerator must approach the zero vector faster than the denominator goes to zero. The numerator is the difference between two vectors, one between points on the surface and the other a particular scalar multiple of a tangent vector. Since the limit must be zero, *that particular scalar multiple* of the tangent vector must approximate the surface pretty well. That's the basic idea...the notation is not for the faint of heart, but here it is.

Accepting the fact that the parametric surface S is differentiable at $(0, \pi/2)$, the limit in Definition 3.5.3 gives

$$\lim_{(\Delta\theta,\Delta\phi)\to(0,0)} \frac{S(0+\Delta\theta,\pi/2+\Delta\phi) - S(0,\pi/2) - DS(0,\pi/2)\cdot\begin{bmatrix}\Delta\theta\\\Delta\phi\end{bmatrix}}{\sqrt{\Delta\theta^2+\Delta\phi^2}} = \langle 0,0\rangle.$$

Now think of the numerator as the difference of the two vectors

$$S(0+\Delta\theta,\pi/2+\Delta\phi) - S(0,\pi/2), \quad\text{and}\quad DS(0,\pi/2)\cdot\begin{bmatrix}\Delta\theta\\\Delta\phi\end{bmatrix}.$$

The limit being zero says that the difference of these two vectors is really close to the zero vector, or that

$$S(0+\Delta\theta,\pi/2+\Delta\phi) - S(0,\pi/2) \approx DS(0,\pi/2)\cdot\begin{bmatrix}\Delta\theta\\\Delta\phi\end{bmatrix}.$$

In particular, when $\Delta\phi = 0$ we see

$$S(0+\Delta\theta,\pi/2) - S(0,\pi/2) \approx DS(0,\pi/2)\cdot\begin{bmatrix}\Delta\theta\\0\end{bmatrix}. \tag{3.5.1}$$

Now recall that DS is the matrix whose columns are S_θ and S_ϕ, and multiplying a matrix by the vector $\begin{bmatrix}\Delta\theta\\0\end{bmatrix}$ gives $\Delta\theta$ times the first column, so the right-hand side of Equation 3.5.1 becomes

$$DS(0,\pi/2)\cdot\begin{bmatrix}\Delta\theta\\0\end{bmatrix} = S_\theta(0,\pi/2)\Delta\theta. \tag{3.5.2}$$

Combining Equations 3.5.1 and 3.5.2 gives the approximation

$$S(0+\Delta\theta,\pi/2) - S(0,\pi/2) \approx S_\theta(0,\pi/2)\Delta\theta,$$

and this is the big deal! This says that the vector $S_\theta(0,\pi/2)\Delta\theta$ is close to the vector from $S(0,\pi/2)$ to $S(0+\Delta\theta,\pi/2)$.

Go back through the notation, keeping in mind the outline of the argument, and determine which equations correspond to which steps in the argument. It may help to refer to Figure 3.4.5 in the process. (Note that the vector $S(0+\Delta\theta,\pi/2) - S(0,\pi/2)$ is not pictured there, where should it be?)

The same argument with $\Delta\theta = 0$ shows that

$$S(0,\pi/2 + \Delta\phi) - S(0,\pi/2) \approx S_\phi(0,\pi/2)\Delta\phi.$$

Combining these approximations justifies using a parallelogram to approximate the surface S. Of course, the smaller $\Delta\theta$ and $\Delta\phi$ are, the better the parallelogram approximates the surface. This observation is very helpful when calculating surface area and defining integrals of functions over surfaces. ▲

The definition of differentiability is one important use of derivative matrices in the multivariable setting. It turns out that derivative matrices also provide the perfect format for describing the multivariable chain rule. The derivative matrix of a composition of functions is the product of their corresponding derivative matrices!

Motivating the Chain Rule: Before stating the chain rule, we consider a familiar example from single-variable calculus that motivates the formula.

Example 3.5.7. *Related rates*

You thought you were done thinking about related rates problems after your first calculus course...unfortunately, they are back! In this context they provide a nice motivation for the multivariable chain rule. We consider the following example:

A conveyor belt pours sand into a conical pile so that its volume increases at a rate of $10\text{ft}^3/\text{min}$ and its radius increases at a rate of $1\text{ft}/\text{min}$. Determine how fast the height of the pile is increasing when $r = h = 50\text{ft}$.

To solve this problem, one differentiates the relationship $V = \frac{\pi}{3}r^2 h$ implicitly with respect to t, and substitutes the appropriate values. Differentiating gives

$$\frac{dV}{dt} = \frac{\pi}{3}2rh\frac{dr}{dt} + \frac{\pi}{3}r^2\frac{dh}{dt}, \qquad (3.5.3)$$

and substituting the given constants allows one to solve the problem. Our purpose here, however, is not to solve the problem, but to put it in a multivariable context and motivate the chain rule.

First note that volume is a function of two variables r and h. These, in turn, are functions of time t. Composing them with the volume formula $V = \frac{\pi}{3}r^2 h$ makes V a function of time. Since V is a composition and since it involves several variables, differentiating it requires a multivariable chain rule! Let's look more closely at Equation 3.5.3 to see what that chain rule is.

The function $V = \frac{\pi}{3}r^2 h$, as a function of two variables, has the gradient

$$\nabla V(r,h) = \left\langle \frac{2\pi}{3}2rh, \frac{\pi}{3}r^2 \right\rangle.$$

Further, since r and h are functions of t, we can consider the vector-valued function $\mathbf{x}(t) = \langle r(t), h(t) \rangle$. Differentiating as with parametric curves, we have

$$\mathbf{x}'(t) = \left\langle \frac{dr}{dt}, \frac{dh}{dt} \right\rangle.$$

The components of both $\nabla V(r,h)$ and $\mathbf{x}'(t)$ appear in certain parts of Equation 3.5.3. In fact, careful inspection shows that to get dV/dt you just take the dot product of $\nabla V(r,h)$ and $\mathbf{x}'(t)$! Indeed, we calculate

$$\nabla V(r,h) \cdot \mathbf{x}'(t) = \left\langle \frac{\pi}{3} 2rh, \frac{\pi}{3} r^2 \right\rangle \cdot \left\langle \frac{dr}{dt}, \frac{dh}{dt} \right\rangle$$

$$= \frac{\pi}{3} 2rh \frac{dr}{dt} + \frac{\pi}{3} r^2 \frac{dh}{dt} = \frac{dV}{dt}. \;\blacktriangle \qquad (3.5.4)$$

Thus the derivative of the composition of $V(r,h)$ and $\mathbf{x}(t)$ is the dot product of their derivatives!

That's the big deal, and that's what you want to remember. Now I'll ramble on about notation a bit. It is important to be precise, and the following describes Equation 3.5.4 while taking a little more care with notation. When we calculate the volume it is at a given time, using the values of r and h at that time—in other words, using $\mathbf{x}(t)$. Thus the volume we are interested in is

$$V(\mathbf{x}(t)) = V \circ \mathbf{x}(t),$$

and our function indeed looks like a composition now! Using composition notation, every occurrence of the ordered pair (r,h) should be replaced by the function $\mathbf{x}(t)$. Thus the gradient $\nabla V(r,h)$ of Equation 3.5.4 is the same as $\nabla V(\mathbf{x}(t))$. Eliminating the middle terms in Equation 3.5.4, and using composition notation, we have

$$\frac{d}{dt} V(\mathbf{x}(t)) = \nabla V(\mathbf{x}(t)) \cdot \mathbf{x}'(t).$$

Chain Rule on a Curve: The above example generalizes, giving a chain rule for functions evaluated on parametric curves (or, equivalently, vector-valued functions). We study the chain rule on curves here, and generalize it to arbitrary compositions in the next paragraph.

Theorem 3.5.1. *Let $f(x,y,z)$ be a function of several variables with continuous partial derivatives and $\mathbf{C}(t) = (x(t), y(t), z(t))$ a differentiable parametric curve. Then $f \circ \mathbf{C}(t)$ is a function of one variable and the derivative of $f(\mathbf{C}(t))$ is given by*

$$\frac{d}{dt} f(\mathbf{C}(t)) = \nabla f(\mathbf{C}(t)) \cdot \mathbf{C}'(t).$$

Proof. To simplify notation we prove the case when f is a function of two variables. Let $\mathbf{C}(t) = (x(t), y(t))$. Then $f \circ \mathbf{C}(t) = f(x(t), y(t))$, and by the definition of the derivative

$$\frac{d}{dt}f(\mathbf{C}(t)) = \lim_{h\to 0} \frac{f(x(t+h),y(t+h)) - f(x(t),y(t))}{h}$$

$$= \lim_{h\to 0} \frac{f(x(t+h),y(t+h)) - f(x(t),y(t+h))}{h} \qquad (3.5.5)$$

$$+ \lim_{h\to 0} \frac{f(x(t),y(t+h)) - f(x(t),y(t))}{h}.$$

The second equality follows from adding $-f\big(x(t),y(t+h)\big) + f\big(x(t),y(t+h)\big)$ to the numerator and separating the fraction to a sum of two limits. We will show that the second term in this sum is $f_y(\mathbf{C}(t)) \cdot y'(t)$. The analysis is similar to show the first term is $f_x(\mathbf{C}(t)) \cdot x'(t)$. The second term of Equation 3.5.5 can be multiplied by $\big(y(t+h) - y(t)\big) / \big(y(t+h) - y(t)\big)$ and rearranged to give

$$\lim_{h\to 0} \frac{f(x(t),y(t+h)) - f(x(t),y(t))}{y(t+h) - y(t)} \frac{y(t+h) - y(t)}{h}. \qquad (3.5.6)$$

The first factor of Expression 3.5.6 holds the x-coordinate constant, so by the mean value theorem in single-variable calculus, there is a c between $y(t)$ and $y(t+h)$ such that

$$\frac{f(x(t),y(t+h)) - f(x(t),y(t))}{y(t+h) - y(t)} = \frac{\partial f}{\partial y}(x(t),c).$$

As $h \to 0$ we see that $y(t+h) \to y(t)$, which implies that $c \to y(t)$ as well. Since we assume f_y is continuous, the limit of the first factor in Expression 3.5.6 is $\frac{\partial f}{\partial y}(x(t),y(t)) = f_y(\mathbf{C}(t))$. The second factor is the definition of the derivative $y'(t)$ of the component function $y(t)$. Therefore the second term in the sum of Equation 3.5.5 is $f_y(\mathbf{C}(t)) \cdot y'(t)$. A similar analysis on the first term shows that it is $f_x(\mathbf{C}(t)) \cdot x'(t)$, showing

$$\frac{d}{dt}f(\mathbf{C}(t)) = f_x(\mathbf{C}(t)) \cdot x'(t) + f_y(\mathbf{C}(t)) \cdot y'(t) = \nabla f(\mathbf{C}(t)) \cdot \mathbf{C}'(t). \ \square$$

Remark: The formula suggested by Example 3.5.7 has now been shown to be true in general. We make two observations before continuing.

1. We can interpret the derivative in terms of the projections and lengths of the vectors involved. Precisely, we can manipulate the result of Theorem 3.5.1 algebraically as follows:

$$\frac{d}{dt}f(\mathbf{C}(t)) = \nabla f(\mathbf{C}(t)) \cdot \mathbf{C}'(t) = \left(\nabla f(\mathbf{C}(t)) \cdot \frac{\mathbf{C}'(t)}{\|\mathbf{C}'(t)\|}\right) \|\mathbf{C}'(t)\|$$

$$= \big(comp_{\mathbf{C}'(t)} \nabla f(\mathbf{C}(t))\big) \|\mathbf{C}'(t)\|$$

$$= \big(comp_{\mathbf{C}'} \nabla f\big) \|\mathbf{C}'(t)\|. \qquad (3.5.7)$$

Some of the notation is suppressed in the last line to lend better insight into the formula. Let's describe it further. According to Equation 3.5.7, the derivative $\frac{d}{dt}f(\mathbf{C}(t))$ is the product of two factors. The second factor, $\|\mathbf{C}'(t)\|$, is the speed of the curve and depends specifically on the parameterization. The first factor, $\text{comp}_{\mathbf{C}'}\nabla f$, depends only on the direction of the curve $\mathbf{C}'$ and the function f. In Section 3.6 we will see that $\text{comp}_{\mathbf{C}'}\nabla f$ is the rate of change of f in the $\mathbf{C}'$ direction.

2. One more remark is in order before moving on to examples. In single-variable calculus, differentiable functions achieved extreme values only when the derivative was zero. Since $f(\mathbf{C}(t))$ is a single-variable function, its extreme values can occur when its derivative vanishes. By Theorem 3.5.1, this requires the dot product $\nabla f \cdot \mathbf{C}'$ to be zero. Thus extreme values of $f(\mathbf{C}(t))$ can occur only when ∇f is perpendicular to $\mathbf{C}'$.

Example 3.5.8. *Verifying the chain rule*

Verify the chain rule for the function $f(x,y) = xe^{2y-3x}$ and the curve $\mathbf{C}(t) = (t, 3-2t)$.

Of course we have a *proof* that the chain rule is true, so in some sense there is no need to verify it! However, for someone seeing the notation for the first time the verification process can help with understanding. To verify

$$\frac{d}{dt}f(\mathbf{C}(t)) = \nabla f(\mathbf{C}) \cdot \mathbf{C}'(t),$$

we simply calculate both sides separately, step back, and say "Hey, they're equal." To calculate $\frac{d}{dt}f(\mathbf{C}(t))$ we first form the composite function $f(\mathbf{C}(t))$ then take its derivative as in Single-variable calculus. In our case we see

$$f(\mathbf{C}(t)) = f(t, 3-2t) = te^{2(3-2t)-3t} = te^{6-7t},$$

so the derivative is

$$\frac{d}{dt}f(\mathbf{C}(t)) = e^{6-7t} - 7te^{6-7t}. \tag{3.5.8}$$

On the other hand we compute $\nabla f(\mathbf{C}) \cdot \mathbf{C}'(t)$ by finding $\nabla f(x,y)$ and evaluating it on $\mathbf{C}(t)$, then dotting the result with the tangent vector $\mathbf{C}'(t)$. In our case

$$\nabla f(x,y) = \left\langle e^{2y-3x} - 3xe^{2y-3x}, 2xe^{2y-3x} \right\rangle,$$

so the needed dot product is

$$\nabla f(\mathbf{C}) \cdot \mathbf{C}'(t) = \left\langle e^{6-7t} - 3te^{6-7t}, 2te^{6-7t} \right\rangle \cdot \langle 1, -2 \rangle = e^{6-7t} - 7te^{6-7t}. \tag{3.5.9}$$

Now we step back and say "Hey, Equations 3.5.8 and 3.5.9 are equal!" So we have verified the chain rule in this case. ▲

Example 3.5.9. *A minimization problem*

Find the point on the parametric curve $\mathbf{C}(t) = (3 - 2t, 4 + t, -3t)$, $-\infty < t < \infty$ closest to the origin.

We could use single-variable calculus techniques, but we want to cast the problem in multivariable language. To minimize distance one frequently minimizes the distance squared, avoiding the square root in calculations. The square of the distance from the point (x, y, z) to the origin is given by $f(x, y, z) = x^2 + y^2 + z^2$. The squared distance from the point $\mathbf{C}(t)$ on the curve to the origin is given by $f(\mathbf{C}(t))$, so this is the function we want to minimize. A minimum will occur when the derivative is zero, which is when the gradient $\nabla f(\mathbf{C}(t))$ is perpendicular to the tangent vector $\mathbf{C}'(t)$. Thus to solve the problem, we need to solve

$$\nabla f(\mathbf{C}(t)) \cdot \mathbf{C}'(t) = 0.$$

Now that we've cast the problem in chain rule language, we commence with the calculations. We see that $\nabla f(x, y, z) = (2x, 2y, 2z)$ and $\mathbf{C}'(t) = \langle -2, 1, -3 \rangle$, so we need to solve

$$\nabla f(\mathbf{C}(t)) \cdot \mathbf{C}'(t) = \langle 6 - 4t, 8 + 2t, -6t \rangle \cdot \langle -2, 1, -3 \rangle = -4 + 28t = 0.$$

The solution is $t = 1/7$, so the point on the line $\mathbf{C}(t)$ that is closest to the origin is $\mathbf{C}(1/7) = (19/7, 29/7, -3/7)$. In general, one has to verify that the result is a minimum, but the context of this problem makes that clear. ▲

Example 3.5.10. *Extreme temperatures*

The temperature at the point (x, y) is given by

$$T(x, y) = 10 - \frac{5}{1 + x^2 - 2xy + y^2}.$$

Find the maximum and minimum temperatures experienced by an ant crawling on the circle $\mathbf{C}(t) = (2 \cos t, 2 \sin t)$, $0 \le t \le 2\pi$.

In this problem we want to optimize the function $T(\mathbf{C}(t))$, so we'll take its derivative using the chain rule. The necessary ingredients are

$$\nabla T(x, y) = \left(\frac{10(x - y)}{\left(1 + x^2 - 2xy + y^2\right)^2}, \frac{-10(x - y)}{\left(1 + x^2 - 2xy + y^2\right)^2} \right)$$

$$\nabla T(\mathbf{C}(t)) = \frac{20}{(5 - 8 \sin t \cos t)^2} \langle \cos t - \sin t, \sin t - \cos t \rangle$$

$$\mathbf{C}'(t) = \langle -2 \sin t, 2 \cos t \rangle.$$

We now solve $\nabla T\left(\mathbf{C}(t)\right) \cdot \mathbf{C}'(t) = 0$:

$$\nabla T(\mathbf{C}(t)) \cdot \mathbf{C}'(t) = \frac{40}{(5 - 8\sin t \cos t)^2}\left(\sin^2 t - \cos^2 t\right)$$

$$= \frac{-40\cos 2t}{5 - 4\sin 2t} = 0.$$

The denominator is never zero, so the derivative is always defined. It equals zero when $\cos 2t = 0$, which is when $t = \pi/4,\ 3\pi/4,\ 5\pi/4,\ 7\pi/4$. Thus the possible extreme temperatures occur at the points $(\sqrt{2}, \sqrt{2}), (-\sqrt{2}, \sqrt{2}), (\sqrt{2}, -\sqrt{2})$, and $(-\sqrt{2}, -\sqrt{2})$. Evaluating T at these points gives

$$T(\sqrt{2}, \sqrt{2}) = T(-\sqrt{2}, -\sqrt{2}) = 5$$

as a minimum temperature and

$$T(-\sqrt{2}, \sqrt{2}) = T(\sqrt{2}, -\sqrt{2}) = 85/9$$

as a maximum. ▲

The General Chain Rule: Theorem 3.5.1 generalizes to vector-valued functions of several variables. More precisely, Theorem 3.5.1 describes how to differentiate $f \circ g$ when $g : \mathbb{R} \to \mathbb{R}^m$ and $f : \mathbb{R}^m \to \mathbb{R}$. We'd like to generalize it to when the domain of g and range of f are arbitrary dimensions. This will be done in two steps. First, partial derivatives quickly allow us to generalize to the case $g : \mathbb{R}^n \to \mathbb{R}^m$ and $f : \mathbb{R}^m \to \mathbb{R}$. Finally, matrix multiplication will provide the most general chain rule.

Example 3.5.11. *Partial differentiation and the chain rule*

Let $\mathbf{S}(s,t) = (s\cos t, s\sin t, s^2)$ and $f(x,y,z) = x^2 - yz$. Then $f(\mathbf{S}(s,t))$ is a function of two variables. Find f_s and f_t.

One straightforward approach is to evaluate $f(\mathbf{S}(s,t))$, simplify, and take partial derivatives. Carrying this out we find

$$f(\mathbf{S}(s,t)) = (s\cos t)^2 - (s\sin t)s^2 = s^2\cos^2 t - s^3\sin t.$$

Taking partial derivatives we get

$$f_s = 2s\cos^2 t - 3s^2\sin t, \text{ and } f_t = -2s^2\cos t\sin t - s^3\cos t.$$

A second approach uses Theorem 3.5.1, which only applies when the first function is a vector-valued function of one variable. However, when computing the partial derivative f_s we consider t constant. This means we are *thinking of $f(\mathbf{S}(s,t))$ as a single-variable function*, and can apply the theorem! The only difference is that instead of dotting ∇f

with $C'(t)$, we'll be dotting it with the partial derivative vector $S_s(s,t)$. With this reasoning we see

$$f_s = \nabla f\left(S(s,t)\right) \cdot S_s(s,t) = \nabla f \cdot S_s.$$

For completeness, we'll compute the partial derivatives of $f\left(S(s,t)\right)$ using this method. First we get the necessary ingredients:

$$\nabla f(x,y,z) = \langle 2x, -z, -y \rangle$$
$$\nabla f\left(S(s,t)\right) = \nabla f(s\cos t, s\sin t, s^2) = \langle 2s\cos t, -s^2, -s\sin t \rangle$$
$$S_s(s,t) = \langle \cos t, \sin t, 2s \rangle$$
$$S_t(s,t) = \langle -s\sin t, s\cos t, 0 \rangle.$$

Taking appropriate dot products gives

$$f_s = \langle 2s\cos t, -s^2, -s\sin t \rangle \cdot \langle \cos t, \sin t, 2s \rangle = 2s\cos^2 t - s^2 \sin t - 2s^2 \sin t$$
$$= 2s\cos^2 t - 3s^2 \sin t$$
$$f_t = \langle 2s\cos t, -s^2, -s\sin t \rangle \cdot \langle -s\sin t, s\cos t, 0 \rangle = -2s^2 \cos t\sin t - s^3 \cos t,$$

which agrees with our previous calculations. ▲

The reasoning in the previous example generalizes to give a chain rule for $f \circ g$ when $g : \mathbb{R}^n \to \mathbb{R}^m$ and $f : \mathbb{R}^m \to \mathbb{R}$. In this case $f \circ g : \mathbb{R}^n \to \mathbb{R}$, and partial derivatives can be taken. Since partial derivatives think of other variables as constants Theorem 3.5.1 applies, giving

$$\frac{\partial (f \circ g)}{\partial x_j}(x) = \nabla f(g(x)) \cdot \left(\frac{\partial g_1}{\partial x_j}(x), \ldots, \frac{\partial g_m}{\partial x_j}(x) \right) = \nabla f \cdot g_{x_j}. \qquad (3.5.10)$$

To get a better feel for this, consider the two-variable case. If $f(x,y)$ is a function of intermediate variables x and y, which are in turn functions of independent variables u and v, then f is a function of u and v, and Equation 3.5.10 tells us

$$f_u = f_x x_u + f_y y_u$$
$$\frac{\partial f}{\partial v} = \frac{\partial f}{\partial x}\frac{\partial x}{\partial v} + \frac{\partial f}{\partial y}\frac{\partial y}{\partial v}.$$

Here we used Leibnitz notation for the second equation and subscript notation in the first, and you should be comfortable with either.

Example 3.5.12. *Quick differentiation*

Suppose $f(x,y) = x^2 - 3xy$, $x = 2u - 3v$ and $y = u^2 v$. Use the chain rule to find the partial derivatives of f with respect to u and v.

By the above remarks,

$$f_u = f_x x_u + f_y y_u$$
$$= 2x \cdot 2 - 3y \cdot 2uv = 4(2u - 3v) - 3(u^2 v)2uv$$
$$= 8u - 12v - 6u^3 v^2.$$

and

$$f_v = f_x x_v + f_y y_v$$
$$= 2x \cdot (-3) - 3y \cdot u^2 = -6(2u - 3v) - 3(u^2 v)u^2$$
$$= -12u + 18v - 3u^4 v.$$

Note that the partials of f are evaluated at the appropriate u- and v-values. ▲

The previous example illustrates that composition of functions is equivalent to multivariable substitution. On the one hand, we can think of substituting $x = 2u - 3v$ and $y = u^2 v$ into $f(x, y) = x^2 - 3xy$. This can also be thought of as the composition $f(g(u, v))$ of $f(x, y)$ with the function $g(u, v) = (2u - 3v, u^2 v)$. You should be comfortable with both notations, and know how to apply the chain rule given either.

At this point we have a chain rule for composing any function $g : \mathbb{R}^n \to \mathbb{R}^m$ with $f : \mathbb{R}^m \to \mathbb{R}$, and are poised to prove the general chain rule. The general chain rule states that the derivative matrix of a composition is the product of the individual derivative matrices. This is analogous to the single-variable setting where the derivative of a composition is the product of the derivatives. It is remarkable that matrix multiplication exactly describes how multivariable derivatives behave under composition.

Theorem 3.5.2. *Let $f : \mathbb{R}^m \to \mathbb{R}^p$ and $g : \mathbb{R}^n \to \mathbb{R}^m$ be differentiable functions and let $\mathbf{x} \in \mathbb{R}^n$. Then $f \circ g : \mathbb{R}^n \to \mathbb{R}^p$ and*

$$D(f \circ g)(\mathbf{x}) = Df(g(\mathbf{x})) \cdot Dg(\mathbf{x}).$$

Proof. The composition $f \circ g$ has p component functions, each of which is a function of n variables. Let $(f \circ g)_i$ denote the ith component function, and x_j the jth variable. By definition, the entry in the ith row and j^{th} column of $D(f \circ g)(\mathbf{x})$ is $\frac{\partial (f \circ g)_i}{\partial x_j}$. Since this is partial differentiation, we fix all coordinates other than x_j, and can think of g as a function of one variable. By Equation 3.5.10, we have

$$\frac{\partial (f \circ g)_i}{\partial x_j} = \nabla f_i(g(\mathbf{x})) \cdot \left\langle \frac{\partial g_1}{\partial x_j}(\mathbf{x}), \dots, \frac{\partial g_m}{\partial x_j}(\mathbf{x}) \right\rangle. \tag{3.5.11}$$

The vector $\nabla f_i(g(\mathbf{x}))$ is the ith row of $Df(g(\mathbf{x}))$, while the jth column of $Dg(\mathbf{x})$ is the vector $\left\langle \frac{\partial g_1}{\partial x_j}(\mathbf{x}), \dots, \frac{\partial g_m}{\partial x_j}(\mathbf{x}) \right\rangle$. Thus the ij entry of $D(f \circ g)$ is the dot product of the ith row of Df with the jth column of Dg. By definition of matrix multiplication, that's just the ij^{th} entry of the product of the matrices! Thus we have

$$D(f \circ g)(\mathbf{x}) = Df(g(\mathbf{x})) \cdot Dg(\mathbf{x}). \;\square$$

We illustrate the theorem with an example.

Example 3.5.13. *General chain rule*

Let $f : \mathbb{R}^2 \to \mathbb{R}^3$ be given by $f(x,y) = (x^2, 5x^2 + y^3, y)$, and $g : \mathbb{R}^2 \to \mathbb{R}^2$ by $g(s,t) = (st, s^2 + 2t)$. We can compute the derivative matrices of both f and g, to find

$$Df = \begin{bmatrix} 2x & 0 \\ 10x & 3y^2 \\ 0 & 1 \end{bmatrix} \qquad Dg = \begin{bmatrix} t & s \\ 2s & 2 \end{bmatrix}$$

Evaluating Df at $g(s,t)$, and multiplying matrices we have:

$$Df(g(s,t)) \cdot Dg(s,t) = \begin{bmatrix} 2(st) & 0 \\ 10(st) & 3(s^2 + 2t)^2 \\ 0 & 1 \end{bmatrix} \begin{bmatrix} t & s \\ 2s & 2 \end{bmatrix}$$

$$= \begin{bmatrix} 2st^2 & 2s^2 t \\ 10st^2 + 6(s^2 + 2t)^2 s & 10s^2 t + 6(s^2 + 2t)^2 \\ 2s & 2 \end{bmatrix}.$$

Alternatively, we could find explicit formulas for the composition $f \circ g$. We have $f \circ g : \mathbb{R}^2 \to \mathbb{R}^3$ is given analytically by $f \circ g(s,t) = f(st, s^2 + 2t) = ((st)^2, 5(st)^2 + (s^2 + 2t)^3, s^2 + 2t)$. The ij^{th} entry is the partial derivative of the ith component function with respect to the jth variable. For example, the entry $(Df \cdot Dg)_{22}$ is the partial derivative $\frac{\partial (f \circ g)_2}{\partial x_2}$, which is

$$\frac{\partial (f \circ g)_2}{\partial t}(s,t) = 10(st) \cdot s + 6(s^2 + 2t)^2.$$

Given enough time and energy, we can verify that the derivative matrix $D(f \circ g)(s,t)$ is the same as $Df(g(s,t)) \cdot Dg(s,t)$ in this case. The upshot is:

The derivative of a composition is the product of the derivatives! ▲

One geometric interpretation of the derivative matrix is that it maps tangent vectors of a curve to tangent vectors of its image. This is a consequence of the chain rule, as the following example illustrates.

Example 3.5.14. *Geometric interpretation of the derivative matrix*

Let $\mathbf{C}(t) = (\cos t, \sin t)$ be the unit circle, and let $f(x,y) = (x - 3y, x + 3y)$ be the map of the plane to itself studied in the first section of this chapter. Then the curve $f(\mathbf{C}(t))$ is

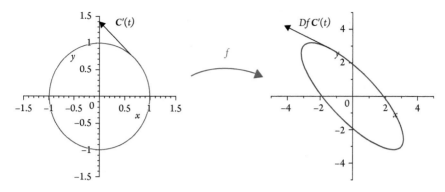

Figure 3.5.3 The geometry of the derivative matrix

an ellipse in the plane, given by $f(\mathbf{C}(t)) = (\cos t - 3\sin t, \cos t + 3\sin t)$ as in Figure 3.1.2. We know that $\frac{d}{dt}f(\mathbf{C}(t))$ is tangent to the ellipse, and by the chain rule:

$$\frac{d}{dt}f(\mathbf{C}(t)) = Df(\mathbf{C}(t)) \cdot \mathbf{C}'(t) = \begin{bmatrix} 1 & -3 \\ 1 & 3 \end{bmatrix}\begin{bmatrix} -\sin t \\ \cos t \end{bmatrix} = \begin{bmatrix} -\sin t - 3\cos t \\ -\sin t + 3\cos t \end{bmatrix}.$$

For example, if $t = \pi/4$, then $\mathbf{C}'(t) = \left(-\frac{\sqrt{2}}{2}, \frac{\sqrt{2}}{2}\right)$ is tangent to the unit circle at $\left(\frac{\sqrt{2}}{2}, \frac{\sqrt{2}}{2}\right)$. To find the tangent to its image under f, multiply it by Df, giving

$$\frac{d}{dt}f(\mathbf{C}(\pi/4)) = Df(\mathbf{C}(\pi/4)) \cdot \mathbf{C}'(\pi/4) = \begin{bmatrix} 1 & -3 \\ 1 & 3 \end{bmatrix}\begin{bmatrix} -\frac{\sqrt{2}}{2} \\ \frac{\sqrt{2}}{2} \end{bmatrix} = \begin{bmatrix} -2\sqrt{2} \\ \sqrt{2} \end{bmatrix}.$$

Thus the vector $\left(-2\sqrt{2}, \sqrt{2}\right)$ is tangent to the ellipse at the point $f(\mathbf{C}(\pi/4)) = (-\sqrt{2}, 2\sqrt{2})$. This is illustrated in Figure 3.5.3. ▲

Exercises

1. Find the gradients of the following functions of several variables.
   (a) $f(x,y) = x^3y - 2\sin(xy)$
   (b) $g(x,y) = 2x - \sin^{-1}(xy)$
   (c) $f(x,y,z) = \frac{x^2+2zy}{y\cos x}$
   (d) $h(x,y,z) = e^{y^2z}\sec xy$

2. Let $f(x,y,z) = x^3y^2z^4$. Compare $\nabla f(\mathbf{x}) \cdot \mathbf{x}$ with f. Describe why your comparison occurs.

3. Let $f(x,y,z) = x^2y + 3xyz - 2yz^2$.

    (a) Show that $f(\lambda\mathbf{x}) = \lambda^3 f(\mathbf{x})$ for all $\lambda \in \mathbb{R}$ (we say that f is homogeneous of degree 3).

    (b) Compare $\nabla f(\mathbf{x}) \cdot \mathbf{x}$ with $f(\mathbf{x})$.

4. The function $f(x,y,z)$ is called homogeneous of degree n if it satisfies $f(\lambda x, \lambda y, \lambda z) = \lambda^n f(x,y,z)$ for all $\lambda \in \mathbb{R}$. If f is homogeneous of degree n, then differentiate both sides of $f(\lambda x, \lambda y, \lambda z) = \lambda^n f(x,y,z)$ with respect to λ to relate $\nabla f(x,y,z) \cdot (x,y,z)$ to $f(x,y,z)$.

5. Find the derivative matrices of the following vector-valued functions of several variables.

    (a) $\mathbf{x}(t) = (t\cos t, t\sin t)$ (is there one row or one column?)

    (b) $f(x,y) = (2x+4, x-2y+7, y-3)$

    (c) $f(x,y,z) = (xz^2, x+zy^2)$

6. Find the derivative matrices for the change-of-coordinate functions, then find their determinants!

    (a) $f(r,\theta) = (r\cos\theta, r\sin\theta)$

    (b) $f(r,\theta,z) = (r\cos\theta, r\sin\theta, z)$

    (c) $f(\rho,\theta,\phi) = (\rho\sin\phi\cos\theta, \rho\sin\phi\sin\theta, \rho\cos\phi)$

7. Verify the chain rule as in Example 3.5.8 for the function $f(x,y) = x^2 + y^2$ and the curve $\mathbf{C}(t) = (3-2t, 5t)$, $-\infty < t < \infty$.

8. Verify the chain rule as in Example 3.5.8 for the function $f(x,y) = xe^y$ and the curve $\mathbf{C}(t) = (2\sin t, 3\cos t)$, $0 \le t \le 2\pi$.

9. Verify the chain rule as in Example 3.5.8 for the function $f(x,y,z) = xy^2 - z^3y$ and the curve $\mathbf{C}(t) = (3t, t^3, 2t)$.

10. Let $f(x,y) = x^2 - 3xy$ and $x = 2u - 3v, y = u + 2v$. Use the chain rule to find f_u and f_v.

11. Let $f(x,y) = xy^2 - 2x^3$ and $x = 2uv, y = 9u^2v^2$. Use the chain rule to find f_u and f_v.

12. Let $f(x,y) = e^{x-y}$ and $x = 2u - 3v, y = 2u - 2v$. Use the chain rule to find f_u and f_v.

13. Let $f(x,y) = \sin(x^2 + y^2)$ and $x = u\cos(v), y = u\sin(v)$. Use the chain rule to find f_u and f_v.

14. Suppose $f(x,y) = xy$, $x = u^2v$, y is a function of u, v, and $f_u = 0$. Find an equation relating y and y_u.

15. Let $f(x,y) = x^2 + y^2$ denote the temperature at the point (x,y), and let $\mathbf{C}(t) = (t, t^2)$ be the path of an ant crawling on the plane. Find how fast the temperature is changing at time $t = 2$.

16. An ant crawls on the unit circle $\mathbf{C}(t) = (\cos t, \sin t)$, and $f(x,y) = \sqrt{3}x - y$ is the temperature at points in the plane. What is the highest temperature the ant encounters? What can you say about the vectors ∇f and $\mathbf{C}'(t)$ at this point?

17. Find all points on the helix $\mathbf{C}(t) = (\cos t, \sin t, t)$ where the gradient of $f(x,y,z) = xy + z$ is normal to the tangent vector $\mathbf{C}'(t)$. Does f have a global extreme value on $\mathbf{C}(t)$?

18. The stream function for a point source $\psi(r,\theta) = K\theta$, for a constant K. Verify that $\psi_x = \psi_r r_x + \psi_\theta \theta_x$.

19. The stream function for a point vortex is $\psi(r,\theta) = K \ln r$, for a constant K. Find ψ_y using the multivariable chain rule.

20. Let $f(x,y,z)$ be a function on $\mathbb{R}^3$, and $\mathbf{C}(t)$ a parametric curve. Describe why ∇f is orthogonal to $\mathbf{C}(t)$ at extreme values of f on $\mathbf{C}$.

21. Sketch the ellipse $\mathbf{C}(t) = (2 \cos t, \sin t)$, for $0 \le t \le 2\pi$. Include the tangent vector $\mathbf{C}'(\pi/4)$ in your sketch. Now consider the map $f(x,y) = (x, 2y)$. Verify the chain rule for $(f \circ \mathbf{C})(t)$ by calculating:
    (a) The matrix Df, and multiplying it by the tangent vector $\mathbf{C}'(\pi/4)$.
    (b) Parametric equations for $(f \circ \mathbf{C})(t)$, and finding the tangent vector $(f \circ \mathbf{C})(\pi/4)$ directly.
    (c) Sketch the curve in part (b), together with its tangent vector.

3.6 Chain Rule Applications

In this section we introduce two applications of Theorem 3.5.1. The first is the notion of directional derivatives, or the rate of change of a function in a given direction. Partial derivatives are rates of change in the positive x, y, and z directions, and directional derivatives allow us to calculate the rate of change in an arbitrary direction. Secondly, we show how gradients are normal to level sets. This allows us to calculate equations for tangent planes, and will be the basis for the method of Lagrange multipliers.

Directional Derivatives: Before defining the directional derivative, we mention some terminology. The *direction* of the vector $\mathbf{v}$ is a *unit* vector pointing in the same direction as $\mathbf{v}$. Thus the direction of $\mathbf{v}$ is just the normalized vector $\frac{\mathbf{v}}{\|\mathbf{v}\|}$. Recall the remark immediately following Theorem 3.5.1, where Equation 3.5.7 shows that the derivative of a function on a curve is

$$\frac{d}{dt}f\left(\mathbf{C}(t)\right) = comp_{\mathbf{C}'}\nabla f \left\|\mathbf{C}'\right\|.$$

Note that any two curves with the same tangent line have the same $comp_{\mathbf{C}'}\nabla f$, so this quantity depends only on f and the direction of $\mathbf{C}'$. The component, then, is how fast f is changing in the $\frac{\mathbf{C}'}{\|\mathbf{C}'\|}$ direction. Since this rate is independent of the choice of parameterization, it is natural to eliminate the mention of the curve $\mathbf{C}$ when defining the rate of change of f in a given direction. We make this more formal.

Definition 3.6.1. *Let $\mathbf{u}$ be a unit vector, P a point, and f a function of several variables. The directional derivative $D_{\mathbf{u}}f$ of f at the point P in the $\mathbf{u}$ direction is*

$$D_{\mathbf{u}}f(P) = comp_{\mathbf{u}}\nabla f(P) = \nabla f(P)\cdot\mathbf{u}.$$

Remark: The directional derivative of $f(x,y,z)$ in the $\mathbf{i}$ direction is

$$D_{\mathbf{i}}f(x,y,z) = \nabla f(x,y,z)\cdot\mathbf{i} = \langle f_x, f_y, f_z\rangle\cdot\langle 1,0,0\rangle = f_x.$$

Similarly, we have $D_{\mathbf{j}}f = f_y$ and $D_{\mathbf{k}}f = f_z$. In other words, the directional derivatives of f in the coordinate directions are just its partial derivatives. Thus directional derivatives generalize partial differentiation, allowing us to calculate rates of change of f in any direction, not just the coordinate directions.

Math App 3.6.1. *Directional derivatives*

The following Math App provides a nice conceptual interpretation of the directional derivative, prior to more analytic investigations.

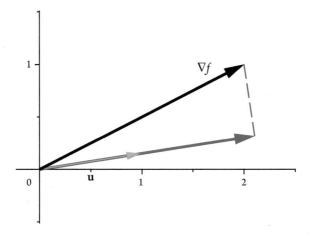

One geometric interpretation of the directional derivative follows from the chain rule, and justifies the definition. Let **u** be any unit vector and P any point. Then the line through P in the direction of **u** is parameterized by $\mathbf{C}(t) = \mathbf{P} + t\mathbf{u}$, and $\mathbf{C}'(t) = \mathbf{u}$. If we evaluate f on the curve $\mathbf{C}(t)$, we get a function of one variable, and can find its rate of change at P by taking the derivative and setting $t = 0$. The chain rule states that

$$\frac{d}{dt}f(\mathbf{C}(0)) = \nabla f(\mathbf{C}(0)) \cdot \mathbf{C}'(t) = \nabla f(P) \cdot \mathbf{u}.$$

Thus the directional derivative is the rate of change of f along a unit speed curve in the **u** direction.

Example 3.6.1. *Calculating directional derivatives*

We begin with some sample calculations.

(a) Calculate the directional derivative of $f(x,y) = x^3y - 2x^2y^3$ at the point $(1,2)$ in the direction of the vector $\langle 3,4 \rangle$.

First we get the ingredients necessary to calculate the directional derivative: the gradient and direction. Normalizing the vector, we see that the direction is $\mathbf{u} = \left(\frac{3}{5}, \frac{4}{5}\right)$. The gradient of f is $\nabla f(x,y) = \left(3x^2y - 4xy^3, x^3 - 6x^2y^2\right)$, and evaluating at $(1,2)$ we have $\nabla f(1,2) = \langle -26, -23 \rangle$. By the definition of the directional derivative we have

$$D_{\mathbf{u}}f(1,2) = \nabla f(1,2) \cdot \mathbf{u} = \langle -26, -23 \rangle \cdot \left(\frac{3}{5}, \frac{4}{5}\right) = -\frac{170}{5} = -34.$$

(b) Find the directional derivative of $f(x,y,z) = xy^3z^2$ at $(3,-1,2)$ in the direction of $\mathbf{u} = \frac{1}{\sqrt{3}}\langle 1, 1, -1 \rangle$.

Since the gradient is $\nabla f(x,y,z) = \langle y^3 z^2, 3xy^2 z^2, 2xy^3 z \rangle$, we have

$$D_{\mathbf{u}} f(3,-1,2) = \langle -4, 36, -12 \rangle \cdot \left(\frac{1}{\sqrt{3}}, \frac{1}{\sqrt{3}}, -\frac{1}{\sqrt{3}} \right) = \frac{44}{\sqrt{3}}. \ \blacktriangle$$

Example 3.6.2. *Directions of zero change*

The temperature on a plate is given by $T(x,y) = x^2 + y^2 - 2x + 4y + 5$. Which direction should an ant at the origin walk so the temperature remains constant?

In this example, the function T and the point $(0,0)$ are fixed, and we're looking for the direction $\mathbf{u}$ that keeps constant temperature. Thus the directional derivative should vanish. The gradient of T is $\nabla T = \langle 2x - 2, 2y + 4 \rangle$, and the goal is to solve

$$D_{\mathbf{u}} T(0,0) = \langle -2, 4 \rangle \cdot \mathbf{u} = 0$$

for $\mathbf{u}$. Switching coordinates and negating one of them gives $\mathbf{v} = \langle 4, 2 \rangle$ is normal to $\nabla T(0,0)$, and normalizing $\mathbf{v}$ gives $\mathbf{u} = \frac{\pm 1}{\sqrt{5}} \langle 2, 1 \rangle$. $\blacktriangle$

Example 3.6.2 is an instance where one specifies a rate of change (zero in that case), and finds a direction in which f changes at that rate. If the ant were cold, it might be interested in the direction for which T is increasing the fastest. The next fact shows that the gradient determines directions and rates of maximal change.

Theorem 3.6.1. *Let f be a differentiable function of several variables with gradient vector ∇f. Then*

1. *The vector ∇f points in the direction of maximal increase of f, and the maximal rate of increase is $\|\nabla f\|$.*

2. *The vector $-\nabla f$ points in the direction of maximal decrease of f. The maximal rate of decrease is $\|\nabla f\|$.*

Proof. For any direction (i.e. unit vector) $\mathbf{u}$ and point P, we have

$$D_{\mathbf{u}} f(P) = \nabla f(P) \cdot \mathbf{u} = \|\nabla f(P)\| \, \|\mathbf{u}\| \cos\theta = \|\nabla f(P)\| \cos\theta.$$

This function is maximized when $\theta = 0$, or when $\nabla f(P)$ points in the same direction as $\mathbf{u}$. Moreover, in that direction $D_{\mathbf{u}} f(P) = \|\nabla f(P)\|$. This proves the first statement of the theorem. The second follows similarly, with $\theta = \pi$. $\square$

Example 3.6.3. *Rates of maximal increase*

In a certain coordinate system, the altitude on a mountain above the point (x,y) is given by $f(x,y) = xy - 2x + y + 2$ (where units are 100 ft above sea level). Which direction should a person standing above $(0,0)$ walk to climb the fastest?

First of all, please don't say "up". Secondly, the direction the person should walk is really a direction in the plane. Typically one might say "north-east", or use compass directions. These really are directions in the plane, or the domain of f. Thus it makes sense that the answer will be a direction (i.e. unit vector) in the plane.

By the previous theorem $\nabla f(x,y) = \langle y-2, x+1 \rangle$ evaluated at $(0,0)$ points in the direction of maximal increase. Normalizing, we have that the direction of maximal increase is

$$\frac{\nabla f(0,0)}{\|\nabla f(0,0)\|} = \frac{1}{\sqrt{5}} \langle -2, 1 \rangle,$$

and the maximal rate of increase is $\|\nabla f(0,0)\| = \sqrt{5}$. This means the hiker climbs $\sqrt{5}$ feet for each horizontal foot traveled in the ∇f direction. ▲

Gradients and Level Sets: We come now to our second application of the chain rule, that the gradient vector is always orthogonal to a level set. Recall that the level surfaces of $f(x,y,z)$ at level c is the solution set to $f(x,y,z) = c$. As an example, recall that the level surfaces of $f(x,y,z) = x^2 + y^2 - z^2$ at levels $c = 1, 0, -1$ are the one-sheeted hyperboloid, the cone, and the two-sheeted hyperboloid, respectively (see Figure 3.6.1).

Theorem 3.6.2. *Let f be a differentiable function of several variables. Then ∇f is orthogonal to the level set $f = c$.*

Proof. We focus on the case where $f(x,y,z)$ is a function of three variables, although the proof works in general. We have to show that $\nabla f(x,y,z)$ is perpendicular to tangent vectors to the level surface $f(x,y,z) = c$. These tangent vectors are of the form $\mathbf{C}'(t)$ for curves $\mathbf{C}(t)$ on the surface.

Let $\mathbf{C}(t)$ be a curve on the surface $f(x,y,z) = c$, so that f evaluated on $\mathbf{C}(t)$ is $f(\mathbf{C}(t)) = c$, a constant function. Since $f(\mathbf{C}(t))$ is constant, its derivative vanishes. By the chain rule we have

$$\frac{d}{dt} f(\mathbf{C}(t)) = \nabla f(\mathbf{C}(t)) \cdot \mathbf{C}'(t) = 0.$$

Thus the vectors $\nabla f(\mathbf{C}(t))$ and $\mathbf{C}'(t)$ are orthogonal, and ∇f is perpendicular to tangent vectors to the surface $f(x,y,z) = c$. □

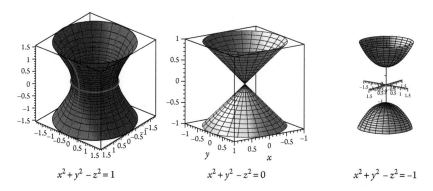

$x^2 + y^2 - z^2 = 1$ $x^2 + y^2 - z^2 = 0$ $x^2 + y^2 - z^2 = -1$

Figure 3.6.1 Level surfaces

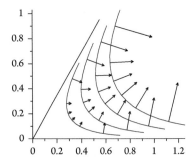

Figure 3.6.2 Streamlines orthogonal to $\nabla\psi$

This theorem is nicely illustrated by a fluid flow example. Recall that streamlines are level curves of the stream function ψ, so Theorem 3.6.2 implies that $\nabla\psi$ is orthogonal to the streamlines of a flow.

Example 3.6.4. *A $\pi/3$-corner flow*

In Example 1.2.11 it was shown that the stream function for an ideal flow around a $\pi/3$ corner is

$$\psi = 3x^2y - y^3.$$

The gradient is $\nabla\psi = \langle 6xy, 3x^2 - 3y^2 \rangle$. Figure 3.6.2 depicts several streamlines together with gradient vectors. The vector field $\nabla\psi$ is everywhere orthogonal to the streamlines. ▲

Math App 3.6.2. *Gradients and level surfaces*

The following Math App plots level surfaces together with their gradient fields as an illustration of Theorem 3.6.2.

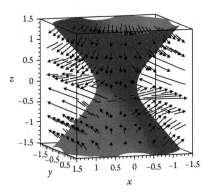

Theorem 3.6.2 allows us to find equations for tangent planes to level surfaces, which we will illustrate in Example 3.6.6 below. Coupled with the techniques of Section 3.4 we

now know how to find tangent planes to surfaces given parametrically, by functions of two variables, and as solution sets to equations in three variables.

Example 3.6.5. *Tangent line to level curve*

Find an equation for the tangent line to the ellipse $5x^2 - 6xy + 5y^2 = 4$ at the point $(-0.5, -1.1)$.

The trick to solving this problem is thinking of the ellipse as a level curve of the function $f(x,y) = 5x^2 - 6xy + 5y^2$ at level $c = 4$. By Theorem 3.6.2 we know $\nabla f(-0.5, -1.1)$ is perpendicular to the level curve $5x^2 - 6xy + 5y^2 = 4$, and therefore perpendicular to the tangent line there. In our case

$$\nabla f(x,y) = \langle 10x - 6y, -6x + 10y \rangle,$$

so the normal vector is $\nabla f(-.5, -1.1) = \langle 1.6, -8 \rangle$.

We can use the normal vector $\langle 1.6, -8 \rangle$ to find a Cartesian equation for the tangent line in the same way we found Cartesian equations for planes in $\mathbb{R}^3$. Let (x,y) be any point on the line, then the vector $\mathbf{v} = \langle x,y \rangle - \langle -.5, -1.1 \rangle$ is perpendicular to $\nabla f(-.5, -1.1)$ (see Figure 3.6.3). Thus the dot product

$$\begin{aligned}
\nabla f(-.5, -1.1) \cdot \mathbf{v} &= \langle 1.6, -8 \rangle \cdot \left(\langle x,y \rangle - \langle -.5, -1.1 \rangle \right) \\
&= \langle 1.6, -8 \rangle \cdot \langle x,y \rangle - \langle 1.6, -8 \rangle \cdot \langle -.5, -1.1 \rangle \\
&= 1.6x - 8y - 8,
\end{aligned}$$

must be zero! Therefore an equation for the tangent line is $1.6x - 8y = 8$. If you don't like decimals, multiplying by $5/8$ gives $x - 5y = 5$. ▲

Example 3.6.6. *Tangent planes to level surfaces*

Find a Cartesian equation for the tangent plane to $x^2 + y^2 - z^2 = 1$ at the point $(1,1,1)$.

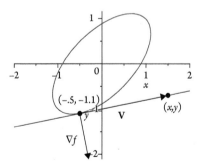

Figure 3.6.3 Gradient is orthogonal to ellipse

We need a point, which is given, and a normal direction. Thinking of the surface as a level surface of the function $f(x,y,z) = x^2 + y^2 - z^2$, the previous theorem tells us that $\nabla f(1,1,1)$ is normal to the surface. Thus $\nabla f(1,1,1) = \langle 2,2,-2 \rangle$ is normal to the tangent plane. To find an equation, we use the scalar multiple $\langle 1,1,-1 \rangle$, and get

$$x + y - z = \langle 1,1,-1 \rangle \cdot \langle 1,1,1 \rangle = 1. \ \blacktriangle$$

Tangent Planes to Level Surfaces

An equation for the tangent plane to the level surface $f(x,y,z) = c$ at the point P on it can be found as follows:

Step 1: Find $\nabla f(P)$.

Step 2: P is on the plane and $\nabla f(P)$ is normal to it, so an equation is

$$\nabla f(P) \cdot \langle x,y,z \rangle = \nabla f(P) \cdot \mathbf{P}.$$

Example 3.6.7. *Tangent to an ellipsoid*

Find an equation for the tangent plane to the ellipsoid

$$9x^2 + y^2 + 4z^2 = 36$$

at the point $(1, -\sqrt{11}, -2)$ (see Figure 3.6.4(*a*)).

Thinking of the ellipsoid as a level surface of $f(x,y,z) = 9x^2 + y^2 + 4z^2$, we find the gradient to be $\nabla f(x,y,z) = \langle 18x, 2y, 8z \rangle$. Thus a normal to the plane is $\nabla f(1, -\sqrt{11}, -2) = \langle 18, -2\sqrt{11}, -16 \rangle$, and an equation for it is

$$18x - 2\sqrt{11}y - 16z = \langle 18, -2\sqrt{11}, -16 \rangle \cdot \langle 1, -\sqrt{11}, -2 \rangle = 72. \ \blacktriangle$$

Example 3.6.8. *Curves of intersection*

Find an equation for the tangent line to the curve of intersection of the surface $z = xy$ and the unit sphere at the point $(1,0,0)$ (see Figure 3.6.4(*b*)).

For an equation of a line we need a point, which is given, and a vector in the line. The key is to notice that the tangent line to the curve of intersection is in the tangent planes to each surface, and hence orthogonal to both normal vectors. Think of the sphere as a level surface of the function $f(x,y,z) = x^2 + y^2 + z^2$, and the other surface as a level surface of $g(x,y,z) = xy - z$. A vector in the line, then, is the cross product $\nabla f(1,0,0) \times \nabla g(1,0,0)$. Calculating we have

$$\nabla f(1,0,0) \times \nabla g(1,0,0) = \langle 2,0,0 \rangle \times \langle 0,1,-1 \rangle = \langle 0,2,2 \rangle.$$

Parametric equations for the line are $\mathbf{C}(t) = \langle 1,0,0 \rangle + t \langle 0,2,2 \rangle = (1,2t,2t). \ \blacktriangle$

(a)

(b)

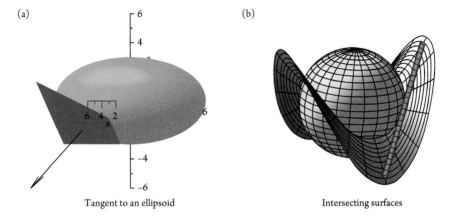

Tangent to an ellipsoid

Intersecting surfaces

Figure 3.6.4 Geometric gradient applications

Exercises

1. Find the directional derivatives of the following functions at the given point in the directions of the given vectors
   (a) $f(x,y) = x^3 - 4xy^2 + 3y$, $P(1,2)$, and $\mathbf{v} = \langle 2,3 \rangle$
   (b) $f(x,y) = ye^x$, $P(\ln 4, 3)$, and $\mathbf{v} = \langle 2,-1 \rangle$
   (c) $f(x,y,z) = x^2y + yz^2 - 2x$, $P(-1,2,1)$, and $\mathbf{v} = \langle 2,-1,1 \rangle$

2. A hiker is standing above $(0,0)$ on the mountain whose altitude is given by $f(x,y) = 8 - x^2 - y^2 - 4x + 2y$ (in thousands of feet). Which direction should she walk to get down the fastest? How fast will she be descending?

3. The temperature near a heat source is $T(x,y,z) = \frac{5}{1+9x^2+4y^2+z^2}$. What direction should a particle at $(2,1,1)$ travel to heat up the fastest? To cool down as quickly as possible?

4. The temperature at the point (x,y) on a plate is given by $T(x,y) = x^2 + 4xy + y^2$. What direction should an ant at $(-2,1)$ crawl in order to warm up the fastest? Is there a direction that the ant could crawl to stay the same temperature?

5. Find an equation for the tangent line to the ellipse $10x^2 - 16xy + 10y^2 = 4$ at the point $(1,1)$.

6. Let $f(x,y) = y - \sin x$.
   (a) Sketch the level curves of f at levels $c = -3,0,5$.
   (b) Include $\nabla f(\pi,0)$ in your sketch.
   (c) Use the chain rule to describe why ∇f is always perpendicular to level curves $f(x,y) = c$.

7. Find the points on $x^2 + 4y^2 + 9z^2 = 1$ where the tangent plane is perpendicular to the line $\mathbf{C}(t) = (3 - t, 4 + t, 4 - 6t)$.

8. Find a Cartesian equation for the tangent plane to $2xy + yz + 5xz = 8$ at the point $(1,1,1)$.

9. Find a Cartesian equation for the tangent plane to the two-sheeted hyperboloid $x^2 + y^2 - z^2 = -1$ at the point $(0, 1, -\sqrt{2})$.

10. Find an equation for the tangent plane to the surface $x^2 + 9y^2 - z^2 = 1$ at the point $(1, 2, 6)$.

11. Find where the tangent plane to the surface $x^2 + 4(y + z)^2 + 9(y - z)^2 = 1$ is horizontal.

12. Find an equation for the tangent line to the curve of intersection of the surfaces $x^2 = 2yz$ and the unit sphere at the point $(\sqrt{2}/2, 1/2, 1/2)$.

13. Show that the tangent vector to $\mathbf{C}(t) = (t \cos t, t \sin t, t)$ is always orthogonal to the normal vector to $x^2 + y^2 = z^2$.

14. The stream function for a uniform flow is $\psi(x, y) = 3x - 4y$.
    (a) Sketch some streamlines for the flow (i.e. level curves of ψ).
    (b) Find $\nabla \psi(x, y)$.
    (c) Argue that $\nabla \psi$ is indeed orthogonal to level curves.

15. The stream function for a point source $\psi(r, \theta) = K\theta$, for a constant K.
    (a) Sketch some streamlines for the flow (i.e. level curves of ψ).
    (b) Find $\nabla \psi(x, y)$ (recall that $\theta = \tan^{-1}(y/x)$, or use the chain rule).
    (c) Argue that $\nabla \psi$ is indeed orthogonal to level curves.

16. The stream function for a point vortex is $\psi(r, \theta) = K \ln r$, for a constant K.
    (a) Sketch some streamlines for the flow (i.e. level curves of ψ).
    (b) Find $\nabla \psi(x, y)$ (recall that $r = \sqrt{x^2 + y^2}$, or use the chain rule).
    (c) Argue that $\nabla \psi$ is indeed orthogonal to level curves.

17. A point charge Q placed at the origin has potential function $V = \frac{kQ}{\rho}$.
    (a) Find level surfaces for V.
    (b) Find $\nabla V = (V_x, V_y, V_z)$.
    (c) Describe why ∇V is orthogonal to level surfaces in this case.

3.7 The Hessian Test

This section and the next develop techniques for optimizing functions of several variables. These techniques apply to a variety of settings, and we now describe one setting which combines both. This may help put two-variable optimization in a familiar context.

One way to think of these optimization techniques is that they generalize optimizing a single-variable function on a closed interval. To find extreme values of $f(x)$ on the closed interval $[a,b]$ one finds all critical points of $f(x)$ in (a,b), then compares function values at the critical points *and* the endpoints. The largest value is the maximum and the smallest value a minimum for f on the interval $[a,b]$.

To generalize this to the multivariable setting, the analog of a closed interval must be discussed. These notions were introduced in Section 3.2, but are recalled here for convenience. The endpoints of the interval are considered its *boundary*, while the open interval (a,b) is the *interior* of $[a,b]$. Analogously, the closed unit disk $\overline{D}$ defined by $x^2 + y^2 \leq 1$ has interior $x^2 + y^2 < 1$ (the open unit disk), and the unit circle is its boundary ∂D. Thus the boundary of a region R is composed of curves, and its interior is the portion of the plane between the curves (see Figure 3.7.1). The boundary of R is denoted ∂R, which is not a derivative! If the region doesn't extend indefinitely; we say that it's bounded. Closed and bounded regions in the plane are an analog of closed intervals of real numbers.

The strategy for optimizing a continuous function $f(x,y)$ on the closed and bounded region R in the plane is analogous to the single-variable strategy outlined above. The steps are:

1. First locate all critical points interior to R,

2. Find the extreme values of f on ∂R,

3. The largest value f attains on all these points is its maximum, and its minimum is the smallest.

This section introduces the Hessian test, which includes techniques for locating and classifying critical points of $f(x,y)$ on the interior of R (Step 1 above). The next section discusses the method of Lagrange multipliers, which in two-dimensions is a technique for optimizing functions on curves (Step 2 above).

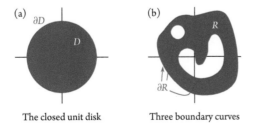

(a) ∂D D (b) R ∂R

The closed unit disk   Three boundary curves

Figure 3.7.1 Closed and bounded regions in the plane

While the Hessian test and the method of Lagrange multipliers can be applied in more general settings, the two sections combined provide an approach for optimizing $f(x,y)$ on closed and bounded planar regions.[1]

Classifying Critical Points: We now describe the Hessian test for classifying critical points of functions of several variables. It is analogous to the second derivative test for extreme values in single-variable calculus. Before considering the two-variable setting, we review the single-variable rule using Taylor polynomials.

Recall that the second degree Taylor polynomial

$$P_2(x) = f(c) + f'(c)(x-c) + \frac{f''(c)}{2!}(x-c)^2$$

at a point $x = c$ is the "best" quadratic approximation of $f(x)$ near $x = c$ (the notion of "best" used here is that $f(x)$ and $P_2(x)$ have the same derivatives at $x = c$ up to second order). The graph of $f(x) = x^5 - 15x^2$ is depicted, together with its second degree Taylor polynomials at $x = -3, 0, 3$, in Figure 3.7.3(b). A close-up of the function and its polynomial at $x = -3$ is in Figure 3.7.3(a), while Figure 3.7.3(c) illustrates the function and polynomial near $x = 0$. Note that near $x = 0$ the best quadratic approximation is actually a line (so $f''(0) = 0$ in P_2).

The idea behind the second derivative test is that for small intervals around $x = c$, the function $f(x)$ looks like its Taylor polynomial P_2. If we know what P_2 does near a

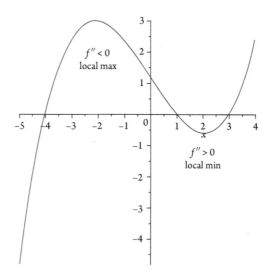

Figure 3.7.2 The second derivative test

[1] In these sections we refrain from formal proofs, choosing instead to provide intuitive arguments. The interested reader is referred to [3] for a careful discussion of these ideas.

(a) (b) (c)

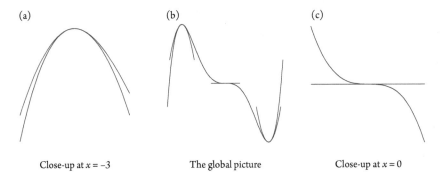

Close-up at $x = -3$ The global picture   Close-up at $x = 0$

Figure 3.7.3 Taylor polynomials at critical points

critical point, then we know what f does. Moreover, the parabola $P_2(x)$ points up if the coefficient of x^2 is positive and down if its negative. Since the coefficient of x^2 is $f''(c)/2$, when $f''(c) > 0$ the Taylor polynomial is a parabola with a minimum. On the other hand, if $f''(x) < 0$ then P_2 has a maximum value (as in Figure 3.7.3(a)). Using the fact that $P_2(x)$ and $f(x)$ behave the same near $x = c$ yields the second derivative test for single-variable calculus:

Suppose $x = c$ is a critical point of $f(x)$. If $f''(c) < 0$ then $f(x)$ has a local maximum at $x = c$, and $f(x)$ attains a local minimum there if $f''(c) > 0$.

Recall that the test is inconclusive if $f''(c) = 0$. This is because the critical point could be a maximum, a minimum, or neither (as in Figure 3.7.3(c)).

This test generalizes to functions of two variables. The generalization requires knowing about conic sections, multivariable Taylor polynomials, and critical points. A thorough discussion of these topics would take us too far afield, so we simply review the results here. The first order of business is defining critical points for functions of two variables.

Definition 3.7.1. *The point (a, b) is a critical point of the function $f(x, y)$ if both first-order partial derivatives of f are zero at (a, b).*

Thus to find critical points you must solve the system of equations

$$f_x(x, y) = 0; \qquad f_y(x, y) = 0,$$

and we illustrate with an example.

Example 3.7.1. *Find critical points*

Find the critical points of $f(x, y) = x^3 + 6xy^2 - 6x$.
This amounts to solving the system of equations

$$f_x(x, y) = 3x^2 + 6y^2 - 6 = 0$$
$$f_y(x, y) = 12xy = 0$$

The second equation implies that either $x = 0$ or $y = 0$, and we treat each case separately. If $x = 0$, the first equation becomes $6y^2 - 6 = 0$, so $y = \pm 1$. Thus $(0, 1)$ and $(0, -1)$ are critical points. If $y = 0$, then the first equation becomes $3x^2 - 6 = 0$, so $x = \pm\sqrt{2}$. Therefore the critical points of f are $(0, \pm 1)$ and $(\pm\sqrt{2}, 0)$. ▲

The multivariable second derivative test is analogous to the single-variable case in that it stems from using the second-order Taylor polynomial to approximate the function. In the single-variable case the leading term of P_2 was $\frac{f''(c)}{2}x^2$, so the second derivative dictated whether P_2 pointed up or down near a critical point. The leading terms of the two-variable Taylor polynomial at a critical point (a, b) are

$$\frac{1}{2}\left(f_{xx}(a,b)x^2 + 2f_{xy}(a,b)xy + f_{yy}(a,b)y^2\right). \tag{3.7.1}$$

If we understand surfaces of the form $P(x,y) = Ax^2 + Bxy + Cy^2$, we will understand $P_2(x,y)$, and therefore $f(x,y)$ near critical points. With absolutely no proof we state that the surface $P(x,y) = Ax^2 + Bxy + Cy^2$ can be qualitatively described by its discriminant $D = 4AC - B^2$. The proof of this result requires some knowledge of quadratic forms; however, a quick visit to Math App 1.2.2 provides experimental evidence for this. It wouldn't hurt to look at that App again before continuing.

Surfaces with positive discriminant are elliptical paraboloids as in Figure 3.7.4 parts (a) and (b). Surfaces with negative discriminant must be saddle points (see Figure 3.7.4(c)),

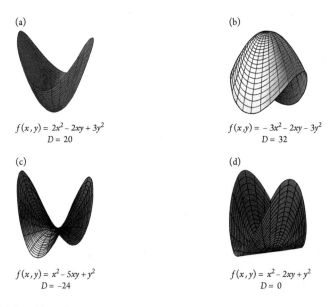

(a) $f(x,y) = 2x^2 - 2xy + 3y^2$, $D = 20$

(b) $f(x,y) = -3x^2 - 2xy - 3y^2$, $D = 32$

(c) $f(x,y) = x^2 - 5xy + y^2$, $D = -24$

(d) $f(x,y) = x^2 - 2xy + y^2$, $D = 0$

Figure 3.7.4 Quadric surfaces

while those with zero discriminant have a whole line of critical points and look like a hard taco shell (see Figure 3.7.4(d)).

The leading terms of P_2 given in Expression 3.7.1 allow us to calculate its discriminant:

$$D = 4AC - B^2 = 4\frac{f_{xx}\,f_{yy}}{2\;2} - f_{xy}^2 = f_{xx}\,f_{yy} - f_{xy}^2.$$

Since P_2 approximates f, the above criterion for P_2 yields a second derivative test for $f(x,y)$:

Near a critical point (a,b) of $f(x,y)$, the surface $z = f(x,y)$

1. Has an extreme value if $f_{xx}\,f_{yy} - f_{xy}^2 > 0$,

2. Is a saddle if $f_{xx}\,f_{yy} - f_{xy}^2 < 0$,

3. Is mysterious if $f_{xx}\,f_{yy} - f_{xy}^2 = 0$.

The discriminant D above can be conveniently defined as the determinant of the Hessian matrix for the function $f(x,y)$. The Hessian matrix for $f(x,y)$ is the matrix of second-order partial derivatives, and it makes remembering the discriminant easier.

Definition 3.7.2. *The Hessian matrix of $f(x,y)$ is*

$$H(f) = \begin{pmatrix} \frac{\partial^2 f}{\partial x^2} & \frac{\partial^2 f}{\partial x \partial y} \\ \frac{\partial^2 f}{\partial y \partial x} & \frac{\partial^2 f}{\partial y^2} \end{pmatrix}$$

The discriminant D is the determinant of the Hessian matrix:

$$D = \begin{vmatrix} \frac{\partial^2 f}{\partial x^2} & \frac{\partial^2 f}{\partial x \partial y} \\ \frac{\partial^2 f}{\partial y \partial x} & \frac{\partial^2 f}{\partial y^2} \end{vmatrix} = f_{xx}\,f_{yy} - f_{xy}^2.$$

Example 3.7.2. *Calculating Hessian matrices*

Find the Hessian matrix $f(x,y) == x^3 + 6xy^2 - 6x$, and its discriminant.

The first partial derivatives were taken in the previous example. Differentiating those we find:

$$H(f)(x,y) = \begin{pmatrix} 6x & 12y \\ 12y & 12x \end{pmatrix}, \text{ and } D = \begin{vmatrix} 6x & 12y \\ 12y & 12x \end{vmatrix} = 72x^2 - 144y^2. \; \blacktriangle$$

We now have introduced everything we need to state the Hessian test for classifying critical points of $f(x,y)$.

The Hessian Test

Let (a,b) be a critical point of $f(x,y)$, and D be the discriminant of f evaluated at (a,b). There are four cases to consider:

1. If $D > 0$ and $f_{xx}(a,b) > 0$, then (a,b) is a local minimum for f.
2. If $D > 0$ and $f_{xx}(a,b) < 0$, then (a,b) is a local maximum for f.
3. If $D < 0$, then f has a saddle point at (a,b).
4. Otherwise, the test is inconclusive.

This is, of course, more than we need to solve the problem stated at the beginning of this section. All that was required to optimize $f(x,y)$ on a closed and bounded region R was to *find* the critical points of f in R. This amounts to finding when the partial derivatives f_x and f_y vanish simultaneously. The Hessian test does more, as it determines the type of critical point under consideration. Another added feature is that the Hessian test does not require the domain R to be closed and bounded. It could be the entire plane—which is certainly unbounded! In any case, some examples are in order.

Example 3.7.3. *Classifying critical points*

Classify the critical points of $f(x,y) = x^3 + 6xy^2 - 6x$. We've determined the critical points are $(0,\pm1)$ and $(\pm\sqrt{2},0)$. We substitute the critical point $(\sqrt{2},0)$ into the discriminant we find $D(\sqrt{2},0) = 72(\sqrt{2})^2 - 144(0)^2 = 144 > 0$. Since the discriminant is positive, we check $f_{xx}(\sqrt{2},0) = 6\sqrt{2} > 0$ and conclude that f has a local minimum at $(\sqrt{2},0)$. Similar analyses give the table below. See if you can find the respective critical points in Figure 3.7.5.

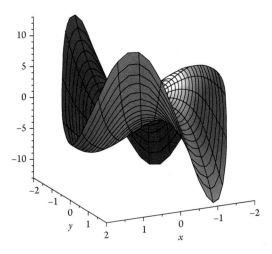

Figure 3.7.5 The surface $f(x,y) = x^3 + 6xy^2 - 6x$

Critical point	Discriminant	f_{xx}	Type
$(\sqrt{2},0)$	144	$6\sqrt{2}$	Minimum
$(-\sqrt{2},0)$	144	$-6\sqrt{2}$	Maximum
$(0,1)$	-144	NA	Saddle
$(0,-1)$	-144	NA	Saddle

Example 3.7.4. *An inconclusive test*

Classify the critical points of $f(x,y) = x^2 - 2xy + y^2$. To do so, we note that the solution set of the system

$$f_x(x,y) = 2x - 2y = 0$$
$$f_y(x,y) = -2x + 2y = 0$$

is the line $y = x$. Moreover, the Hessian and discriminant are

$$H(f)(x,y) = \begin{pmatrix} 2 & -2 \\ -2 & 2 \end{pmatrix}, \text{ and } D = \begin{vmatrix} 2 & -2 \\ -2 & 2 \end{vmatrix} = 0.$$

Since the discriminant is always zero, the Hessian test is inconclusive in this situation. In this case, it's because the critical points of f are not isolated: there is an entire line of critical points! The reason for this is easier to see if one factors $f(x,y) = x^2 - 2xy + y^2 = (x - y)^2$. Thus the graph is just a parabola translated back and forth on the line $y = x$ (see Figure 3.7.4(d)). The line of critical points is the line containing the vertices of the parabola. ▲

Example 3.7.5. *An exponential example*

Classify the critical points of $f(x,y) = xe^{x-y^2}$ as local maxima, minima, or saddle points.

To do this we find all second-order partial derivatives, finding

$$f_x = (1+x)e^{x-y^2} \qquad f_y = -2xye^{x-y^2}$$

$$f_{xx} = (2+x)e^{x-y^2} \qquad f_{xy} = -2y(1+x)e^{x-y^2} \qquad f_{yy} = -2x(1-2y^2)e^{x-y^2}$$

and the discriminant is

$$D = -2x(2+x)(1-2y^2)e^{2x-2y^2} - 4y^2(1+x)^2e^{2x-2y^2}.$$

Since exponential functions are never zero, $x = -1$ is the only value making $f_x = 0$, and f_y vanishes exactly when x or y does. Thus $(-1,0)$ is the only critical point of f. The discriminant is $D(-1,0) = 2e^{-2} > 0$, and $f_{xx}(-1,0) = e^{-1} > 0$, so the point $(-1,0)$ is a local minimum of the function. ▲

Example 3.7.6. *Minimizing distance*

Find the point on the parametric surface $S(s,t) = (t+s, t-s, t^2)$ closest to $(12,4,1)$.

To do this, we will minimize the square of the distance from $(12,4,1)$ to any point on the surface. We square the distance to make our calculations easier. The square of the distance is given by

$$f(s,t) = (12-t-s)^2 + (4-t+s)^2 + \left(1-t^2\right)^2.$$

The problem of finding the closest point on the surface to $(12,4,1)$ is reduced to minimizing $f(s,t)$, a function of two variables. Using the chain rule to take partial derivatives, we have the system of equations

$$f_s(s,t) = -2(12-t-s) + 2(4-t+s) = -16 + 4s = 0$$
$$f_t(s,t) = -2(12-t-s) - 2(4-t+s) - 4t(1-t^2) = -32 + 4t^3 = 0.$$

From these we deduce that the only critical point is $(s_0, t_0) = (4,2)$. We calculate the Hessian to verify that this is a minimum of the function. We find

$$H(f)(s,t) = \begin{pmatrix} 4 & 0 \\ 0 & 12t^2 \end{pmatrix}, \text{ and } D = \begin{vmatrix} 4 & 0 \\ 0 & 48 \end{vmatrix} = 192 > 0.$$

Since $f_s(s,t) = 4 > 0$, we know that $(4,2)$ is a minimum of $f(s,t)$, and the closest point on the surface to $(12,4,1)$ is $S(4,2) = (6,-2,4)$. ▲

The Hessian test works because $f(x,y)$ looks locally like its second-degree Taylor polynomial, and the discriminant classifies those. The only part of the test not yet discussed is the additional check of f_{xx} when $D > 0$. To gain some intuition for this, remember that taking partial derivatives of f is equivalent to intersecting the surface with planes, and considering the curves you get. If $D = f_{xx} f_{yy} - f_{xy}^2 > 0$, then f_{xx} and f_{yy} must have the same sign. This means the curves of intersection are either both concave up or both concave down, yielding a minimum or a maximum respectively (see Figure 3.7.6(a)).

Another situation that is easy to see intuitively is the case where f_{xx} and f_{yy} have opposite sign. In this setting it is clear that the discriminant will be negative, so (a,b) is a saddle point. This makes sense because if the second partials have opposite sign, then the surface is concave up in one direction and concave down in the other, yielding a saddle point (see Figure 3.7.6(b)).

Example 3.7.7. *Least squares regression*

One interesting application of the Hessian test is that it finds the formula for the least squares regression line for a collection of paired data. We illustrate this with an oversimplified example.

Find the least squares regression line for the data set $(3,7)$, $(5,4)$, and $(7,4)$.

Before getting started, let's review least squares regression. The goal is to find a line that "best" fits the data. The definition of "best" might vary, but we describe the sense in which the least squares regression is the best fit. First, let $y = mx + b$ be a line approximating

(a)

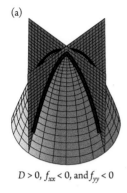

(b)

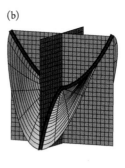

$D > 0$, $f_{xx} < 0$, and $f_{yy} < 0$ f_{xx} and f_{yy} have opposite sign

Figure 3.7.6 The geometry of the Hessian test

data. A *residual* is the vertical distance between the observed data (x_i, y_i) and the point on the line $(x_i, mx_i + b)$ (see Figure 3.7.7(a)). The least squares regression line is the choice of m and b that minimizes the sum of the squared residuals.

Let's make this very explicit in our example. Given any choice of slope and intercept, the sum of the squared residuals for our points is

$$f(m, b) = (7 - (3m + b))^2 + (4 - (5m + b))^2 + (4 - (7m + b))^2.$$

Our goal is to find the slope m and y-intercept b that minimizes $f(m, b)$. Thus finding the least squares regression line amounts to optimizing a function of two variables! We use the Hessian test to do this. Taking partial derivatives gives

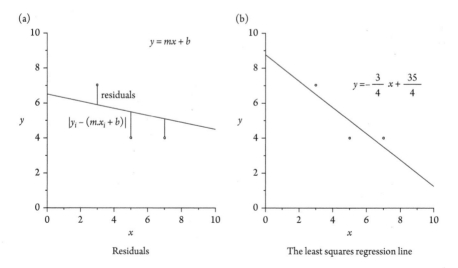

Residuals   The least squares regression line

Figure 3.7.7 Least squares regression

$f_m = -6(7 - 3m - b) - 10(4 - 5m - b) - 14(4 - 7m - b) = -138 + 166m + 30b$

$f_b = -2(7 - 3m - b) - 2(4 - 5m - b) - 2(4 - 7m - b) = -30 + 30m + 6b$

After some algebra, one finds that $(m_0, b_0) = \left(-\frac{3}{4}, \frac{35}{4}\right)$ is a critical point. In this case, the discriminant is

$$D = \begin{vmatrix} 166 & 30 \\ 30 & 6 \end{vmatrix} = 996 - 900 > 0,$$

and $f_{mm} > 0$, so the critical point is a minimum. Thus the least squares regression line for $\{(3,7), (5,4), (7,4)\}$ is $y = -\frac{3}{4}x + \frac{35}{4}$. ▲

Exercises

1. Classify the critical points of the following functions
   (a) $f(x,y) = x^2 - y^2 + 6x + 2y + 8$
   (b) $f(x,y) = x^3 + 4xy^2 - 12x + 12y$
   (c) $f(x,y) = (x^3 - x)e^{-y^2}$
   (d) $f(x,y) = \frac{x}{1+x^2+y^2}$

2. Classify the critical points of $f(x,y) = 2x^3 + xy^2 + 5x^2 + y^2$ as local maxima, minima, or saddle points.

3. Classify the critical points of $f(x,y) = x^3 - 3x^2y + 3y^2 + 9y$ as local maxima, minima, or saddle points.

4. Classify the critical points of $f(x,y) = (1 - y^2)e^{-x^2+4x}$ as local maxima, minima, or saddle points.

5. Classify the critical points of $f(x,y) = 2 - y^2 + e^{-x^2+4x}$ as local maxima, minima, or saddle points.

6. Classify the critical points of $f(x,y) = \dfrac{x - x^2}{1+y^2}$ as local maxima, minima, or saddle points.

7. Classify the critical points of $f(x,y) = \dfrac{x^3 - 3x}{1+y^2}$ as local maxima, minima, or saddle points.

8. Apply the Hessian test to the function $f(x,y) = x^2 + 4xy + 4y^2$. What do you find?

9. Let $f(x,y) = Ax^2 + Bxy + Cy^2$ and suppose $f_{xx} + f_{yy} = 0$ (assume A, B, C are non-zero constants).

(a) What can you say about the constants A, B, C?

(b) What can you say about any critical points of $f(x,y)$? Justify your answer.

10. What can you say about the constants A through F in $f(x,y) = Ax^2 + Bxy + Cy^2 + Dx + Ey + F$ when the Hessian test is inconclusive?

11. Without calculating, what is the result of applying the Hessian test to the monkey saddle $f(x,y) = x(x - y)(x + y)$ (see Figure 1.2.8)? Justify your answer.

12. Find the point on the parametric surface $\mathbf{S}(s,t) = (s, t, t - s)$ closest to $(2, 1, 0)$.

13. Find the point on the parametric surface $\mathbf{S}(s,t) = (s, t + s, t - s)$ closest to $(2, 1, 1)$.

14. Find the least squares regression line for $(5, 2)$, $(10, 12)$, and $(12, 10)$.

15. Derive the formula for the slope and intercept of the least squares regression line for the general data set $\{(x_1, y_1), \ldots, (x_n, y_n)\}$. (Hint: use summation notation and the fact that $n\bar{x} = \sum_{i=1}^{n} x_i$, where $\bar{x}$ denotes the mean of the x_i).

3.8 Constrained Optimization

There are many situations where you want to optimize a function subject to some constraint. Formally, these types of questions look like: maximize the function $f(x,y,z)$ subject to the constraint $g(x,y,z) = c$. The function $f(x,y,z)$ is called the *objective function*, and $g(x,y,z) = c$ the *constraint equation*, in this context.

We actually met this type of question in Example 1.2.7 of Section 1.2 for the two-variable case. We maximized the function $f(x,y) = x^2 - 2x + y^2 + 4y + 1$ subject to the constraint $x^2 + y^2 = 1$. In that instance we argued that $f(x,y)$ would achieve extreme values on the unit circle where the level curves were tangent to it. The reasoning went as follows: if a level curve crosses the unit circle, then nearby level curves will too. The nearby levels will be different: greater in one direction, smaller in the other (see Figure 3.8.1(a)). This means f can't attain an extreme value where level curves cross the constraint curve. Thus any extreme values occur where level curves are tangent to the constraint curve.

Combining this observation with what we know about gradients and level curves will lead to the method of Lagrange multipliers. Recall that ∇f is perpendicular to the level curves of f. Moreover, we can think of the constraint curve $x^2 + y^2 = 1$ as the level curve of $g(x,y) = x^2 + y^2$ at level 1. Thought of in this way, we see that ∇g will be perpendicular to the constraint curve. Of course, when we say a vector is orthogonal to a curve, we really mean it's orthogonal to the tangent line to the curve.

Now consider what happens when the level curve $f(x,y) = c$ is tangent to the constraint curve $g(x,y) = 1$. At this point, the tangent lines to both curves are the same. Therefore ∇f and ∇g are perpendicular to the same tangent line, and must be parallel (see Figure 3.8.1(b)). This means they are scalar multiples, and there must be a scalar λ such that $\nabla f = \lambda \nabla g$. We remark that when $\nabla f = \mathbf{0}$ the level curves of f can cross $g(x,y) = c$ while f-values increase in both directions unlike Figure 3.8.1(a) (see Example 3.8.2).

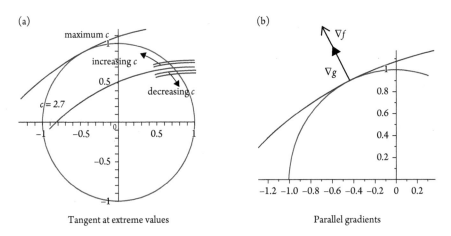

Tangent at extreme values   Parallel gradients

Figure 3.8.1 Level curves, gradients and extreme values

An alternate justification will be given later showing that there is a scalar λ satisfying $\nabla f = \lambda \nabla g$ at extreme points.

Math App 3.8.1. *Justifying* $\nabla f = \lambda \nabla g$

The following Math App allows you to experiment with the ideas just discussed.

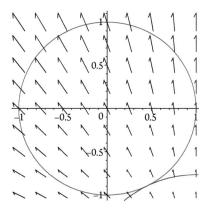

These observations lead to the method of Lagrange multipliers, which is used to optimize functions of several variables subject to a constraint. We outline the method below and follow with examples.

The Method of Lagrange Multipliers

To optimize the function $f(x,y,z)$ subject to the constraint $g(x,y,z) = c$, solve the system of equations

$$\nabla f(x,y,z) = \lambda \nabla g(x,y,z)$$
$$g(x,y,z) = c. \qquad (3.8.1)$$

Evaluate f at the solutions. The largest value is the maximum, the least is the minimum.

Notice that the first equation in the Lagrange system is a *vector* equation. Since two vectors are equal if and only if *all* their components are, the one vector equation $\nabla f(x,y,z) = \lambda \nabla g(x,y,z)$ represents *three* scalar equations, one for each coordinate. This method works since, as described above, at extreme values of f on the constraint $g(x,y,z) = c$ the gradients are parallel.

Before illustrating this procedure in several examples, we note that solving the system of Equations 3.8.1 can involve a fair amount of trickery and ingenuity. For example, we might factor a given equation, divide consecutive equations, or simply use substitution.

We mention that λ could be zero when ∇f is the zero vector, which happens at critical points of f.

Example 3.8.1. *A two-variable example*

Find the extreme values that the function $f(x,y) = x^2 + y^2 - 2x + 2y$ takes on the unit circle.

This can be stated in Lagrange format: Optimize $f(x,y) = x^2 + y^2 - 2x + 2y$ subject to the constraint $x^2 + y^2 = 1$. The Lagrange system of equations, in vector format, is

$$\langle 2x - 2, 2y + 2 \rangle = \lambda \langle 2x, 2y \rangle$$
$$x^2 + y^2 = 1$$

Separating the vector equation into a system of component equations gives:

$$2x - 2 = \lambda 2x$$
$$2y + 2 = \lambda 2y$$
$$x^2 + y^2 = 1$$

Rearranging the first equation gives $(1 - \lambda)x = 1$, which implies that $1 - \lambda \neq 0$. Similarly, the second component equation becomes $(1 - \lambda)y = -1$. Combining these, we get $(1 - \lambda)x = -(1 - \lambda)y$ and dividing by $1 - \lambda$ gives $x = -y$ (we can do this division since we know $1 - \lambda \neq 0$). Substituting $x = -y$ into the constraint equation $x^2 + y^2 = 1$ and solving gives $x = \pm\frac{\sqrt{2}}{2}$. Thus the points where f can take on extreme values are $(\frac{\sqrt{2}}{2}, -\frac{\sqrt{2}}{2})$ and $(-\frac{\sqrt{2}}{2}, \frac{\sqrt{2}}{2})$. To find the extreme values, we evaluate f at those points:

$$f\left(\frac{\sqrt{2}}{2}, -\frac{\sqrt{2}}{2}\right) = \frac{1}{2} + \frac{1}{2} - \sqrt{2} - \sqrt{2} = 1 - 2\sqrt{2}$$

$$f\left(-\frac{\sqrt{2}}{2}, \frac{\sqrt{2}}{2}\right) = \frac{1}{2} + \frac{1}{2} + \sqrt{2} + \sqrt{2} = 1 + 2\sqrt{2}$$

Thus the maximum value that f attains on the unit circle is $1 + 2\sqrt{2}$, and the minimum value is $1 - 2\sqrt{2}$. We can interpret this geometrically as follows. The graph of $f(x,y)$ is a paraboloid. Restricting the domain to the unit circle is equivalent to looking at the curve on the paraboloid above the unit circle. We found the highest and lowest point on that curve (see Figure 3.8.2). ▲

Example 3.8.2. *Extreme values when $\nabla f = 0$*

Find the extreme values that the function $f(x,y) = x^2 + y^2 - 2xy$ takes on the unit circle.

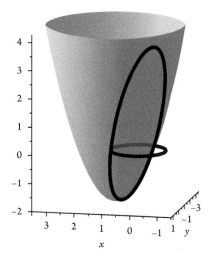

Figure 3.8.2 Geometric interpretation of constrained optimization

The Lagrange system of equations for this problem is

$$\langle 2x - 2y, 2y - 2x \rangle = \lambda \langle 2x, 2y \rangle$$
$$x^2 + y^2 = 1$$

Separating the vector equation into a system of component equations gives:

$$2x - 2y = \lambda 2x$$
$$2y - 2x = \lambda 2y$$
$$x^2 + y^2 = 1$$

Solving the first equation for y and the second for x gives the system

$$y = (1 - \lambda)x$$
$$x = (1 - \lambda)y$$
$$x^2 + y^2 = 1$$

On the unit circle x and y can't both be zero, so the above system implies $x \neq 0$, $y \neq 0$, and $\lambda \neq 1$ (convince yourself that any of these cases implies both x and y are zero—a contradiction). Dividing the first two equations gives

$$\frac{y}{x} = \frac{(1 - \lambda)x}{(1 - \lambda)y},$$

which reduces to $y^2 = x^2$. Combining this with the constraint $x^2 + y^2 = 1$ yields the four points $(\pm\frac{\sqrt{2}}{2}, \pm\frac{\sqrt{2}}{2})$. Direct calculation gives $f(\frac{\sqrt{2}}{2}, -\frac{\sqrt{2}}{2}) = f(-\frac{\sqrt{2}}{2}, \frac{\sqrt{2}}{2}) = 2$ and $f(-\frac{\sqrt{2}}{2}, -\frac{\sqrt{2}}{2}) = f(\frac{\sqrt{2}}{2}, \frac{\sqrt{2}}{2}) = 0$.

Thus f has a minimum value of 0 and maximum value of 2 on the unit circle. The reader can verify that ∇f vanishes at the points $\pm(\sqrt{2}/2, \sqrt{2}/2)$, so this is an instance where extreme values occur where $\nabla f = \mathbf{0}$.

Once again, let's interpret this geometrically. Factoring the function gives $f(x,y) = (x-y)^2$, which means that f is always at least 0 and above the unit disk is the taco shell pictured in Figure 3.8.3. The fold of the taco shell is along the line $y = x$, and is strange because it corresponds to an entire line of minima for f. As in Section 3.7, these are points where $\nabla f = \mathbf{0}$. The Lagrange vector equation $\nabla f = \lambda \nabla g$ is then satisfied when $\lambda = 0$. Thus critical points of f correspond to solutions to the Lagrange system corresponding to $\lambda = 0$. ▲

Example 3.8.3. *Local and global extremes*

Find the extreme values of $f(x,y) = x^3 - 5x + y^2$ subject to the constraint $x^2 + y^2 = 4$.

The system of equations to solve is:

$$3x^2 - 5 = 2\lambda x$$
$$2y = 2\lambda y$$
$$x^2 + y^2 = 4. \qquad (3.8.2)$$

The second equation implies either $y = 0$ or $\lambda = 1$, and the cases are treated separately.

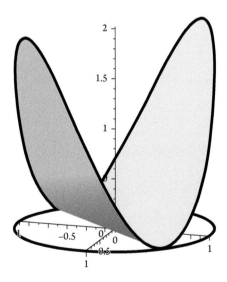

Figure 3.8.3 Extreme values when $\nabla f = 0$

If $y = 0$, the constraint equation yields $x = \pm 2$, so we get the points $(\pm 2, 0)$.

If $\lambda = 1$, then the first equation simplifies to $3x^2 - 2x - 5 = (3x - 5)(x + 1) = 0$. Thus x is -1 or $5/3$, and combining with the constraint equation yields the points $(-1, \pm\sqrt{3}), (5/3, \pm\sqrt{11/9})$.

Points satisfying the system of equations 3.8.2 are all the *potential* extreme points for f on $x^2 + y^2 = 4$. No other point on the circle can yield a max or min. To find which points give extreme values, just evaluate f at all six points. Doing so we find

$$f(2,0) = -2 \qquad f(-2,0) = 2$$
$$f(-1, \pm\sqrt{3}) = 7 \quad f(5/3, \pm\sqrt{11/9}) = -67/27.$$

Since $-67/24 < -2$, that is the minimum of f on the circle, and f attains a maximum value of 7 at points $(-1, \pm\sqrt{3})$. Notice that the points $(\pm 2, 0)$ yield only relative extreme values for f, which is why they are still solutions to System 3.8.2. Figure 3.8.4 is a nice illustration of the difference between local and global extreme values. Be sure to convince yourself from the figure that $f(\pm 2, 0)$ are local extreme values, but that over the entire circle the function achieves more extreme values than these. ▲

Example 3.8.4. *Maximizing area for fixed perimeter*

Find the triangle of perimeter two with maximal area.

To solve this problem, let's denote the lengths of the sides of the triangle by a, b, and c. Since we know information about the lengths of sides and want to maximize the area, we need to find a formula that gives the area of a triangle as a function of its side lengths. Toward this end, we recall Heron's formula. Let $s = \frac{a+b+c}{2}$ be the semiperimeter of the triangle, then Heron's formula says the area A of the triangle is $A = \sqrt{s(s-a)(s-b)(s-c)}$.

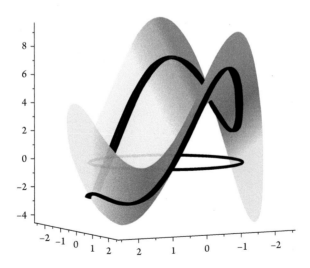

Figure 3.8.4 A function with both local and global extreme values

We will maximize the square of the area, since it gives the same triangle and simplifies our calculations considerably. Since the perimeter is two, in our case $s = 1$ and we have

$$f(a,b,c) = A^2 = (1-a)(1-b)(1-c) = 1 - a - b - c + ab + ac + bc - abc$$
$$= -1 + ab + ac + bc - abc,$$

where the last equality follows from the fact that $a + b + c = 2$. So we want to maximize f, subject to the constraint that the perimeter is two. The Lagrange system of equations, component-wise rather than in vector form, is

$$b + c - bc = \lambda$$
$$a + c - ac = \lambda$$
$$a + b - ab = \lambda$$
$$a + b + c = 2.$$

Equating the first two equations gives $b + c - bc = a + c - ac$, which simplifies to $(a-b)c = (a-b)$. This implies either $c = 1$ or $a - b = 0$. We consider each case separately. If $c = 1$, the system of equations becomes

$$b + 1 - b = \lambda$$
$$a + 1 - a = \lambda$$
$$a + b - ab = \lambda$$
$$a + b + 1 = 2.$$

The first implies that $\lambda = 1$ and the fourth that $a + b = 1$. Substituting these into the third equation gives $1 - ab = 1$, or $-ab = 0$. This, of course, implies that at least one of a or b is zero, which can't happen because they represent lengths of the sides of a triangle. We conclude that $c \neq 1$, which leaves that $a = b$ as the only possible case.

Identical reasoning, using the second and third equations yields $b = c$. Thus the triangle of perimeter two with maximum area is an equilateral triangle, with edge length $\frac{2}{3}$. ▲

Example 3.8.5. *Velocity potential along streamlines*

Show that for an ideal flow around a right-angled corner, the velocity potential never attains extreme values along streamlines.

Recall that the velocity potential and stream function are $\varphi(x,y) = x^2 - y^2$ and $\psi(x,y) = 2xy$, respectively. A streamline is a level curve $2xy = c$ of the stream function. Any extreme value of φ along $\psi = c$ occurs when $\nabla \varphi$ is parallel to $\nabla \psi$, by the reasoning that opened this section. Since Example 3.5.3 shows that the gradients are perpendicular everywhere, φ doesn't achieve any extreme values on $\psi = c$. ▲

We now give an alternate argument for the existence of a scalar satisfying $\nabla f = \lambda \nabla g$ at extreme points of f on $g = c$. This argument works for hypersurfaces in all dimensions, and makes use of directional derivatives. Recall that $D_{\mathbf{u}}f(P) = comp_{\mathbf{u}}\nabla f(P)$ is the rate of change of f in the $\mathbf{u}$ direction at P.

Let P be a point satisfying $g(P) = c$, so that P is on the constraint, and we are interested in whether or not $f(P)$ is an extreme value for f on the constraint. Suppose there is a direction $\mathbf{u}$ which is tangent to $g = c$ at P, and for which $D_{\mathbf{u}}f(P) \neq 0$. We will show that P cannot be an extreme value for f on $g = c$.

Since $\mathbf{u}$ is tangent to the constraint, there is a path C on $g = c$ with tangent vector at P equal to $\mathbf{u}$. Moreover, the rate of change of f along C is not zero because it is $D_{\mathbf{u}}f(P) \neq 0$. In fact, f-values will increase in one direction along C and decrease in the other. Indeed, if $D_{\mathbf{u}}f(P) > 0$, then f-values increase moving forward along $C(t)$ (this is the case depicted in Figure 3.8.5). Further, since $D_{-\mathbf{u}}f(P) = -D_{\mathbf{u}}f(P)$, f-values decrease moving backwards along C. Thus f achieves larger and smaller values on $g = c$ near the point P, and $f(P)$ cannot be an extreme point (if $D_{\mathbf{u}}f(P) < 0$ the same argument justifies the claim).

We have shown that if there is a tangent direction $\mathbf{u}$ to the constraint at P with $D_{\mathbf{u}}f(P) \neq 0$, then $f(P)$ is not an extreme value of f on $g = c$. Equivalently, if $f(P)$ is an extreme value then $D_{\mathbf{u}}f(P) = 0$ for all tangent vectors to $g = c$. This implies that $\nabla f(P)$ is orthogonal to $g = c$, or that it is the zero vector. In either case, a scalar λ exists satisfying $\nabla f = \lambda \nabla g$.

We proceed with some three-variable examples. The only difference between two and three variables is the level of difficulty in solving the system of equations.

Example 3.8.6. *A three-variable example*

Find the maximum and minimum values that $f(x, y, z) = 8x - 6y + 10z$ attain on the sphere $x^2 + y^2 + z^2 = 2$.

Setting up the Lagrange system of equations gives:

$$\langle 8, -6, 10 \rangle = \lambda \langle 2x, 2y, 2z \rangle$$
$$x^2 + y^2 + z^2 = 2$$

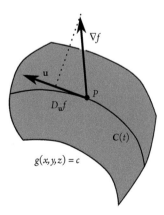

Figure 3.8.5 The f-values not extreme at P

Solving the component equations for x, y, and z gives $x = \frac{4}{\lambda}$, $y = -\frac{3}{\lambda}$, and $z = \frac{5}{\lambda}$. Substituting these into the constraint equations and solving yields

$$\left(\frac{4}{\lambda}\right)^2 + \left(-\frac{3}{\lambda}\right)^2 + \left(\frac{5}{\lambda}\right)^2 = 2$$
$$\frac{50}{\lambda^2} = 2$$
$$25 = \lambda^2$$

Therefore $\lambda = \pm 5$. We've already solved for x, y, and z in terms of λ. If $\lambda = 5$ we get $(\frac{4}{5}, -\frac{3}{5}, 1)$ and if $\lambda = -5$ we get $(-\frac{4}{5}, \frac{3}{5}, -1)$. Substituting into f, the extreme values of f on the sphere are:

$$f\left(\frac{4}{5}, -\frac{3}{5}, 1\right) = \frac{32}{5} - \frac{18}{5} + 10 = 12.8$$
$$f\left(-\frac{4}{5}, \frac{3}{5}, -1\right) = -12.8. \ \blacktriangle$$

Example 3.8.7. *An ellipsoidal constraint*

Maximize $f(x,y,z) = xyz$ on the ellipsoid $16x^2 + 9y^2 + 25z^2 = 96$. Computing gradients we get $\nabla f(x,y,z) = \langle yz, xz, xy \rangle$ and $\nabla g(x,y,z) = \langle 32x, 18y, 50z \rangle$. We set up the gradient and constraint equations:

$$yz = 32\lambda x$$
$$xz = 18\lambda y$$
$$xy = 50\lambda z$$
$$16x^2 + 9y^2 + 25z^2 = 96$$

This system is most easily solved by taking quotients of consecutive equations. Taking the quotient of the first two equations and simplifying gives

$$\frac{yz}{xz} = \frac{32\lambda x}{18\lambda y}$$
$$\frac{y}{x} = \frac{16x}{9y}$$
$$9y^2 = 16x^2$$

Similar calculations show that $25z^2 = 16x^2$. Substituting these into the constraint equation gives

$$16x^2 + 9y^2 + 25z^2 = 16x^2 + 16x^2 + 16x^2 = 48x^2 = 96.$$

Therefore we find $x = \pm\sqrt{2}$. Solving for y^2 instead will give $27y^2 = 96$, or $y = \pm\frac{4\sqrt{2}}{3}$. Analogously, one calculates $z = \pm\frac{4\sqrt{2}}{5}$. The product is maximized when all are positive, so it is $f(\sqrt{2}, \frac{4\sqrt{2}}{3}, \frac{4\sqrt{2}}{5}) = \sqrt{2}\frac{4\sqrt{2}}{3}\frac{4\sqrt{2}}{5} = \frac{32\sqrt{2}}{15}$. ▲

Closed and Bounded Regions: The introduction to Section 3.7 outlined how the techniques of that section can combine with Lagrange multipliers to optimize functions on closed and bounded regions. Indeed, any extrema on the interior of a region occur at critical points, while the method of Lagrange multipliers optimizes f on the boundary. We now combine these techniques.

Example 3.8.8. *Optimizing on a disk*

Find the absolute extreme values of $f(x,y) = x^2 + y^2 - 6y + 8x$ on the disk $x^2 + y^2 \le 36$.

First find the critical points of f interior to the disk by solving

$$2x + 8 = 0$$
$$2y - 6 = 0,$$

yielding the point $(-4, 3)$, which is interior to the disk.

To find potential extrema on the boundary circle $x^2 + y^2 = 36$, solve the Lagrange system:

$$2x + 8 = 2\lambda x$$
$$2y - 6 = 2\lambda y$$
$$x^2 + y^2 = 36.$$

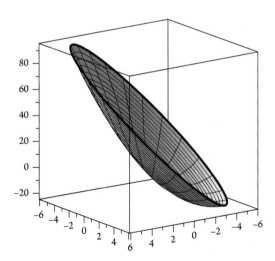

Figure 3.8.6 Absolute extrema on closed and bounded region

If x or y is zero the first two equations yield a contradiction, and if λ is zero the solution to the first two doesn't satisfy the third. Therefore we can divide by x, y, λ at will. Dividing the first two equations and simplifying yields

$$\frac{x+4}{y-3} = \frac{x}{y}$$
$$xy + 4y = xy - 3x,$$

so $y = \frac{-3}{4}x$. Substituting this into the boundary equation gives

$$x^2 + y^2 = x^2 + \frac{9}{16}x^2 = \frac{25}{16}x^2 = 36,$$

which yields $x = \pm\frac{24}{5}$. The substitution $y = \frac{-3}{4}x$ gives the points $\pm(24/5, -18/5)$.

At this point any extreme values of f occur at one of the three points $(-4, 3)$, $\pm(24/5, -18/5)$, so evaluating f there and comparing values completes the task. We obtain

$$f(-4,3) = -25; \quad f(24/5, -18/5) = 96; \quad f(-24/5, 18/5) = -24.$$

The minimum value of f on the disk is -25, occurring at the interior point, and the maximum value is 96. ▲

We conclude this section by optimizing a function on an annulus (a ring with two boundary components).

Example 3.8.9. *Two boundary components*

Optimize $f(x,y) = xe^{-x^2-y^2+x}$ on the ring $\frac{9}{16} \le x^2 + y^2 \le \frac{9}{4}$.

First note that the boundary of the region consists of two circles, those of radius $3/4$ and $3/2$ centered at the origin. The region itself is the ring between these circles.

To find absolute extreme values on closed and bounded regions evaluate f at all critical points interior to the region and all boundary points found using the method of Lagrange multipliers. The partial derivatives are necessary for both of these processes, after some simplification they are

$$f_x = -(2x+1)(x-1)e^{-x^2-y^2+x}, \text{ and } f_y = -2yxe^{-x^2-y^2+x}.$$

Critical points are where these vanish simultaneously, and recall that exponential functions are never zero. It follows that $x = -1/2, 1$ are the only values that make $f_x = 0$. The only way f_y can vanish for those x-values is for $y = 0$. Thus the critical points of f are $(-1/2, 0)$ and $(1, 0)$. Since the first point is not in the region, the only critical point of interest is $(1, 0)$.

Now apply the method of Lagrange multipliers to the boundary curves of the region. Since both boundary curves are level curves of the function $g(x,y) = x^2 + y^2$, the gra-

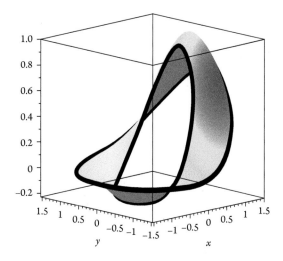

1.0

0.8

0.6

0.4

0.2

0

−0.2

1.5 1 0.5 0 −0.5 −1 −1.5 −1 −0.5 0 0.5 1 1.5

y x

Figure 3.8.7 A bounded region with two boundary curves

dient ∇g will be the same for both curves. The only difference between the Lagrange systems of equations will be the constant in the constraint. Initially we focus on solving $\nabla f = \lambda \nabla g$:

$$-(2x+1)(x-1)e^{-x^2-y^2+x} = 2\lambda x$$

$$-2yxe^{-x^2-y^2+x} = 2\lambda y.$$

If $y = 0$ the second equation is satisfied, and we can use the constraint equations $x^2 + y^2 = 9/16$ and $x^2 + y^2 = 9/4$ to find x-values. This gives the points $(\pm 3/4, 0)$ and $(\pm 3/2, 0)$.

If $y \neq 0$, a tedious check shows that x and λ don't vanish either (do this—don't take my word for it!). Knowing this we can divide the equations to get

$$\frac{-(2x+1)(x-1)e^{-x^2-y^2+x}}{-2yxe^{-x^2-y^2+x}} = \frac{2\lambda x}{2\lambda y}$$

$$-y(2x+1)(x-1) = -2yx^2$$

$$(x+1)y = 0.$$

We've assumed y isn't zero, so $x = -1$. Since no point on the inner circle has x-coordinate -1, the only additional points to check are $(-1, \pm\sqrt{1/2})$.

In summary, the points we must check are: $(1,0)$ (the critical point interior to the region), and the boundary points $(\pm 3/4, 0), (\pm 3/2, 0), (-1, \pm\sqrt{1/2})$. Evaluating f at these points gives

$$f(1,0) = 1 \qquad f(3/4,0) \approx 0.9047 \qquad f(-3/4) \approx -0.2019$$
$$f(3/2,0) \approx 0.7085 \quad f(-3/2,0) \approx -0.0353 \quad f(-1,\pm\sqrt{1/2}) \approx -0.0821$$

Thus f attains a maximum value of 1 at the point $(1,0)$ and a minimum value of -0.2019 at the point $(-3/4,0)$, in the ring $\dfrac{9}{16} \le x^2 + y^2 \le \dfrac{9}{4}$. ▲

Exercises

1. Use the method of Lagrange multipliers to optimize the following functions f subject to the given constraints.

   (a) $f(x,y) = x^2 + y^2 + 4x - 4y$, $x^2 + y^2 = 1$.

   (b) $f(x,y) = xy$, $x^2 + 4y^2 = 1$.

   (c) $f(x,y) = e^{-x^3-y^3+12x+3y}$, $x^3 + y^3 = 1$.

   (d) $f(x,y,z) = x + y + z$, $x^2 + y^2 + z^2 = 1$.

   (e) $f(x,y,z) = xyz^2$, $x + y + z = 3$, with all x, y, z positive.

2. Let $f(x,y) = x^3 - x + y^2$.

   (a) Find the extreme values of $f(x,y)$ on the unit circle.

   (b) Find the extreme values of $f(x,y)$ on $x^2 + y^2 = 4$.

   (c) Are there local extreme values in either case? Analyze the graph of f over $x^2 + y^2 = r^2$ for various radii r. Can you determine radii for which f has local extreme values? Where do they occur?

3. Find the rectangle with maximum area and perimeter 8.

4. Find the rectangle of maximal area that can be inscribed in the ellipse $\frac{x^2}{9} + y^2 = 1$.

5. Recall that the Cobb–Douglas production function is $Q = K^\alpha L^{1-\alpha}$, where K is capital and L is labor. In economics, one can model the total cost C of production by $C = aK + bL$, where a is the cost per unit capital (the interest rate) and b the cost per unit labor. Given a fixed cost C_0, it is reasonable to ask how to allocate resources to maximize production. In other words, maximize $Q = K^\alpha L^{1-\alpha}$ subject to the constraint $aK + bL = C_0$. This is the setting of Lagrange multipliers.

   (a) Maximize the Cobb–Douglas production function $Q = K^{1/4} L^{3/4}$ subject to the constraint $3K + 2L = 7$.

   (b) Maximize $Q = K^\alpha L^{1-\alpha}$ subject to $aK + bL = C_0$.

6. Find three positive numbers whose sum is S and whose product is maximized. What is the maximum value of the product?

7. Find n positive numbers whose sum is S and whose product is maximized.

8. Find the dimensions of the box of greatest volume that can be made from 24 square feet of material. Assume no waste in construction.

9. Find the dimensions of the largest open-top box that can be made from 24 square feet of material.

10. The curve $3x^2y - y^3 = 1$ is a streamline for an ideal flow around a $\pi/3$-corner. Show that the velocity potential $\varphi = x^3 - 3xy^2$ does not attain a maximum or minimum along $3x^2y - y^3 = 1$.

11. Show that the velocity potential $\varphi = \ln\sqrt{x^2 + y^2}$ for a source does not achieve extreme values on the streamline $y = 3x, x > 0$.

12. Find the absolute extreme values of $f(x,y) = x^2 + y^2 + 4x - 2y$ on the disk $x^2 + y^2 \leq 9$.

13. Find the absolute extreme values of $f(x,y) = x^2 + y^2 - 2x$ on the disk $x^2 + y^2 \leq 4$.

14. Find the absolute extreme values of $f(x,y) = e^{x^2+y^2+4x-2y}$ on the ring $4 \leq x^2 + y^2 \leq 9$.

15. Find the absolute extreme values of $f(x,y) = x^4 - 8x^2 + y^2$ on the ring $1 \leq x^2 + y^2 \leq 3$.

Problem Solving

We have learned a powerful technique in Lagrange multipliers. It turns out that Problem 7 has far-reaching consequences for problem solving, and we discuss a piece of Polya's book *Mathematics and Plausible Reasoning, Volume I: Induction and Analogy in Mathematics.* One notices that the solution to Problem 7 implies that if S is fixed and $x_1 + \cdots + x_n = S$, then

$$x_1 x_2 \cdots x_n \leq \left(\frac{S}{n}\right)^n = \left(\frac{x_1 + \cdots + x_n}{n}\right)^n. \qquad (3.8.3)$$

Equality holds if and only if each x_i is S/n. Taking the nth root of both sides gives the famous relationship between geometric and arithmetic means:

$$\sqrt[n]{x_1 x_2 \cdots x_n} \leq \frac{x_1 + \cdots + x_n}{n}.$$

Moreover, this result can be used to easily solve Problem 8, when looked at the right way. If we let x, y, and z denote the dimensions of the box, and S the surface area (which is 24 in the problem), then the constraint is $2xy + 2yz + 2xz = S$. Thus xy, yz and xz are three positive numbers whose sum is $S/2$. By Inequality 3.8.3, we see

$$(xyz)^2 = (xy)(yz)(xz) \leq \left(\frac{S/2}{3}\right)^3 = \left(\frac{S}{6}\right)^3,$$

with equality holding exactly when $xy = yz = xz$ (implying $x = y = z$). Now substitute $V = xyz$ to see that for any choice of dimensions

$$V^2 \le \left(\frac{S}{6}\right)^3,$$

with equality holding exactly when the box is a cube. So one clever application of the relationship between arithmetic and geometric means shows that a cube maximizes the volume of a box with fixed surface area, and that the relationship is $V^2 = \left(\frac{S}{6}\right)^3$.

Let's take a minute to discuss the application a little more closely. To apply Inequality 3.8.3, we thought of the constraint $xy + yz + xz = S/2$ as a statement about the arithmetic mean of the three numbers xy, yz, and xz. Even though we want to know x, y, and z individually, we pushed on, applying the inequality for the numbers xy, yz, and xz. The next key observation is this. We want to maximize the volume. While the product of xy, yz, and xz is not the volume, it *is* the square of the volume! With appropriate substitutions, the inequality ends up relating the volume and surface area of a box. Knowing that equality is reached only when the original numbers xy, yz, and xz are equal solves the optimization problem.

Notice that the same reasoning also proves the box with least surface area for a fixed volume is the cube, without any more calculation! Can you solve the open-box problem in a similar fashion?

Polya describes more examples of using a problem-solving approach to solve optimization problems. He outlines what he calls the pattern of partial variation to prove the geometric mean is at most the arithmetic mean. He poses many other geometric optimization problems that can be solved in a manner similar to the above reasoning: Think of the constraint as a statement about a mean, apply the geometric–arithmetic mean inequality 3.8.3, and note that the other side of the inequality is (a function of) the quantity you want to maximize or minimize. The interested reader is referred to Chapter VIII of [6] for more examples and problems.

4 Integration

In this chapter we introduce several extensions of the single-variable notion of integration. In the single-variable setting, the notation $\int_a^b f(x)\, dx$ involves three ingredients: the interval $[a,b]$ of integration, the integrand $f(x)$, and the differential dx (an infinitesimal change in length). We first extend the integrand to functions $f(x,y)$ of two variables, and the interval $[a,b]$ of the x-axis to a region R of the xy-plane. The infinitesimal length dx extends to an infinitesimal area dA, resulting in the definition of a double integral $\iint_R f(x,y)\, dA$ in Section 4.1. The definition of $\iint_R f(x,y)\, dA$ is difficult to use for calculation, so Section 4.2 introduces techniques for setting up and calculating double integrals. The skill, introduced in Section 1.6, of describing regions R using inequalities is reviewed, and gives rise to limits of integration. Section 4.3 extends our abilities by another dimension, to triple integrals $\iiint_W f(x,y,z)\, dV$. The region of integration is a solid W in $\mathbb{R}^3$, the integrand a function of three variables, and dV an infinitesimal volume. Section 4.4 shows how to handle integration using coordinate systems other than Cartesian, and Section 4.5 considers two applications of integration. Finally, in Section 4.6, we show how to integrate functions along curves and surfaces. By the end of the chapter, then, you will be able to integrate functions of several variables over regions as diverse as surfaces in space, solids in space, regions in the plane, and curves in either $\mathbb{R}^2$ or $\mathbb{R}^3$—quite an increase in sophistication!

4.1 Double Integrals—The Definition

To motivate the definition of the double integral we recall the definition of the definite integral $\int_a^b f(x)\, dx$ from single-variable calculus. Recall that if $f(x) \geq 0$ on the interval $[a,b]$, the integral $\int_a^b f(x)\, dx$ is the area A under the curve, and we focus on this situation to add geometric intuition. The big idea of the definite integral was this:

Approximate A with areas you can calculate, then take a limit as your approximation gets better.

Actually working out this idea is an algebraic nightmare, but it's a recurring nightmare that gets less scary each time you see it. We remind you of it now, without resorting to complete generality for ease of exposition.

To calculate the area A under the graph of $y = f(x)$, first partition the interval $[a, b]$ into n equal subintervals of width $\Delta x = \frac{b-a}{n}$. Label the partition points $a = x_0 < x_1 < \cdots < x_n = b$. With this notation, the ith subinterval is $[x_{i-1}, x_i]$ and the width of each subinterval is $x_i - x_{i-1} = \Delta x$. Now build a rectangle R_i above each subinterval $[x_{i-1}, x_i]$ by choosing the height of the rectangle to be a point on the curve. Analytically this means: arbitrarily pick values h_i in each subinterval $[x_{i-1}, x_i]$ and let $f(h_i)$ be the height of the rectangle R_i (see Figure 4.1.1(a)).

We now have lots of rectangles, and the sum of their areas is approximately the area under the curve. More importantly, we can calculate the areas of the rectangles! The height of R_i is $f(h_i)$, and its width is Δx, so its area is $f(h_i)\Delta x$. Adding up the areas of the rectangles gives the familiar Riemann sum approximation for the area under the curve,

$$A \approx \sum_{i=1}^{n} f(h_i)\Delta x.$$

We have accomplished the first step in the process of finding the area A under the curve $y = f(x)$. We have approximated A with areas we can calculate.

Of course, there is usually error in this approximation, which is seen geometrically as portions of the rectangles that are either above or below the curve. One asks how they might get a better approximation, and an obvious-ish answer is to use more (and hence smaller) rectangles.

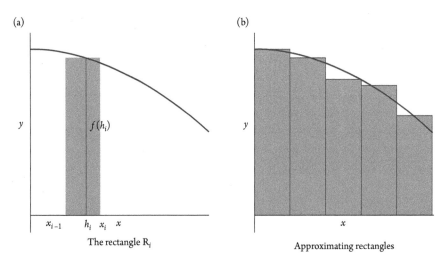

(a)

y

$f(h_i)$

x_{i-1} h_i x_i x

The rectangle R_i

(b)

y

x

Approximating rectangles

Figure 4.1.1 The definite integral

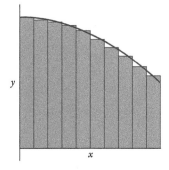

Figure 4.1.2 A better approximation of A

This is the clue to completing the second step in the process—taking a limit as your approximation gets better. Recall that the first thing we did was partition $[a,b]$ into n subintervals of equal length. Since the number of rectangles constructed is n, and we want to use more rectangles, we simply let n tend to infinity. This motivated the definition of the definite integral:

$$\int_a^b f(x)\,dx = \lim_{n\to\infty} \sum_{i=1}^n f(h_i)\Delta x.$$

Let us now turn our attention to a higher-dimensional problem—that of finding the volume under the surface $z = f(x,y)$. Again we make some initial simplifying assumptions, which will be removed later. Let's assume that we want the volume under $z = f(x,y)$ and over the rectangle $[a,b] \times [c,d]$ in the xy-plane. Further, we assume that $f(x,y) \geq 0$ above this rectangle. To calculate the volume V, we use the same basic idea as in the area problem considered above:

Approximate V with volumes you can calculate, then take a limit as your approximation gets better.

The idea is completely analogous to the single-variable case, but there is a little more to keep track of with two variables. As before, partition the x-interval $[a,b]$ into n equal subintervals with partition points $a = x_0 < x_1 < \cdots < x_n = b$. Now do the same with the y-interval $[c,d]$ obtaining partition points $c = y_0 < y_1 < \cdots < y_n = d$. This results in partitioning the rectangle $[a,b] \times [c,d]$ into n^2 sub-rectangles $R_{ij} = [x_{i-1}, x_i] \times [y_{j-1}, y_j]$, each with area $\Delta x \Delta y = \frac{b-a}{n} \cdot \frac{d-c}{n}$. Now build a rectangular box above each rectangle R_{ij} by picking a point $(h_i, k_j) \in R_{ij}$ and letting $f(h_i, k_j)$ be the height of the box. As before, we can calculate the volumes of the boxes, and their sum approximates the volume under the surface. The difference is that to add all the volumes we must sum on both indices i and j, resulting in a double sum.

$$V \approx \sum_{i=1}^n \sum_{j=1}^n f(h_i, k_j)\Delta x \Delta y$$

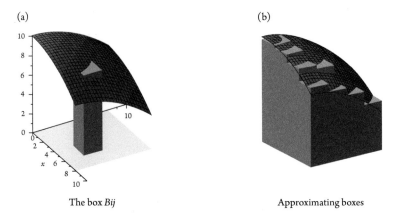

The box B_{ij}

Approximating boxes

Figure 4.1.3 Approximating volume with boxes

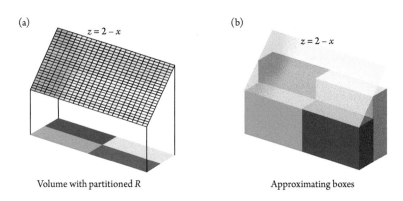

Volume with partitioned R

Approximating boxes

Figure 4.1.4 Approximating volume

Again we observe that our approximation becomes better if we take finer partitions, so we define the double integral of f over the rectangle R to be

$$\iint_R f(x,y)\, dA = \lim_{n\to\infty} \sum_{i=1}^n \sum_{j=1}^n f(h_i, k_j)\,\Delta x \Delta y.$$

Example 4.1.1. *Volume of a prism*

Approximate the volume under $f(x,y) = 2 - x$ above the rectangle $R = [0,1] \times [0,3]$ using four sub-rectangles (see Figure 4.1.4(a)).

We will use the above technique, partitioning each axis into two subintervals (i.e. $n = 2$). In this case $\Delta x = \frac{1-0}{2} = \frac{1}{2}$ and $\Delta y = \frac{3-0}{2} = \frac{3}{2}$. Our partition points are

$$x_0 = 0 \qquad\qquad y_0 = 0$$
$$x_1 = 0 + \Delta x = \tfrac{1}{2} \qquad\qquad y_1 = 0 + \Delta y = \tfrac{3}{2}$$
$$x_2 = 0 + 2\Delta x = 0 + 2\tfrac{1}{2} = 1 \qquad y_2 = 0 + 2\Delta y = 0 + 2\tfrac{3}{2} = 3.$$

It is evident that the general pattern for partition points in this context is

$$x_i = x_0 + i\Delta x \qquad\qquad y_j = y_0 + j\Delta y \qquad\qquad (4.1.1)$$

These partition points divide R into four sub-rectangles, and we use the "upper right" corner of each to determine the height of our box. Thus we must evaluate f at points $(1/2, 3/2)$, $(1/2, 3)$, $(1, 3/2)$, and $(1, 3)$ to get the heights of the boxes (see Figure 4.1.4(b)). Summing the volumes of the boxes approximates the volume under the plane:

$$V \approx \sum_{i=1}^{2} \sum_{j=1}^{2} f(x_i, y_j) \Delta x \Delta y$$

$$= f(x_1, y_1) \Delta x \Delta y + f(x_1, y_2) \Delta x \Delta y + f(x_2, y_1) \Delta x \Delta y + f(x_2, y_2) \Delta x \Delta y$$

$$= \left(2 - \frac{1}{2}\right) \frac{3}{4} + \left(2 - \frac{1}{2}\right) \frac{3}{4} + (2 - 1) \frac{3}{4} + (2 - 1) \frac{3}{4} = \frac{15}{4}. \ \blacktriangle$$

Math App 4.1.1. *Double Integration*

In this Math App you will be able to control the number of partitions in a double integral, and see the corresponding improvements in volume approximation. As a warning, however, for a 10×10 partition it takes a bit to plot the graphs, so be patient!

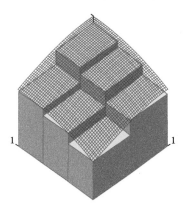

The previous example approximated volume using a finite subdivision. The next example calculates the volume approximation for an arbitrary subdivision then takes a limit as the approximation improves, thereby calculating the integral.

Example 4.1.2. *Volume of a prism—again!*

This time we want to approximate the volume under $f(x,y) = 2 - x$ above the rectangle $R = [0,1] \times [0,3]$ using n partition points, then taking a limit as $n \to \infty$.

Recall that for any constant k we have $\sum_{j=1}^{n} k = kn$, and that the sum of the first n integers is given by $\sum_{i=1}^{n} i = \frac{n(n+1)}{2}$.

When using n partition poins we have $\Delta x = \frac{1}{n}$ and $\Delta y = \frac{3}{n}$. Moreover, from Equation 4.1.1 we get the general formula for the partition points:

$$x_i = i\Delta x = \frac{i}{n} \qquad\qquad y_j = j\Delta y = \frac{3j}{n}.$$

Again using the "upper right" corner of each sub-rectangle we see the height of the box above the ij^{th} rectangle is $f(x_i, y_j) = 2 - \frac{i}{n}$. Thus the sum of the volumes of the boxes can be calculated

$$V \approx \sum_{i=1}^{n}\sum_{j=1}^{n} f(x_i, y_j)\Delta x \Delta y$$

$$= \sum_{i=1}^{n}\sum_{j=1}^{n}\left(2 - \frac{i}{n}\right)\frac{1}{n}\cdot\frac{3}{n} = \sum_{i=1}^{n}\sum_{j=1}^{n}\left(\frac{6}{n^2} - \frac{3i}{n^3}\right)$$

$$= \sum_{i=1}^{n}\left(\frac{6}{n} - \frac{3i}{n^2}\right) \qquad \text{(multiply by } n \text{ since terms constant wrt } j\text{)}$$

$$= \sum_{i=1}^{n}\frac{6}{n} - \frac{3}{n^2}\sum_{i=1}^{n} i \qquad \text{(algebra with summations)}$$

$$= 6 - \frac{3}{n^2}\cdot\frac{n(n+1)}{2} = 6 - \frac{3(n+1)}{2n}.$$

Of course, the volume is the limit as our approximation gets finer, and it is also the double integral. Thus we can calculate

$$V = \iint_R 2 - x\, dA = \lim_{n\to\infty} 6 - \frac{3(n+1)}{2n} = 6 - \frac{3}{2} = \frac{9}{2}.$$

We have calculated our first double integral! Luckily, we won't calculate many double integrals this way. In the next section we introduce iterated integrals and Fubini's theorem. These allow us to integrate functions one variable at a time, much like partial derivatives allow us to differentiate one variable at a time.

Before leaving this example, notice that we can use geometry to calculate the volume of the solid. In fact, the volume will just be the area of the trapezoidal face times the length of the prism. Since the area of the face is $\frac{3}{2}$ and the length is 3, we have the volume is $V = \frac{3}{2}\cdot 3 = \frac{9}{2}$, which agrees with our integral calculation. ▲

We complete one final example for good measure. The properties of sums to be used are

$$\sum_{k=1}^{n} c = nc; \qquad \sum_{k=1}^{n} k = \frac{n(n+1)}{2}; \qquad \sum_{k=1}^{n} k^2 = \frac{n(n+1)(2n+1)}{6}.$$

Example 4.1.3. *Another limit of Riemann sums*

Evaluate $\iint_R 4 - x - y^2 \, dA$ using the definition of the double integral, where $R = [0,2] \times [0,1]$.

Subdividing each edge of R into n subintervals gives $\Delta x = 2/n$, $\Delta y = 1/n$, which leads to $x_i = 2i/n$ and $y_j = j/n$.

From this we calculate the height of the ij^{th} box to be $f(2i/n, j/n) = 4 - 2i/n - (j/n)^2$, so that the sum of the volumes of the boxes is

$$V \approx \sum_{i=1}^{n} \sum_{j=1}^{n} f(x_i, y_j) \Delta x \Delta y$$

$$= \sum_{i=1}^{n} \sum_{j=1}^{n} \left(4 - \frac{2i}{n} - \left(\frac{j}{n}\right)^2 \right) \frac{2}{n} \cdot \frac{1}{n} = \sum_{i=1}^{n} \sum_{j=1}^{n} \left(\frac{8}{n^2} - \frac{4i}{n^3} - \frac{2}{n^4} j^2 \right)$$

$$= \sum_{i=1}^{n} \left(\sum_{j=1}^{n} \left(\frac{8}{n^2} - \frac{4i}{n^3} \right) - \frac{2}{n^4} \sum_{j=1}^{n} j^2 \right) \qquad \text{(algebra with summations)}$$

$$= \sum_{i=1}^{n} \left(\frac{8}{n} - \frac{4i}{n^2} - \frac{2}{n^4} \frac{n(n+1)(2n+1)}{6} \right) \qquad \text{using } \sum_{j=1}^{n} c \text{ and } \sum_{j=1}^{n} j^2$$

$$= \sum_{i=1}^{n} \left(\frac{8}{n} - \frac{(n+1)(2n+1)}{3n^3} \right) - \frac{4}{n^2} \sum_{i=1}^{n} i \qquad \text{(summation algebra)}$$

$$= 8 - \frac{(n+1)(2n+1)}{3n^2} - \frac{4}{n^2} \cdot \frac{n(n+1)}{2} \qquad \text{using } \sum_{i=1}^{n} c \text{ and } \sum_{i=1}^{n} i$$

$$= 8 - \frac{8n^2 + 9n + 1}{3n^2}.$$

Taking the limit as $n \to \infty$ gives the double integral

$$\iint_R 4 - x - y^2 \, dA = \lim_{n \to \infty} 8 - \frac{8n^2 + 9n + 1}{3n^2} = 8 - \frac{8}{3} = \frac{16}{3}. \; \blacktriangle$$

Technical Remarks about R: At this point we have a definition of the double integral $\iint_R f(x,y) \, dA$ when R is a rectangle in the xy-plane. We'd like to integrate functions over much more general regions than these. You can extend to other regions R by taking a union of rectangles that approximate R and summing the integrals of $f(x,y)$ over those. Once that is done, you take a limit as your rectangle approximations of R improve. This

(a)   (b)

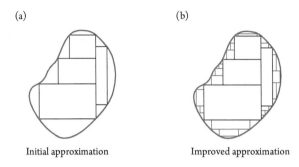

Initial approximation   Improved approximation

Figure 4.1.5 Partitioning a general region

is illustrated in Figure 4.1.5, where the left subdivision of R into rectangles is improved upon in the right.

So, for more general regions, the integral $\iint_R f(x,y)dA$ is a limit of limits (one limit being the integral over a single rectangle, the second being as the rectangles approach R). Should you decide to pursue mathematics further you may have the pleasure of verifying that this can be done! However, these are details we won't concern ourselves with here. Suffice it to say that if the boundary of R is a union of piecewise smooth curves, you can define the double integral $\iint_R f(x,y)\,dA$.

Exercises

1. Use four sub-rectangles to approximate the volume under $f(x,y) = 5 - y$ over the rectangle $R = [0,2] \times [0,3]$.

2. Use four sub-rectangles to approximate the volume under $f(x,y) = 4 - x^2$ over the rectangle $R = [0,1] \times [0,3]$.

3. Use four sub-rectangles to approximate the volume under $f(x,y) = 3 - x - y$ over the rectangle $R = [0,2] \times [0,1]$.

4. Use four sub-rectangles to approximate the volume under $f(x,y) = 4 - x^2 - y^2$ over the square $R = [0,1] \times [0,1]$.

5. Using the definition of the double integral, evaluate $\iint_R 5 - y\, dA$ where $R = [0,2] \times [0,3]$.

6. Using the definition of the double integral, evaluate $\iint_R 4 - x^2\ dA$ where $R = [0,2] \times [0,3]$.

7. Using the definition of the double integral, evaluate $\iint_R 4 - x^2 - y^2\ dA$ where $R = [0,1] \times [0,1]$.

8. Using the definition of the double integral, evaluate $\iint_R 3 - x - y\ dA$ where $R = [0,2] \times [0,1]$.

9. Why doesn't $\iint_R 2 - y\,dA$, where $R = [0,3] \times [0,3]$, give the volume of a solid?

10. Use the definition of $\iint_R f(x,y)\,dA$ to show that

$$\iint_R f(x,y) + g(x,y)\,dA = \iint_R f(x,y)\,dA + \iint_R g(x,y)\,dA.$$

11. Use the definition of $\iint_R f(x,y)\,dA$ to show that

$$\iint_R cf(x,y)\,dA = c\iint_R f(x,y)\,dA.$$

12. Do you think that

$$\iint_R f(x,y) \cdot g(x,y)\,dA = \iint_R f(x,y)\,dA \cdot \iint_R g(x,y)\,dA?$$

Why or why not?

4.2 Calculating Double Integrals

Now that we know what a double integral is, we need to get comfortable calculating them. In the single-variable case this was accomplished by the fundamental theorem of calculus. Remember that this theorem relates the definite integral with taking an antiderivative, so you never have to take limits of Riemann sums again! In the two variable case, Fubini's theorem states that the double integral we just defined can be calculated by integrating one variable at a time—partial integration, if you will. The fancy math term for this is an iterated integral, and we introduce the notation. An iterated integral is denoted

$$\int_a^b \int_c^d f(x,y)\ dydx,$$

and means first evaluate $\int_c^d f(x,y)\ dy$, thinking of x as a constant. The result will be a function of x, since evaluating at the endpoints c and d gets rid of the y's. Then integrate that with respect to x. An example illustrates this.

Example 4.2.1. *Iterated integrals*

Evaluate $\int_{-1}^2 \int_0^2 4y + x^2 y^2\ dydx$. Since this is the first example we've seen, we'll separate the steps. Very quickly you will become familiar with the notation and we will carry the double integral notation along as we go. Thinking of x as a constant, we evaluate

$$\int_0^2 4y + x^2 y^2\ dy = 2y^2 + x^2 \frac{y^3}{3}\Big|_{y=0}^2 = 8 - \frac{8}{3}x^2,$$

which is a function of x. Note that when evaluating at the endpoints, we only substitute the values for y, leaving x alone. This is because our integration was with respect to y at this stage. We finish the iterated integral by integrating the result with respect to x:

$$\int_{-1}^2 \int_0^2 4y + x^2 y^2\ dydx = \int_{-1}^2 \left(\int_0^2 4y + x^2 y^2\ dy \right) dx$$

$$= \int_{-1}^2 8 - \frac{8}{3}x^2\ dx$$

$$= 8x - \frac{8}{9}x^3 \Big|_{-1}^2 = 16 - \frac{64}{9} - \left(-8 + \frac{8}{9} \right) = 16.\ \blacktriangle$$

We note that there is nothing special about integrating with respect to y first, then with respect to x. Iterated integrals can be calculated in the other order as well, as long as some care is taken with the limits of integration. When the region of integration is a rectangle, the limits merely switch as follows:

$$\int_{-1}^{2}\int_{0}^{2} 4y + x^2 y^2 \, dy dx = \int_{0}^{2}\int_{-1}^{2} 4y + x^2 y^2 \, dx dy$$

$$= \int_{0}^{2}\left(4yx - y^2\frac{x^3}{3}\Big|_{x=-1}^{2}\right) dy$$

$$= \int_{0}^{2} 8y - \frac{8}{3}y^2 - \left(-4y + \frac{1}{3}y^2\right) dy = \int_{0}^{2} 12y - 3y^2 \, dy$$

$$= 6y^2 - y^3\big|_{0}^{2} = 24 - 8 = 16.$$

Before leaving this example we mention again that the limits on the inside integral sign correspond to the first variable of integration, while the limits on the outer integral sign correspond to the second. Take some time to familiarize yourself with the notation of an iterated integral; it will be worth the effort. ▲

Now that we know what an iterated integral is, we can state Fubini's theorem. Before doing so, remember that our definition of the double integral is a limit of Riemann sums. It's very ugly, and difficult to calculate. For this reason, Fubini's theorem is quite nice.

Theorem 4.2.1. *Let $f(x,y)$ be a continuous function defined on the rectangle $R = [a,b] \times [c,d]$ in the xy-plane. Then*

$$\iint_{R} f(x,y) \, dA = \int_{a}^{b}\int_{c}^{d} f(x,y) \, dy dx = \int_{c}^{d}\int_{a}^{b} f(x,y) \, dx dy$$

This theorem implies that we needn't take limits like we did in Section 4.1, but can perform iterated integrals instead. Rather than giving a formal proof, we provide a plausible reason. When using limits to calculate

$$\iint_{R} f(x,y) dA = \lim_{n\to\infty} \sum_{i=1}^{n}\sum_{j=1}^{n} f(h_i, k_j)\Delta x \Delta y,$$

we simplified the sum first. In the process we evaluated the double-sum one variable at a time, thinking of the other variable as constant. This is analogous to integrating with respect to one variable, thinking of the other as constant—which is iterated integration! Therefore Fubini's theorem *should* be true...and it is.

Here is one more example of an iterated integral over a rectangle, before considering more general regions of integration.

Example 4.2.2. *Integrating over a rectangle*

Evaluate $\iint_{R} \dfrac{y}{1+x^2} dA$ where R is the rectangle $[0,1] \times [1,2]$. By Fubini's theorem we see that the double integral can be evaluated as the iterated integral

$$\int_0^1 \int_1^2 \frac{y}{1+x^2}\,dy\,dx = \int_0^1 \frac{1}{1+x^2}\left(\int_1^2 y\,dy\right)dx$$

$$= \int_0^1 \frac{1}{1+x^2}\left(\left.\frac{y^2}{2}\right|_1^2\right)dx$$

$$= \int_0^1 \frac{3}{2(1+x^2)}\,dx$$

$$= \left.\frac{3}{2}\tan^{-1}x\right|_0^1 = \frac{3}{2}\tan^{-1}(1) - \frac{3}{2}\tan^{-1}(0) = \frac{3\pi}{8}. \ \blacktriangle$$

Limits of Integration from R: There are three aspects of the double integral $\iint_R f(x,y)\,dA$ to consider when setting up and evaluating it: the region R, the integrand $f(x,y)$, and the differential dA. The infinitesimal area dA will depend on the coordinates we use to describe the region R. Different interpretations of the function $f(x,y)$ will yield different interpretations of the integral. For example, if $f(x,y)$ is the density of a thin plate R, then $\iint_R f(x,y)\,dA$ is the mass of R. The region R dictates the limits of integration in iterated integrals. Thus all aspects of the double integral are vital for setting up, evaluating, and interpreting it. We now wish to investigate how the region R determines the limits of integration.

So far we have focused on integrating functions over rectangles. We want to be able to integrate over more general regions in the plane. There are two particularly nice types of regions for which it is easy (sort of) to find limits of integration. We describe them in terms of which variable you will integrate with respect to first.

A *dy*-region in the plane is one in which the top and bottom are functions of x, and whose shadow on the x-axis is an interval (see Figure 4.2.1(a)). A *dx*-region is one in which the right and left are functions of y and whose shadow on the y-axis is an interval (see Figure 4.2.1(b)). Some regions in the plane are both *dx*- and *dy*-regions, and some

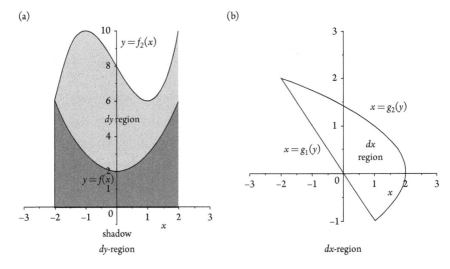

Figure 4.2.1 Elementary regions

are neither. It is relatively easy to describe such regions with a system of inequalities. This is an important skill, needed to determine the limits of integration for iterated integrals.

To Describe a *dy*-Region:

1. Solve the equations of boundary curves for y, so that the top curve is $y = f_2(x)$ and the bottom curve is $y = f_1(x)$.
2. Find the shadow on the x-axis. This should be an interval $[a, b]$.
3. The system of inequalities is:

$$f_1(x) \le y \le f_2(x)$$
$$a \le x \le b$$

To integrate a function $f(x, y)$ over a *dy*-region, the system of inequalities gives the limits of integration. We get

$$\iint_R f(x, y)\, dA = \int_a^b \int_{f_1(x)}^{f_2(x)} f(x, y)\, dy dx.$$

This technique is now illustrated in an animated Math App intended to strengthen your intuition.

Math App 4.2.1. *Limits in double integrals*

In the following Math App the limits of integration are found for a *dy*-region. The animation assists you in viewing them.

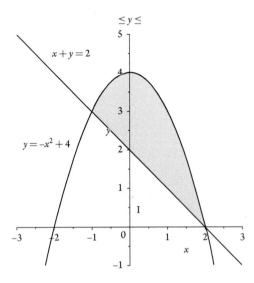

The technique for finding limits of integration for a *dx*-region is analogous to this. We include the description for completeness.

To Describe a *dx*-Region:

1. Solve the equations of boundary curves for x, so that the right curve is $x = g_2(y)$ and the left curve is $x = g_1(y)$.

2. Find the shadow on the y-axis. This should be an interval $[c, d]$.

3. The system of inequalities is:

$$g_1(y) \leq x \leq g_2(y)$$
$$c \leq y \leq d$$

To integrate a function $f(x, y)$ over a *dx*-region, the system of inequalities gives the limits of integration. You get

$$\iint_R f(x, y)\, dA = \int_a^b \int_{g_1(x)}^{g_2(x)} f(x, y)\, dxdy.$$

Some regions can be both *dx*- and *dy*-regions, as the following Math App demonstrates.

Math App 4.2.2. *A region that is both dx and dy*

In this example we use animations to compare the *dx*- and *dy*-limits of integration for the same region R. They turn out to be surprisingly different!

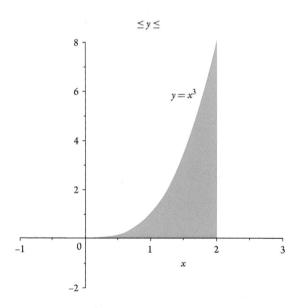

Example 4.2.3. *Describing regions in the plane*

Use inequalities to describe the triangle T in the first quadrant bounded by the axes and the line $3x + 2y = 6$. Thinking of the triangle as a dy-region, so the line is the top of the region, we solve the boundary equation for y to get $y = -\frac{3}{2}x + 3$. The line $y = 0$ is the bottom of T and the line $y = -\frac{3}{2}x + 3$ is the top. Looking at the shadow of T on the x-axis gives $0 \leq x \leq 2$. Thus, as a dy-region, we can describe T by the system of inequalities

$$0 \leq y \leq -\frac{3}{2}x + 3$$
$$0 \leq x \leq 2.$$

To integrate any function $f(x,y)$ over the triangle, you get the integral

$$\int_0^2 \int_0^{-\frac{3}{2}x+3} f(x,y)\, dy dx.$$

An intuitive way of finding the above limits of integration is the following. Walk toward ∞ along a line parallel to the y-axis that goes through your region. The equation of the curve where you enter the region is the lower limit, and the equation for the curve where you leave the region is your upper limit.

If one thinks of the line $3x + 2y = 6$ as the *right* boundary of the region, rather than the top, T becomes a dx-region. The left boundary is the y-axis, or the line $x = 0$, and one solves the equation $3x + 2y = 6$ for x to find the right boundary. The shadow on the y-axis is the interval $[0,3]$. So we can describe T via the system of inequalities

$$0 \leq x \leq -\frac{2}{3}y + 2$$
$$0 \leq y \leq 3.$$

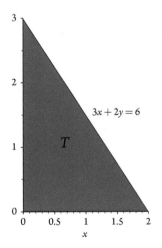

Figure 4.2.2 The triangle T

Thus another way to integrate $f(x,y)$ over the triangle is

$$\int_0^3 \int_0^{-\frac{2}{3}y+3} f(x,y) \; dxdy.$$

Again, an intuitive way to find the limits on x in this example is to walk toward ∞ on a line through the region and parallel to the x-axis. The lower limit is the equation of the curve where you enter the region and the upper limit is the equation of the curve where you leave it. Both these equations are solved for x when you're integrating with respect to x. ▲

Example 4.2.4. *A dx-region*

Describe the region R bounded by the coordinate axes and the lines $y = 2$ and $x + 2y = 3$ both as a dy-region and as a dx-region. To describe R as a dy-region, notice that the top function changes at $x = 2$. This means we have to split our description there, and use two systems of inequaties. Since we're thinking dy first, we solve our boundary equations for y and use the "bottom-top" approach. This gives the systems:

$$0 \le y \le 2 \qquad\qquad\qquad 0 \le y \le 3 - \frac{1}{2}x$$

$$0 \le x \le 2 \qquad\qquad\qquad 2 \le x \le 6$$

It's actually easier to think of the region as a dx-region, because there are single-boundary curves on the left and right sides of the region. Walking along a line through R and parallel to the x-axis, you enter the region at the curve $x = 0$ and leave R at the curve $x = 6 - 2y$. The shadow on the y-axis is the interval $[0,2]$, yielding the system of inequalities

$$0 \le x \le 6 - 2y$$
$$0 \le y \le 2.$$

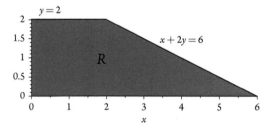

Figure 4.2.3 The region R

These inequalities provide the limits of integration when integrating any function over R. ▲

So far we have seen that describing R using a system of inequalities gives limits of integration for the region R. Conversely, given limits of integration we can find the region R. This will be useful when we want to switch the order of integration in certain examples.

Example 4.2.5. *The region from limits*

Determine the region R of integration for the double integral

$$\int_{-1}^{0} \int_{-y}^{2-y^2} f(x,y) \, dxdy.$$

To accomplish this, we write down the system of inequalities corresponding to the given limits, then sketch the region. The system is

$$-y \le x \le 2 - y^2$$
$$-1 \le y \le 0.$$

Sketching the boundary curves gives the region in Figure 4.2.4. ▲

Example 4.2.6. *Integrals over more general regions*

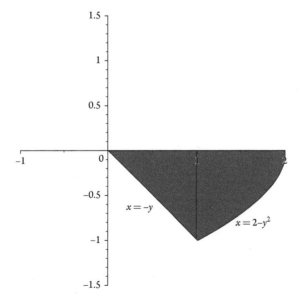

Figure 4.2.4 The region of integration

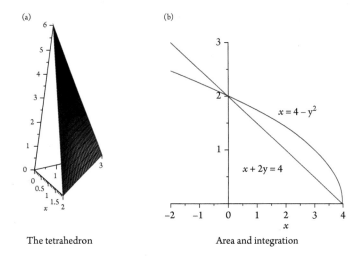

(a) The tetrahedron

(b) Area and integration

Figure 4.2.5 Double integral applications

Find the volume of the tetrahedron bounded by the coordinate planes and the plane $3x + 2y + z = 6$ (see Figure 4.2.5(a)). We can think of the tetrahedron as a solid under the graph $z = f(x,y)$. This solid is the region under the graph of $z = 6 - 3x - 2y$, and above the triangle T in the xy-plane from the previous example. To find the volume, then, we just need to integrate using the limits found in the previous example.

$$\iint_T 6 - 3x - 2y \, dA = \int_0^2 \int_0^{-\frac{3}{2}x+3} 6 - 3x - 2y \, dy dx$$

$$= \int_0^2 (6 - 3x)y - y^2 \Big|_0^{-\frac{3}{2}x+3} dx$$

$$= \int_0^2 (6 - 3x)\left(-\frac{3}{2}x + 3\right) - \left(-\frac{3}{2}x + 3\right)^2 dx$$

$$= \int_0^2 \frac{9}{4}x^2 - 9x + 9 \, dx$$

$$= \frac{3}{4}x^3 - \frac{9}{2}x^2 + 9x \Big|_0^2 = 6. \blacktriangle$$

Example 4.2.7. *Double integrals and area*

Recall that if $f(x,y) \geq 0$ over the region R in the plane, then $\iint_R f(x,y) dA$ is the volume of the solid under $z = f(x,y)$ and above R. If we use the constant function $f(x,y) = c$, then the volume of the solid is the area of R times the height c. More specifically, if $f(x,y) = 1$ then $\iint_R dA$ is the area of R. We use this observation now.

Find the area of the region bounded by $x = 4 - y^2$ and $x + 2y = 4$. Sketching the region in Figure 4.2.5(b) we see that the line is the left side and the parabola the right. Moreover, the points of intersection of the curves can be found by substituting $4 - y^2$ for x in the equation for the line. One obtains

$$4 - y^2 + 2y = 4$$
$$y^2 - 2y = 0,$$

from which we see that $y = 0$ or $y = 2$. This implies that the region can be described by

$$-2y + 4 \leq x \leq 4 - y^2$$
$$0 \leq y \leq 2.$$

Thus the area is given by

$$\int_0^2 \int_{4-2y}^{4-y^2} dx\, dy = \int_0^2 x\Big|_{4-2y}^{4-y^2} dy$$
$$= \int_0^2 (4 - y^2) - (4 - 2y)\, dy = \int_0^2 2y - y^2\, dy$$
$$= y^2 - \frac{y^3}{3}\Big|_0^2 = \frac{4}{3}. \; \blacktriangle$$

We also observe that after integrating with respect to x we are left with the integral $\int_0^2 (4 - y^2) - (4 - 2y)\, dy$. This is precisely the integral you get using single variable techniques to find area! The area of a region between two curves is the integral of the "top" curve minus the "bottom" curve. In this case you use the right curve minus the left curve because everything is sideways.

Example 4.2.8. *Switching order of integration*

Evaluate $\int_0^{\sqrt{\pi/2}} \int_x^{\sqrt{\pi/2}} \sin y^2 \, dy\, dx$.

To do this, notice we don't really know how to integrate $\int \sin y^2 \, dy$, but $\int \sin y^2 \, dx = x \sin y^2 + C$ since $\sin y^2$ is a constant when integrating with respect to x. With this as motivation, we switch the order of integration and hope for the best. To do so, we use the given limits of integration to find the region R of integration. The limits describe the system of inequalities

$$x \leq y \leq \sqrt{\pi/2}$$
$$0 \leq x \leq \sqrt{\pi/2},$$

which is the triangle pictured in Figure 4.2.6. Thinking of it as a dx-region, the left side is $x = 0$, the right is $x = y$, and the shadow on the y-axis is the interval $[0, \sqrt{\pi/2}]$. Thus, if we want to integrate with respect to x first, the integral becomes:

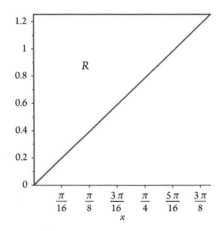

Figure 4.2.6 The triangle R

$$\iint_R \sin y^2\, dA = \int_0^{\sqrt{\pi/2}} \int_0^y \sin y^2\, dx dy$$

$$= \int_0^{\sqrt{\pi/2}} x \sin y^2 \big|_0^y\, dy$$

$$= \int_0^{\sqrt{\pi/2}} y \sin y^2\, dy$$

$$= \frac{1}{2} \sin y^2 \Big|_0^{\sqrt{\pi/2}} = \frac{1}{2}.$$

The final integration is a simple substitution. This is an example of an integral that is quite direct when viewed one way, but originally very difficult to do. ▲

Exercises

In Problems 1–9 describe the given regions first as a dx-region, then as a dy-region.

1. The triangle in the first quadrant bounded by the coordinate axes and the line $4x + 3y = 12$.

2. The triangle in the fourth quadrant bounded by the coordinate axes and the line $7x - 2y = 8$.

3. The trapezoid in the first quadrant bounded by the coordinate axes together with the lines $y = 2$ and $3x + y = 6$.

4. The bounded region between the x-axis and the parabola $y = 4 - x^2$.

5. The quarter circle in the first quadrant and under $x^2 + y^2 = 1$.

6. The unit disk $x^2 + y^2 \le 1$.

7. The region in the first quadrant between the unit circle and the circle $x^2 + y^2 = 4$.

8. The bounded region between the y-axis and the parabola $x = 1 - y^2$.

9. The bounded region between $x = 1 - y^2$ and $x + 2y = 1$.

In problems 10–17 sketch the regions of integration for the iterated integral.

10. $\displaystyle\int_2^3 \int_{-1}^4 f(x,y)\,dx\,dy$

11. $\displaystyle\int_2^3 \int_{-1}^4 f(x,y)\,dy\,dx$

12. $\displaystyle\int_{-1}^1 \int_0^{3-x} f(x,y)\,dy\,dx$

13. $\displaystyle\int_0^2 \int_{y-2}^0 f(x,y)\,dx\,dy$

14. $\displaystyle\int_{-1}^2 \int_{-x}^{2-x^2} f(x,y)\,dy\,dx$

15. $\displaystyle\int_{-1/2}^1 \int_{1-y}^{2-2y^2} f(x,y)\,dx\,dy$

16. $\displaystyle\int_{-1}^3 \int_{x^2}^{2x+3} f(x,y)\,dy\,dx$

17. $\displaystyle\int_{-1}^1 \int_{y^2-1}^{2-2y^2} f(x,y)\,dx\,dy$

In Problems 18–23 evaluate the iterated integrals.

18. $\displaystyle\int_{-1}^0 \int_0^2 2x - 3y^2\,dy\,dx$

19. $\displaystyle\int_0^1 \int_0^{\pi/2} xy\sin y\,dy\,dx$

20. $\displaystyle\int_0^1 \int_0^{1/2} \frac{y}{\sqrt{1-x^2}}\,dx\,dy$

21. $\displaystyle\int_1^2 \int_0^x 2x - y\,dy\,dx$

22. $\displaystyle\int_0^{\sqrt{\pi/2}} \int_0^x \sin x^2\,dy\,dx$

23. $\displaystyle\int_{-2}^2 \int_{x^2-4}^{4-x^2} dy\,dx$

24. Find the volume of the tetrahedron bounded by the coordinate planes and $3x + 2y + z = 6$.

25. Find the volume of the solid under the paraboloid $z = 4 - x^2 - y^2$ and above the rectangle $[0,1] \times [-1,1]$.

26. Find the volume of the solid under the plane $2x + y + z = 10$ and over the rectangle $[1,3] \times [-1,0]$.

27. Find the volume of the solid under the plane $x + y + z = 20$ in $\mathbb{R}^3$ and over the region in the xy-plane bounded by the curves $x = 2 - y^2$ and $y = -x$.

28. Find the area between $y = x^2 - 2$ and $y = x$ using double integration.

29. Find the area between $x + y = 1$ and $x = y^2 - 1$ using double integration.

30. Evaluate $\displaystyle\int_0^1 \int_x^1 e^{y^2} \, dy dx$ by first switching the order of integration.

31. Evaluate $\displaystyle\int_0^1 \int_0^{\sqrt[3]{1-x^2}} \sqrt{1 - y^3} \, dy dx$ by first switching the order of integration.

32. Evaluate $\iint_R x - y \, dA$, where R is the triangle in the plane with vertices $(0,0)$, $(1,1)$, and $(2,1)$.

33. Evaluate $\iint_R \, dA$ where R is the bounded region between the parabolas $y = 1 - x^2$ and $y = x^2 - 1$.

4.3 Triple Integrals

The integral of a function $f(x,y,z)$ of three variables over a solid W is a triple integral, and denoted $\iiint_W f(x,y,z)\, dV$. It is defined completely analogously to the double integral. Partition the solid W into rectangular boxes with length Δx_i, width Δy_j and height Δz_k. The volume of the ijk^{th} box is $\Delta V_{ijk} = \Delta x_i \Delta y_j \Delta z_k$. Choose a point (a_i, b_j, c_k) in each box, evaluate f there, form the product $f(a_i, b_j, c_k)\Delta V_{ijk}$ and sum the values to estimate the triple integral. Taking a limit as the partition gets finer yields

$$\iiint_W f(x,y,z)\, dV = \lim_{n\to\infty} \sum_{i=1}^{n}\sum_{j=1}^{n}\sum_{k=1}^{n} f(a_i, b_j, c_k)\Delta x_i \Delta y_j \Delta z_k.$$

Technically this integral would be defined on boxes, then generalized to more complicated solids in the same way double integrals on rectangles are extended to more general regions (revisit Figure 4.1.5).

Motivating the Triple Integral—Mass from Density: This process can seem pretty abstract, so we discuss a physical example in which integration arises naturally: calculating mass from density. Recall that the density of an object is mass per unit volume. If two objects are exactly the same size, the more dense object is heavier.

Example 4.3.1. *Constant density*

If the density of an object is constant, then the mass is simply density times volume. Suppose the tetrahedron of Example 4.2.6 is made from material with constant density 10 grams per cubic centimeter, then the mass m is

$$M = \text{volume} \cdot \text{density} = 6 \cdot 10 = 60 \text{ gm.}$$

This follows because in Example 4.2.6 we calculated the volume of the tetrahedron to be 6 cm^3, assuming the units on each axis to be centimeters. ▲

Example 4.3.2. *Mass from variable density*

Now suppose the tetrahedron is more dense at its base than at its apex, and we still want to calculate its mass. More precisely, suppose the density function is given by $\delta(x,y,z) = 40 - z^2$. Our approach could be the standard calculus approach to such problems: Approximate the quantity using things you can calculate, then take a limit as your approximation gets better.

We describe an example and then the more general, yet still finite, setting. The actual calculations are tedious, and not that important. It is significant that:

The process of approximating mass, then taking a limit as approximations improve, leads naturally to integration.

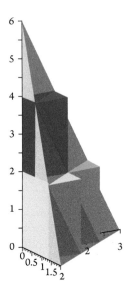

Figure 4.3.1 Approximating the mass of a tetrahedron

A particularly simple approximation would be the following. We could say that the tetrahedron is approximated by the three boxes

$$B_1 = [0, \tfrac{4}{3}] \times [0, 1] \times [0, 2],$$
$$B_2 = [0, \tfrac{2}{3}] \times [1, 2] \times [0, 2], \text{ and}$$
$$B_3 = [0, \tfrac{2}{3}] \times [0, 1] \times [2, 4].$$

Note the corresponding volumes of the boxes are $V_1 = 8/3$, $V_2 = 4/3$, and $V_3 = 4/3$. To further simplify our calculation we could assume that the box has constant density, the density $\delta(x, y, z) = 40 - z^2$ of its geometric center. The sum of the masses of the boxes with these assumptions approximates the mass of the tetrahedron. Then our approximate mass of the tetrahedron is

$$M \approx V_1 \delta(2/3, 1/2, 1) + V_2 \delta(1/3, 3/2, 1) + V_3 \delta(1/3, 1/2, 3)$$
$$= \frac{8}{3} \cdot 39 + \frac{4}{3} \cdot 39 + \frac{4}{3} \cdot 36 = 204 \text{ gm.} \blacktriangle$$

The assumptions we made were that our object was made up of boxes (so we can easily calculate volume) and that the density of each box was constant (which facilitates the mass calculation). If we use many small boxes to approximate the tetrahedron, with constant densities which are close to that given by $\delta(x, y, z)$ for each box, we could add up the masses of the boxes to approximate the mass of the tetrahedron.

In our case, we could partition the tetrahedron into tiny boxes

$$B_{ijk} = [x_i, x_{i+1}] \times [y_j, y_{j+1}] \times [z_k, z_{k+1}]$$

and assume the density is constant on each box. If the boxes are small, this assumption is reasonable. To determine the approximate density on each cube, pick a point (a_i, b_j, c_k) in B_{ijk} and use its density $\delta(a_i, b_j, c_k)$. Again, if the boxes are small, then $\delta(a_i, b_j, c_k)$ should be close to the density of the material at each point in box B_{ijk}. The mass M_{ijk} of the box B_{ijk} is approximated by

$$M_{ijk} \approx (\text{Density at } (a_i, b_j, c_k)) \cdot (\text{Volume of } B_{ijk})$$
$$= \delta(a_i, b_j, c_k) \Delta x_i \Delta y_j \Delta z_k, \tag{4.3.1}$$

where $\Delta x_i = x_{i+1} - x_i$, and similarly for Δy_j and Δz_k. The sum of the masses of the boxes M_{ijk} approximates the total mass M of the tetrahedron. Each M_{ijk} can be approximated as in Equation 4.3.1, so we can approximate M by

$$M \approx \sum_{i=1}^{n} \sum_{j=1}^{n} \sum_{k=1}^{n} M_{ijk} \approx \sum_{i=1}^{n} \sum_{j=1}^{n} \sum_{k=1}^{n} \delta(a_i, b_j, c_k) \Delta x_i \Delta y_j \Delta z_k. \tag{4.3.2}$$

Since the approximation gets better as we use smaller and smaller boxes, we say

$$M = \lim_{n \to \infty} \sum_{i=1}^{n} \sum_{j=1}^{n} \sum_{k=1}^{n} \delta(a_i, b_j, c_k) \Delta x_i \Delta y_j \Delta z_k. \tag{4.3.3}$$

This limit is what we defined to be the triple integral, so we find that the integral of the density function is the mass of a solid. We summarize this discussion in the following:

Calculating Mass from Density

Let W be a solid, and let the density at each point (x, y, z) of W be given by the function $\delta(x, y, z)$. The mass M of W is given by

$$M = \iiint_W \delta(x, y, z)\, dV.$$

Thus far we have introduced the notion of a triple integral and motivated the definition using a physical application. We now turn our attention to calculating triple integrals.

It turns out that Fubini's theorem generalizes, and we can compute triple integrals via iterated integrals. Thus over the box $W = [a,b] \times [c,d] \times [e,f]$ we have

$$\iiint_W f(x,y,z)\, dV = \int_a^b \int_c^d \int_e^f f(x,y,z)\, dz\, dy\, dx.$$

Example 4.3.3. *Mass of a box*

The density of the box $W = [0,2] \times [0,3] \times [0,1]$ is given by $\delta(x,y,z) = x^2 y + 3z$. Find the mass of W.

$$\int_0^2 \int_0^3 \int_0^1 x^2 y + 3z\, dz\, dy\, dx = \int_0^2 \int_0^3 x^2 yz + \frac{3}{2}z^2 \Big|_{z=0}^{1} dy\, dx$$

$$= \int_0^2 \int_0^3 x^2 y + \frac{3}{2}\, dy\, dx = \int_0^2 \frac{1}{2}x^2 y^2 + \frac{3}{2}y \Big|_{y=0}^{3} dy\, dx$$

$$= \int_0^2 \frac{9}{2}x^2 + \frac{9}{2}\, dx = \frac{3}{2}x^3 + \frac{9}{2}x \Big|_{x=0}^{2} = 21. \blacktriangle$$

Solids from Limits: Just as in double integrals, there are three features of triple integrals that are important to consider: the solid W of integration, the integrand $f(x,y,z)$, and the differential (or volume element) dV. We have already seen that the integral of density is mass. We will see many additional interpretations of the integrand that yield other interesting integrals as we proceed. For now, however, we wish to focus on the relationship between the limits and the solid of integration. The first order of business is to use the limits of a triple integral to sketch the solid of integration.

Recall from double integrals that the region R of integration in $\iint_R f(x,y)\, dA$ is a rectangle when the limits of integration are all constants. Analogously, the solid W of integration in $\iiint_W f(x,y,z)\, dV$ is a rectangular prism (a box) when all the limits are constant. One obtains the solid, as in the double integral setting, by first translating the limits into a system of inequalities.

Example 4.3.4. *Constant limits*

Sketch the solid of integration for

$$\int_{-1}^0 \int_0^3 \int_3^4 x^2 y + 2z\, dx\, dy\, dz.$$

Since we only care about the solid of integration, we can ignore the integrand entirely. Recalling that the notation for iterated integrals is nested, so that the innermost limits correspond to the first variable of integration, we get the system of inequalities:

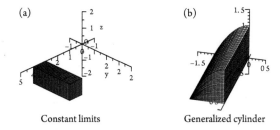

(a) Constant limits   (b) Generalized cylinder

Figure 4.3.2 Solid of integration from limits

$$3 \leq x \leq 4; \quad 0 \leq y \leq 3; \quad -1 \leq z \leq 0.$$

The box is pictured in Figure 4.3.2(a). ▲

We now look at solids of integration where two sets of limits are constant and one set contains variable expressions. The variable expression in the first example only has the variable y, which makes the bounding surface a generalized cylinder. The second example has a variable expression in x and z, making one bounding surface the graph of a function (a paraboloid in this case).

Example 4.3.5. *A generalized cylinder*

Sketch the solid of integration for

$$\int_0^3 \int_{-1}^0 \int_0^{1-y^2} f(x,y,z)\, dzdydx.$$

We get the corresponding system of inequalities from the limits:

$$0 \leq z \leq 1 - y^2$$
$$-1 \leq y \leq 0$$
$$0 \leq x \leq 3.$$

The shadow of the solid W in the xy-plane, then, is the rectangle $R = [0,3] \times [-1,0]$. The inequality $0 \leq z \leq 1 - y^2$ indicates that the solid itself lies above the plane $z = 0$ and below the generalized cylinder $z = 1 - y^2$. The solid is pictured in Figure 4.3.2(b). ▲

Example 4.3.6. *The graph of a function*

Sketch the solid of integration for

$$\int_{-1}^1 \int_{-1}^1 \int_0^{4-z^2-x^2} f(x,y,z)\, dydxdz.$$

(a)

(b)

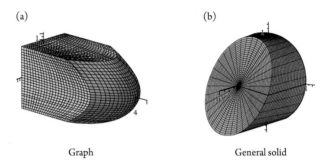

Graph   General solid

Figure 4.3.3 Solid of integration from Limits

We get the corresponding system of inequalities from the limits:

$$0 \le y \le 4 - z^2 - x^2$$
$$-1 \le x \le 1$$
$$-1 \le z \le 1.$$

The shadow of the solid W in the xz-plane, then, is the rectangle $R = [-1, 1] \times [-1, 1]$. The inequality $0 \le y \le 4 - z^2 - x^2$ indicates that the solid itself lies to the right of the plane $y = 0$ and to the left of the paraboloid $y = 4 - z^2 - x^2$. The solid is pictured in Figure 4.3.3(a). ▲

We finish the "solids from limits" paragraph with an example in which two sets of limits have variable expressions, and only the last set is constant. This is, of course, the most general case.

Example 4.3.7. *The general case*

Sketch the solid of integration for

$$\int_{-1}^{1} \int_{-\sqrt{1-z^2}}^{\sqrt{1-z^2}} \int_{0}^{6+2y-3z} f(x, y, z) \, dxdydz.$$

We get the corresponding system of inequalities from the limits:

$$0 \le x \le 6 + 2y - 3z$$
$$-\sqrt{1-z^2} \le y \le \sqrt{1-z^2}$$
$$-1 \le z \le 1.$$

Since the last two inequalities describe the unit disk in the yz-plane, that is the shadow of the solid W. The inequality $0 \le x \le 6 + 2y - 3z$ indicates that the solid itself lies in front of the plane $x = 0$ and behind the plane $x = 6 + 2y - 3z$. The solid is pictured in Figure 4.3.3(b). ▲

(a)   (b)   (c)

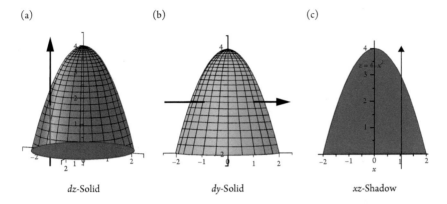

dz-Solid   dy-Solid   xz-Shadow

Figure 4.3.4 The solid W

Limits from Solids: In the previous paragraph we looked at finding solids of integration from given limits. We now reverse the process, and discuss finding limits of integration from given solids. This can be slightly challenging, simply because it's harder to visualize solids in three dimensions. The goal, then, is to reduce the problem to two dimensions as quickly as possible. We make the convention that the x-axis is pointing out of the paper at you, the y-axis to your right, and the z-axis points straight up.

A dz-solid is one that has a nice top and bottom surface. When integrating over such a solid, you integrate with respect to z first. To find limits of integration, walk along a line parallel to the z-axis which pierces the solid. The equation for the surface where you enter the solid is the lower limit of integration and the equation for the surface where you leave is the upper limit (see Figure 4.3.4). Both equations must be solved for z, since that is the variable you are integrating with respect to. Once you have the limits on z, you immediately take the shadow R of the solid W in the xy-plane. Now follow double-integral techniques to find limits on x and y.

A dy-solid has a nice left surface and right surface. To find limits, walk along a line parallel to the y-axis. The equation for the surface (solved for y of course) where you enter the solid is the lower limit, and where you leave is the upper limit (a nice example is pictured in Figure 4.3.8). A dx-solid is similar, but with nice front and back surfaces (see Figure 4.3.6).

Math App 4.3.1. *Finding limits for a dx-solid*

In the following Math App, limits for a dx-solid are illustrated using animations. Click the hyperlink below, or print users open manually, to investigate the process.

Example 4.3.8. *Finding limits for a dz-solid*

Let W be the solid below the paraboloid $x^2 + y^2 + z = 4$ and above the xy-plane (i.e. the plane $z = 0$). Describe the solid as a dz-solid.

A vertical line piercing the solid enters at the plane $z = 0$ and leaves at the paraboloid $z = 4 - x^2 - y^2$. Thus limits on z are $0 \le z \le 4 - x^2 - y^2$. Once these have been determined, immediately reduce to a two-dimensional problem by taking the shadow R of W

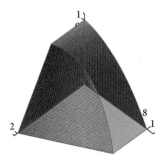

in the xy-plane. The boundary of the region R can easily be seen to be the disk of radius 2 and centered at the origin. Using techniques from the previous section we can describe the solid by the system of inequalities

$$0 \leq z \leq 4 - x^2 - y^2$$
$$-\sqrt{4 - x^2} \leq y \leq \sqrt{4 - x^2}$$
$$-2 \leq x \leq 2. \ \blacktriangle$$

In more complicated examples one can sometimes find the boundary of the region R from the equations for the top and bottom surfaces. The equations for the surfaces should already be solved for z, so setting them equal to each other gives a single equation in just x and y. This is the equation for the boundary of R when the top and bottom surfaces intersect on the edge of the solid.

Example 4.3.9. *Limits as a dy-solid.*

Express the same solid as a *dy*-solid.

Now walk along a line parallel to the y-axis that pierces the solid. The equation for the surface where you enter the solid is your lower limit on y, the surface where you leave is your upper limit. This line enters and leaves at the paraboloid $x^2 + y^2 + z = 4$ (see Figure 4.3.4(b)). Since we're integrating dy first we solve for y, yielding $y = \pm\sqrt{4 - z - x^2}$, and limits on y are

$$-\sqrt{4 - z - x^2} \leq y \leq \sqrt{4 - z - x^2}.$$

The shadow in the xz-plane is the region R under the curve $z = 4 - x^2$ (see Figure 4.3.4(c)). A line parallel to the z-axis enters R at the curve $z = 0$ and leaves at $z = 4 - x^2$. The shadow on the x-axis is the interval $[-2, 2]$. Thus the system of inequalities describing W as a *dy*-solid are:

$$-\sqrt{4 - z - x^2} \leq y \leq \sqrt{4 - z - x^2}$$
$$0 \leq z \leq 4 - x^2$$
$$-2 \leq x \leq 2. \ \blacktriangle$$

(a)

(b)

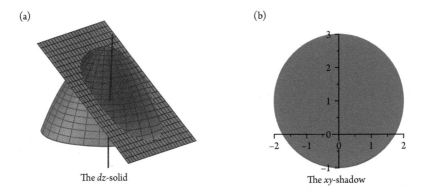

The dz-solid

The xy-shadow

Figure 4.3.5 The solid of integration

Example 4.3.10. *A slight variation*

Let W be the solid between the same paraboloid $z = 4 - x^2 - y^2$ and the plane $2y + z = 1$. Describe W as a dz-solid (see Figure 4.3.5).

A line parallel to the z-axis enters W at the plane and leaves at the paraboloid. Thus limits on z are $1 - 2y \le z \le 4 - x^2 - y^2$. We reduce to two dimensions, and the equation for the boundary of the shadow is obtained by setting the two surface equations equal to each other. Thus we get the boundary curve

$$-2y + 1 = 4 - x^2 - y^2$$
$$x^2 + y^2 - 2y + 1 = 4$$
$$x^2 + (y - 1)^2 = 4.$$

Thus the region R is the disk radius 2 and centered at $(0, 1)$. Solving the circle equation for x, then taking the shadow on the y-axis gives that W is described by:

$$1 - 2y \le z \le 4 - x^2 - y^2$$
$$-\sqrt{4 - (y - 1)^2} \le x \le \sqrt{4 - (y - 1)^2}$$
$$-1 \le y \le 3$$

For the limits on x, note that a line parallel to the x-axis enters the disk at the left semicircle (the negative square root) and leaves at the right one (the positive square root).

We finish this example by noting that any integration problem with these limits would be rather messy. This example is just intended to illustrate how to set up the limits of integration. ▲

Math App 4.3.2. *Finding limits for a tetrahedron*

The following Math App is a repeat of Math App 1.6.1. In that Math App we found a system of inequalities to describe certain tetrahedra. In particular, we consider tetrahedra with vertices $(0,0,0)$, $(x,0,0)$, $(0,y,0)$, and $(0,0,z)$ where you specify values for the x-, y-, and z-intercepts. In the present context, this process gives the limits of integration for the tetrahedron. See if you can determine the corresponding order of integration!

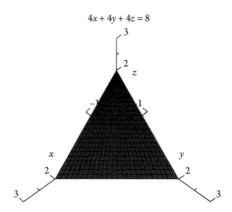

Some Examples: The remainder of this section consists of several examples illustrating different aspects of triple integration. The first two involve using triple integrals to find the volume of a solid. To derive the formula, recall that if the integrand of a double integral is 1, then $\iint_R dA = $ Area of R. Analogously, the triple integral of $f(x,y,z) = 1$ over the solid W is the volume of W, or

$$\text{Volume of } W = \iiint_W dV.$$

This can also be seen by assuming unit density using the mass calculation given at the beginning of this section, because for unit density the mass is the same magnitude as the volume (different units, of course).

Example 4.3.11. *Volume*

Find the volume of the solid bounded by the coordinate planes, the plane $y+z=1$ and $x+2y+z=6$. Note that the plane $y+z=1$ is a cylinder, since the equation is void of x, and the plane $x+2y+z=6$ slices through the first octant since its intercepts are $(6,0,0)$, $(0,3,0)$, and $(0,0,6)$. The solid bounded by the planes, then, is a dx-solid. A line parallel to the x-axis enters the solid at $x=0$ and leaves at $x=6-z-2y$. The shadow of the solid in the yz-plane is the triangle bounded by the coordinate axes and the line $y+z=1$. Thus we get the volume by calculating the iterated integral

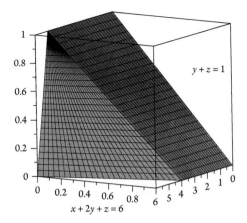

Figure 4.3.6 The solid bounded by planes

$$V = \int_0^1 \int_0^{1-y} \int_0^{6-z-2y} dxdzdy$$

$$= \int_0^1 \int_0^{1-y} x\Big|_{x=0}^{6-z-2y} dzdy = \int_0^1 \int_0^{1-y} 6 - z - 2y \, dzdy$$

$$= \int_0^1 (6-2y)z - \frac{z^2}{2}\Big|_{z=0}^{1-y} dy = \int_0^1 \frac{11}{2} - 7y + \frac{3}{2}y^2 \, dy$$

$$= \frac{11}{2}y - \frac{7}{2}y^2 + \frac{1}{2}y^3\Big|_0^1 = \frac{5}{2}. \; \blacktriangle$$

Example 4.3.12. *A generalized cylinder*

Find the volume of the solid bounded by the generalized cylinder $y = 4 - x^2$, the *xy*-plane and the plane $z = 2y$.

Step 1. Determine limits of integration.

The equation $y = 4 - x^2$ is missing a *z*, so the surface is obtained by translating the parabola $y = 4 - x^2$ vertically. The planes then cut a wedge out of the generalized cylinder as in Figure 4.3.7. The plane $z = 2y$ is the top, while the *xy*-plane $(z = 0)$ is the bottom of the wedge. This implies that the limits on *z* are

$$0 \le z \le 2y.$$

Since we have limits for *z*, we take the shadow in the *xy*-plane to find limits on *x* and *y*. The shadow is the region between the parabola $y = 4 - x^2$ and the *x*-axis. Thinking of this as a *dy*-region we get

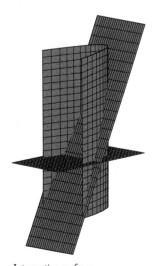

Intersecting surfaces

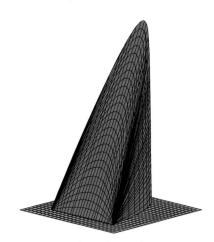

The resulting wedge

Figure 4.3.7 Volume of a wedge

$$0 \le y \le 4 - x^2$$
$$-2 \le x \le 2.$$

Step 2. Integrate

The volume is given by

$$
\begin{aligned}
V &= \int_{-2}^{2} \int_{0}^{4-x^2} \int_{0}^{2y} dz\,dy\,dz = \int_{-2}^{2} \int_{0}^{4-x^2} z\big|_{z=0}^{2y} dy\,dx \\
&= \int_{-2}^{2} \int_{0}^{4-x^2} 2y\,dy\,dx = \int_{-2}^{2} y^2\big|_{y=0}^{4-x^2} dx \\
&= \int_{-2}^{2} (4-x^2)^2 dx = \int_{-2}^{2} 16 - 8x^2 + x^4 dx \\
&= 16x - \frac{8}{3}x^3 + \frac{1}{5}x^5 \Big|_{-2}^{2} = \frac{512}{15}. \; \blacktriangle
\end{aligned}
$$

We take time to find the exact mass of the tetrahedron that we approximated in Example 4.3.2.

Example 4.3.13. *Mass of the tetrahedron*

Calculate the mass of the tetrahedron of example 4.2.6 if the density is given by $\delta(x,y,z) = 40 - z^2$.

We need to evaluate $\iiint_W \delta(x,y,z)\, dV$.

Step 1. Determine the limits of integration

Considering the tetrahedron to be a dz-solid, we walk through it parallel the z-axis. The equation for the surface where we enter the solid is the lower limit and where we leave is the upper limit (the equations must be solved for z). We enter at the xy-plane, or $z = 0$, and leave at the plane $3x + 2y + z = 6$. Solving for z gives

$$0 \leq z \leq 6 - 3x - 2y.$$

After finding limits on z, we find limits on the projection in the xy-plane. Notice that we reduce the number of dimensions as soon as we take care of limits on z. In this case, the projection is the triangle in the first quadrant bounded by the line $3x + 2y = 6$. Thinking of this as a dy-region in the plane gives the limits

$$0 \leq y \leq (6 - 3x)/2$$
$$0 \leq x \leq 2.$$

Step 2. Integrate

We see that the mass is given by

$$
\begin{aligned}
m &= \int_0^2 \int_0^{(6-3x)/2} \int_0^{6-3x-2y} 40 - z^2 \, dz dy dz \\
&= \int_0^2 \int_0^{(6-3x)/2} 40z - \frac{z^3}{3} \bigg|_{z=0}^{6-3x-2y} dy dx \\
&= \int_0^2 \int_0^{(6-3x)/2} 40(6 - 3x - 2y) - \frac{(6 - 3x - 2y)^3}{3} \, dy dx \\
&= \int_0^2 -10(6 - 3x - 2y)^2 + \frac{(6 - 3x - 2y)^4}{24} \bigg|_{y=0}^{(6-3x)/2} dx \\
&= \int_0^2 10(6 - 3x)^2 - \frac{(6 - 3x)^4}{24} \, dx \\
&= -\frac{10(6 - 3x)^3}{9} + \frac{(6 - 3x)^5}{360} \bigg|_0^2 = \frac{1092}{5} = 218.4 \, gm. \; \blacktriangle
\end{aligned}
$$

When working with double integrals, we encountered problems in which we switched the limits of integration. To do so, we had to sketch the region of integration given the limits. Similar problems can arise with triple integrals.

Example 4.3.14. *The solid of integration*

Sketch the solid of integration corresponding to the following iterated integral, then evaluate the integral.

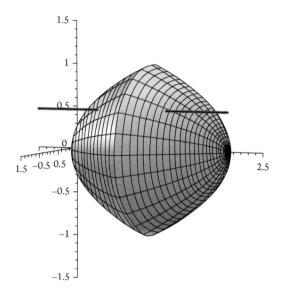

Figure 4.3.8 The solid of integration

$$\int_{-1}^{1} \int_{-\sqrt{1-x^2}}^{\sqrt{1-x^2}} \int_{x^2+z^2}^{2-x^2-z^2} dy\,dz\,dx.$$

The first limits indicate that $x^2 + z^2 \le y \le 2 - x^2 - z^2$. Thus y is between the paraboloids $y = x^2 + z^2$ and $y = 2 - x^2 - z^2$. The last two sets of limits are for the unit circle in the xz-plane, so the shadow of the solid lies inside $x^2 + z^2 = 1$. Since the curve of intersection of the paraboloids is the circle $x^2 + z^2 = 1$ lying in the plane $y = 1$, the solid of integration is the intersection of the paraboloids $y = x^2 + z^2$ and $y = 2 - x^2 - z^2$ (see Figure 4.3.8).

We now proceed to calculate the integral. The triple integral gives the volume, and by symmetry it is

$$\int_{-1}^{1} \int_{-\sqrt{1-x^2}}^{\sqrt{1-x^2}} \int_{x^2+z^2}^{2-x^2-z^2} dy\,dz\,dx = 4 \int_{0}^{1} \int_{0}^{\sqrt{1-x^2}} \int_{x^2+z^2}^{2-x^2-z^2} dy\,dz\,dx$$

$$= 4 \int_{0}^{1} \int_{0}^{\sqrt{1-x^2}} 2 - 2x^2 - 2z^2 \, dz\,dx = 4 \int_{0}^{1} \left. 2\left(1-x^2\right)z - \frac{2}{3}z^3 \right|_{z=0}^{\sqrt{1-x^2}} dx$$

$$= 4 \int_{0}^{1} \frac{4}{3}\left(1-x^2\right)^{3/2} dx = \frac{16}{3} \int_{0}^{1} \left(1-x^2\right)^{3/2} dx.$$

Now make the trigonometric substitution $x = \sin\theta$, $dx = \cos\theta\,d\theta$, and change the limits of integration to get

$$\frac{16}{3}\int_0^1 (1-x^2)^{3/2}\,dx = \frac{16}{3}\int_0^{\pi/2}(\cos^2\theta)^{3/2}\cos\theta\,d\theta$$

$$= \frac{16}{3}\int_0^{\pi/2}\cos^4\theta\,d\theta = \frac{16}{3}\int_0^{\pi/2}\left(\frac{1}{2}(1+\cos 2\theta)\right)^2 d\theta,$$

where the last equality follows from the half-angle formula $\cos^2\theta = \frac{1}{2}(1+\cos 2\theta)$. Expanding the integrand, and another application of the half-angle formula gives

$$\frac{16}{3}\int_0^{\pi/2}\left(\frac{1}{2}(1+\cos 2\theta)\right)^2 d\theta = \frac{4}{3}\int_0^{\pi/2}1+\cos 2\theta + \cos^2 2\theta\,d\theta$$

$$= \frac{4}{3}\int_0^{\pi/2}1+\cos 2\theta + \frac{1}{2}(1+\cos 4\theta)\,d\theta = \frac{4}{3}\int_0^{\pi/2}\frac{3}{2}+\cos 2\theta + \frac{1}{2}\cos 4\theta\,d\theta$$

$$= \frac{4}{3}\left(\frac{3}{2}\theta + \sin 2\theta + \frac{1}{8}\sin 4\theta\right)\Big|_0^{\pi/2} = \pi.\ \blacktriangle$$

This was a fairly complicated integral. In the next section we'll see a much easier way to evaluate it. Stay tuned!

Exercises

1. The vertices of tetrahedra are given below. Express each as a dx-, dy-, and dz-solid.
   (a) $(0,0,0)$, $(-2,0,0)$, $(0,1,0)$, $(0,0,-3)$.
   (b) $(0,0,0)$, $(1,0,0)$, $(0,1,0)$, $(0,0,1)$.
   (c) $(-1,0,0)$, $(1,0,0)$, $(0,1,0)$, $(0,0,1)$.

2. Express the tetrahedron in the first octant bounded by $x + 5y + 2z = 10$ as a dy-solid.

3. Express the tetrahedron in the first octant bounded by $7x+y+3z = 21$ as a dx-solid.

4. Express the tetrahedron in the first octant bounded by $2x + 3y + z = 6$ as a dz-solid. In exercises 5–17, express each solid as a dx-, dy-, or dz-solid. Choose the most convenient direction.

5. The solid inside the cone $z = \sqrt{x^2 + y^2}$ and under the plane $z = 3$.

6. The solid between the paraboloid $z = 4 - x^2 - y^2$ and the plane $2x + z = 4$.

7. The solid inside $x^2 + z^2 = 4$ and between $y = 0$ and $2y + z = 2$.

8. The solid inside $x^2 + y^2 = 9$ and between the xy-plane and the plane $9x + 4y + 6z = 36$.

9. The solid bounded by the xy-plane, the cylinder $x^2 + y^2 = 4$, and the plane $5x + 10y + z = 50$.

10. The solid bounded by the yz-plane, the cylinder $z^2 + y^2 = 1$, and the plane $x + y + z = 4$.

11. The solid bounded by the xz-plane, the cylinder $x^2 + z^2 = 4$, and the plane $x - 2y + z = 6$.

12. The solid bounded by the xz-plane and the paraboloid $y = x^2 + z^2 - 4$.

13. The solid bounded by the xz-plane and the paraboloid $y = 4 - x^2 - z^2$.

14. The solid bounded by the $x^2 + y^2 + z = 0$ and the plane $2x + 2y - z = 0$.

15. The solid inside the sphere $x^2 + y^2 + z^2 = 1$ and above the cone $x^2 + y^2 - z^2 = 0$.

16. The solid inside the sphere $x^2 + y^2 + z^2 = 1$ and to the right of the plane $y = 1/2$.

17. The solid inside the sphere $x^2 + y^2 + z^2 = 1$ and below the plane $z = -1/2$.

18. Let T be the tetrahedron bounded by the coordinate planes and $3x + y + 5z = 15$. Evaluate $\iiint_T dV$.

19. Sketch the solid of integration, then evaluate the following triple integrals:

    (a) $\int_0^2 \int_0^{2-y} \int_0^{2-z-y} dx dz dy$

    (b) $\int_0^1 \int_{4x^2}^4 \int_0^{\sqrt{z-4x^2}} x \, dy dz dx$

    (c) $\int_0^1 \int_{-x}^x \int_{-\sqrt{x^2-z^2}}^{\sqrt{x^2-z^2}} y \, dy dz dx$

20. Find the volume of the solid bounded by the coordinate planes, the plane $x + 2y = 4$, and the plane $x + y + z = 5$.

21. Find the volume of the solid inside the parabolic cylinder $z = x^2$ and the planes $y = 0$ and $z + y = 1$.

22. Let W be the solid in the first octant which is inside the paraboloid $z = x^2 + y^2$ and under the plane $z = 1$. Find the mass of W if the density is given by $\delta(x,y,z) = y$.

23. Compute the volume of the solid inside the cylinders $y^2 + z^2 = 1$ and $x^2 + z^2 = 1$.

4.4 Integration in Different Coordinate Systems

In this section we describe how to change coordinates in multiple integration. Each facet of multiple integrals—limits, integrand, and differential—needs to change when translating an integral to a different coordinate system. We'll begin by studying double integrals in polar coordinates, then move on to treat triple integrals in cylindrical and spherical coordinates in turn. You might want to review the other coordinate systems in Section 1.1, because familiarity with them is essential for this section. We will also find that the area form dA and volume form dV are more subtle and significant than you might think! Without further ado, let's begin.

Polar Double Integrals: We begin by discussing how polar coordinates can simplify limits of integration. Consider example 4.3.14 from the previous section. The shadow in the xz-plane was the unit disk $x^2 + z^2 \leq 1$, so limits in Cartesian coordinates were

$$-\sqrt{1-x^2} \leq z \leq \sqrt{1-x^2}; \quad -1 \leq x \leq 1.$$

These limits made the integration difficult, requiring a trigonometric substitution. The region of integration, however, has a very simple description using polar coordinates in the xz-plane, where r is the distance to the origin in the xz-plane and θ the angle with the positive x-axis. The unit disk is all points within one unit of the origin (regardless of the angle θ), so the following system of inequalities describes the region in polar coordinates:

$$0 \leq r \leq 1; \quad 0 \leq \theta \leq 2\pi.$$

In this coordinate system the limits of integration are constants, as opposed to unpleasant square roots. Instead of integrating with respect to Cartesian coordinates, then, we'd rather integrate with respect to polar. The next few examples introduce the mechanics of integration in polar coordinates, while the polar differential is justified after gaining some experience with the technique.

Example 4.4.1. *Double integral in polar coordinates*

Integrate the function $f(x,y) = x^2$ over the portion R of the unit disk in the first quadrant. In Cartesian coordinates we have

$$\iint_R f(x,y)\, dA = \int_0^1 \int_0^{\sqrt{1-x^2}} x^2\, dy dx = \int_0^1 \int_0^{\sqrt{1-y^2}} x^2\, dx dy.$$

Either order of integration gets messy because of the square root, so we switch to polar coordinates. The following steps outline the necessary adjustments.

Step 1. Limits in polar coordinates
   Using polar coordinates, we see that the quarter circle R is described by the system of inequalities

$$0 \leq r \leq 1 \quad 0 \leq \theta \leq \frac{\pi}{2}.$$

Step 2. Integrand in polar coordinates

Since we're working in polar coordinates, we have to substitute $x = r\cos\theta$ and $y = r\sin\theta$ in the integrand $f(x, y) = x^2$. This gives the integrand

$$f(r\cos\theta, r\sin\theta) = r^2 \cos^2\theta.$$

Step 3. dA in polar coordinates

One final change is required, and it may be surprising at first glance. Historically it represented an infinitesimal area, coming from the limit in the partitioning process of the Riemann sum. We'll go deeper into the description later, for now it suffices to say that in polar coordinates

$$dA = r \, dr d\theta.$$

Do not forget the "extra" r that pops up out of nowhere! Using

- the limits we found on polar coordinates,
- our integrand translated into polar coordinates, and
- the polar version of dA,

we have

$$\iint_R f(x, y) \, dA = \int_0^{\pi/2} \int_0^1 r^2 \cos^2\theta \, r dr d\theta = \int_0^1 \int_0^1 r^3 \cos^2\theta \, dr d\theta$$

$$= \int_0^{\pi/2} \cos^2\theta \left.\frac{r^4}{4}\right|_{r=0}^1 d\theta = \frac{1}{4} \int_0^{\pi/2} \cos^2\theta \, d\theta$$

$$= \frac{1}{8} \int_0^{\pi/2} 1 + \cos 2\theta \, d\theta = \frac{1}{8} \left.\left(\theta + \frac{\sin 2\theta}{2}\right)\right|_0^{\pi/2} = \frac{\pi}{16}.$$

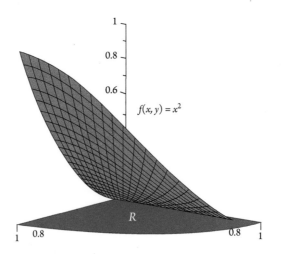

Figure 4.4.1 The region of integration suggests polar coordinates

Since $f(x,y) \geq 0$ on R in this case, the integral has a geometric interpretation. Namely, it is the volume of the solid under the surface $z = x^2$ and above the quarter unit disk R (see Figure 4.4.1). ▲

This example illustrates the essential features for evaluating integrals in polar coordinates. We summarize the process below.

Double Integrals in Polar Coordinates

To evaluate $\iint_R f(x,y)\,dA$ using polar coordinates:

1. **Limits:** Find limits for R in terms of r and θ.

2. **Integrand:** Evaluate and simplify $f(r\cos\theta, r\sin\theta)$.

3. **Integrate:** Let $dA = r\,dr\,d\theta$ and integrate.

Math App 4.4.1. *Constant limits of integration*

When using Cartesian coordinates we found that the region of integration for $\int_a^b \int_c^d f(x,y)\,dx\,dy$ is the rectangle $R = [a,b] \times [c,d]$. One wonders what types of regions arise in polar coordinates when the limits are constants, i.e. when evaluating $\int_a^b \int_c^d f(r\cos\theta, r\sin\theta)\,r\,dr\,d\theta$. Click the following hyperlink, or print users open manually, to investigate the constant limit setting in polar coordinates.

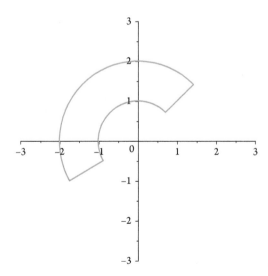

Another situation in which polar integrals are useful is when the region of integration is given by polar equations. Here is such an example.

Example 4.4.2. *Regions given by polar equations*

Find the area inside the cardioid $r = 1 + \cos\theta$.

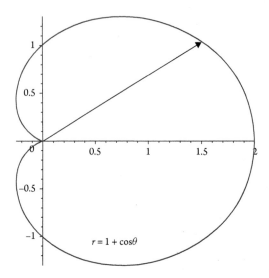

Figure 4.4.2 Finding the area of a cardioid

Step 1. Polar limits

To find limits of integration on r, sketch a ray emanating from the origin that intersects the region. The equation for the curve where it enters your region is the lower limit, and the equation for the curve where your ray leaves the region is your upper limit (both equations should be solved for r). Limits on θ are those angles for which the ray intersects the region. In this case, the region is described by

$$0 \le r \le 1 + \cos\theta$$
$$0 \le \theta \le 2\pi.$$

Step 2. Polar integrand

The integrand is $f(x,y) = 1$ since we're trying to find area, which is the same in polar coordinates.

Step 3. Polar dA

Since $dA = r\,dr d\theta$, the area enclosed by the cardioid is given by

$$\iint_R dA = \int_0^{2\pi} \int_0^{1+\cos\theta} r\,dr d\theta = \int_0^{2\pi} \frac{r^2}{2}\Big|_{r=0}^{1+\cos\theta} d\theta$$

$$= \frac{1}{2} \int_0^{2\pi} 1 + 2\cos\theta + \cos^2\theta\, d\theta = \frac{1}{2} \int_0^{2\pi} \frac{3}{2} + 2\cos\theta + \frac{\cos 2\theta}{2}\, d\theta$$

$$= \frac{1}{2} \left(\frac{3}{2}\theta + 2\sin\theta + \frac{\sin 2\theta}{4} \right)\Big|_0^{2\pi} = \frac{3\pi}{2}. \ \blacktriangle$$

This is the first example where we used functions to find limits of integration in polar coordinates. For your reference we summarize the process in the following box.

Finding Limits in Polar Coordinates

- Limits on r: Draw a ray emanating from the origin that intersects the region R. The equation for the curve where it enters R is your lower limit and the equation for the curve where it leaves R is your upper limit (both equations solved for r, of course).
- Limits on θ: Determine for which angles θ the rays go through R.

Example 4.4.3. *Limits in polar coordinates*

Let R be the region inside the unit circle and outside the cardioid $r = 1 - \sin\theta$. Describe R using a system of inequalities in polar coordinates.

Draw a ray from the origin that passes through R, as in Figure 4.4.3. It enters the region at the cardioid $r = 1 - \sin\theta$ and leaves at the unit circle $r = 1$. Moreover, rays pass through R for angles θ between 0 and π. Thus R is described by the system

$$1 - \sin\theta \le r \le 1$$
$$0 \le \theta \le \pi. \blacktriangle$$

Justification that $dA = rdrd\theta$: We now know the mechanics of integration in polar coordinates, so let's look at where the extra r comes from in the area form. Recall that

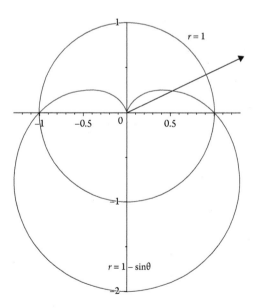

Figure 4.4.3 Limits in polar coordinates

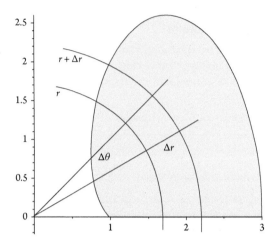

Figure 4.4.4 Polar partitions yield $dA = r\,dr\,d\theta$

to define the double integral over a rectangle R, we started by partitioning R into small sub-rectangles. These came from using partitions in the x and y directions. Alternatively, one could use the r and θ directions to partition the region as in Figure 4.4.4. The result is a partitioning of the region into portions of circular sectors.

To integrate a function $f(x,y)$ in polar coordinates, then we approximate the integral $\iint_R f(x,y)\,dA$ as follows:

1. Using the above partition, find the area of each sub-sector. To do this, recall that the area that an angle $\Delta\theta$ cuts out of a circle radius r is given by $\frac{\Delta\theta}{2\pi}\pi r^2 = \frac{\Delta\theta}{2}r^2$. This follows from the fact that the ratio of the areas of the sector to the circle is the same as the ratio of the angle to 2π. Now let the inner radius be r, and Δr be the change in radius, so the outer radius is $r + \Delta r$. Then the area of the sub-sector is the outer area minus the inner:

$$\frac{\Delta\theta}{2}(r+\Delta r)^2 - \frac{\Delta\theta}{2}r^2 = \frac{\Delta\theta}{2}(2r+\Delta r)\,\Delta r = \left(r + \frac{\Delta r}{2}\right)\Delta r\Delta\theta = r_a\Delta r\Delta\theta$$

where r_a is the average of the two boundary radii.

2. Pick a point $(r_i\cos\theta_j, r_i\sin\theta_j)$ in each sub-sector, and evaluate f there. It turns out the limit is independent of choice here, so we make the choice that $r_i = r_a$. For each sub-sector we get the contribution

$$f(r_i\cos\theta_j, r_i\sin\theta_j)r_i\Delta r_i\Delta\theta_j$$

to the Riemann sum.

3. Approximate the integral with the double-sum of the contributions from each sector

$$\iint_R f(x,y)\,dA \approx \sum_{i=1}^{n}\sum_{j=1}^{n} f(r_i\cos\theta_j, r_i\sin\theta_j)r_i\Delta r_i\Delta\theta_j.$$

4. Take a limit as the partition gets finer, to find the integral. Notice that the Riemann sum in the previous step is formally the same as our original double integral, but where the function is $f(r\cos\theta, r\sin\theta)r$ (not *just* $f(r\cos\theta, r\sin\theta)$), followed by the change in variables. Therefore in the limit we get the extra r:

$$\iint_R f(x,y)\ dA = \int_a^b \int_c^d f(r\cos\theta, r\sin\theta)r\,dr\,d\theta.$$

Thus we've seen the "extra" r in the polar differential dA arises naturally when calculating areas of circular sectors, and in fact is a necessary component of polar integration.

Cylindrical Triple Integrals: We now introduce triple integrals in cylindrical and spherical coordinates. The techniques are similar to double integrals in polar coordinates, with minor modifications.

Triple Integrals in Cylindrical Coordinates

To evaluate $\iiint_W f(x,y,z)\ dV$ using cylindrical coordinates:

1. Find limits for W in terms of z, r and θ.
2. Evaluate and simplify $f(r\cos\theta, r\sin\theta, z)$.
3. Let $dV = rdzdrd\theta$ and integrate.

Math App 4.4.2. *Constant limits in cylindrical coordinates*

You may have noticed that the previous Math App on constant limits in polar coordinates had cylindrical and spherical as well. Feel free to review the solids you obtain in those coordinate systems now.

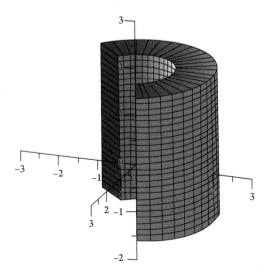

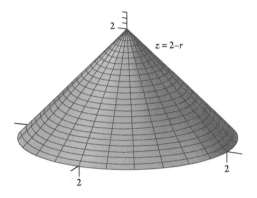

$z = 2-r$

2

2

2

Figure 4.4.5 The cone $z = 2 - \sqrt{x^2 + y^2}$

Example 4.4.4. *Mass of a cone*

Find the mass of the cone below $z = 2 - \sqrt{x^2 + y^2}$ and above the xy-plane if the density is given by $\delta(x, y, z) = 3 - z$ (see Figure 4.4.5).

Before we begin the process, note that the equation of the cone in cylindrical coordinates is $z = 2 - r$, which is considerably nicer than its Cartesian equation. This is one hint that using cylindrical coordinates might make calculations easier.

Step 1. Cylindrical limits

We find limits of integration on z first, then project into the xy-plane and use polar coordinates there. The cone is above the plane $z = 0$ and below the surface $z = 2 - r$, so limits on z are

$$0 \le z \le 2 - r.$$

Now project into the xy-plane and use polar coordinates. The projection into the xy-plane is the disk radius 2, centered at the origin. Using polar coordinates, any point (r, θ) in the region has r at most 2 and no restriction on θ, so the disk is described by the system of inequalities

$$0 \le r \le 2; \quad 0 \le \theta \le 2\pi.$$

Step 2. Cylindrical integrand

We must translate the integrand into cylindrical coordinates. Since the integrand is $\delta(x, y, z) = 3 - z$, and z is a cylindrical coordinate, there is no change required.

Step 3. Cylindrical dV

Now that we have the limits of integration, and the integrand, we use the appropriate volume form and integrate.

$$\iiint_W \delta \, dV = \int_0^{2\pi} \int_0^2 \int_0^{2-r} (3-z) r \, dz \, dr \, d\theta$$

$$= \int_0^{2\pi} \int_0^2 \left. 3rz - \frac{r}{2}z^2 \right|_{z=0}^{2-r} dr \, d\theta = \int_0^{2\pi} \int_0^2 4r - r^2 - \frac{1}{2}r^3 \, dr \, d\theta$$

$$= \int_0^{2\pi} \left. 2r^2 - \frac{1}{3}r^3 - \frac{1}{8}r^4 \right|_{r=0}^2 d\theta = \int_0^{2\pi} \frac{10}{3} \, d\theta = \frac{20\pi}{3}. \; \blacktriangle$$

Remark: In cylindrical coordinates we usually use polar coordinates in the *xy*-plane, together with the Cartesian coordinate *z*. There is nothing special, however, about which coordinate is Cartesian and which two are polar. We could just as easily have used polar coordinates in the *yz*-plane and *x* for the Cartesian. The appropriate adjustment in the volume form would be $dV = r \, dx \, dr \, d\theta$. The following example uses *y* as the Cartesian coordinate.

Example 4.4.5. *Paraboloids revisited*

Recall Example 4.3.14, where we calculated the volume of the solid between the paraboloids $x^2 + y + z^2 = 2$ and $y = x^2 + z^2$. The integration was fairly complex, but it will be greatly simplified by using cylindrical coordinates r, θ, y. Notice that we're not using *z* as the third axis, but we're using polar coordinates in the *xz*-plane. Thus we'll let $x = r\cos\theta, y = y,$ and $z = r\sin\theta$.

Step 0. Cylindrical surface equations

Before starting the problem, let's get all the equations in the appropriate coordinates. Straightforward substitutions give the equations $y = 2 - r^2$ and $y = r^2$.

Step 1. Cylindrical limits

To find limits of integration, first find limits on *y*, then project into the *xz*-plane as before. In cylindrical coordinates we see $r^2 \le y \le 2 - r^2$. The shadow in the *xz*-plane is still the unit disk so limits are $0 \le r \le 1$ and $0 \le \theta \le 2\pi$.

Step 2. Cylindrical integrand

Since we are finding volume the integrand is the constant one—in any coordinate system. So no substitutions are necessary to adjust the integrand.

Step 3. Cylindrical dV

The volume form is $dV = r \, dy \, dr \, d\theta$, so the volume is given by

$$\iiint_W dV = \int_0^{2\pi} \int_0^1 \int_{r^2}^{2-r^2} r \, dy \, dr \, d\theta$$

$$= \int_0^{2\pi} \int_0^1 \left. ry \right|_{y=r^2}^{2-r^2} dr \, d\theta = \int_0^{2\pi} \int_0^1 2r - 2r^3 \, dr \, d\theta$$

$$= \int_0^{2\pi} \left. r^2 - \frac{1}{2}r^4 \right|_{r=0}^1 d\theta = \int_0^{2\pi} \frac{1}{2} \, d\theta = \pi.$$

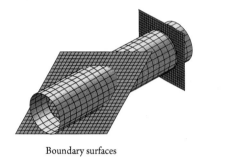

Boundary surfaces   The solid

Figure 4.4.6 A cylindrical solid along the x-axis

Notice how much easier cylindrical coordinates were than Cartesian in this problem. ▲

Example 4.4.6. *A cylinder in the x-direction*

Find the volume of the solid inside the cylinder $y^2 + z^2 = 4$ and between the yz-plane and $x + y + z = 10$.

Step 0. Cylindrical surface equations

The cylinder is the circle radius 2 in the yz-plane translated along the x-axis. The plane $x + y + z = 10$ cuts the cylinder in front of the yz-plane as in Figure 4.4.6.

Using cylindrical coordinates in the yz-plane, letting $y = r\cos\theta$ and $z = r\sin\theta$, and x be the coordinate, which remains Cartesian, we can translate the equations for the bounding surfaces into cylindrical coordinates. The cylinder $y^2 + z^2 = 4$ becomes $r = 2$, the yz-plane remains $x = 0$, while the slanted plane becomes $x = 10 - r\cos\theta - r\sin\theta$.

Step 1. Cylindrical limits

The solid has a nice front and back surface, so it is a dx-solid. The back surface is $x = 0$, while the front is the slanted plane, yielding the limits on x:

$$0 \leq x \leq 10 - r\cos\theta - r\sin\theta.$$

The projection of the solid in the yz-plane is the disk centered at the origin with radius 2. Using polar coordinates in the yz-plane, the disk can be described by $0 \leq r \leq 2, 0 \leq \theta \leq 2\pi$, resulting in the system

$$0 \leq x \leq 10 - r\cos\theta - r\sin\theta; \quad 0 \leq r \leq 2; \quad 0 \leq \theta \leq 2\pi.$$

Step 2. Cylindrical integrand

We want volume, so the integrand is the constant function 1, which is the same in any coordinate system.

Step 3. Cylindrical dV.

Using the volume form $dV = r\,dx\,dr\,d\theta$, we calculate

$$\iiint_W dV = \int_0^{2\pi} \int_0^2 \int_0^{10-r\cos\theta-r\sin\theta} r\,dx\,dr\,d\theta$$

$$= \int_0^{2\pi} \int_0^2 rx\Big|_{x=0}^{10-r\cos\theta-r\sin\theta}\,dr\,d\theta = \int_0^{2\pi} \int_0^2 10r - r^2\cos\theta - r^2\sin\theta\,dr\,d\theta$$

$$= \int_0^{2\pi} \int_0^2 5r^2 - \frac{r^3}{3}\cos\theta - \frac{r^3}{3}\sin\theta\Big|_{r=0}^{2}\,dr\,d\theta$$

$$= \int_0^{2\pi} 20 - \frac{8}{3}\cos\theta - \frac{8}{3}\sin\theta\,d\theta = 20\theta - \frac{8}{3}\sin\theta + \frac{8}{3}\cos\theta\Big|_0^{2\pi} = 40\pi. \; \blacktriangle$$

It seems natural that if the polar differential is $dA = r\,dr\,d\theta$, then the cylindrical differential should be $dV = r\,dz\,dr\,d\theta$, but we give an intuitive justification anyway. Recall that the differential dV of the integral $\iiint_W f\,dV$ represents an infinitesimal change in volume resulting from infinitesimal changes in your variables. To find dV, first consider the solid resulting from small changes $\Delta r, \Delta\theta, \Delta z$ in your variables (see Figure 4.4.7). Since r is the distance to the z-axis in $\mathbb{R}^3$, changing r by Δr moves you away from the z-axis. Changing z by Δz results in a vertical change by Δz units. Finally a change of $\Delta\theta$ in the θ-variable increases the angle with the z-axis.

The resulting solid is a portion of a cylindrical shell, and we must approximate its volume ΔV. The solid is not a box, but for small changes $\Delta r, \Delta\theta, \Delta z$ let's pretend it is. If it's a box, then its volume is length times width times height. The width and height of the

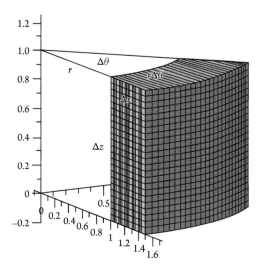

Figure 4.4.7 Cylindrical solid with volume ΔV

solid in Figure 4.4.7 are Δr and Δz, respectively. The "length" of the cylindrical shell is approximately the length of the innermost circular arc.

The length of the innermost arc is *not* $\Delta\theta$ since $\Delta\theta$ is a change in the angle, and is not even a distance! However, $\Delta\theta$ *can be used* to calculate the length. To do so, recall that if the angle α cuts an arc of length s out of a circle radius r, then the radian measure of α is given by $\alpha = s/r$ (draw a quick sketch of this scenario to see where all the pieces fit). To find the length of the innermost arc, then, note that the angle is $\Delta\theta$ and the radius is r (the initial radius, before changing by Δr). Thus the arclength is $s = r\Delta\theta$.

OK, read the last two paragraphs again, looking at Figure 4.4.7 as you do. This is not beyond you, and helps develop some intuition!

At this point we can approximate ΔV, the volume of the cylindrical shell, using the assumption that $\Delta r, \Delta\theta, \Delta z$ are small enough that the shell is close to a box. In this case, ΔV is the product of its dimensions, and we've just found those, so

$$\Delta V \approx r\Delta\theta\,\Delta r\Delta z.$$

This approximation improves as $\Delta r, \Delta\theta, \Delta z$ get smaller, and in the limit the changes in values become differentials, so

$$dV = r d\theta\, dr dz.$$

This is not a rigorous discussion with epsilons and deltas, but it emphasizes the rule-of-thumb in calculus that things become linear when you look close enough.

Spherical Triple Integrals: We've investigated several examples of integration in polar and cylindrical coordinates, and it is time to turn our attention to spherical coordinates. As in the previous cases, it turns out that certain solids of integration, and certain integrands, are best described using spherical coordinates. Moreover, the steps to integrating in spherical coordinates are very similar to those we've already encountered.

Triple Integrals in Spherical Coordinates

To evaluate $\iiint_W f(x,y,z)\,dV$ using spherical coordinates:

1. Find limits for W in terms of ρ, ϕ and θ.
2. Evaluate and simplify $f(\rho\sin\phi\cos\theta, \rho\sin\phi\sin\theta, \rho\cos\phi)$.
3. Let $dV = \rho^2\sin\phi d\rho d\phi d\theta$ and integrate.

The differential $dV = \rho^2\sin\phi d\rho d\phi d\theta$ in spherical coordinates is the most complicated yet, but it is derived in much the same manner as the cylindrical differential. Let ΔV be the volume of the solid obtained by making small changes $\Delta\rho, \Delta\phi, \Delta\theta$ in the spherical coordinates. This solid is part of a spherical shell, and pictured in Figure 4.4.8. This spherical shell has one straight edge (corresponding to $\Delta\rho$), and two edges that are

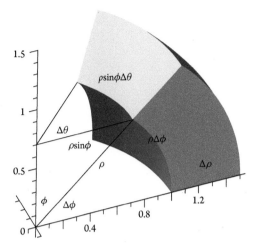

Figure 4.4.8 Spherical solid with volume ΔV

circular arcs (corresponding to $\Delta\phi$ and $\Delta\theta$). As in the cylindrical coordinate example, we approximate the volume ΔV by assuming all edges are straight and just multiplying their lengths.

The length of the straight edge is $\Delta\rho$. The vertical circular arc is opposite angle $\Delta\phi$, and lies in a circle radius ρ. Since arclength is the angle times radius (as described in the cylindrical differential justification above), the length of the vertical arc is $\rho\Delta\phi$. The horizontal circular arc subtends an angle $\Delta\theta$ in a circle radius $\rho\sin\phi$ (take two— three?—minutes to believe this), so its length is $\rho\sin\phi\Delta\theta$. Multiplying these lengths approximates the volume with

$$\Delta V \approx \Delta\rho \cdot \rho\Delta\phi \cdot \rho\sin\phi\Delta\theta = \rho^2\sin\phi\Delta\rho\Delta\phi\Delta\theta.$$

This approximation improves the smaller $\Delta\rho, \Delta\phi, \Delta\theta$ get, so in the limit we have the differential

$$dV = \rho^2\sin\phi\, d\rho\, d\phi\, d\theta.$$

With this intuition on the spherical differential, some examples are in order.

Example 4.4.7. *Volume of a snowcone*

Find the volume of the region above the cone $z^2 = x^2 + y^2$ and inside the unit sphere (see Figure 4.4.9).

Step 0. Spherical surface equations

Recall that the cone has spherical equation $\phi = \frac{\pi}{4}$ and the sphere $\rho = 1$. The cone equation can be obtained making substitutions $x = \rho\sin\phi\cos\theta, y = \rho\sin\phi\sin\theta$,

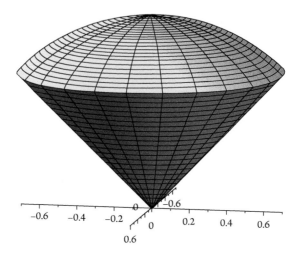

Figure 4.4.9 A snowcone

$z = \rho \cos\phi$, and simplifying, but that is rather laborious. Hopefully you remember that the cone is a constant coordinate surface, as described in Section 1.1!

Step 1. Spherical limits

To find limits on ρ, draw a ray emanating from the origin that intersects your solid. The equation for the surface where it enters the solid is your lower limit, and the equation for the surface where it leaves is your upper limit. In our case, the ray starts in the solid and leaves at $\rho = 1$, so limits are $0 \le \rho \le 1$. To find limits on ϕ, start with the positive z-axis $\phi = 0$, and let the ray drop until it reaches the boundary of your solid. In our case, the ray starts in our solid, and leaves at $\phi = \frac{\pi}{4}$. Thus limits are $0 \le \phi \le \frac{\pi}{4}$. Finally, ask for which angles θ the rays intersect the solid, leaving $0 \le \theta \le 2\pi$.

Step 2. Spherical integrand

Since volume is obtained by integrating the constant function $f(x,y,z) = 1$, the integrand is unchanged by switching to spherical coordinates.

Step 3. Spherical dV

The volume is given by

$$\iiint_W dV = \int_0^{2\pi} \int_0^{\pi/4} \int_0^1 \rho^2 \sin\phi \, d\rho \, d\phi \, d\theta$$

$$= \int_0^{2\pi} \int_0^{\pi/4} \frac{\rho^3}{3} \sin\phi \Big|_{\rho=0}^1 d\phi \, d\theta = \int_0^{2\pi} \int_0^{\pi/4} \frac{1}{3} \sin\phi \, d\phi \, d\theta$$

$$= \int_0^{2\pi} -\frac{1}{3} \cos\phi \Big|_0^{\pi/4} d\theta = \int_0^{2\pi} \frac{2 - \sqrt{2}}{6} d\theta = \frac{(2 - \sqrt{2})\pi}{3}. \ \blacktriangle$$

Math App 4.4.3. *Spherical limits of integration*

In this Math App we explore finding limits of integration in spherical coordinates. The limits split into two sets of inequalities because the surfaces where rays from the origin leave the solid change for different values of ϕ.

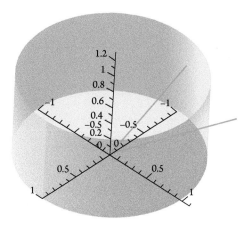

Example 4.4.8. *Another volume example*

Find the volume of the solid inside $\rho = 1 - \cos\phi$, for $0 \le \phi \le \pi$ (see Figure 4.4.10).

Step 1. Spherical limits

A ray emanating from the origin enters the solid at $\rho = 0$ and leaves at the surface $\rho = 1 - \cos\phi$. The limits for ϕ are given, and there is no restriction on θ. Thus the solid is described by

$$0 \le \rho \le 1 - \cos\phi$$
$$0 \le \phi \le \pi$$
$$0 \le \theta \le 2\pi.$$

Step 2. Spherical integrand

We want volume, so the integrand is 1.

Step 3. Spherical dV

Letting $dV = \rho^2 \sin\phi\, d\rho\, d\phi\, d\theta$ we get

$$\iiint_W dV = \int_0^{2\pi} \int_0^\pi \int_0^{1-\cos\phi} \rho^2 \sin\phi\, d\rho\, d\phi\, d\theta$$

$$= \int_0^{2\pi} \int_0^\pi \frac{\rho^3}{3} \sin\phi \bigg|_{\rho=0}^{1-\cos\phi} d\phi\, d\theta = \int_0^{2\pi} \int_0^\pi \frac{(1-\cos\phi)^3}{3} \sin\phi\, d\phi\, d\theta.$$

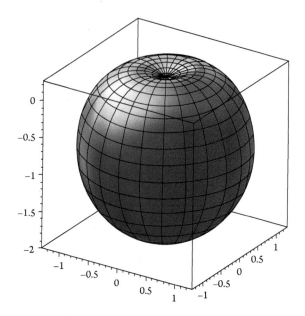

Figure 4.4.10 The surface $\rho = 1 - \cos\phi$

Using the substitution $u = 1 - \cos\phi$, $du = \sin\phi \, d\phi$, and changing the limits of integration gives

$$\iiint_W dV = \int_0^{2\pi} \int_0^2 \frac{u^3}{3} \, du \, d\theta = \int_0^{2\pi} \frac{u^4}{12}\bigg|_0^2 \, d\theta = \int_0^{2\pi} \frac{4}{3} \, d\theta = \frac{8\pi}{3}. \; \blacktriangle$$

Example 4.4.9. *Mass with variable density*

Determine the mass of the unit ball B^3 if the density (in gm/cm^3) is

$$\delta(x,y,z) = \frac{1}{1 + x^2 + y^2 + z^2}.$$

Step 1. Spherical limits

Find limits of integration in spherical coordinates.

In this case, the solid of integration is the unit ball in $\mathbb{R}^3$, which is described using spherical coordinates by the system of inequalities

$$0 \leq \rho \leq 1; \quad 0 \leq \phi \leq \pi; \quad 0 \leq \theta \leq 2\pi.$$

Step 2. Spherical integrand

To find mass, we integrate density, so we must translate the density function into spherical coordinates. Since $\rho^2 = x^2 + y^2 + z^2$, the density function is

$$\delta = \frac{1}{1 + \rho^2}.$$

Step 3. Spherical dV

We integrate in spherical coordinates:

$$M = \iiint_{B^3} \delta(x,y,z)\,dV = \int_0^{2\pi}\int_0^{\pi}\int_0^1 \frac{1}{1+\rho^2}\rho^2\sin\phi\,d\rho\,d\phi\,d\theta$$

Now $\frac{\rho^2}{1+\rho^2} = 1 - \frac{1}{1+\rho^2}$, so we have

$$M = \int_0^{2\pi}\int_0^{\pi}\int_0^1 \left(1 - \frac{1}{1+\rho^2}\right)\sin\phi\,d\rho\,d\phi\,d\theta$$

$$= \int_0^{2\pi}\int_0^{\pi} \rho - \tan^{-1}\rho\Big|_{\rho=0}^{1}\sin\phi\,d\phi\,d\theta$$

$$= \left(1 - \frac{\pi}{4}\right)\int_0^{2\pi}\int_0^{\pi}\sin\phi\,d\phi\,d\theta = \left(1 - \frac{\pi}{4}\right)4\pi = 4\pi - \pi^2. \; \blacktriangle$$

Change of Variables Formula: Thus far we've seen how to integrate functions in Cartesian, polar, cylindrical, and spherical coordinates. In each case we had to pay attention to limits of integration, the integrand, and the differential. The differential for each coordinate system was found using approximations and geometry, but all of them follow from a more general concept—the change of variables formula—which we now describe. The main take-away from this paragraph is that if $g : \mathbb{R}^n \to \mathbb{R}^n$, then the determinant of the derivative matrix Dg is a local measure of how much g stretches volume. This paragraph can be skipped with little consequence for the remainder of the text.

As a bit of notation, let $\det Dg$ denote the determinant of the derivative matrix Dg. Let's verify the take-away statement for the function $g : \mathbb{R}^2 \to \mathbb{R}^2$ from the $r\theta$-plane to the xy-plane given by $g(r,\theta) = (r\cos\theta, r\sin\theta)$ (this is the standard change of coordinates). Consider a small rectangle U in the $r\theta$-plane determined by three vertices (r,θ), $(r+\Delta r, \theta), (r, \theta+\Delta\theta)$. The function g maps U to a small sector R in the xy-plane determined by the vertices $g(r,\theta), g(r+\Delta r, \theta), g(r, \theta+\Delta\theta)$ (see Figure 4.4.11). In our discussion on differentiability we saw that

$$g(r+\Delta r, \theta) - g(r,\theta) \approx g_r(r,\theta)\Delta r \text{ and } g(r,\theta+\Delta\theta) - g(r,\theta) \approx g_\theta(r,\theta)\Delta\theta$$

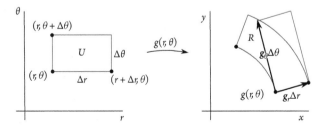

Figure 4.4.11 Area locally stretched by $|\det Dg|$

(see Example 3.5.6 for the idea). Thus the area of R is approximately the area of the parallelogram determined by the vectors $g_r(r,\theta)\Delta r$ and $g_\theta(r,\theta)\Delta\theta$.

In Section 2.4 we saw that given two vectors in the plane, the area of the parallelogram they determine is the determinant of the matrix with those vectors as rows (or columns). Letting the columns of our matrix be $g_r(r,\theta)\Delta r$ and $g_\theta(r,\theta)\Delta\theta$, we calculate

$$\begin{vmatrix} \cos\theta\,\Delta r & -r\sin\theta\,\Delta\theta \\ \sin\theta\,\Delta r & r\cos\theta\,\Delta\theta \end{vmatrix} = r\cos^2\theta\,\Delta r\Delta\theta + r\sin^2\theta\,\Delta r\Delta\theta = \begin{vmatrix} \cos\theta & -r\sin\theta \\ \sin\theta & r\cos\theta \end{vmatrix}\Delta r\Delta\theta.$$

$$(4.4.1)$$

Now the area of U is $\Delta r\Delta\theta$, and for our g we have

$$\det Dg = \begin{vmatrix} \cos\theta & -r\sin\theta \\ \sin\theta & r\cos\theta \end{vmatrix}.$$

Thus the calculation of Equation 4.4.1 shows that the area of the parallelogram (the left-hand expression) is $|\det Dg|$ times the area of U (the right-hand expression). Since the area of $R = g(U)$ is almost the area of the parallelogram, we see

$$\text{Area of } g(U) \approx |\det Dg| \cdot (\text{Area of } U).$$

This shows that $|\det Dg|$ measures how much g stretches area.

The above argument depended on approximating changes in function values using derivatives, and the fact that determinants give area. The derivative approximations hold in any dimension (given appropriate assumptions on the function g), and it is true that determinants of larger matrices give n-dimensional volumes. Thus the intuitive arguments above hold in much more general circumstances. We proceed with the general theorem.

Let $f : \mathbb{R}^2 \to \mathbb{R}$ be an integrable function defined on a region R in the xy-plane, and let $g : \mathbb{R}^2 \to \mathbb{R}^2$ be a continuously differentiable function from the vw-plane to the xy-plane. Further, suppose that U is a subset of the vw-plane such that $g(U) = R$. (There are several technical assumptions on the functions f and g; the interested reader is referred to [8] page 326.) The statement of the change of variables theorem is

Theorem 4.4.1. *Let f, R, g, and U be as above. Then*

$$\iint_R f(x,y)\,dxdy = \iint_U f(g(v,w))\,|\det Dg|\,dvdw.$$

Note that we are taking the absolute value of the determinant of Dg in the right-hand side of this equation. This is because it comes from area considerations, so it needs to be positive. Before indicating why this result is true, we illustrate that the formula of Theorem 4.4.1 agrees with techniques already presented.

Example 4.4.10. *Change of variables*

Let R be the unit disk in the xy-plane, and $f(x,y) = x^2 - 2xy + y^2$. Set up the integral $\iint_U f(g(v,w)) \left| \det Dg \right| dr d\theta$ using the change-of-variables function $g(r,\theta) = (r\cos\theta, r\sin\theta)$.

Example 3.1.2 shows that the rectangle $U = [0,1] \times [0,2\pi]$ gets mapped to the unit disk by g (see Figure 3.1.3), so limits for r and θ are given by

$$0 \leq r \leq 1; \quad 0 \leq \theta \leq 2\pi.$$

Direct substitution shows that $f(g(r,\theta)) = r^2 - 2r^2\cos\theta\sin\theta$, and direct computation shows

$$\det Dg = \begin{vmatrix} \cos\theta & -r\sin\theta \\ \sin\theta & r\cos\theta \end{vmatrix} = r\cos^2\theta + r\sin^2\theta = r. \tag{4.4.2}$$

Placing all these calculations in the right-hand side of Theorem 4.4.1 gives

$$\iint_R f(x,y)\,dxdy = \int_0^{2\pi} \int_0^1 \left(r^2 - 2r^2\cos\theta\sin\theta \right) r\,dr d\theta.$$

The reader can confirm that the polar integral techniques described earlier yield the same result. Note that the extra r in the differential $dA = r dr d\theta$ is the result of the determinant in this context, as opposed to coming from approximating ΔA_{ij}. More on this to come. ▲

The point of this example was to show that the formula of Theorem 4.4.1 gives the expected integral. Moreover, each component of the polar double integral—the limits of integration, the integrand, the differential—can be interpreted in terms of the map $g : \mathbb{R}^2 \to \mathbb{R}^2$.

To argue for Theorem 4.4.1 in the case of polar coordinates, we will return to the definition of the double integral. Recall that to evaluate $\iint_R f(x,y)dA$ you first partition R into sub-regions of area ΔA_{ij}. Then evaluate $f(x_i,y_j)$ at a point (x_i,y_j) in each sub-region, and sum the products $f(x_i,y_j)\Delta A_{ij}$ to approximate the integral. The integral itself is the limit of these sums as the partition you choose becomes finer. The discussion in Section 4.1 made some simplifying assumptions (R is a rectangle, equal partitions, etc.) and arrived at the formula

$$\iint_R f(x,y)dA = \lim_{n\to\infty} \sum_{i=1}^n \sum_{j=1}^n f(x_i,y_j)\Delta A_{ij}.$$

The key to understanding Theorem 4.4.1 is to use a partition of U and the function g to create the partition of R. We illustrate this using the previous example.

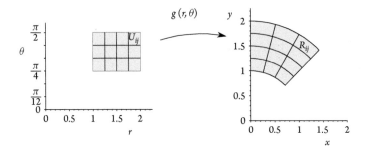

Figure 4.4.12 The induced partition of $g(r, \theta)$

We know that $g : U \to R$ where U is the rectangle $[0, 1] \times [0, 2\pi]$ in the $r\theta$-plane and R is the unit disk in the xy-plane. Partition U into sub-rectangles U_{ij} with width Δr_i and length $\Delta \theta_j$. Letting $R_{ij} = g(U_{ij})$ partitions R into sub-regions—not necessarily sub-rectangles (see Figure 4.4.12 for a sample illustration). The partition points in R are the images $(x_i, y_j) = g(r_i, \theta_j)$ of the partition points in U. This describes the partition of R induced by that of U.

Reviewing the discussion at the beginning of this paragraph, the region R_{ij} can be approximated by the parallelogram determined by the vectors $g_r \Delta r_i, g_\theta \Delta \theta_j$. The area ΔA_{ij} of R_{ij} is nicely approximated by the area of the parallelogram, which is the area of U_{ij} scaled by the factor $|\det Dg|$. Translating the last sentence into symbols, and remembering to evaluate Dg at partition points, gives

$$\Delta A_{ij} \approx |\det Dg(r_i, \theta_j)| \, \Delta r_i \Delta \theta_j.$$

Putting this together with the fact that partition points of R are given by $(x_i, y_j) = g(r_i, \theta_j)$, we have

$$\iint_R f(x, y) \, dA = \lim_{n \to \infty} \sum_{i=1}^{n} \sum_{j=1}^{n} f(x_i, y_j) \Delta A_{ij}$$

$$= \lim_{n \to \infty} \sum_{i=1}^{n} \sum_{j=1}^{n} f(g(r_i, \theta_j)) \, |\det Dg(r_i, \theta_j)| \, \Delta r_i \Delta \theta_j$$

$$= \iint_U f(g(r, \theta)) \, |\det Dg| \, dr d\theta.$$

The plausible argument just given can be summarized as follows: Use the function g and a partition U_{ij} of U to partition R into sub-regions R_{ij}. Now note that the area of the R_{ij} is approximately the area of U_{ij} stretched by a factor of $|\det Dg(r_i, \theta_j)|$.

This reasoning applies to more general functions $g : \mathbb{R}^2 \to \mathbb{R}^2$, as well as to higher dimensions. The factor $|\det Dg|$ measures how g changes n-dimensional volume locally, and Theorem 4.4.1 is valid in arbitrary dimensions. A careful statement of the theorem, together with necessary differentiability hypotheses, can be found in [8].

The change of variables theorem is important in many theoretical applications. In particular, it can be used to show that integrals of functions or vector fields over surfaces are independent of the parameterization. This brief, intuitive discussion was intended to place integrals in other coordinate systems in a broader context, as well as to present a theorem that will be used in subsequent sections.

Exercises

1. Sketch the region of integration for the following polar integrals:

a) $\displaystyle\int_0^{\pi/4} \int_1^2 r\,dr\,d\theta$

b) $\displaystyle\int_{-\pi/2}^{\pi/3} \int_0^3 r\,dr\,d\theta$

c) $\displaystyle\int_{\pi/6}^{5\pi/6} \int_{1/2}^{\sin\theta} r\,dr\,d\theta$

d) $\displaystyle\int_{-\pi/2}^{\pi/2} \int_1^{1+\cos\theta} r\,dr\,d\theta$

2. Evaluate $\iint_R \sqrt{1+x^2+y^2}\,dA$ where R is the quarter unit disk in the first quadrant.

3. Evaluate $\iint_R xy\,dA$ where R is the region in the first quadrant between $x^2+y^2=1$ and $x^2+y^2=2$.

4. Evaluate $\iint_R e^{-(x^2+y^2)}\,dA$ where R is the disk centered at the origin and radius 10. How does your answer change if R has radius a? What happens to the integral as $a \to \infty$?

5. Find the area between the unit circle and the spiral $r = e^\theta$ for $0 \le \theta \le 2\pi$.

6. Find the area of the region inside the cardioid $r = 1 + \cos\theta$ and outside the unit circle.

7. Find the area enclosed by one petal of the four-leafed rose $r = \cos 2\theta$.

8. Find the area enclosed by the lemniscate $r^2 = \cos 2\theta$. (Hint: find the area for one side and double it)

9. Find the volume of the solid between the xy-plane and the cone $z = 10 - 2\sqrt{x^2+y^2}$.

10. Find the volume between the paraboloid $x = y^2 + z^2 - 1$ and the yz-plane.

11. Find the volume inside the cylinder $x^2 + z^2 = 1$, and between the plane $y = 0$ and the paraboloid $y = x^2 + z^2$.

12. Find the volume between the paraboloids $x = 2 - 2y^2 - 2z^2$ and $x = y^2 + z^2 - 1$.

13. Find the volume of the solid under $z = \sqrt{x^2+y^2}$ and over the circle $x^2 + (y-1)^2 = 1$ in the xy-plane.

14. Find the volume of the solid between the paraboloid $z = x^2 + y^2$ and the plane $z = 4$.

15. Find the volume of the solid inside the cylinder $y^2 + z^2 = 1$ and between the planes $x = 0$ and $2x + y + 3z = 6$.

16. Find the volume of the solid inside the cylinder $x^2 + z^2 = 1$ and between the planes $y + z = 1$ and $y - z = -2$.

17. Find the mass of the cylinder inside $x^2 + y^2 = 1$ and between $z = \pm 1$ if the density is $\delta(x,y,z) = 4 - y - z$.

18. Sketch the solids of integration for the following spherical integrals:

$$a) \int_0^{2\pi} \int_0^{\pi/4} \int_0^1 \rho^2 \sin\phi \, d\rho \, d\phi \, d\theta \qquad b) \int_0^{2\pi} \int_{\pi/2}^{\pi} \int_0^1 \rho^2 \sin\phi \, d\rho \, d\phi \, d\theta$$

$$c) \int_0^{\pi/2} \int_0^{\pi/2} \int_0^1 \rho^2 \sin\phi \, d\rho \, d\phi \, d\theta \qquad d) \int_0^{2\pi} \int_0^{\pi} \int_0^1 \rho^2 \sin\phi \, d\rho \, d\phi \, d\theta$$

19. Find the volume of the solid outside the cone $z^2 = x^2 + y^2$ and inside the unit sphere.

20. Find the volume of the solid inside the sphere $\rho = 1$ and outside the cylinder $r = 1/2$.

21. Find the mass of the unit ball if the density is given by $\delta(x,y,z) = z^2$.

22. Find the volume inside $\rho = 1 + \cos\phi$ and outside $\rho = 1$.

23. Use integration to show that the volume of a sphere of radius R is $\frac{4\pi}{3} R^3$.
    Set up the triple integrals that give the solids in exercises 24–27 in a) Cartesian, b) cylindrical, and c) spherical coordinates.

24. The solid above $\phi = \pi/6$ and below the unit sphere.

25. The solid under $z = 4 - x^2 - y^2$ and above the xy-plane.

26. The solid inside $\rho = 1$ and above $z = 1/2$.

27. The solid outside the cylinder $r = 1/2$ and inside the unit sphere.

28. Let $g(r,\theta,z) = (r\cos\theta, r\sin\theta, z)$. Find $\left|\det Dg\right|$ and verify that $dxdydz = \left|\det Dg\right| dzdrd\theta$.

29. Let $g(\rho,\theta,\phi) = (\rho\sin\phi\cos\theta, \rho\sin\phi\sin\theta, \rho\cos\phi)$. Find $\left|\det Dg\right|$ and verify that $dxdydz = \left|\det Dg\right| d\rho d\theta d\phi$.

4.5 Applications of Integration

In this section we introduce two applications of integration. We begin with calculating the area of parametric surfaces, then use this to derive the formula for the surface area of the graph of a function of two variables. The technique for calculating surface area is intuitive geometrically, and will reappear in Section 4.6 when defining how to integrate a function on a surface. This section concludes with finding centers of mass of solids and planar regions.

Surface Area: Recall that a parametric surface $S(s,t)$ takes a region R in the st-plane and bends, twists, and stretches it into a surface in $\mathbb{R}^3$. For example, the surface $S(s,t) = (s,t,1-s^2-t^2)$ takes the rectangle $R = [0,1] \times [0,1]$ in the st-plane and maps it to a portion of a paraboloid (see Figure 4.5.1).

Since parameterizations can stretch or shrink, the area of the image $S(R)$ is not necessarily the same as that of the domain R. It is our goal to develop a method for calculating the area of $S(R)$. Not surprisingly, we use the typical calculus approach:

Approximate the surface area with things you can calculate, then take a limit as your approximation gets better.

To approximate the surface area, first partition R into small rectangles R_{ij} of width Δs_i and height Δt_j. As in Example 3.4.6, the area of $S(R_{ij})$ (the image of R_{ij} under the map S) is approximately the area of the parallelogram spanned by the vectors $S_s(s_i,t_j)\Delta s_i$ and $S_t(s_i,t_j)\Delta t_j$ (see Figure 4.5.2). The area of that parallelogram is the length of the cross product

$$\left\| \left(S_s(s_i,t_j)\Delta s_i \right) \times \left(S_t(s_i,t_j)\Delta t_j \right) \right\|.$$

Scalars can factor out of cross products and out of length calculations, giving

$$\text{Area of } S(R_{ij}) \approx \left\| S_s(s_i,t_j)\Delta s_i \times S_t(s_i,t_j)\Delta t_j \right\|$$
$$= \left\| S_s(s_i,t_j) \times S_t(s_i,t_j) \right\| \Delta s_i \Delta t_j.$$

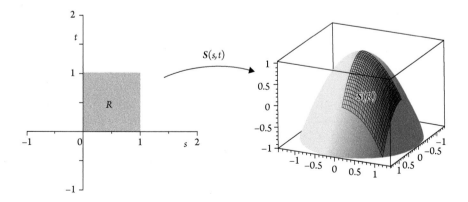

Figure 4.5.1 A parametric paraboloid

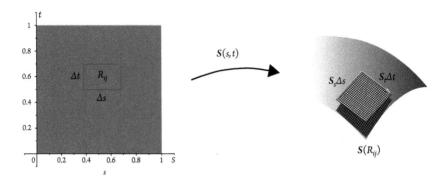

Figure 4.5.2 Approximating surface area

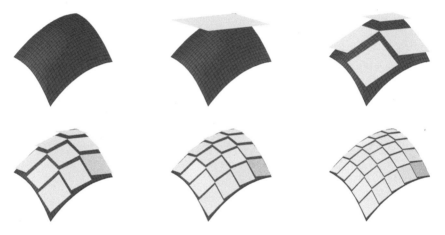

Figure 4.5.3 Improving approximations of surface area

The area of $\mathbf{S}(R)$ is approximated by summing these approximate areas of $\mathbf{S}(R_{ij})$, so

$$\text{Area of } \mathbf{S}(R) \approx \sum_{i=1}^{n} \sum_{j=1}^{n} \left\| \mathbf{S}_s(s_i, t_j) \times \mathbf{S}_t(s_i, t_j) \right\| \Delta s_i \Delta t_j. \tag{4.5.1}$$

Taking finer partitions improves the approximation. This is illustrated for the parametric surface $\mathbf{S}(s,t) = (s, t, 1 - s^2 - t^2)$, $0 \le s, t \le 1$ in Figure 4.5.3. The original surface is pictured first, then the approximating parallelograms for different partitions of the domain $[0,1] \times [0,1]$ into sub-rectangles. It is evident from the picture that the areas of the parallelograms approach the area of the surface as the partition gets finer. Taking the limit of Approximation 4.5.1 gives an integral formula for the area of a parametric surface.

Parametric Surface Area

Let $S(s,t)$ be a parametric surface over the region R in the st-plane. The image $S(R)$ has area

$$\iint_R \|S_s(s,t) \times S_t(s,t)\| \, dsdt = \iint_R dS.$$

Thus we introduce the notation $dS = \|S_s(s,t) \times S_t(s,t)\| \, dsdt$ for a parametric surface $S(s,t)$.

Notation: The notation for integrals is important, and can be daunting. For example, dA represents an infinitesimal area in a plane. Depending on which coordinate system you use, it could mean $dxdy$ or $rdrd\theta$. We are now introduced to the area element dS which represents an infinitesimal area on the surface S. If S is given parametrically by $S(s,t)$, then $dS = \|S_s \times S_t\| \, dsdt$. In Chapter 5 we will want to integrate vector fields over surfaces, and the notation $d\mathbf{S}$ will be introduced. While the symbols dS and $d\mathbf{S}$ are similar, they mean quite different things. We have seen that dS is a scalar, and it will turn out that $d\mathbf{S}$ is a vector normal to the surface S. Try your best to pay attention to notation; it is subtle.

After such a lengthy discussion, it is high time to illustrate these ideas with some examples.

Example 4.5.1. *Paraboloid patch*

Find the area of the patch of the paraboloid

$$S(s,t) = (s, t, 1 - s^2 - t^2), \ 0 \le s, t \le 1.$$

Step 1. Find the integrand $\|S_s \times S_t\|$.
We first compute the cross product:

$$S_s \times S_t = \begin{vmatrix} \mathbf{i} & \mathbf{j} & \mathbf{k} \\ 1 & 0 & -2s \\ 0 & 1 & -2t \end{vmatrix} = \langle 2s, 2t, 1 \rangle,$$

whose length is $\|S_s \times S_t\| = \sqrt{4s^2 + 4t^2 + 1}$.
Step 2. Evaluate the integral.

$$\iint_R dS = \int_0^1 \int_0^1 \sqrt{4s^2 + 4t^2 + 1} \, dsdt = \frac{7}{12} \ln(5) + 1 - \frac{1}{12} \arctan\left(\frac{4}{3}\right),$$

where the integral is done by inspection. OK, OK, we're kidding. The integral is really hard and we used software to calculate it! However, if we change the domain R in the st-plane, we end up with an integral we can calculate.

Let S be the same parametric surface, $\mathbf{S}(s,t) = (s,t,1 - s^2 - t^2)$ over the unit disk in the st-plane. Find the area of S.

Step 1. The parameteric equations are the same, so the integrand is $\|\mathbf{S}_s \times \mathbf{S}_t\| = \sqrt{4s^2 + 4t^2 + 1}$.

Step 2. Evaluate the integral over the unit disk in the st-plane. Since R is the unit disk, we switch to polar coordinates to evaluate below.

$$
\iint_R dS = \iint_{unit\ disk} \sqrt{4s^2 + 4t^2 + 1}\, dA
$$
$$
= \int_0^{2\pi} \int_0^1 \sqrt{1 + 4r^2}\, r\, dr\, d\theta = \frac{1}{8} \int_0^{2\pi} \int_1^{\sqrt{5}} \sqrt{u}\, du\, d\theta
$$
$$
= \frac{1}{12} \int_0^{2\pi} u^{3/2}\Big|_{u=1}^{\sqrt{5}}\, d\theta = \frac{5\sqrt{5} - 1}{12} \int_0^{2\pi} d\theta = \frac{(5\sqrt{5} - 1)\pi}{6}. \ \blacktriangle
$$

Math App 4.5.1. *Surface area*

The following Math App provides the opportunity to interact with surface area approximations.

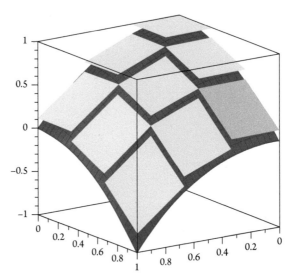

Example 4.5.2. *Area of a torus*

Let $\mathbf{S}(s,t) = ((\cos t + 2) \cos s, (\cos t + 2) \sin s, \sin t)$ for $0 \le s, t \le 2\pi$. In Example 3.4.2 we analyzed this torus (see Figure 3.4.3), and found that

$$
\mathbf{S}_s \times \mathbf{S}_t = \langle (\cos t + 2) \cos s \cos t, (\cos t + 2) \sin s \cos t, (\cos t + 2) \sin t \rangle.
$$

To find the area of the surface of the torus, we integrate $\|\mathbf{S}_s \times \mathbf{S}_t\|$ over the domain. Calculating the length gives us

$$\|\mathbf{S}_s \times \mathbf{S}_t\| = \|\langle(\cos t + 2)\cos s \cos t, (\cos t + 2)\sin s \cos t, (\cos t + 2)\sin t\rangle\|$$
$$= |\cos t + 2| \, \|\langle\cos s \cos t, \sin s \cos t, \sin t\rangle\| = \cos t + 2.$$

The area of S is

$$\iint_R dS = \int_0^{2\pi} \int_0^{2\pi} \cos t + 2 \, ds \, dt = \int_0^{2\pi} (\cos t + 2)s \Big|_0^{2\pi} dt$$
$$= 2\pi \int_0^{2\pi} \cos t + 2 \, dt = 2\pi (\sin t + 2t) \Big|_0^{2\pi} = 8\pi^2. \ \blacktriangle$$

Recall that one way to parameterize the graph of $z = f(x,y)$ is to let the independent variables be parameters and use the function to parameterize z, as in $\mathbf{S}(x,y) = (x,y,f(x,y))$. This, combined with the formula for the area of a parametric surface, yields a formula for the area of the graph of a function. One readily computes

$$\|\mathbf{S}_x \times \mathbf{S}_y\| = \|\langle 1,0,f_x\rangle \times \langle 0,1,f_y\rangle\| = \sqrt{1 + f_x^2 + f_y^2}.$$

Applying the above formula for surface area we get the following:

Surface Area of a Graph

The area of the graph of $z = f(x,y)$ over the region R in the xy-plane is

$$\iint_R dS = \iint_R \sqrt{1 + f_x^2 + f_y^2} \, dx \, dy.$$

Thus if S is the graph of $z = f(x,y)$, then the area element dS is given by

$$dS = \sqrt{1 + f_x^2 + f_y^2} \, dx \, dy,$$

and represents the area of an infinitesimally small patch on S. It is recommended that you remember two area formulas, one for the area of a parametric surface and the other for the area of a graph. Two graph area calculations follow.

Example 4.5.3. *Area of a plane*

Find the area of the portion of $2x + 3y + z = 6$ in the first octant. This is the triangle with vertices $(3,0,0)$, $(0,2,0)$ and $(0,0,6)$ so we could use geometry to find the area. However, we can also think of the triangle as the portion of the graph of $z = 6 - 2x - 3y$ in the first octant. We need to find the region R in the xy-plane that it lies above. Of course, R is just the triangle in the first quadrant below the line $2x + 3y = 6$.

$$\iint_R \sqrt{1+f_x^2+f_y^2}\,dxdy = \int_0^2 \int_0^{(6-3y)/2} \sqrt{1+(-2)^2+(-3)^2}\,dxdy$$

$$= \int_0^2 \sqrt{14}x\Big|_0^{(6-3y)/2}\,dy = \frac{\sqrt{14}}{2}\int_0^2 6-3y\,dy$$

$$= \frac{\sqrt{14}}{2}\left(6y-\frac{3}{2}y^2\right)\Big|_0^2 = 3\sqrt{14}.\ \blacktriangle$$

The integrand $\sqrt{1+f_x^2+f_y^2}$ measures how the function $f(x,y)$ changes area near each point. The fact that the integrand is constant in the previous example indicates that the function multiplies area by a factor of $\sqrt{14}$ everywhere. If R is any region in the xy-plane with area A, then the area of the portion of the plane $2x+3y+z=6$ above R is $\iint_R dS = \sqrt{14}A$. This situation is not common, as the following example illustrates.

Example 4.5.4. *Area of a hemisphere*

Find the area of the hemisphere $z = \sqrt{9-x^2-y^2}$.

In this case the hemisphere is given as the graph of a function whose domain R is all points satisfying $9-x^2-y^2 \geq 0$, or the disk centered at the origin of radius 3. Direct calculation shows

$$f_x = \frac{-x}{\sqrt{9-x^2-y^2}} \text{ and } f_y = \frac{-y}{\sqrt{9-x^2-y^2}},$$

so the integrand in the area calculation is

$$\sqrt{1+f_x^2+f_y^2} = \sqrt{1+\frac{x^2}{9-x^2-y^2}+\frac{y^2}{9-x^2-y^2}} = \sqrt{\frac{9}{9-x^2-y^2}}.$$

Both the integrand and the region of integration strongly suggest using polar coordinates to evaluate the integral. Doing so yields

$$\iint_R \sqrt{\frac{9}{9-x^2-y^2}}\,dA = \int_0^{2\pi} \int_0^3 \frac{3}{\sqrt{9-r^2}}rdrd\theta$$

$$= -\frac{3}{2}\int_0^{2\pi}\int_9^0 u^{-1/2}\,dud\theta \qquad \text{(letting } u = 9-r^2\text{)}$$

$$= \int_0^{2\pi} 9d\theta = 18\pi.$$

Of course, the area of a sphere is $4\pi r^2$, so a hemisphere has half that area. In our case the radius is three, so the formula confirms our integration. $\blacktriangle$

Centers of Mass: We now look at another application of integration. We've seen that integrating the density function over a solid gives the mass of the solid. Since the center of mass is the weighted average position, we have

---

Let W be a solid with density $\delta(x,y,z)$, mass M, and center of mass $(\bar{x}, \bar{y}, \bar{z})$. They are related by the following formulas:

$$M = \iiint_W \delta(x,y,z)\,dV$$

$$\bar{x} = \frac{\iiint_W x\delta(x,y,z)\,dV}{M} \qquad \bar{y} = \frac{\iiint_W y\delta(x,y,z)\,dV}{M} \qquad \bar{z} = \frac{\iiint_W z\delta(x,y,z)\,dV}{M}$$

There are analogous formulas in two dimensions. We continue with a three- and a two-dimensional example.

Example 4.5.5. *Center of mass of tetrahedron*

Find the center of mass of the tetrahedron with vertices $(0,0,0)$, $(2,0,0)$, $(0,3,0)$, and $(0,0,6)$ if the density function is given by $\delta(x,y,z) = 40 - z^2$.

This is the tetrahedron of Examples 4.2.6 and 4.3.13, which is bounded by the coordinate planes and $3x + 2y + z = 6$. We found our limits of integration to be

$$0 \le z \le 6 - 3x - 2y$$
$$0 \le y \le (6 - 3x)/2$$
$$0 \le x \le 2.$$

And we found the mass to be $M = 1092/5$. To find the center of mass, we have to calculate the numerators of each fraction above. We have

$$\iiint_W x\delta(x,y,z)\,dV = \int_0^2 \int_0^{(6-3x)/2} \int_0^{6-3x-2y} x(40 - z^2)\,dz\,dy\,dx$$

$$= \int_0^2 \int_0^{(6-3x)/2} x\left(40(6 - 3x - 2y) - \frac{(6 - 3x - 2y)^3}{3}\right)dy\,dx$$

$$= \int_0^2 x\left(-10(6 - 3x)^2 + \frac{(6 - 3x)^4}{24}\right)dx = \frac{564}{5}.$$

So the x-coordinate of the center of mass is

$$\bar{x} = \frac{\frac{564}{5}}{\frac{1092}{5}} = \frac{47}{91}.$$

Similar brute force calculations yield

$$\iiint_W y\delta(x,y,z)dV = \frac{846}{5}$$

$$\iiint_W z\delta(x,y,z)dV = \frac{1476}{5},$$

which gives $\bar{y} = \frac{141}{182}$ and $\bar{z} = \frac{123}{91}$. ▲

Example 4.5.6. *A thin plate*

A thin plate covering the quarter of the unit disk in the first quadrant has density $\delta(x,y) = \sqrt{x^2 + y^2}$. Find its center of mass.

First, since the region *and* the density function are symmetric in x and y, we know $\bar{x} = \bar{y}$. The mass is most easily calculated using polar coordinates, so

$$M = \iint_R \sqrt{x^2 + y^2}\, dA = \int_0^{\pi/2}\int_0^1 r^2\, dr d\theta = \int_0^{\pi/2} \frac{r^3}{3}\Big|_0^1 d\theta = \int_0^{\pi/2} \frac{1}{3}\, d\theta = \frac{\pi}{6}.$$

Again using polar coordinates, we compute the numerator of $\bar{x}$.

$$\iint_R x\sqrt{x^2 + y^2}\, dA = \int_0^{\pi/2}\int_0^1 r\cos\theta\, r(r dr d\theta) = \int_0^{\pi/2}\int_0^1 r^3 \cos\theta\, dr d\theta$$

$$= \int_0^{\pi/2} \frac{r^4}{4}\cos\theta\Big|_0^1 d\theta = \frac{1}{4}\int_0^{\pi/2}\cos\theta\, d\theta$$

$$= \frac{1}{4}\sin\theta\Big|_0^{\pi/2} = \frac{1}{4}.$$

Combining the calculations gives $\bar{x} = \bar{y} = \frac{1/4}{\pi/6} = \frac{3}{2\pi}$. ▲

In the previous example we used the two-dimensional formulas for mass and centers of mass, and we now take a closer look at them. Let R be a region in the plane with density $\delta(x,y)$, mass M, and center of mass $(\bar{x},\bar{y})$. They are given by the following formulas:

$$M = \iint_R \delta(x,y)dA$$

$$\bar{x} = \frac{\iint_R x\delta(x,y)dA}{M} \qquad \bar{y} = \frac{\iint_R y\delta(x,y)dA}{M}. \qquad (4.5.2)$$

Notice the symmetry in the formulas. To find the numerator of $\bar{x}$, integrate x times the density while the integral of $y\delta(x,y)$ gives the numerator of $\bar{y}$. In order to truly appreciate this symmetry, let's compare the double integral formulas with those from single-variable calculus. We'll consider the constant density case where $\delta(x,y) = 1$, when the center of mass is sometimes called the centroid.

Example 4.5.7. *Derivation of single-variable formulas*

Let R be the region under $y = f(x)$ on the interval $[a,b]$, and suppose it has constant unit density. Find single-variable formulas for the centroid of R.

To do so, note that the mass of R is $\iint_R dA$, which is the area A of R. Since R is the region under $y = f(x)$ on the interval $[a,b]$, it is described by the system of inequalities

$$0 \leq y \leq f(x)$$
$$a \leq x \leq b.$$

Hence the numerator of $\overline{x}$ is given by

$$\iint_R x\,dA = \int_a^b \int_0^{f(x)} x\,dy\,dx = \int_a^b xy\big|_{y=0}^{f(x)} dx = \int_a^b xf(x)\,dx,$$

and the numerator of $\overline{y}$ is

$$\iint_R y\,dA = \int_a^b \int_0^{f(x)} y\,dy\,dx = \int_a^b \frac{y^2}{2}\bigg|_0^{f(x)} dx = \int_a^b \frac{f^2(x)}{2}\,dx.$$

Thus if A is the area of R, then the centroid of R is given by

$$\overline{x} = \frac{\int_a^b xf(x)\,dx}{A} \qquad \overline{y} = \frac{\frac{1}{2}\int_a^b f^2(x)\,dx}{A}.$$

The derivation of these formulas using single-variable techniques is tedious. Moreover, they aren't intuitive at all so you probably just memorized them. Thirdly, they only apply when the density is constant!

On the other hand, the double-integral formulas of Equation 4.5.2 can be used for variable density $\delta(x,y)$ and apply to more general regions than just those under the graph of a function. Finally, and most importantly, the double-integral formulas make intuitive sense. The center of mass is the average position, where position is weighted by mass. For a finite system of particles in a line, this means the denominator should be the total mass of the system and the numerator should be the sum of the positions times the masses of the individual particles:

$$\frac{x_1 \cdot m_1 + \cdots + x_n \cdot m_n}{m_1 + \cdots + m_n}$$

For continuous systems (like regions or solids) the sum turns into an integral, and masses translate into the density function, yielding the formulas of Equation 4.5.2. For a nice transition from the finite to continuous case from basic principles of physics, see Section 9.4 of *Calculus II*, by Marsden and Weinstein, where they consider the single-variable situation (see [4]). ▲

Exercises

1. Find the center of mass of the rectangle $R = [0,1] \times [0,1]$ if the density is $\delta(x,y) = xy$. Use symmetry where possible.

2. Find the center of mass of the rectangle $R = [0,\pi/4] \times [0,\pi/4]$ if the density is $\delta(x,y) = \sin(x+y)$. Use symmetry where possible.

3. Find the center of mass of the region between $y = x^2$ and $y = 1$ if the density function is $\delta(x,y) = y$. Use symmetry where possible.

4. Find the center of mass of the quarter unit disk in the first quadrant if the density function is $\delta(x,y) = xy$. Use symmetry where possible.

5. Find the center of mass of the region inside the unit circle and above $y = |x|$ if the density function is $\delta(x,y) = y$. Use symmetry where possible.

6. Find the center of mass of the constant density $\rho(x,y,z) = 1$ tetrahedron with vertices at $(0,0,0)$, $(3,0,0)$, $(0,2,0)$, and $(0,0,6)$.

7. Let W be the solid in the first octant under $x + z = 1$ and between the planes $y = 0$ and $y = 2$. Find the center of mass of W if the density is $\delta(x,y,z) = y$.

8. Find the center of mass of the upper hemisphere of the unit ball with density function $\delta(x,y,z) = x^2 + y^2 + z^2$. Use symmetry where possible. (The solid is the region inside the unit sphere and above the xy-plane.)

9. Find the center of mass of the solid inside the unit sphere and above the cone $\phi = \pi/4$ with density function $\delta(x,y,z) = \sqrt{x^2 + y^2}$. Use symmetry where possible.

10. Let W be the solid in the first octant that is inside $x^2 + y^2 = 1$ and between the planes $z = 0$ and $z = 2$. Find the center of mass of W if the density of W is $\delta(x,y,z) = xy(2 - z)$. Use symmetry where possible.

11. Let W be the solid above the unit disk and under the cone $z = \sqrt{x^2 + y^2}$ with density function $\delta(x,y,z) = (1 - \sqrt{x^2 + y^2})z$. Find the center of mass of W, using symmetry when possible.

12. Find the area of the portion of the plane $3x - 2y + z = 5$ above the rectangle $[0,1] \times [0,1]$.

13. Find the area of the portion of the plane $2x - 3y + z = 40$ above the rectangle $[0,3] \times [1,2]$.

14. Find the area of the portion of the plane $x + 2y - z = 2$ above the region R between the y-axis and $x = 1 - y^2$.

15. Find the area of the cone $z = 10 - \sqrt{x^2 + y^2}$ above the unit disk.

16. Find the area of the portion of the cone $z = 2 + 4\sqrt{x^2 + y^2}$ over the rectangle $[-1,1] \times [-1,1]$.

17. Find the area of the portion of $z = \sqrt{1 - x^2}$ lying above the rectangle $[0,1] \times [-2,2]$.

18. Find the area of the portion of the surface $z = x^2 + y^2$ below the plane $z = 9$.

19. Find the area of the parametric surface

$$S(s,t) = (2s + t - 1, 3s + 5t + 2, -s + 2t + 4),$$

for $0 \le s, t \le 1$.

20. Find the area of the parametric surface

$$S(s,t) = (s + t - 2, 2s - t - 3, s + 3t + 5),$$

for $0 \le s, t \le 1$.

21. Find the area of the parametric surface $S(s,t) = (s \cos t, 4 - s^2, s \sin t)$, for $0 \le s \le 2$ and $0 \le t \le 2\pi$.

22. Find the area of $S(s,t) = (s \cos t, s \sin t, s^2 \cos 2t)$, for $0 \le s \le 1$ and $0 \le t \le \pi$.

23. Find the area of $S(s,t) = (s \cos t, 3 - 2s, s \sin t)$, for $0 \le s \le 1$ and $0 \le t \le \pi$.

24. Find the area of the parametric surface $S(s,t) = (3s, s \cos t, s \sin t)$, for $0 \le s \le 2$ and $0 \le t \le 2\pi$.

25. Find the area of the portion of the sphere $S(s,t) = (\sin s \cos t, \sin s \sin t, \cos s)$, for $0 \le s, t \le \pi/6$.

26. Find the area of a sphere radius R. Recall that one parameterization of the sphere radius R is $S(s,t) = (R \sin s \cos t, R \sin s \sin t, R \cos s)$, for $0 \le s \le \pi$ and $0 \le t \le 2\pi$.

27. Find the surface area of the torus $S(s,t) = ((2 \cos t + 3) \cos s, (2 \cos t + 3) \sin s, 2 \sin t)$ for $0 \le s, t \le 2\pi$. Note that this torus is obtained by revolving the circle $(y - 3)^2 + z^2 = 4$ around the z-axis.

4.6 Curve and Surface Integrals

In first-year calculus you learned how to integrate functions of a single variable $\int_a^b f(x)dx$ defined on intervals $[a,b]$ of real numbers. In this chapter we defined double integrals $\iint_R f(x,y)dA$ of two-variable functions whose domain was the region R in the xy-plane. We now generalize the domain of integration for both of these settings. Integrals of functions defined on intervals $[a,b]$ are extended to integrals of functions defined on curves C in $\mathbb{R}^3$. Analogously, double integrals of functions defined on regions R in $\mathbb{R}^2$ will be generalized to double integrals of functions defined on surfaces S in $\mathbb{R}^3$. The strategy in each case will be to define the generalized integral via ordinary single and double integrals using a parameterization for C or S. Then we show that the value of the integral is independent of our choice of parameterization. Since the integral is independent of how we parameterize our curves or surfaces, we think of it as integrating the function *on the curve or surface itself*! Our notation, $\int_C f\, dC$ and $\iint_S f\, dS$, is chosen to emphasize this point of view.

Integrals along a curve: Let C be a curve in $\mathbb{R}^3$ and f a function defined on C. We wish to define the integral of f over C, denoted $\int_C f dC$. To do so, we recall how the definite integral $\int_a^b f(x)dx$ was defined in single-variable calculus. First, partition the interval $[a,b]$ into subintervals of length Δx_i, and choose a point x_i^* in each subinterval. Then evaluate f at the chosen points x_i^* and form the approximation

$$\int_a^b f(x)dx \approx \sum_{i=1}^n f(x_i^*)\Delta x_i.$$

Finally, take the limit as the partition becomes finer and define

$$\int_a^b f(x)dx = \lim_{n\to\infty} \sum_{i=1}^n f(x_i^*)\Delta x_i.$$

The idea for defining $\int_C f dC$ is completely analogous. Partition the curve C into sub-curves C_i of length ΔC_i, and choose representative points P_i from each sub-curve (see Figure 4.6.1). For each sub-curve evaluate $f(P_i)\Delta C_i$, and take a limit as the partition of C becomes finer to define

$$\int_C f dC = \lim_{n\to\infty} \sum_{i=1}^n f(P_i)\Delta C_i. \tag{4.6.1}$$

Conceptually this generalization is quite natural, but how does one calculate it? To answer this question, recall that curves are usually given to us parametrically. Our first task is to calculate the ingredients of Equation 4.6.1 given a parameterization $C(t)$, $a \le t \le b$ of the curve C. We will make several approximations along the way.

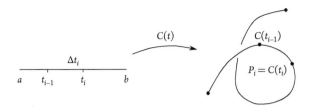

Figure 4.6.1 Partitioning a parametric curve

A partition $a = t_0 < t_1 < \cdots < t_n = b$ of the interval $[a,b]$ yields a partition of the curve C into sub-curves. Indeed, just let C_i be the portion of C from $\mathbf{C}(t_{i-1})$ to $\mathbf{C}(t_i)$. Since our choice of P_i was arbitrary above, we might as well choose $P_i = \mathbf{C}(t_i)$. Now that we've made the choices necessary, we need to calculate lengths and function values. For small Δt, the length ΔC_i of the sub-curve C_i is approximated by $\|\mathbf{C}'(t_i)\| \Delta t_i$ (recall that $\|\mathbf{C}'(t_i)\|$ is the speed of the curve, so distance traveled, ΔC_i, is speed times time, or approximately $\|\mathbf{C}'(t_i)\| \Delta t_i$). We then get the approximation

$$\int_C f dC \approx \sum_{i=1}^{n} f(P_i) \Delta C_i \approx \sum_{i=1}^{n} f(\mathbf{C}(t_i)) \|\mathbf{C}'(t_i)\| \Delta t_i.$$

The sum on the right is a Riemann sum, so taking the limit as $n \to \infty$ gives a single-variable integral. In the limit, the Δt_i becomes the differential dt, and the rest becomes the integrand $f(\mathbf{C}(t)) \|\mathbf{C}'(t)\|$. With this motivation, we make the following definition.

Definition 4.6.1. *Let $\mathbf{C}(t)$, $a \le t \le b$, be a parametric curve, and f a function defined on it. The integral of f along $\mathbf{C}$ is*

$$\int_C f dC = \int_a^b f(\mathbf{C}(t)) \|\mathbf{C}'(t)\| dt.$$

At this point we have defined the integral of a function along a given parameterization. Theorem 4.6.1 will show that any two parameterizations of the same geometric curve C lead to the same value of the integral, and hence we *think* of it as integrating f along C. Before we do, let's calculate some examples.

Example 4.6.1. *Integrating along a line.*

Integrate the function $f(x,y,z) = x^2 + yz$ along the curve $\mathbf{C}(t) = (1 - t, 2 + t, 3t)$, $0 \le t \le 1$.

We first find the ingredients for the integrand, then put them together and integrate. We calculate

$$f(\mathbf{C}(t)) = f(1 - t, 2 + t, 3t) = (1 - t)^2 + (2 + t)3t = 4t^2 + 4t + 1,$$
$$\|\mathbf{C}'(t)\| = \|\langle -1, 1, 3 \rangle\| = \sqrt{11}.$$

Putting these together, we get that the integral of f over the line segment is

$$\int_0^1 (4t^2 + 4t + 1)\sqrt{11}\,dt = \sqrt{11}\left(\frac{4}{3}t^3 + 2t^2 + t\right)\Big|_0^1 = \frac{13\sqrt{11}}{3}. \quad \blacktriangle$$

Example 4.6.2. *Mass from density*

Let $C(t) = (3\cos t, 3\sin t, t)$, $0 \le t \le \pi/2$, represent a thin wire in space with density function $\delta(x,y,z) = x^2 + y^2 + z^2$. Find the mass of the wire.

We know that integrating density gives mass, so we compute the ingredients for the integrand

$$\delta(C(t)) = \delta(3\cos t, 3\sin t, t) = (3\cos t)^2 + (3\sin t)^2 + t^2 = 9 + t^2,$$
$$\|C'(t)\| = \|\langle -3\sin t, 3\cos t, 1\rangle\| = \sqrt{10}.$$

The mass of the wire M is then

$$M = \int_0^{\pi/2}\left(9 + t^2\right)\sqrt{10}\,dt = \sqrt{10}\left(9t + \frac{t^3}{3}\right)\Big|_0^{\pi/2} = \sqrt{10}\left(\frac{9\pi}{2} + \frac{\pi^3}{24}\right). \quad \blacktriangle$$

Math App 4.6.1. *The area of a vase*

In this Math App we provide another context for a line integral. In particular, the curve C is planar, and the function f represents the height of a vase lying above the curve. Then the integral $\int_C f\,dC$ represents the surface area of the vase.

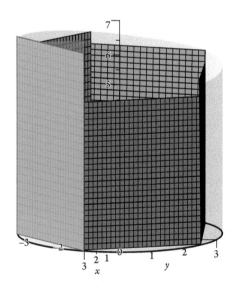

Recall that the same curve C in $\mathbb{R}^3$ can have many different parameterizations. For example, both $\mathbf{C}_1(t) = (\cos t, \sin t)$, $0 \le t \le \pi$, and $\mathbf{C}_2(t) = (\cos 2t, \sin 2t)$, $0 \le t \le \pi/2$ parameterize the top half of the unit circle oriented counterclockwise. Definition 5.1.1 does not guarantee that using two different parameterizations for the same curve will yield the same integral, but we prove that now.

Theorem 4.6.1. *The integral of f along a continuously differentiable curve C is independent of the parameterization chosen for C.*

Proof. The proof relies on a fact relating two parameterizations of the same curve, together with the chain rule for differentiation. Let $\mathbf{C}_1(t)$, $a \le t \le b$ be a parameterization of C. It turns out that any other (regular) parameterization of C can be obtained by composing with a function $g : [c,d] \to [a,b]$ such that $g'(s) > 0$. Let $\mathbf{C}_2(s) = \mathbf{C}_1(g(s))$, $c \le s \le d$, be any other parameterization of C. Using the chain rule we have

$$\mathbf{C}_2'(s) = \frac{d}{ds}\mathbf{C}_1(g(s)) = \mathbf{C}_1'(g(s))g'(s).$$

This calculation shows that the line integral is

$$\int_c^d f(\mathbf{C}_2(s)) \cdot \mathbf{C}_2'(s)\, ds = \int_c^d f(\mathbf{C}_1(g(s))) \cdot \mathbf{C}_1'(g(s))g'(s)\, ds$$

$$= \int_a^b f(\mathbf{C}_1(t)) \cdot \mathbf{C}_1'(t)\, dt,$$

where the last equality follows by substituting $t = g(s)$ and $dt = g'(s)\, ds$ and using the fact that $a = g(c)$ and $b = g(d)$. $\square$

Notation: Because these integrals are independent of parameterization, we use the notation $\int_C f\, dC$ which emphasizes the curve C without denoting any specific parameterization $\mathbf{C}(t)$. Notice that, given a parameterization $\mathbf{C}(t)$ for C, the differentials are related by $dC = \|\mathbf{C}'(t)\|\, dt$. It will be important to keep track of the differential notations as we proceed. We summarize a method for integrating functions along curves:

Integrating a Function Along a Curve

To evaluate $\int_C f dC$:

1. Parameterize the curve $\mathbf{C}(t)$, for $a \le t \le b$, and find $\mathbf{C}'(t)$ (be sure to find the limits a and b on your parameterization since they become the limits of integration).

2. Evaluate f on the curve $\mathbf{C}(t)$ and find $f(\mathbf{C}(t)) \|\mathbf{C}'(t)\|$.

3. Integrate: $\int_a^b f(\mathbf{C}(t)) \|\mathbf{C}'(t)\|\, dt$.

Example 4.6.3. *Integrating along a spiral*

Let $f(x,y,z) = \sqrt{x^2 + y^2 + z^2}$ and let C be the parametric curve $\mathbf{C}(t) = (t\cos t, t\sin t, t)$, $0 \le t \le 2\pi$. Evaluate $\int_C f dC$.

Since the parameterization is given, we can begin by finding the integrand. One computes that

$$f(\mathbf{C}(t))\,\|\mathbf{C}'(t)\| = f(t\cos t, t\sin t, t)\,\|\langle \cos t - t\sin t, \sin t + t\cos t, 1\rangle\|$$
$$= \sqrt{t^2\cos^2 t + t^2\sin^2 t + t^2} \cdot \sqrt{2 + t^2}$$
$$= \sqrt{2} \cdot t\sqrt{2 + t^2},$$

where the calculation $\|\langle \cos t - t\sin t, \sin t + t\cos t, 1\rangle\| = \sqrt{2 + t^2}$ should be confirmed by the reader! Integrating we get

$$\int_C f dC = \sqrt{2}\int_0^{2\pi} t\sqrt{2 + t^2}\,dt = \frac{\sqrt{2}}{2}\int_2^{2+4\pi^2} u^{1/2}\,du \qquad (\text{letting } u = 2 + t^2)$$
$$= \frac{\sqrt{2}}{3}u^{3/2}\Big|_2^{2+4\pi^2} = \frac{\sqrt{2}}{3}\left(2 + 4\pi^2\right)^{3/2} - \frac{4}{3}. \;\blacktriangle$$

Thus far we have seen how to integrate functions along curves. In Chapter 5 we will see how to integrate *vector fields* along curves! Before that, we discuss integrating functions on surfaces.

Surface Integrals: To define the integral $\iint_S f\,dS$ of a function f over a surface S, we generalize double integrals. This is analogous to the integral of a function along a curve generalizing that along an interval. Recall that $\iint_R f\,dA$ was defined by first partitioning the region R into sub-rectangles of area ΔA_{ij}, and choosing a point (h_i, k_j) in each sub-rectangle. Then evaluate $f(h_i, k_j)$ and form the approximation

$$\iint_R f(x,y)dA \approx \sum_{i=1}^{n}\sum_{j=1}^{n} f(h_i, k_j)\Delta A_{ij}.$$

Finally, take the limit as the partition becomes finer and define

$$\iint_R f(x,y)dA = \lim_{n\to\infty} \sum_{i=1}^{n}\sum_{j=1}^{n} f(h_i, k_j)\Delta A_{ij}.$$

To define $\iint_S f dS$, partition the surface into sub-surfaces S_{ij} of area ΔS_{ij}, and choose representative points P_{ij} from each sub-surface. Evaluate $f(P_{ij})\Delta S_{ij}$, and take a limit as the partition of S becomes finer to define

$$\iint_S f dS = \lim_{n \to \infty} \sum_{i=1}^{n} \sum_{j=1}^{n} f(P_{ij}) \Delta S_{ij}. \qquad (4.6.2)$$

Our goal is to obtain a formula for $\iint_S f dS$ given a parametric surface $S(s,t)$ over some region R in the st-plane. Partitioning R into sub-rectangles R_{ij} leads to a partition of the surface, letting S_{ij} be the image of R_{ij} under S. As in Section 4.5, the area of $S_{ij} = S(R_{ij})$ is approximately $\|S_s \times S_t\|$ times the area of R_{ij} (see Figure 4.5.2). Thus ΔS_{ij} of Equation 4.6.2 is approximated by $\|S_s \times S_t\| \Delta s_i \Delta t_j$. Further, the choice of P_{ij} is arbitrary, so let $P_{ij} = S(s_i, t_j)$ and we get the approximation

$$\iint_S f dS \approx \sum_{i=1}^{n} \sum_{j=1}^{n} f(S(s_i, t_j)) \|S_s \times S_t\| \Delta s_i \Delta t_j.$$

Taking a limit, the $\Delta s_j \Delta t_i$ becomes $dsdt$, and the rest becomes the integrand $f(S(s,t)) \|S_s \times S_t\|$. With this motivation, we make the following definition.

Definition 4.6.2. *Let S be parametrized by $S(s,t)$ over the region R in the st-plane, and let f be a function defined on S. The integral of f on S is*

$$\iint_S f dS = \iint_R f(S(s,t)) \|S_s \times S_t\| \, dsdt.$$

Remark:

1. Notice from the formula for surface area, if f is the constant function 1, the surface integral gives the area of the surface S. Symbolically,

$$\text{Area of } S = \iint_S dS.$$

2. The differential is $dS = \|S_s \times S_t\| \, dsdt$, and $\|S_s \times S_t\|$ is the length of the normal vector to $S(s,t)$. Thus one can think that integrating a function on a parametric surface is like double integration, with the additional factor of the length of the normal vector in the integrand.

Example 4.6.4. *Mass of a hemisphere*

The upper hemisphere of the unit sphere has density $\delta(x,y,z) = 1 - z^2$, find its mass. We know that mass is the integral of density, so we have to integrate the density function on the hemisphere.

Step 1. Parameterize the surface.

Recall that the unit sphere has spherical equation $\rho = 1$, so the coordinate transformations give the parameterization

$$S(s,t) = (\sin s \cos t, \sin s \sin t, \cos s), \ 0 \le s \le \pi/2, \ 0 \le t \le 2\pi.$$

Step 2. Find the integrand. The cross product of the partial derivatives of $S(s,t)$ is

$$S_s \times S_t = \begin{vmatrix} \mathbf{i} & \mathbf{j} & \mathbf{k} \\ \cos s \cos t & \cos s \sin t & -\sin s \\ -\sin s \sin t & \sin s \cos t & 0 \end{vmatrix}$$

$$= \left\langle \sin^2 s \cos t, \sin^2 s \sin t, \sin s \cos s \right\rangle,$$

so the length of the normal vector is

$$\|S_s \times S_t\| = \left\| \left\langle \sin^2 s \cos t, \sin^2 s \sin t, \sin s \cos s \right\rangle \right\|$$

$$= \sin s \, \| \langle \sin s \cos t, \sin s \sin t, \cos s \rangle \| = \sin s.$$

Observing that the density function evaluated on the surface is

$$\delta(S(s,t)) = \delta(\sin s \cos t, \sin s \sin t, \cos s) = 1 - \cos^2 s,$$

the integrand is

$$\delta(S(s,t)) \|S_s \times S_t\| = \left(1 - \cos^2 s\right) \sin s.$$

Step 3. Integrate.

$$\int_0^{2\pi} \int_0^{\pi/2} \left(1 - \cos^2 s\right) \sin s \, ds dt = \int_0^{2\pi} \left(-\int_1^0 1 - u^2 du \right) dt \qquad (u = \cos s)$$

$$= \int_0^{2\pi} \left. \frac{u^3}{3} - u \right|_1^0 dt = \frac{2}{3} \int_0^{2\pi} dt = \frac{4\pi}{3}. \; \blacktriangle$$

Example 4.6.5. *Mass of a generalized cylinder*

Let S be the parametric surface $S(s,t) = (\sin t, s, t)$ for $0 \le s \le 1$ and $0 \le t \le \pi/2$ with density function $\delta(x,y,z) = x$. Find the mass of S.

Evaluating the function on the surface we have

$$\delta(S(s,t)) = \delta(\sin t, s, t) = \sin t.$$

The normal vector to the surface is

$$S_s \times S_t = \begin{vmatrix} \mathbf{i} & \mathbf{j} & \mathbf{k} \\ 0 & 1 & 0 \\ \cos t & 0 & 1 \end{vmatrix} = \langle 1, 0, -\cos t \rangle,$$

so the length of the normal vector is $\sqrt{1 + \cos^2 t}$. The mass of the surface is then given by the integral

$$\iint_S \delta \, dS = \int_0^{\pi/2} \int_0^1 \sin t \sqrt{1 + \cos^2 t} \, ds dt$$

$$= \int_0^{\pi/2} \sin t \sqrt{1 + \cos^2 t} \, dt = -\int_1^0 \sqrt{1 + u^2} du$$

$$= -\left(\frac{u}{2}\sqrt{1+u^2} + \frac{1}{2}\ln\left|u + \sqrt{1+u^2}\right| \right)\Big|_1^0 = \frac{\sqrt{2}}{2} + \frac{1}{2}\ln\left(1 + \sqrt{2}\right).$$

We made the substitution $u = \cos t$, then used an integral table for the rest! ▲

Integrating a Function on a Parametric Surface

The integral of f over S with parameterization $S(s,t)$, over the region R in the st-plane, is found as follows:

1. Find the integrand $f(S(s,t)) \, \|S_s \times S_t\|$. This amounts to

   a. Evaluating f on the surface (i.e find $f(S(s,t))$).

   b. Find the length of the normal vector to the surface (i.e. find $\|S_s \times S_t\|$).

2. Integrate the function from Step 1 over the region R in the st-plane.

$$\iint_S f dS = \iint_R f(S(s,t)) \, \|S_s \times S_t\| \, dA.$$

Just as for curves, surfaces have many different parameterizations, which can be thought of as a change of variables with the right assumptions. The change of variables formula in Theorem 4.4.1 can then be used to prove that the integral of a function over a surface is independent of the parameterization. We omit the details, but the result justifies calling $\iint_S f dS$ an integral over S since it doesn't matter which parameterization of S you choose.

Integration over $z = g(x,y)$: We now know how to integrate functions over parametric surfaces, and can apply that knowledge to integrating functions over graphs of $z = g(x,y)$. As you know, the graph of $z = g(x,y)$ can be parameterized by $S(x,y) = (x, y, g(x,y))$ and the techniques for parametric surfaces apply. As in our surface area calculations in Section 4.5, one calculates that the length of the normal vector is $\|\mathbf{n}\| = \sqrt{g_x^2 + g_y^2 + 1}$, and evaluating f on the surface gives

$$\iint_S f dS = \iint_R f(x,y,g(x,y)) \sqrt{g_x^2 + g_y^2 + 1} \, dA.$$

We illustrate how to apply this formula with an example.

Example 4.6.6. *Integrating a function over a graph*

Integrate $f(x,y,z) = x^2 + y^2 + z$ over the portion of the paraboloid $z = x^2 + y^2$ lying over the unit disk.

Step 1. Find the integrand.

The function evaluated on the surface is

$$f(x,y,x^2+y^2) = x^2 + y^2 + (x^2+y^2) = 2(x^2+y^2),$$

while the length of the normal vector is

$$\|\mathbf{n}\| = \|\langle -2x, -2y, 1\rangle\| = \sqrt{4x^2 + 4y^2 + 1}.$$

Step 2. Integrate. The surface integral is given by

$$\iint_S f dS = \iint_{\text{unit disk}} 2(x^2+y^2)\sqrt{4x^2+4y^2+1}\, dA,$$

which is best calculated in polar coordinates. The limits for the unit disk are $0 \le \theta \le 2\pi$, $0 \le r \le 1$, and $dA = r dr d\theta$. Using $r^2 = x^2 + y^2$, the surface integral becomes

$$\iint_S f dS = \int_0^{2\pi} \int_0^1 2r^2\sqrt{1+4r^2}\, r dr d\theta.$$

Letting $u = 1 + 4r^2$, one can solve for r^2 to get $r^2 = (u-1)/4$. Substituting using the differential $du = 8r dr$ yields

$$\iint_S f dS = \int_0^{2\pi} \frac{1}{16} \int_1^5 (u-1)\sqrt{u}\, du d\theta = \int_0^{2\pi} \frac{1}{16} \int_1^5 u^{3/2} - u^{1/2}\, du d\theta$$

$$= \int_0^{2\pi} \frac{1}{16} \left(\frac{2}{5}5^{5/2} - \frac{2}{3}5^{3/2} - \left(\frac{2}{5} - \frac{2}{3}\right)\right) d\theta$$

$$= \int_0^{2\pi} \frac{1}{16}\left(\frac{20\sqrt{5}}{3} + \frac{4}{15}\right) d\theta = \frac{\pi}{8}\left(\frac{20\sqrt{5}}{3} + \frac{4}{15}\right). \ \blacktriangle$$

Exercises

1. A fence lies above the curve $\mathbf{C}(t) = (t, t^2)$ for $1 \le t \le 4$. Its height above any point (x,y) on the curve is given by $f(x,y) = x + \sqrt{y}$. Find the area of the fence.

2. A fence lies above the curve $\mathbf{C}(t) = (2t+1, 3t-5)$ for $0 \le t \le 2$. Its height above any point (x,y) on the curve is given by $f(x,y) = 3x^2 - 4y$. Find the area of the fence.

3. Integrate the function $f(x,y) = 5x^3 + 4y$ along the curve $C(t) = (t, t^3)$, $0 \leq t \leq 1$.

4. Evaluate the integral of $f(x,y,z) = xy + z$ along the line segment from the origin to $(1, 2, 3)$.

5. Evaluate the integral of $f(x,y,z) = x^2 + y^2 + z^2$ along the curve $C(t) = (\cos t, \sin t, t)$, $0 \leq t \leq 2\pi$.

6. Evaluate the integral of $f(x,y,z) = 2x - y + z$ along the line segment $C(t) = (t, 1 + 2t, 1 + t)$, $0 \leq t \leq 3$.

7. Find the mass of the wire $C(t) = (t, t^2)$, $0 \leq t \leq 1$ if the density is given by $\delta(x,y) = xy$.

8. Find the mass of the unit circle if the density is given by $\delta(x,y) = x^2 - 2xy + y^2$.

9. Find the mass of the wire $C(t) = (2 - t, 3 + 2t, -1 + t)$, $-1 \leq t \leq 1$ if the density is given by $\delta(x,y,z) = x^2 + z^2$.

10. Find the mass of the wire $C(t) = (t\cos t, t\sin t, t)$, $0 \leq t \leq 2\pi$ if the density is given by $\delta(x,y,z) = \sqrt{x^2 + y^2 + z^2}$.

11. Evaluate the integral of $f(x,y,z) = xyz$ over the cylinder $S(s,t) = (\cos t, s, \sin t)$, for $0 \leq s \leq 1$ and $0 \leq t \leq 2\pi$.

12. Evaluate the integral of $f(x,y,z) = x + y + z$ on the surface $S(s,t) = (2s - t + 1, 3 + 2t - 1, 3s + 2t - 3)$, $0 \leq s, t \leq 1$.

13. The density of $S(s,t) = (s, t, s^2)$, $0 \leq s, t \leq 1$ is given by $\delta(x,y,z) = xy$. Find the mass of the surface.

14. Evaluate the integral of $f(x,y,z) = yz$ on the surface $S(s,t) = (2t^3, 3t^2, s)$, $0 \leq s, t \leq 1$.

15. Evaluate the integral of $f(x,y,z) = x^2 + y + z$ on the surface $S(s,t) = (2t^3, 3t^2, s)$, $0 \leq s, t \leq 1$.

16. Find the mass of the surface $S(s,t) = (2t^3, 3t^2, s)$, $0 \leq s, t \leq 1$, if the density is $\delta(x,y,z) = z^2 + 2z$.

17. Evaluate the integral of $f(x,y,z) = x^2yz$ on the surface $S(s,t) = (\cos t, \sin t, s)$, $0 \leq s \leq 1$, $0 \leq t \leq \pi/2$.

18. Evaluate the integral of $f(x,y,z) = e^z$ on the surface $S(s,t) = (s\cos t, s\sin t, s)$, $0 \leq s \leq 1$, $0 \leq t \leq \pi/2$.

19. The density of the surface $S(s,t) = (\cos t, \sin t, s)$, $0 \leq s \leq 1$, $0 \leq t \leq \pi/2$ is $\delta(x,y,z) = 2x^2 + 3y^2 + z + 5$. Find its mass.

20. Evaluate the integral of $f(x,y,z) = x^2 + y^2 + z^2$ on the surface $S(s,t) = (s\cos t, s\sin t, s)$, $0 \leq s \leq 1$, $0 \leq t \leq \pi/2$.

21. Evaluate the integral of $f(x,y,z) = xy$ on the surface $S(s,t) = (s\cos t, s\sin t, s^2)$, $0 \leq s \leq 1$, $0 \leq t \leq \pi/2$.

22. Evaluate the integral of $f(x,y,z) = \frac{xy}{z}$ on the surface $S(s,t) = (s\cos t, s\sin t, s^2)$, $1 \leq s \leq 2$, $0 \leq t \leq \pi/2$.

23. The density of the surface $S(s,t) = (s\cos t, s\sin t, s^2)$, $0 \le s \le 1$, $0 \le t \le \pi/2$ is $\delta(x,y,z) = x^2 + y^2$. Find its mass.

24. Evaluate the integral of $f(x,y,z) = z^2$ on the surface $S(s,t) = (\sin s \cos t, \sin s \sin t, \cos s)$, $0 \le s \le \pi/2$, $0 \le t \le \pi/2$.

25. Evaluate the integral of $f(x,y,z) = xyz$ on the surface $S(s,t) = (\sin s \cos t, \sin s \sin t, \cos s)$, $0 \le s \le \pi/2$, $0 \le t \le \pi/2$.

26. Find the mass of the portion of the cone $z = \sqrt{x^2 + y^2}$ that lies over the unit disk if the density function is $\delta(x,y,z) = 1 - z$.

27. Evaluate the integral of $f(x,y,z) = yz$ over the portion of $2x + y + z = 6$ that lies above the rectangle $R = [0,2] \times [0,1]$.

28. The density of the portion of $z = 4 - x^2 - y^2$ that lies above the xy-plane is $\delta(x,y,z) = 4 - z$. Find the mass of the surface.

5 Vector Analysis

This chapter embarks on the study of vector analysis. Vector analysis has significant applications to a wide range of scientific disciplines, so this is one topic that has an answer to the age-old question "When will I ever use this?" In Sections 5.1, 5.2, and 5.3, the integral of a vector field along a curve is investigated. Section 5.1 motivates the integral of a vector field along a curve using the notion of work introduced in Section 2.3, and derives a formula for the integral given a parameterization of the curve. Sections 5.2 and 5.3 provide other techniques for computing line integrals in certain situations. The integral of a vector field across a surface is defined in Section 5.4, setting the stage for two important theorems in vector analysis—Stokes' and the divergence theorem. Section 5.5 discusses derivatives of vector fields. In particular, curl and divergence are introduced. With these preliminaries taken care of, the next two sections cover Stokes' theorem and the divergence theorem. The chapter concludes by applying these theorems to the problem of finding formulas for curl, divergence and the Laplacian in cylindrical and spherical coordinates.

5.1 Line Integrals

In this section we describe how to integrate a vector field along a curve. Such an integral is called a line integral, even though it's really an integral along a curve. Line integrals have many applications, and can be motivated by physical considerations. We do that now.

Example 5.1.1. *Work by constant force*

Recall that, in physics, the work done by a force **F** in moving an object from one point to another is defined to be the component of force in the direction of movement times the distance traveled. In short, work is force times distance. We saw in Section 2.3 that the work w done by a constant force **F** on an object moving through the displacement vector **d** is given by

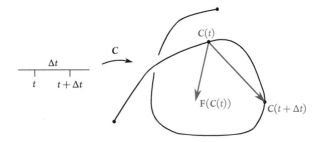

Figure 5.1.1 Work by variable force along a curve

$$w = \left(comp_{\mathbf{d}}\mathbf{F} \right) \|\mathbf{d}\| = \frac{\mathbf{F} \cdot \mathbf{d}}{\|\mathbf{d}\|} \|\mathbf{d}\| = \mathbf{F} \cdot \mathbf{d}.$$

Given a constant force on an object moving in a straight line, then, work is merely the dot product of the force and displacement vectors.

We now want to consider what happens if we try to calculate work for a variable force along a path that is not straight. The answer is just what you would expect in a calculus class:

Approximate the solution with work you can calculate, then take a limit as your approximation becomes better.

In order to approximate the work done we will approximate the path with small line segments, and assume the force is constant on each. Keep these assumptions in mind during the following discussion.

Let the parametric curve $\mathbf{C}$ describe the position of a particle in space, and $\mathbf{F}$ a vector field describing a force. Our initial goal is to approximate the work done by $\mathbf{F}$ on the particle as it moves between nearby points on $\mathbf{C}$. For small Δt, the points $\mathbf{C}(t)$ and $\mathbf{C}(t + \Delta t)$ are close and the curve between them is almost straight, which means the vector $\mathbf{C}(t + \Delta t) - \mathbf{C}(t)$ approximates the path (see Figure 5.1.1). Again for small Δt, the force is almost constant along the curve and can be approximated by $\mathbf{F}(\mathbf{C}(t))$. Since work is force dotted with displacement, the work done by $\mathbf{F}$ on the particle as it travels along this small arc of $\mathbf{C}$ is approximated by

$$w \approx \mathbf{F}(\mathbf{C}(t)) \cdot (\mathbf{C}(t + \Delta t) - \mathbf{C}(t)). \tag{5.1.1}$$

This gives a reasonable approximation for work over small arcs. To approximate the work done by $\mathbf{F}$ on all of $\mathbf{C}$, partition the interval $[a, b]$ into small subintervals $[t_i, t_{i+1}]$, and use this to partition the curve into sub-arcs from $\mathbf{C}(t_i)$ to $\mathbf{C}(t_{i+1})$ (here $t_{i+1} = t_i + \Delta t_i$). On each sub-arc use the approximation of Equation 5.1.1, then add the estimates for all sub-arcs to estimate the total work. The resulting estimate is

$$w \approx \sum_{i=0}^{n-1} \mathbf{F}(\mathbf{C}(t_i)) \cdot (\mathbf{C}(t_i + \Delta t_i) - \mathbf{C}(t_i)).$$

This approximation improves as the partition becomes finer, and we'll say the work *is* the limit as $n \to \infty$.

The sum in this approximation doesn't look like a Riemann sum yet, but an algebraic trick can change that. Multiplying and dividing by Δt we get

$$w = \lim_{n \to \infty} \sum_{i=0}^{n-1} \mathbf{F}(\mathbf{C}(t_i)) \cdot \frac{(\mathbf{C}(t_i + \Delta t_i) - \mathbf{C}(t_i))}{\Delta t_i} \Delta t_i, \tag{5.1.2}$$

which is a Riemann sum because each term is a function times Δt. In taking the limit, then, we arrive at a single-variable integral whose integrand is easy to describe.

To do so, we consider what each vector in the dot product of Equation 5.1.2 approaches as $\Delta t \to 0$. Note that by definition

$$\lim_{\Delta t \to 0} \frac{(\mathbf{C}(t + \Delta t) - \mathbf{C}(t))}{\Delta t} = \mathbf{C}'(t),$$

so the second vector in the dot product just becomes $\mathbf{C}'(t)$ in the integrand! The first becomes $\mathbf{F}(\mathbf{C}(t))$, the Δt becomes the differential dt, and the limit of Riemann sums gives work as the integral

$$w = \int_a^b \mathbf{F}(\mathbf{C}(t)) \cdot \mathbf{C}'(t) dt.$$

This discussion motivates the following definition:

Definition 5.1.1. *The* line integral *of the vector field* $\mathbf{F}$ *along the parametric curve* $\mathbf{C}(t)$, *for* $a \le t \le b$, *is*

$$\int_a^b \mathbf{F}(\mathbf{C}(t)) \cdot \mathbf{C}'(t) \, dt.$$

The integrand of Definition 5.1.1 is the dot product of two vectors, so it is a scalar function, and the integral is a single-variable integral. Implicit in the notation $\mathbf{F}(\mathbf{C}(t))$ is the assumption that you have a parameterization, and the notation means that you substitute the parametric equations into the component functions of the vector field. This is outlined in the following procedure for evaluating line integrals.

Parametric Method for Evaluating Line Integrals

To evaluate $\int_a^b \mathbf{F}(\mathbf{C}(t)) \cdot \mathbf{C}'(t) \, dt$:

1. Parameterize the curve $\mathbf{C}(t)$, for $a \le t \le b$, and find $\mathbf{C}'(t)$ (be sure to find the limits a and b on your parameterization since they become the limits of integration).

2. Evaluate $\mathbf{F}$ on the curve $\mathbf{C}(t)$ and find $\mathbf{F}(\mathbf{C}(t)) \cdot \mathbf{C}'(t)$.

3. Integrate: $\int_a^b \mathbf{F}(\mathbf{C}(t)) \cdot \mathbf{C}'(t) \, dt$.

We carry this procedure out in a few examples. It will be helpful to recall that

$$C(t) = (1-t)\mathbf{P} + t\mathbf{Q} \text{ for } 0 \le t \le 1.$$

parameterizes the line segment from P to Q.

Example 5.1.2. *Work by variable force*

Let $\mathbf{F}(x,y,z) = \left(\dfrac{-x}{\sqrt{x^2+y^2+z^2}}, \dfrac{-y}{\sqrt{x^2+y^2+z^2}}, \dfrac{-z}{\sqrt{x^2+y^2+z^2}} \right)$, and find the work done by $\mathbf{F}$ on a particle that travels along the line segment from the origin to the point $(1,2,3)$.

Step 1. Parameterize the curve and differentiate it.

In our case, the line segment from $(0,0,0)$ to $(1,2,3)$ can be parameterized by $C(t) = (1-t)(0,0,0) + t(1,2,3) = (t,2t,3t)$ for $0 \le t \le 1$. So $C'(t) = \langle 1,2,3 \rangle$.

Step 2. Find $\mathbf{F}(C(t)) \cdot C'(t)$.

Evaluating $\mathbf{F}$ on the curve gives

$$\mathbf{F}(C(t)) = \mathbf{F}(t,2t,3t)$$

$$= \left(\frac{-t}{\sqrt{t^2+(2t)^2+(3t)^2}}, \frac{-2t}{\sqrt{t^2+(2t)^2+(3t)^2}}, \frac{-3t}{\sqrt{t^2+(2t)^2+(3t)^2}} \right)$$

$$= \left(\frac{-t}{\sqrt{10t}}, \frac{-2t}{\sqrt{10t}}, \frac{-3t}{\sqrt{10t}} \right) = \left(\frac{-1}{\sqrt{10}}, \frac{-2}{\sqrt{10}}, \frac{-3}{\sqrt{10}} \right).$$

We remark that it is not common for $\mathbf{F}(C(t))$ to be a constant vector; that just happens in this example. The integrand of the line integral is

$$\mathbf{F}(C(t)) \cdot C'(t) = \left(\frac{-1}{\sqrt{10}}, \frac{-2}{\sqrt{10}}, \frac{-3}{\sqrt{10}} \right) \cdot \langle 1,2,3 \rangle = -\frac{10}{\sqrt{10}} = -\sqrt{10}.$$

Step 3. Integrate.

$$\int_a^b \mathbf{F}(C(t)) \cdot C'(t)\, dt = \int_0^1 -\sqrt{10}\, dt = -\sqrt{10}t \Big|_0^1 = -\sqrt{10}.$$

The fact that the work done by $\mathbf{F}$ is negative means that the force was hindering movement rather than helping it. Geometrically, it means that $\mathbf{F}$ and C' point in opposite directions. ▲

Independence of Parameterization The first step in integrating a vector field along a curve is to parameterize the curve. Recall that there are many parameterizations for the same curve. For example both

$$C_1(t) = (\cos t, \sin t), \ 0 \le t \le \pi, \text{ and}$$
$$C_2(t) = (\cos 2t, \sin 2t), \ 0 \le t \le \pi/2,$$

parameterize the top half of the unit circle oriented counterclockwise. A priori, using different parameterizations of the same curve C could result in different values of the line integral. It turns out they don't (see Theorem 5.1.1). This allows us to talk about the integral of a vector field $\mathbf{F}$ along a curve C, rather than having to specify the parameterization we use. Before stating the theorem, we illustrate it with an example to get more familiar with the parameteric method of integrating vector fields along curves.

Example 5.1.3. *Independence of parameterization*

Evaluate the line integral of $\mathbf{F}(x,y) = \langle xy, 2x \rangle$ along the top half of the unit circle using the parameterizations:

(a) $\mathbf{C}(t) = (\cos t, \sin t), 0 \le t \le \pi$, and
(b) $\mathbf{C}(t) = (\cos 2t, \sin 2t), 0 \le t \le \pi/2$

We follow the above steps in each case:
(a) Using the standard parameterization of the unit circle.

1. $\mathbf{C}(t) = (\cos t, \sin t)$, for $0 \le t \le \pi$, and $\mathbf{C}'(t) = \langle -\sin t, \cos t \rangle$.
2. $\mathbf{F}(\cos t, \sin t) = \langle \cos t \sin t, 2 \cos t \rangle$, and

$$\mathbf{F}(\mathbf{C}(t)) \cdot \mathbf{C}'(t) = \langle \cos t \sin t, 2 \cos t \rangle \cdot \langle -\sin t, \cos t \rangle$$
$$= -\cos t \sin^2 t + 2 \cos^2 t.$$

3. $\int_a^b \mathbf{F}(\mathbf{C}(t)) \cdot \mathbf{C}'(t) \, dt = \int_0^\pi -\cos t \sin^2 t + 2 \cos^2 t \, dt$. The first sum can be integrated using the substitution $u = \sin t$, and the second using the half-angle formula $\cos^2 t = \frac{1}{2}(1 + \cos 2t)$.

Explicitly, we have:

$$\int_0^\pi -\cos t \sin^2 t + 2 \cos^2 t \, dt = \int_0^\pi -\cos t \sin^2 t \, dt + \int_0^\pi 2 \cos^2 t \, dt$$
$$= -\frac{\sin^3 t}{3} + t + \frac{\sin 2t}{2} \Big|_0^\pi = \pi.$$

(b) Using the parameterization that is twice as fast.

1. $\mathbf{C}(t) = (\cos 2t, \sin 2t)$, for $0 \le t \le \pi/2$, and $\mathbf{C}'(t) = \langle -2 \sin 2t, 2 \cos 2t \rangle$.
2. $\mathbf{F}(\cos 2t, \sin 2t) = \langle \cos 2t \sin 2t, 2 \cos 2t \rangle$, so

$$\mathbf{F}(\mathbf{C}(t)) \cdot \mathbf{C}'(t) = \langle \cos 2t \sin 2t, 2 \cos 2t \rangle \cdot \langle -2 \sin 2t, 2 \cos 2t \rangle$$
$$= -2 \cos 2t \sin^2 2t + 4 \cos^2 2t.$$

3. $\int_a^b \mathbf{F}(\mathbf{C}(t)) \cdot \mathbf{C}'(t) \, dt = \int_0^{\pi/2} -2 \cos 2t \sin^2 2t + 4 \cos^2 2t \, dt$. Making the substitution $u = 2t$, $du = 2dt$, and changing the limits gives

$$\int_0^{\pi} -\cos u \sin^2 u + 2\cos^2 u \, du.$$

Since this is the same integral as part (a), the value of the line integral of $\mathbf{F}$ for either parameterization is π. ▲

As mentioned above, this is not an isolated phenomenon, but a general fact. Let C be any geometric curve from point A to point B (for example, the semicircle from $(1,0)$ to $(-1,0)$ in the plane). As long as we parameterize C with the correct orientation (from A to B) and don't backtrack while traversing C, the line integral is independent of the parameterization we choose. We state it as a theorem:

Theorem 5.1.1. *The line integral of a vector field $\mathbf{F}$ with continuous component functions along a continuously differentiable curve C is independent of the parameterization chosen for C.*

Proof. The proof relies on a fact relating two parameterizations of the same curve, together with the chain rule for differentiation. Let $\mathbf{C}_1(t)$, $a \le t \le b$ be a parameterization of C. Any other regular parameterization of C can be obtained by composing with a function $g : [c,d] \to [a,b]$ such that $g'(s) > 0$ (recall Definition 1.4.1). Let $\mathbf{C}_2(s) = \mathbf{C}_1(g(s))$, $c \le s \le d$, be any other parameterization of C. Using the chain rule we have

$$\frac{d}{ds}\mathbf{C}_2(s) = \frac{d}{ds}\mathbf{C}_1(g(s)) = \mathbf{C}_1'(g(s))g'(s).$$

This calculation shows that the line integral is

$$\int_c^d \mathbf{F}(\mathbf{C}_2(s)) \cdot \mathbf{C}_2'(s)\, ds = \int_c^d \mathbf{F}(\mathbf{C}_1(g(s))) \cdot \mathbf{C}_1'(g(s))g'(s)\, ds$$

$$= \int_a^b \mathbf{F}(\mathbf{C}_1(t)) \cdot \mathbf{C}_1'(t)\, dt,$$

where the last equality follows by substituting $t = g(s)$ and $dt = g'(s)\, ds$ and using the fact that $a = g(c)$ and $b = g(d)$. □

Because the line integral depends only on the curve C, and not on a choice of parameterization $\mathbf{C}(t)$, we often use the following notation for line integrals:

$$\int_a^b \mathbf{F}(\mathbf{C}(t)) \cdot \mathbf{C}'(t)\, dt = \int_C \mathbf{F} \cdot d\mathbf{C}.$$

The parameterization $\mathbf{C}(t)$ is suppressed in the notation $\int_C \mathbf{F} \cdot d\mathbf{C}$, and we've also introduced the differential $d\mathbf{C}$ to represent $\mathbf{C}'(t)dt$. This is analogous to the use of differential notation in single-variable calculus, where if $x = f(t)$ we know $dx = f'(t)dt$.

Remark: It is worth noting that the line integral, or integral of a vector field along a curve, can also be viewed as integrating a function along the curve as in Section 4.6. Given a vector field **F** along a curve C parameterized by **C**, the component of **F** in the tangential direction is a *function* defined on the curve C by the formula

$$comp_{C'}\mathbf{F} = \mathbf{F}(\mathbf{C}(t)) \cdot \frac{\mathbf{C}'(t)}{\|\mathbf{C}'(t)\|}.$$

Since this is a function on C, the techniques of section 4.6 can be used to evaluate

$$\int_C comp_{C'}\mathbf{F}dC = \int_a^b \left(\mathbf{F}(\mathbf{C}(t)) \cdot \frac{\mathbf{C}'(t)}{\|\mathbf{C}'(t)\|} \right) \|\mathbf{C}'(t)\| \, dt$$

$$= \int_a^b \mathbf{F}(\mathbf{C}(t)) \cdot \mathbf{C}'(t) \, dt$$

$$= \int_C \mathbf{F} \cdot d\mathbf{C}.$$

The first equality follows from the fact that $dC = \|\mathbf{C}'(t)\| \, dt$, and the factors of $\|\mathbf{C}'(t)\|$ in numerator and denominator cancel to verify that

$$\int_C \mathbf{F} \cdot d\mathbf{C} = \int_C comp_{C'}\mathbf{F}dC.$$

Thus the integral of a vector field along a curve is simply the integral its tangential component (a scalar function) along the curve. This interpretation justifies some intuition. If **F** and **C**$'$ point roughly the same direction most of the time, the line integral $\int_C \mathbf{F} \cdot d\mathbf{C}$ will be positive. If the line integral is negative, **F** points in the opposite direction from **C**$'$ more often than not.

Math App 5.1.1. *Work and line integrals*

The following Math App provides an interactive look at integrating the tangential component of a vector field along a curve.

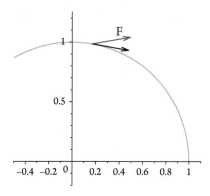

Curve Notation: Sometimes we encounter curves that are made by concatenating pieces of smooth curves. For example, the boundary of the triangle with vertices $(0,0)$, $(1,0)$, and $(0,1)$ consists of three line segments. Let C denote the boundary of the triangle, oriented counterclockwise. Then C is made up of the line segments C_1 from $(0,0)$ to $(1,0)$, C_2 from $(1,0)$ to $(0,1)$, and C_3 from $(0,1)$ back to $(0,0)$. In situations like this, we say that the line integral of a vector field $\mathbf{F}$ along the boundary of the triangle C is the sum of the line integrals along the curves that combine to make C. Symbolically we have

$$\int_C \mathbf{F} \cdot d\mathbf{C} = \int_{C_1 + C_2 + C_3} \mathbf{F} \cdot d\mathbf{C} = \int_{C_1} \mathbf{F} \cdot d\mathbf{C} + \int_{C_2} \mathbf{F} \cdot d\mathbf{C} + \int_{C_3} \mathbf{F} \cdot d\mathbf{C}.$$

Similarly we let $-C$ denote the curve C backwards. More precisely any parameterization of the curve C induces an orientation of C, in other words, a direction of travel along C. The orientation is the direction of increasing t-values. For example $\mathbf{C}(t) = (\cos t, \sin t)$ for $0 \le t \le 2\pi$ starts at $(1,0)$ and induces a counterclockwise orientation of the unit circle. On the other hand, $\mathbf{C}(t) = (\cos t, -\sin t)$ for $0 \le t \le 2\pi$ also starts at $(1,0)$, but traverses the unit circle in a clockwise fashion. If C is a curve oriented from A to B, then we denote the same curve from B to A as $-C$. It follows from the chain rule (or rather, from substitution in integration) that

$$\int_{-C} \mathbf{F} \cdot d\mathbf{C} = - \int_C \mathbf{F} \cdot d\mathbf{C},$$

and we will use this fact regularly. This is intuitively obvious as well. The line integral of a vector field $\mathbf{F}$ along C is just the integral of the component of $\mathbf{F}$ in the tangential direction. The curve $-C$ has the opposite tangential direction to C. Since the tangents to C and $-C$ are negatives of each other, the components of $\mathbf{F}$ in tangential directions will be as well. Hence the line integrals will have the same magnitude, but opposite signs.

We illustrate these notions with some examples.

Example 5.1.4. *Integrals along piecewise smooth curves*

Find the line integral $\int_C \mathbf{F} \cdot d\mathbf{C}$ of $\mathbf{F}(x,y) = (xy, -x^2)$ along the boundary C of the triangle with vertices $(0,0)$, $(1,0)$, and $(0,1)$ oriented counterclockwise (see Figure 5.1.2). We still use the approach outlined above of parameterizing, evaluating and taking dot products to find the integrand.

C_1: This segment is parameterized by $\mathbf{C}_1(t) = (t,0)$ for $0 \le t \le 1$, so $\mathbf{C}_1'(t) = (1,0)$. Evaluating the vector field on the curve we have $\mathbf{F}(t,0) = (0, -t^2)$, so the line integral is

$$\int_{C_1} \mathbf{F} \cdot d\mathbf{C} = \int_0^1 (0, -t^2) \cdot (1,0) \, dt = 0.$$

C_2: This curve is parameterized by $\mathbf{C}_2 = (1-t, t)$, for $0 \le t \le 1$, so that $\mathbf{C}_2'(t) = (-1, 1)$. Evaluating the vector field on the curve gives $\mathbf{F}(1-t, t) = ((1-t)t, -(1-t)^2)$, and the line integral becomes

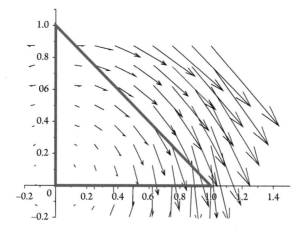

Figure 5.1.2 Line integrals along piecewise defined curves

$$\int_{C_2} \mathbf{F} \cdot d\mathbf{C} = \int_0^1 \left\langle (1-t)t, (1-t)^2 \right\rangle \cdot \langle -1, 1 \rangle \, dt$$

$$= \int_0^1 1 - 3t + 2t^2 dt = t - \frac{3t^2}{2} + \frac{2t^3}{3} \Big|_0^1 = \frac{1}{6}.$$

C_3: This curve is parameterized by $C_3(t) = (0, 1-t)$, for $0 \le t \le 1$ so that $C_3'(t) = \langle 0, -1 \rangle$. Since each component function of $\mathbf{F}$ has an x in it, we see that $\mathbf{F}(0, 1-t) = \langle 0, 0 \rangle$ and the integrand in the line integral will be zero. Thus $\int_{C_3} \mathbf{F} \cdot d\mathbf{C} = 0$. Putting these together we have

$$\int_C \mathbf{F} \cdot d\mathbf{C} = \int_{C_1} \mathbf{F} \cdot d\mathbf{C} + \int_{C_2} \mathbf{F} \cdot d\mathbf{C} + \int_{C_3} \mathbf{F} \cdot d\mathbf{C} = 0 + \frac{1}{6} + 0 = \frac{1}{6}. \blacktriangle$$

Example 5.1.5. *Integrating along* $-C_2$

In the previous example we calculated the line integral of $\mathbf{F}(x, y) = \langle xy, -x^2 \rangle$ along the line segment C_2 from $(1, 0)$ to $(0, 1)$. We found $\int_{C_2} \mathbf{F} \cdot d\mathbf{C} = \frac{1}{6}$. The curve $-C_2$ is the line segment from $(0, 1)$ to $(1, 0)$, which is parameterized by $\mathbf{C}(t) = (t, 1-t)$ for $0 \le t \le 1$ (so $\mathbf{C}'(t) = \langle 1, -1 \rangle$). As $\mathbf{F}(t, (1-t)) = \langle t(1-t), -t^2 \rangle$, we have

$$\int_{-C_2} \mathbf{F} \cdot d\mathbf{C} = \int_0^1 \left\langle t(1-t), -t^2 \right\rangle \cdot \langle 1, -1 \rangle \, dt = \int_0^1 t - 2t^2 dt = \frac{t^2}{2} - \frac{2t^3}{3} \Big|_0^1 = -\frac{1}{6}. \blacktriangle$$

This illustrates that when you integrate backwards along a curve you get the negative of the forward integral.

Differential Notation: There is yet another notation for line integrals that is common, and useful. So far we have denoted the line integral of **F** along a curve C by $\int_C \mathbf{F} \cdot d\mathbf{C}$, which is relatively simple notation but can hide some of the details of what we mean. A more descriptive definition, given initially, is $\int_a^b \mathbf{F}(\mathbf{C}(t)) \cdot \mathbf{C}'(t)\, dt$. This notation indicates that to calculate line integrals we use a parameterization of the curve, and it makes it clearer that we're just integrating the component of **F** in the direction tangent to $\mathbf{C}(t)$. We now introduce differential notation, which will be convenient when discussing some of the main theorems in vector analysis.

Let $\mathbf{F}(x,y,z) = P(x,y,z)\mathbf{i} + Q(x,y,z)\mathbf{j} + R(x,y,z)\mathbf{k}$ be a vector field, and C any curve. We introduce the notation

$$\int_C \mathbf{F} \cdot d\mathbf{C} = \int_C P(x,y,z)\,dx + Q(x,y,z)\,dy + R(x,y,z)\,dz,$$

which will sometimes be shortened to $\int_C P\,dx + Q\,dy + R\,dz$, suppressing the independent variables.

We now describe why this notation makes sense, given our experience with differentials dx, dy, and dz. Let $\mathbf{C}(t) = (x(t), y(t), z(t))$, for $a \le t \le b$, be a parameterization of C. By the definition of the differential $dx = x'(t)\,dt$, and similarly for y and z. Using this notation, we then have

$$\int_C \mathbf{F} \cdot d\mathbf{C} = \int_a^b \langle P(\mathbf{C}(t)), Q(\mathbf{C}(t)), R(\mathbf{C}(t)) \rangle \cdot \langle x'(t), y'(t), z'(t) \rangle\, dt$$

$$= \int_a^b P(\mathbf{C}(t))x'(t)\,dt + \int_a^b Q(\mathbf{C}(t))y'(t)\,dt + \int_a^b R(\mathbf{C}(t))z'(t)\,dt.$$

$$= \int_C P(x,y,z)\,dx + Q(x,y,z)\,dy + R(x,y,z)\,dz$$

We describe how to evaluate line integrals given in differential notation.

Evaluating Line Integrals in Differential Notation

To evaluate $\int_C P\,dx + Q\,dy + R\,dz$,

1. Parameterize C by $\mathbf{C}(t) = (x(t), y(t), z(t))$ for $a \le t \le b$.
2. Evaluate the component functions P, Q, and R on $\mathbf{C}(t)$, and let $dx = x'(t)\,dt$, $dy = y'(t)\,dt$ and $dz = z'(t)\,dt$.
3. Integrate $\int_a^b P(\mathbf{C}(t))x'(t) + Q(\mathbf{C}(t))y'(t) + R(\mathbf{C}(t))z'(t)\,dt$.

Example 5.1.6. *Evaluating a line integral in differential notation*

Let C be the helical arc given by $\mathbf{C}(t) = (\cos t, \sin t, t)$ for $0 \le t \le \pi$. Evaluate $\int_C z\,dx + xy\,dy + xz\,dz$.

Notice that we already have a parameterization, so the first step is done. The differentials are given by $dx = -\sin t dt$, $dy = \cos t dt$ and $dz = dt$. Substituting we get

$$\int_C zdx + xydy + xzdz = \int_0^\pi t(-\sin t \, dt) + \cos t \sin t(\cos t \, dt) + t \cos t \, dt$$

$$= -\int_0^\pi t \sin t dt + \int_0^\pi \cos^2 t \sin t dt + \int_0^\pi t \cos t dt$$

$$= -(-t\cos t + \sin t) + \frac{\cos^3 t}{3} + (t\sin t + \cos t) \Big|_0^\pi$$

$$= -\pi - \frac{8}{3}.$$

Integration by parts and substitution were used to evaluate the integrals. ▲

Things to Know

- The line integral $\int_C \mathbf{F} \cdot d\mathbf{C}$ of a vector field $\mathbf{F}$ along a curve C is the integral of the component of $\mathbf{F}$ in the tangential direction.
- Given a parameterization $\mathbf{C}$ of the curve C,

$$\int_C \mathbf{F} \cdot d\mathbf{C} = \int_a^b \mathbf{F}(\mathbf{C}(t)) \cdot \mathbf{C}'(t) \, dt.$$

- Differential notation:

$$\int_C Pdx + Qdy + Rdz$$

  emphasizes the components of $\mathbf{F}$ and is used in other contexts.
- Know how to calculate $\int_C \mathbf{F} \cdot d\mathbf{C}$ in either notation using parameterizations.
- $\int_C \mathbf{F} \cdot d\mathbf{C}$ is independent of the parameterization of C.

Exercises

1. Let $\mathbf{F}(x,y,z) = \langle -x, -y, -z \rangle$, and let C be the line segment from the origin to the point $(3, -1, 2)$. Do you expect $\int_C \mathbf{F} \cdot d\mathbf{C}$ to be positive or negative, and why?
2. Assume that both curves A and B in the Figure 5.1.3 are oriented counterclockwise.
   (a) Are $\int_A \mathbf{F} \cdot d\mathbf{C}$ and $\int_B \mathbf{F} \cdot d\mathbf{C}$ positive or negative? Why?
   (b) Is $\int_A \mathbf{F} \cdot d\mathbf{C} < \int_B \mathbf{F} \cdot d\mathbf{C}$? Be careful of the signs, and justify your conclusion.
3. Let C be the quarter unit circle in the first quadrant, oriented counterclockwise.

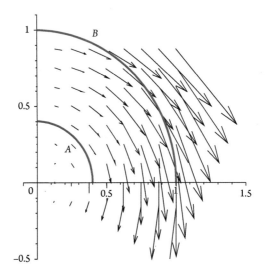

Figure 5.1.3 Relative size of line integrals

    (a) Sketch C and some of the vectors of $\mathbf{F} = \langle x, y \rangle$ along C.

    (b) Without calculating, what can you say about $\int_C \mathbf{F} \cdot d\mathbf{C}$? Why?

4. Let C be the quarter unit circle in the first quadrant oriented counterclockwise.

    (a) Sketch C and some of the vectors of $\mathbf{F} = \langle -y, x \rangle$ along C.

    (b) Without calculating, what can you say about $\int_C \mathbf{F} \cdot d\mathbf{C}$? Why?

5. Let C be the quarter unit circle in the first quadrant oriented counterclockwise.

    (a) Sketch C and some of the vectors of $\mathbf{F} = \langle y, -x \rangle$ along C.

    (b) Without calculating, what can you say about $\int_C \mathbf{F} \cdot d\mathbf{C}$? Why?

6. Let C be the quarter unit circle in the first quadrant oriented counterclockwise. Sketch two vector fields $\mathbf{F}_1$ and $\mathbf{F}_2$ on C for which

$$\int_C \mathbf{F}_1 \cdot d\mathbf{C} > \int_C \mathbf{F}_2 \cdot d\mathbf{C} > 0.$$

7. In each figure below sketch a non-zero vector field $\mathbf{F}$ along the curve C with the desired line integral

(a)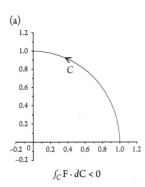

$\int_C \mathbf{F} \cdot d\mathbf{C} < 0$

(b)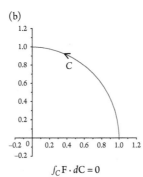

$\int_C \mathbf{F} \cdot d\mathbf{C} = 0$

(c)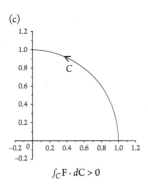

$\int_C \mathbf{F} \cdot d\mathbf{C} > 0$

8. In each plot below sketch an oriented line segment C **not** through the origin, for which $\int_C \mathbf{F} \cdot d\mathbf{C}$ has the stated values.

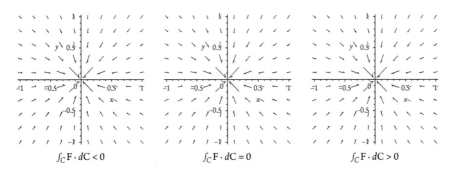

$\int_C \mathbf{F} \cdot d\mathbf{C} < 0$ $\qquad$ $\int_C \mathbf{F} \cdot d\mathbf{C} = 0$ $\qquad$ $\int_C \mathbf{F} \cdot d\mathbf{C} > 0$

Evaluate $\int_C \mathbf{F} \cdot d\mathbf{C}$ in Exercises 9–15 for the given $\mathbf{C}$ and $\mathbf{F}$.

9. $\mathbf{C}(t) = (2 - t, 2t)$, $0 \le t \le 1$ and $\mathbf{F}(x,y) = (x + 2y, y^2)$.

10. $\mathbf{C}(t) = (t, t^2)$, $0 \le t \le 1$ and $\mathbf{F}(x,y) = (3x^2 - y, 2y + x)$.

11. $\mathbf{C}(t) = (1 - t, 1 - t)$, $0 \le t \le 1$ and $\mathbf{F}(x,y) = (3x^2 - y, 2y + x)$.

12. $\mathbf{C}(t) = (2 + t, 3 - 2t, 4t)$, $0 \le t \le 1$ and $\mathbf{F}(x,y,z) = (2x - z, y + 2z, x - 3y)$.

13. $\mathbf{C}(t) = (\pi + t, 2t, t^2)$, $0 \le t \le 1$ and $\mathbf{F}(x,y,z) = (\cos x, \sin y, z)$.

14. $\mathbf{C}(t) = (1 + t, t, 1 - t)$, $0 \le t \le 1$ and $\mathbf{F}(x,y,z) = (xy, xz, yz)$.

15. $\mathbf{C}(t) = (2\cos t, 3\sin t)$, $0 \le t \le \pi/2$ and $\mathbf{F}(x,y) = (y, -x + 2)$.

16. Let $\mathbf{F}(x,y) = (x,y)$ and C be the line segment from $(2,0)$ to $(0,1)$.
    (a) Find $\int_C \mathbf{F} \cdot d\mathbf{C}$ using the parameterization $\mathbf{C}(t) = (2 - 2t, t)$ for $0 \le t \le 1$.
    (b) Find $\int_C \mathbf{F} \cdot d\mathbf{C}$ using the parameterization $\mathbf{C}(t) = (2 - 2t^2, t^2)$ for $0 \le t \le 1$. (you should convince yourself that both parameterize the segment of $x + 2y = 2$ from $(2,0)$ to $(0,1)$.

17. Let $\mathbf{F}(x,y) = \langle y, x^2 \rangle$. Evaluate $\int_C \mathbf{F} \cdot d\mathbf{C}$ where C is
    (a) the line segment from $(1,0)$ to $(-1,0)$
    (b) the upper semicircle from $(1,0)$ to $(-1,0)$.

18. Find the work done by the constant vector field $\mathbf{F}(x,y,z) = \langle 0,0,-9.8 \rangle$ in moving a particle from $(1,0,0)$ to $(1,0,2\pi)$ along
    (a) the vertical line segment
    (b) the helical arc $\mathbf{C}(t) = \langle \cos t, \sin t, t \rangle$, $0 \le t \le 2\pi$. Are your answers the same or different?

19. Find the work done by the force $\mathbf{F}(x,y) = \langle 2y - x, y \rangle$ in moving a particle from the origin to $(1,1)$ along the parabola $y = x^2$.

20. Find the work done by $\mathbf{F}(x,y) = \langle xy, 1+x \rangle$ along the curve $\mathbf{C}(t) = (2 - 3t^{10}, 1 + t^{10})$ for $0 \le t \le 1$. (Hint: find a different parameterization of the same curve that makes your calculation easier.)

21. Let $\mathbf{F}(x,y) = \langle y, -x \rangle$.
    (a) Find $\int_C \mathbf{F} \cdot d\mathbf{C}$ where C is the unit circle oriented counterclockwise.
    (b) Now parameterize the unit circle in a clockwise fashion to calculate $\int_{-C} \mathbf{F} \cdot d\mathbf{C}$. This should verify that $\int_{-C} \mathbf{F} \cdot d\mathbf{C} = -\int_C \mathbf{F} \cdot d\mathbf{C}$.

22. Evaluate both $\int_C \mathbf{F} \cdot d\mathbf{C}$ and $\int_{-C} \mathbf{F} \cdot d\mathbf{C}$ where $\mathbf{F}(x,y) = \langle x^2, x+y \rangle$ and C is the line segment from $(0,3)$ to $(4,3)$.

23. Let C start at $(0,0)$ and travel counterclockwise around the triangle with vertices $(0,0)$, $(1,0)$, and $(0,2)$. Find the line integral of $\mathbf{F}(x,y) = \langle x + 2y, 3x - y \rangle$ around C.

24. Let C be any curve on the cone $x^2 + y^2 - z^2 = 0$, and let $\mathbf{F}(x,y,z) = \langle x, y, -z \rangle$. Describe why $\int_C \mathbf{F} \cdot d\mathbf{C} = 0$.

25. Evaluate $\int_C xy \, dx - 2y \, dy$ where C is the line segment from $(2,1)$ to $(4,0)$.

26. Evaluate $\int_C -y \, dx + x \, dy$ where C is the unit circle.

27. Evaluate $\int_C -y \, dx + x \, dy$ where $\mathbf{C}(t) = (3 \cos t, 2 \sin t)$, $0 \le t \le 2\pi$.

28. Evaluate $\int_C \frac{-y}{x^2+y^2} \, dx + \frac{x}{x^2+y^2} \, dy$ where C is the unit circle.

29. Evaluate $\int_C \frac{-y}{x^2+y^2} \, dx + \frac{x}{x^2+y^2} \, dy$ where C is the circle $x^2 + y^2 = 4$.

30. Evaluate $\int_C 2xy \, dx + x^2 \, dy$ where C is the triangle with vertices $(0,0)$, $(1,0)$, and $(0,2)$ oriented counterclockwise.

31. Evaluate $\int_C z \, dx + x \, dy - y \, dz$ where C is $\mathbf{C}(t) = (\cos t, \sin t, t)$, $0 \le t \le \pi$

32. Let C be a curve parameterized by $\mathbf{C}(t) = (x(t), y(t))$, $a \le t \le b$, where the product $x(t)y(t)$ is a constant. Show that $\int_C y \, dx + x \, dy = 0$.

5.2 Fundamental Theorem for Line Integrals

Recall that the fundamental theorem of calculus says that if $F'(x) = f(x)$, then

$$\int_a^b f(x)dx = F(b) - F(a).$$

To evaluate an integral, then, find an antiderivative and evaluate at endpoints. To evaluate $\int_C \mathbf{F} \cdot d\mathbf{C}$, the integral of a vector field along a curve, one might hope that a similar process might be used: find an antiderivative of $\mathbf{F}$ and evaluate at endpoints. The question is, what is an antiderivative of a vector field? One possible answer to this was given in Section 3.5. If $\mathbf{F} = \nabla f$, then we can think of the function f as an antiderivative of $\mathbf{F}$. In such a case the function f is called a *potential function* for $\mathbf{F}$. The fundamental theorem for line integrals says that when $\mathbf{F} = \nabla f$

$$\int_C \mathbf{F} \cdot d\mathbf{C} = f(B) - f(A), \tag{5.2.1}$$

where A and B are the initial and terminal points of the curve C, respectively. A proof of this theorem is the ultimate goal of this section.

Note that if $\mathbf{F}$ is a vector field for which Equation 5.2.1 holds, then the integral $\int_C \mathbf{F} \cdot d\mathbf{C}$ depends only on the endpoints of C and not on the particular path taken. This section begins by showing that there are vector fields which can't be integrated using the fundamental theorem. Then conservative vector fields are defined, and techniques for finding potential functions introduced. Finally, the fundamental theorem for line integrals is proven.

Path Dependence: We have seen that the integral of a vector field along a curve does not depend on the choice of parameterization. One might also focus on the endpoints of the curve and ask if line integrals depend on the path between two points. In general two paths C_1 and C_2 with the same endpoints yield different line integrals, as we see in the next example. There is an important class of vector fields, called *conservative* vector fields, for which line integrals are independent of the path. Before considering conservative vector fields, we consider one that isn't.

Example 5.2.1. *Path-dependent integrals*

Let $\mathbf{F}(x,y) = \langle -y, x \rangle$. Find the work done by $\mathbf{F}$ on a particle moving from $(1,0)$ to $(-1,0)$ along

(a) the straight line segment connecting them

(b) the top semicircle of the unit circle.

We treat each case separately, using the three-step strategy of parameterizing, finding the integrand, and integrating.

(*a*) The line segment

Step 1. Parameterize and differentiate

A parameterization of the line segment is

$$\mathbf{C}(t) = (1-t)(1,0) + t(-1,0) = (1-2t,0), \text{ for } 0 \le t \le 1.$$

Thus $\mathbf{C}'(t) = \langle -2,0 \rangle$.

Step 2. Find $\mathbf{F}(\mathbf{C}(t)) \cdot \mathbf{C}'(t)$

Evaluating $\mathbf{F}$ on $\mathbf{C}(t)$ gives $\mathbf{F}(1-2t,0) = \langle 0, 1-2t \rangle$. We obtain the integrand:

$$\mathbf{F}(\mathbf{C}(t)) \cdot \mathbf{C}'(t) = \langle 0, 1-2t \rangle \cdot \langle -2,0 \rangle = 0.$$

Step 3. Integrate

$$\int_a^b \mathbf{F}(\mathbf{C}(t)) \cdot \mathbf{C}'(t)\, dt = \int_0^1 0\, dt = 0.$$

Notice that in this example the integrand is always zero. This means that the vector field $\mathbf{F}$ is always perpendicular to the tangent vector of the line segment (see Figure 5.2.1).

(*b*) The integral of $\mathbf{F}$ over the semicircle

Step 1. $\mathbf{C}(t) = (\cos t, \sin t)$ for $0 \le t \le \pi$ and $\mathbf{C}'(t) = \langle -\sin t, \cos t \rangle$.

Step 2. $\mathbf{F}(\cos t, \sin t) = \langle -\sin t, \cos t \rangle$, and

$$\mathbf{F}(\mathbf{C}(t)) \cdot \mathbf{C}'(t) = \langle -\sin t, \cos t \rangle \cdot \langle -\sin t, \cos t \rangle = \sin^2 t + \cos^2 t = 1.$$

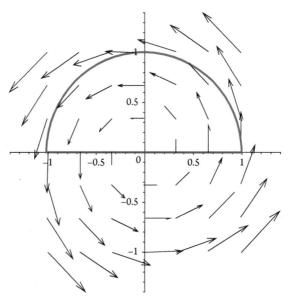

Figure 5.2.1 Work depends on the path

Step 3. Integrate

$$\int_a^b \mathbf{F}(\mathbf{C}(t)) \cdot \mathbf{C}'(t) \, dt = \int_0^\pi 1 \, dt = \pi.$$

Notice that the line integral of $\mathbf{F}$ along the semicircle is different from that along the line, even though the endpoints are the same. We say the that line integral is path-dependent. More precisely, if $\mathbf{F}$ is a vector field and C_1 and C_2 are both paths from point A to point B, then it may be that $\int_{C_1} \mathbf{F} \cdot d\mathbf{C} \neq \int_{C_2} \mathbf{F} \cdot d\mathbf{C}$. ▲

Conservative Fields: The fundamental theorem of line integrals does not apply to the vector field of the previous example. Vector fields for which the fundamental theorem applies are called *conservative*. There are several equivalent ways to define a conservative vector field, and we need some terminology about loops before stating them. A *closed loop* is any curve that starts and stops at the same point. A *simple closed loop* is a closed loop is a curve that doesn't intersect itself. A lemniscate is a closed loop since it starts and stops in the same place, but it is not simple because it intersects itself (see Figure 5.2.2(a)). An ellipse is an example of a simple closed loop (Figure 5.2.2(b)). Simple closed loops will be our focus in Section 5.3. In this section we consider closed loops in general.

We can now define what we mean by a conservative vector field. We then give an equivalent definition using paths rather than loops, and characterize conservative vector fields in terms of antiderivatives.

Definition 5.2.1. *A vector field $\mathbf{F}$ is conservative if $\int_C \mathbf{F} \cdot d\mathbf{C} = 0$ for every closed loop C in the domain of $\mathbf{F}$.*

An equivalent definition is: The vector field $\mathbf{F}$ is conservative if all line integrals of $\mathbf{F}$ are independent of the path taken. More precisely, let C_1 and C_2 be any two curves from point A to point B in the domain of $\mathbf{F}$. Then $\mathbf{F}$ is conservative if for every C_1, C_2

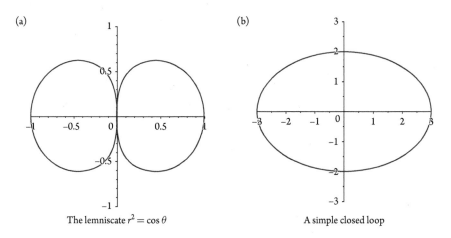

(a)

The lemniscate $r^2 = \cos \theta$

(b)

A simple closed loop

Figure 5.2.2 Simple vs. non-simple closed loops

we have $\int_{C_1} \mathbf{F} \cdot d\mathbf{C} = \int_{C_2} \mathbf{F} \cdot d\mathbf{C}$. To see that this is equivalent to Definition 5.2.1, notice that if C_1 and C_2 both start and stop at the same points, then the piecewise defined curve $C = C_1 - C_2$ is a closed loop. Path independence now implies

$$\int_C \mathbf{F} \cdot d\mathbf{C} = \int_{C_1 - C_2} \mathbf{F} \cdot d\mathbf{C} = \int_{C_1} \mathbf{F} \cdot d\mathbf{C} - \int_{C_2} \mathbf{F} \cdot d\mathbf{C} = 0.$$

Since any closed loop can be expressed as the difference of two paths with the same endpoints, path independence of line integrals implies integrals around closed loops are zero. The same equation implies the converse: if the line integral of $\mathbf{F}$ around any closed loop is zero, then line integrals of $\mathbf{F}$ along depend only on their endpoints.

Math App 5.2.1. *Visualizing non-conservative fields*

This Math App develops some intuition for (non)conservative vector fields by finding paths where the line integrals differ. Feel free to take a closer look.

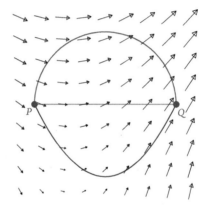

Thus there is a closed loop definition of a conservative vector field, and an equivalent definition in terms of paths. While these are nice intuitive definitions of a conservative field, they aren't very helpful in determining if a given vector field is conservative or not. It turns out that gradient fields ∇f are exactly conservative vector fields, and we state this as a theorem (without proof).

Theorem 5.2.1. *Let $\mathbf{F}$ be a vector field on $\mathbb{R}^3$ with continuously differentiable component functions. Then $\mathbf{F}$ is conservative if and only if there is a function f such that $\mathbf{F} = \nabla f$ on the domain of $\mathbf{F}$.*

This fact gives us a method for testing whether or not a vector field is conservative. If $\mathbf{F} = \langle P, Q \rangle$ is a gradient field, then there is a function f such that $\mathbf{F} = \nabla f$. The function f is called the *potential function* of $\mathbf{F}$, and note that $P = \frac{\partial f}{\partial x}$ and $Q = \frac{\partial f}{\partial y}$. Since mixed partials are equal we notice that

$$\frac{\partial P}{\partial y} = \frac{\partial^2 f}{\partial y \partial x} = \frac{\partial^2 f}{\partial x \partial y} = \frac{\partial Q}{\partial x}.$$

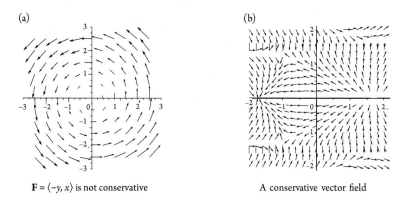

(a)

$\mathbf{F} = \langle -y, x \rangle$ is not conservative

(b)

A conservative vector field

Figure 5.2.3 Conservative vs. non-conservative fields

This shows that if $\mathbf{F}$ is conservative, then $P_y = Q_x$. Equivalently, if $P_y \neq Q_x$ then $\mathbf{F} = \langle P, Q \rangle$ is not conservative.

Unfortunately, if $P_y = Q_x$ that doesn't always mean that $\mathbf{F}$ has a potential function. The classic example is the vector field

$$\mathbf{F}(x,y) = \left\langle \frac{-y}{x^2 + y^2}, \frac{x}{x^2 + y^2} \right\rangle. \tag{5.2.2}$$

You can verify that $P_y = Q_x$, but that the integral around the unit circle doesn't vanish. By Definition 5.2.1, the non-vanishing integral implies $\mathbf{F}$ is not conservative, so it can't have a potential function. For the examples we encounter in this text, if $P_y = Q_x$ then $\mathbf{F}$ will have a potential function. The interested reader can consult sections 7.3 and 8.3 of Lang's *Calculus of Several Variables* ([2]) for a thorough discussion of when a planar vector field admits a potential function.

Example 5.2.2. *Determining conservative vector fields*

Determine if the following vector fields are conservative:

(a) $\mathbf{F} = \langle -y, x \rangle$ is not conservative since $\frac{\partial P}{\partial y} = -1$, but $\frac{\partial Q}{\partial x} = 1$.

(b) $\mathbf{F}(x,y) = \langle x^2 - 3\cos y, \tan y + 3x \sin y \rangle$ is conservative since

$$\frac{\partial P}{\partial y} = 3\sin y = \frac{\partial Q}{\partial x}. \; \blacktriangle$$

Once we know that a vector field is conservative, it will be important to find its potential function. The key to doing this is partial integration. Suppose $\mathbf{F} = \langle P, Q \rangle$ is conservative, then $P = f_x$ for some function f. Thus integrating P with respect to x will give f back, up to a constant. One key fact to remember is that any function of y is a constant when differentiating with respect to x. Thus when integrating $\int P dx$ you have to add not just a constant C, but also an arbitrary function of y. We illustrate with an example.

Example 5.2.3. *Finding a two-dimensional potential function*

Find a potential function for $\mathbf{F}(x,y) = \left(x^2 - 3\cos y, \tan y + 3x\sin y\right)$. First we integrate P with respect to x, adding an arbitrary constant function.

$$f(x,y) = \int P\,dx = \int x^2 - 3\cos y\,dx = \frac{x^3}{3} - 3x\cos y + g(y).$$

Now we also know $Q = f_y$, so we set the partial derivative of the $f(x,y)$ we just found equal to Q in order to find $g(y)$. Since $f_y = 3x\sin y + g'(y)$, we solve

$$3x\sin y + g'(y) = \tan y + 3x\sin y$$

for $g'(y)$. Thus $g'(y) = \tan y$, and integrating gives $g(y) = -\ln|\cos y| + C$. Putting it all together gives the potential function

$$f(x,y) = \frac{x^3}{3} - 3x\cos y - \ln|\cos y|.$$

We ignore the arbitrary constant C since we won't use it for our purposes. ▲

In the three-dimensional setting there are more mixed partials to consider. To show that the vector field $\mathbf{F} = \langle P, Q, R \rangle$ is conservative, verify that the following three equations are satisfied:

$$\frac{\partial P}{\partial y} = \frac{\partial Q}{\partial x}; \quad \frac{\partial P}{\partial z} = \frac{\partial R}{\partial x}; \quad \frac{\partial Q}{\partial z} = \frac{\partial R}{\partial y}.$$

These equations are necessarily satisfied by a conservative vector field because of equality of mixed partial derivatives (in the same way that $P_y = Q_x$ is a consequence of the equality $f_{xy} = f_{yx}$). There is no three-dimensional analog of the vector field in Equation 5.2.2, as the extra dimension changes things. A complete discussion of the three-dimensional situation can be found in Section 8.3 of Marsden and Tromba's *Vector Calculus* ([3]). For the vector fields we encounter, satisfying the above equations will imply conservative. To find a potential function in three dimensions, once you know a vector field is conservative, just apply the two-dimensional process twice.

Example 5.2.4. *A three-dimensional example*

Show that

$$\mathbf{F}(x,y,z) = \left(2xyz - 3y^2 + 1, x^2z - 6xy + 5z^2, x^2y + 10yz\right)$$

is a conservative vector field, and find a potential function for it.

Taking partials we find

$$\frac{\partial P}{\partial y} = 2xz - 6y = \frac{\partial Q}{\partial x}$$

$$\frac{\partial P}{\partial z} = 2xy = \frac{\partial R}{\partial x}$$

$$\frac{\partial Q}{\partial z} = x^2 + 10z = \frac{\partial R}{\partial y},$$

so $\mathbf{F}$ is conservative.

Integrating P with respect to x, and noting that the arbitrary constant is now a function of *both* y and z, we have

$$f(x,y,z) = \int 2xyz - 3y^2 + 1\ dx = x^2yz - 3xy^2 + x + g(y,z).$$

Knowing that $f_y(x,y,z) = Q$, we solve

$$x^2z - 6xy + g_y(y,z) = x^2z - 6xy + 5z^2$$

$$g_y(y,z) = 5z^2.$$

Integrating with respect to y, and adding the arbitrary constant $h(z)$, gives

$$g(y,z) = \int 5z^2\,dy = 5z^2y + h(z).$$

Thus far we have $f(x,y,z) = x^2yz - 3xy^2 + x + 5z^2y + h(z)$. Solving $f_z(x,y,z) = R$ shows that $h(z) = 0$, and we have found a potential function

$$f(x,y,z) = x^2yz - 3xy^2 + x + 5z^2y. \ \blacktriangle$$

Example 5.2.5. *Electric potential*

A point charge Q at the origin generates an electric field given most easily in spherical coordinates

$$\mathbf{E} = \frac{kQ}{\rho^2}\widehat{\rho},$$

and Example 3.5.2 showed that $\mathbf{E} = -\nabla V$ where

$$V = \frac{kQ}{\rho}.$$

Of course, that was cheating because we knew the formula for V and just verified the gradient calculation. Starting with the electric field $\mathbf{E}$ in Cartesian coordinates we can now honestly find V. In Cartesian coordinates we have

$$\mathbf{E} = kQ\left(\frac{x}{(x^2+y^2+z^2)^{3/2}}, \frac{y}{(x^2+y^2+z^2)^{3/2}}, \frac{z}{(x^2+y^2+z^2)^{3/2}}\right),$$

from which you can verify that

$$P_y = kQ\frac{-3xy}{(x^2+y^2+z^2)^{5/2}} = Q_x,$$

and that $P_z = R_x$, $Q_z = R_y$ as well. Thus $\mathbf{E}$ has a potential function, and integrating P we find

$$f(x,y,z) = \int kQ\frac{x}{(x^2+y^2+z^2)^{3/2}}\,dx = -\frac{kQ}{(x^2+y^2+z^2)^{1/2}} = -\frac{kQ}{\rho}.$$

Note that integrating Q with respect to y, or R with respect to z, gives the same result, so $\mathbf{E} = \nabla f$. Physicists define electric potential V of a field $\mathbf{E}$ to be the function such that $\mathbf{E} = -\nabla V$. This example verifies our earlier claim that that the electric potential is $V = kQ/\rho$. ▲

The Fundamental Theorem of Line Integrals: The terminology of conservative vector fields comes from physics, in the context of conservation of kinetic and potential energy. Mathematically, one significance of conservative vector fields is that line integrals become relatively straightforward when the vector field is conservative. We now give the fundamental theorem of calculus for line integrals.

Theorem 5.2.2. *Let $\mathbf{F}$ be a conservative vector field with potential function f, and let C be a piecewise continuously differentiable path from point A to point B. Then*

$$\int_C \mathbf{F}\cdot d\mathbf{C} = f(B) - f(A).$$

Proof. Let $\mathbf{C}(t)$ for $a \le t \le b$ be a parameterization of the curve C. Thus $\mathbf{C}(a) = A$ and $\mathbf{C}(b) = B$. Since $\mathbf{F} = \nabla f$ we have

$$\int_C \mathbf{F}\cdot d\mathbf{C} = \int_a^b \mathbf{F}(\mathbf{C}(t))\cdot \mathbf{C}'(t)\,dt = \int_a^b \nabla f(\mathbf{C}(t))\cdot \mathbf{C}'(t)\,dt.$$

But the chain rule for curves tells us that the integrand $\nabla f(\mathbf{C}(t))\cdot \mathbf{C}'(t) = \frac{d}{dt}f(\mathbf{C}(t))$. The fundamental theorem of calculus for a single variable implies

$$\int_a^b \nabla f(\mathbf{C}(t)) \cdot \mathbf{C}'(t)\,dt = \int_a^b \frac{d}{dt} f(\mathbf{C}(t))\,dt$$

$$= f(\mathbf{C}(t))\Big|_a^b = f(\mathbf{C}(b)) - f(\mathbf{C}(a)) = f(B) - f(A). \;\square$$

Thus to evaluate the line integral of a conservative vector field, we find the potential function and evaluate at the endpoints. More explicitly, we have:

Conservative Fields: The Fundamental Theorem of Line Integrals

Let C be any path from point A to point B, and let $\mathbf{F}$ be a conservative vector field. To evaluate $\int_C \mathbf{F} \cdot d\mathbf{C}$:

1. Find a potential function f for $\mathbf{F}$ (think antiderivative).
2. Use the fact that $\int_C \mathbf{F} \cdot d\mathbf{C} = f(B) - f(A)$.

Example 5.2.6. *Using the fundamental theorem*

Evaluate the integral of $\mathbf{F}(x,y) = \langle 2xy - 2\cos x, x^2 - \sin y \rangle$ along the quarter circle from $(\pi/2, 0)$ to $(0, \pi/2)$. In this instance the vector field is conservative since

$$\frac{\partial P}{\partial y} = 2x = \frac{\partial Q}{\partial x}.$$

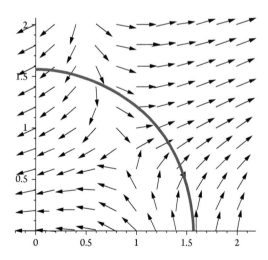

Figure 5.2.4 Integrating a conservative field

So we find a potential function and apply the theorem. Integrating we have

$$f(x,y) = \int P \, dx = \int 2xy - 2\cos x \, dx = x^2 y - 2\sin x + g(y).$$

Taking the partial with respect to y, and equating to Q gives

$$f_y(x,y) = x^2 + g'(y) = x^2 - \sin y.$$

Thus $g'(y) = -\sin y$, and $g(y) = \cos y$. The potential function is $f(x,y) = x^2 y - 2\sin x + \cos y$, and we use it to evaluate the integral of $\mathbf{F}$ over C:

$$\int_C \mathbf{F} \cdot d\mathbf{C} = f(\pi/2, 0) - f(0, \pi/2) = -1. \ \blacktriangle$$

Example 5.2.7. *Work in three dimensions*

Find the work done by $\mathbf{F}(x,y,z) = \langle y - z, x + z, y - x \rangle$ on a particle that travels along the helical arc $\mathbf{C}(t) = (\cos t, \sin t, t), 0 \le t \le \pi$.
Since

$$\frac{\partial P}{\partial y} = 1 = \frac{\partial Q}{\partial x}$$

$$\frac{\partial P}{\partial z} = -1 = \frac{\partial R}{\partial x}$$

$$\frac{\partial Q}{\partial z} = 1 = \frac{\partial R}{\partial y},$$

the vector field is conservative. By integrating we know $f(x,y,z) = xy - xz + g(y,z)$, and inspection shows that $g(y,z) = yz$. Thus a potential function is $f(x,y,z) = xy - xz + yz$. Since $\mathbf{C}(0) = (1,0,0)$ and $\mathbf{C}(\pi) = (-1,0,\pi)$, the fundamental theorem of line integrals implies

$$w = \int_C \mathbf{F} \cdot d\mathbf{C} = f(-1,0,\pi) - f(1,0,0) = \pi. \ \blacktriangle$$

Electric Potential Energy and Electric Potential: Potential energy in physics is defined relative to a chosen fixed position. For example, gravitational potential energy is usually relative to sea level because that's how we tend to measure altitude. The potential energy of a mass at height h is defined to be the work necessary to move the same mass from sea level to that position. Note that potential energy is defined in terms of work, which is a line integral!

In this paragraph we consider a point charge Q placed at the origin, its electric potential, and the electric potential energy that other charges have relative to it. Recall that the electric field generated by Q is the *force per unit charge* that Q exerts on a charge q at a point P. For a single point charge of Q coulombs at the origin, the electric field is

$$\mathbf{E} = \frac{kQ}{\rho^2}\hat{\rho}.$$

Example 5.2.5 showed that $\mathbf{E} = -\nabla V$ where

$$V = \frac{kQ}{\rho}$$

is the electric potential created by Q.

The force that Q exerts on a charge q is also conservative. Suppose a second charge of q coulombs is at the point (x, y, z). Since $\mathbf{E}$ is the force per unit charge that Q exerts, multiplying $\mathbf{E}$ by the charge q gives force, and we have

$$\mathbf{F} = q\mathbf{E}.$$

Since $\mathbf{F}$ is a constant multiple of a conservative vector field, it is conservative as well. Moreover, the preceding remarks imply that $\mathbf{F}$ is the gradient of

$$\mathbf{F} = \nabla\left(-\frac{kQq}{\rho}\right).$$

Thus we can use Theorem 5.2.2 to evaluate work done by Q on particles moving in space. Since work is potential energy, we can use the fundamental theorem of line integrals to calculate electric potential energy.

To define gravitational potential energy a point of reference (sea-level) had to be chosen. The same is true for electric potential energy, and infinity is the natural choice here. Fixing a point charge Q at the origin, the potential energy of a second charge q at a point P is the work required to bring q from infinity to the point P. To calculate potential energy, we first compute the work done by Q on q as it moves from infinity to the point P. Letting C be any curve from infinity to P, Theorem 5.2.2 yields the calculation

$$w = \int_C \mathbf{F} \cdot d\mathbf{C} = -\frac{kQq}{\rho}\Big|_\infty^P = -\frac{kQq}{\rho(P)} + \frac{kQq}{\infty} = -\frac{kQq}{\rho(P)}, \tag{5.2.3}$$

where $\rho(P)$ denotes the distance from P to the origin.

Now observe that the work needed for me to move q from infinity to P is the negative of the work that Q does. To see this, imagine that I push q just a smidge to get it moving toward Q. Once q is moving, I must exert the same force as Q but in the opposite direction to make the net force on q be zero, which maintains constant speed (recall that no net force means no acceleration, implying constant velocity). Thus, after a minuscule push to get q going, the force I place on q is $-\mathbf{F}$, and the work I do is the opposite of that done by Q.

The potential energy of q at P, from the calculation in Equation 5.2.3, is the quantity $\frac{kQq}{\rho}$, where ρ now represents the distance from P to the origin.

This underscores the relationship between electric potential $V = kQ/\rho$ and electric potential energy kQq/ρ. The only difference is a factor of the charge q, so electric potential is the *potential energy per unit charge*, in the same way that the electric field generated by Q is the force per unit charge.

Example 5.2.8. *Work by an electric field*

A positive 2C point charge Q is placed at the origin, and suppose that a positive 3C point charge q is moved from $(0,0,5)$ to $(1,0,0)$. Find the work done by Q on q during this movement.

The preceding discussion shows that $\mathbf{F}$ is the gradient of

$$-3V = -\frac{6k}{\rho}.$$

We can use the fundamental theorem of line integrals to see the work is

$$w = \int_C \mathbf{F} \cdot d\mathbf{C} = -\frac{6k}{\rho}\bigg|_{(0,0,5)}^{(1,0,0)} = -6k + \frac{6k}{5} = -\frac{24k}{5}.$$

This work is negative because the charges have the same sign, so the charge Q is hindering q from moving closer. Note that we never mentioned what path q took to get from $(0,0,5)$ to $(1,0,0)$. In fact we didn't need to, because the integral is independent of the path since $\mathbf{F}$ is conservative. ▲

Things to Know

- In general, different paths C_1 and C_2 from point A to point B lead to different integrals. In other words:

$$\int_{C_1} \mathbf{F} \cdot d\mathbf{C} \neq \int_{C_2} \mathbf{F} \cdot d\mathbf{C}.$$

- The following are equivalent definitions of conservative vector fields:

  1. $\int_C \mathbf{F} \cdot d\mathbf{C} = 0$ for any closed loop C.
  2. $\int_C \mathbf{F} \cdot d\mathbf{C}$ is independent of the path.
  3. There is a potential function f such that $\mathbf{F} = \nabla f$.

- To see if $\mathbf{F} = \langle P, Q, R \rangle$ is conservative, check equality of partials:

$$P_y = Q_x; \ P_z = R_x; \ Q_z = R_y.$$

- To find a potential function for a conservative field $\mathbf{F}$, use partial integration several times.

- When **F** is conservative, with potential function f, you don't have to parameterize C to evaluate $\int_C \mathbf{F} \cdot d\mathbf{C}$! By the fundamental theorem

$$\int_C \mathbf{F} \cdot d\mathbf{C} = f(B) - f(A),$$

where A is the initial and B the terminal point of the curve C.

Exercises

1. Determine which of the following vector fields are conservative. If they are conservative, find a potential function.

   (a) $\mathbf{F}(x,y) = \langle x^2 - 3xy, y^2 - 3xy \rangle$

   (b) $\mathbf{F}(x,y) = \langle 3x^2 y^2 - 2y \cos x, 2x^3 y - 2 \sin x - \sin y \rangle$

   (c) $\mathbf{F}(x,y) = \left\langle \dfrac{x}{\sqrt{x^2+y^2}}, \dfrac{x}{\sqrt{x^2+y^2}} \right\rangle$

   (d) $\mathbf{F}(x,y) = \langle x, y \rangle$.

2. For each conservative vector field in the previous problem, evaluate the line integral along the curve $\mathbf{C}(t) = (t, \sin t)$ for $0 \le t \le 2\pi$.

3. Let $\mathbf{F}(x,y,z) = \langle 2x, \cos y - 2z^3, 6yz^2 \rangle$ and $\mathbf{C}(t) = (\cos t, t, \sin t)$, for $0 \le t \le \pi$. Evaluate $\int_C \mathbf{F} \cdot d\mathbf{C}$.

4. Find the work done by $\mathbf{F}(x,y,z) = \langle 2x - yz, 4 - xz, z^2 - xy \rangle$ on a particle moving along the curve $\mathbf{C}(t) = \left(e^t \cos \pi t, t \sin \pi t, t^2 + \frac{3}{2}t \right)$, $0 \le t \le 1/2$.

5. Evaluate $\int_C \mathbf{F} \cdot d\mathbf{x}$ where $\mathbf{F}(x,y,z) = \langle yz, xz, xy \rangle$ and C is the parametric curve $\mathbf{C}(t) = (\cos t, \sin t, t)$, $0 \le t \le \frac{\pi}{3}$.

6. Evaluate $\int_C \mathbf{F} \cdot d\mathbf{x}$, where $\mathbf{F}(x,y) = \langle x^2 + xy, xy \rangle$ and C is

   (a) the line segment from $(0, -1)$ to $(1, 0)$

   (b) the quarter of the unit circle from $(0, -1)$ to $(1, 0)$.

   (c) Based only on your answers for parts a and b above, can **F** be a conservative vector field? Explain.

7. Let $\mathbf{F} = \langle 2xz, z^3 - 3, x^2 + 3yz^2 \rangle$ and $\mathbf{C}(t) = (\tan(\pi t/4), (t-1)\cos t, 4t)$, $0 \le t \le 1$. Evaluate $\int_C \mathbf{F} \cdot d\mathbf{C}$.

8. Find the work done by $\mathbf{F}(x,y,z) = \langle 2xy - 2y^2 + 3z^2, x^2 - 4xy, 6xz \rangle$ on a particle moving along the path $\mathbf{C}(t) = \left((t-1)e^t, \cos(\pi t), t^3 \right)$, $0 \le t \le 1$.

9. The vector field $\mathbf{F} = \langle -y, x \rangle$ is pictured below. Sketch two curves C_1, C_2 from $(2, 0)$ to $(-2, 0)$ such that $\int_{C_1} \mathbf{F} \cdot d\mathbf{C} \ne \int_{C_2} \mathbf{F} \cdot d\mathbf{C}$.

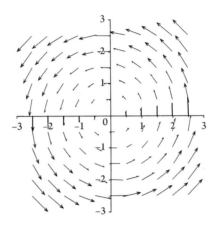

10. Find the work done by $\mathbf{F}(x,y) = \langle \cos y, -x \sin y \rangle$ on a particle moving along the line segment from $(0,0)$ to $(1,\pi)$.

11. Let the curve C be parameterized by

$$\mathbf{C}(t) = \langle e^t \cos t, e^t \sin t \rangle, \; 0 \le t \le \pi/4,$$

and let $\mathbf{F} = \langle 3x^2 y, x^3 \rangle$. Evaluate $\int_C \mathbf{F} \cdot d\mathbf{C}$.

12. The electric force that a certain charge placed at the origin exerts on a unit charge at (x,y,z) is given by

$$\mathbf{F} = \frac{1}{\rho^2}\hat{\rho}.$$

Find the work done by the electric force on a particle traveling from $(0,0,1)$ to $(1,3,2)$.

13. A point charge of 3C at the origin exerts a force on a 5C charge as it is moved from $(1,0,0)$ to $(0,3,4)$. Compute the work done by this force on the 5C charge.

14. A point charge of 3C at the origin exerts a force on a 1C charge that starts at $(1,0,0)$.

   (a) Compute the work done by this force as the unit charge moves to $(0,1,0)$ and to $(3/5,0,4/5)$.

   (b) How much work is done as the 1C charge moves to any point on the unit sphere? Justify your answer.

15. A 2C point charge at the origin exerts a force on a 3C point charge in space. Determine the work done by this force as the 3C charge moves from any point on the radius 3 sphere centered at the origin to any point on that of radius 5.

5.3 Green's Theorem

So far we have two methods for evaluating line integrals: either parameterize the curve and use the original definition or, in the case of a conservative vector field, find a potential function and evaluate at the endpoints. In this section we will consider another method for calculating line integrals, which can be used when the curve(s) involved form the boundary of a region in the plane.

Green's Theorem—circulation Form: We have seen that the line integral arose naturally out of an effort to calculate the work done by a variable force along a general path. Another physical situation, fluid flow, gives a second interpretation of line integrals. The notions of flux and circulation of a fluid flow were introduced in Section 2.6. Letting $\mathbf{F}$ represent the velocity field of a fluid flow and C be a closed loop, the tangential component of $\mathbf{F}$ measured the flow of fluid *along* C while the normal component measured the flow *across* C (a quick look at Figure 2.6.4 would be helpful). The circulation of $\mathbf{F}$ around C was introduced intuitively in Section 2.6 as a measure of how the fluid circulates around C. Locally, the rate at which the fluid flows along C is given by the tangential component $comp_{C'}\mathbf{F}$. The circulation of $\mathbf{F}$ around C, then, is the integral of these local contributions given by

$$\int_C comp_{C'}\mathbf{F} dC,$$

which is just the line integral $\int_C \mathbf{F} \cdot d\mathbf{C}$.

We now have two interpretations of the line integral. When $\mathbf{F}$ is a force, the line integral calculates work. On the other hand, if $\mathbf{F}$ is the velocity field of a fluid flow and C a closed loop, then the line integral $\int_C \mathbf{F} \cdot d\mathbf{C}$ is the circulation of $\mathbf{F}$ around C.

When we want to emphasize that the curve C is a closed loop, the notation $\oint_C \mathbf{F} \cdot d\mathbf{C}$ is sometimes used instead of $\int_C \mathbf{F} \cdot d\mathbf{C}$. Both notations just mean the line integral of $\mathbf{F}$ around C.

Green's theorem shows how to calculate the circulation of $\mathbf{F}$ around a closed loop C using a double integral over the region in the plane that C bounds. We begin by considering when C is the boundary of the rectangle $R = [a,b] \times [c,d]$ oriented counterclockwise. We have the special case of Green's theorem:

Theorem 5.3.1. *Let* $\mathbf{F}(x,y) = P(x,y)\mathbf{i} + Q(x,y)\mathbf{j}$ *be a vector field in the plane where P and Q have continuous partial derivatives in the rectangle* $R = [a,b] \times [c,d]$. *Moreover, let C be the boundary of R oriented counterclockwise. Then*

$$\oint_C Pdx + Qdy = \iint_R \frac{\partial Q}{\partial x} - \frac{\partial P}{\partial y} dA.$$

Proof. The proof amounts to calculating the integrals on both sides of the equation, and noting that they are equal.

Since integration "distributes" across sums, we treat the integrals involving P separately from those involving Q. Let's start by evaluating those involving Q.

To evaluate the line integral $\int_C Q\,dy$ we parameterize C, thinking of $C = C_1 + C_2 + C_3 + C_4$ where

- C_1 is the horizontal line segment from (a,c) to (b,c), parameterized by $\mathbf{C_1}(t) = (t,c)$ for $a \le t \le b$.
- C_2 is the vertical line segment from (b,c) to (b,d), parameterized by $\mathbf{C_2}(t) = (b,t)$ for $c \le t \le d$.
- C_3 is the horizontal line segment from (b,d) to (a,d), parameterized by $\mathbf{C_3}(t) = (-t,d)$ for $-b \le t \le -a$.
- C_4 is the vertical line segment from (a,d) to (a,c), parameterized by $\mathbf{C_4}(t) = (a,-t)$ for $-d \le t \le -c$.

The parameterizations are a little strange to account for the orientation on the different line segments. When evaluating $\int_{C_i} Q\,dy$ we need to find dy, and notice that $dy = 0$ for C_1 and C_3 since they are horizontal line segments. Thus

$$\int_C Q\,dy = \int_{C_2} Q\,dy + \int_{C_4} Q\,dy$$

$$= \int_c^d Q(b,t)\,dt + \int_{-d}^{-c} Q(a,-t)(-dt)$$

$$= \int_c^d Q(b,t)\,dt + \int_d^c Q(a,u)\,du$$

$$= \int_c^d Q(b,t)\,dt - \int_c^d Q(a,t)\,dt, \qquad (5.3.1)$$

where we've made the substitution $u = -t$ to get the third equality. To get the last equality, we changed the label from u back to t and changed the order of the limits by taking the negative of the integral.

Now let's consider the double integral involving Q, and integrate with respect to x first. We calculate

$$\iint_R \frac{\partial Q}{\partial x}\,dA = \int_c^d \int_a^b \frac{\partial Q}{\partial x}\,dx\,dy$$

$$= \int_c^d Q(x,y)\big|_{x=a}^b\,dy$$

$$= \int_c^d Q(b,y) - Q(a,y)\,dy. \qquad (5.3.2)$$

Comparing Equations 5.3.1 and 5.3.2 we see that $\int_C Q\,dy = \iint_R \frac{\partial Q}{\partial x}\,dA$, as desired. The analysis that $\int_C P\,dx = -\iint_R \frac{\partial P}{\partial y}\,dA$ is similar, and left as an exercise for the reader. $\square$

Theorem 5.3.1 gives a third method for calculating line integrals: replacing the line integral with a double integral of a related function (when the curve is the boundary of a rectangle oriented counterclockwise). The integrand of Green's theorem,

$$\frac{\partial Q}{\partial x} - \frac{\partial P}{\partial y},$$

is called the *scalar curl* of the vector field $\mathbf{F} = \langle P, Q \rangle$. Before discussing its significance, we demonstrate how to use Theorem 5.3.1 and its generalizations.

Example 5.3.1. *An initial application*

Let R be the rectangle $[0, 1] \times [2, 3]$, and C be its boundary oriented counterclockwise, and let $\mathbf{F}(x, y) = \langle xe^x - y^2, \tan^{-1} y + x \rangle$. Evaluate $\int_C \mathbf{F} \cdot d\mathbf{C}$.

Using previous techniques to evaluate $\int_C P dx + Q dy$, we'd start by parameterizing the curve, substitute, and integrate. Just along the right side of the rectangle C_2, we'd get $\mathbf{C}(t) = (1, t)$, for $2 \le t \le 3$. Substituting gives

$$\int_{C_2} P dx + Q dy = \int_2^3 (e - t^2) \cdot 0 + (\tan^{-1} t + 1) dt = \int_2^3 \tan^{-1} t + 1 \ dt.$$

This integral alone is mildly unsavory, and we'd have to calculate three others to complete the problem. Luckily Theorem 5.3.1 provides an alternative method of evaluation. To apply it, we first find the integrand, then evaluate the integral. In our case we have

$$\frac{\partial Q}{\partial x} - \frac{\partial P}{\partial y} = \frac{\partial}{\partial x}\left(\tan^{-1} y + x \right) - \frac{\partial}{\partial y}\left(xe^x - y^2 \right) = 1 + 2y.$$

Applying Theorem 5.3.1 gives

$$\oint_C \mathbf{F} \cdot d\mathbf{C} = \iint_R \frac{\partial Q}{\partial x} - \frac{\partial P}{\partial y} dA$$
$$= \int_0^1 \int_2^3 1 - 2y \, dy dx = \int_0^1 y - y^2 \big|_{y=2}^3 \, dx$$
$$= \int_0^1 -4 dx = -4. \ \blacktriangle$$

Green's theorem applies to much more general regions than the rectangles of Theorem 5.3.1. We state a more general form, which still isn't the most general result.

Theorem 5.3.2. *Let R be any region in the plane bounded by simple closed curves $C_1, \ldots, C_n$, oriented so that R is on their left, and let $C = \{C_1, \ldots, C_n\}$. Further, let $\mathbf{F}(x, y) = P(x, y)\mathbf{i} + Q(x, y)\mathbf{j}$ be a vector field where P and Q have continuous partial derivatives. Then*

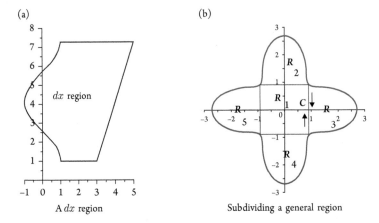

(a)

dx region

A dx region

(b)

R_2

R_1 C

R

R_3

Subdividing a general region

Figure 5.3.1 More general regions for Green's theorem

$$\int_C Pdx + Qdy = \iint_R \frac{\partial Q}{\partial x} - \frac{\partial P}{\partial y} dA.$$

Rather than a formal proof, we provide a plausible argument. Essentially the proof of Theorem 5.3.1 given above generalizes to the case where R is any region with a single boundary curve which is either a dx-region or a dy-region (see Figure 5.3.1(a)). The only difference is that the parameterizations are a little more involved than in the proof of Theorem 5.3.1. For more general regions, subdivide them into sub-regions that are either dx or dy and apply the theorem in each of those. Boundaries of regions must be oriented so that the region lies on your left as you traverse the boundary. Thus if a curve bounds a region on both the left and the right, it will be oriented in opposite directions in the two regions. Since the integrals along C and $-C$ are negatives of each other, the net contribution from one of these curves is zero. For example, the vertical line labeled C in Figure 5.3.1(b) contributes positively to the integral around R_1 and negatively around R_3, thereby cancelling out. Thus the sum of all the double integrals over the sub-regions is equal to the line integral around the boundary.

The theorem extends to the case where R has multiple boundary components, as in Figure 3.7.1(b). The idea of the proof is the same, just divide R into regions on which it applies. The integrals on interior arcs cancel, leaving only integrals on boundary curves on the left-hand side of Theorem 5.3.2. The double integrals add so that the region of integration on the right is all of R.

Example 5.3.2. *Green's theorem on a trapezoid*

Let R be the trapezoid with vertices $(0,0)$, $(1,0)$, $(1,1)$, and $(0,2)$, and let C be its boundary curve oriented counterclockwise. Evaluate $\oint_C xydx - x^2dy$ using Green's theorem. We calculate, setting up the limits on R,

$$\oint_C xydx - x^2dy = \iint_R \frac{\partial}{\partial x}(-x^2) - \frac{\partial}{\partial y}(xy)dA = \iint_R -3xdA$$

$$= \int_0^1 \int_0^{2-x} -3xdydx = \int_0^1 -3xy\Big|_{y=0}^{2-x} dx$$

$$= \int_0^1 3x^2 - 6xdx = x^3 - 3x^2\Big|_0^1 = -2. \blacktriangle$$

The following example is a nice application of Green's theorem to a region with more than one boundary component.

Example 5.3.3. *Equal integrals*

Let C be any simple closed loop encircling the origin in counterclockwise fashion (as in Figure 5.3.3), and consider the vector field

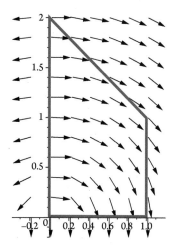

Figure 5.3.2 Green's theorem on a trapezoid

Figure 5.3.3 A simple closed loop C around the origin

$$\mathbf{F}(x,y) = \left\langle \frac{-y}{x^2+y^2}, \frac{x}{x^2+y^2} \right\rangle.$$

Show that the circulation $\oint_C \mathbf{F} \cdot d\mathbf{C} = 2\pi$.

First we show that the statement is true for any circle X centered at the origin. If X has radius r a counterclockwise parameterization is

$$\mathbf{C}(t) = (r\cos t, r\sin t), \ 0 \le t \le 2\pi.$$

In this case, $\mathbf{C}'(t) = \langle -r\sin t, r\cos t\rangle$ and

$$\mathbf{F}(\mathbf{C}(t)) = \left\langle \frac{-r\sin t}{r^2}, \frac{r\cos t}{r^2} \right\rangle = \frac{1}{r}\langle -\sin t, \cos t\rangle.$$

The line integral is

$$\oint_C \mathbf{F} \cdot d\mathbf{C} = \int_0^{2\pi} \frac{1}{r}\langle -\sin t, \cos t\rangle \cdot \langle -r\sin t, r\cos t\rangle \, dt$$

$$= \int_0^{2\pi} \sin^2 t + \cos^2 t \, dt = 2\pi.$$

Now let C be a simple closed loop encircling the origin once, as in Figure 5.3.3, and let X be a circle centered at the origin lying entirely inside C. Let R be the region between X and C, so that R does not contain the origin and has two boundary curves. Orient X and C so that R lies to their left, which implies X is oriented clockwise and C counterclockwise. Since R doesn't contain the origin, the vector field $\mathbf{F}$ has continuous partial derivatives on R and Green's theorem can be applied. Thus

$$\int_{C-X} \mathbf{F} \cdot d\mathbf{C} = \oint_C \mathbf{F} \cdot d\mathbf{C} - \oint_X \mathbf{F} \cdot d\mathbf{C} = \iint_R \frac{\partial Q}{\partial x} - \frac{\partial P}{\partial y} \, dA.$$

Here we've used $-X$ to denote the clockwise orientation on X and the fact that $\int_{-X} \mathbf{F} \cdot d\mathbf{C} = -\int_X \mathbf{F} \cdot d\mathbf{C}$. A tedious computation shows

$$\frac{\partial Q}{\partial x} - \frac{\partial P}{\partial y} = \frac{y^2 - x^2}{(x^2+y^2)^2} - \frac{y^2 - x^2}{(x^2+y^2)^2} = 0,$$

so the integrand of the double integral is zero. But this means

$$\int_C \mathbf{F} \cdot d\mathbf{C} - \int_X \mathbf{F} \cdot d\mathbf{C} = 0,$$

or the circulation around C is the same as that around X oriented counterclockwise. Since X is a circle centered at the origin, our initial computation shows

$$\oint_C \mathbf{F} \cdot d\mathbf{C} = \oint_X \mathbf{F} \cdot d\mathbf{C} = 2\pi.$$

Thus the integral around any counterclockwise simple closed loop encircling the origin once is 2π. ▲

Scalar Curl: The integrand in Green's theorem is at this point unmotivated, and rather unintuitive. The quantity $\frac{\partial Q}{\partial x} - \frac{\partial P}{\partial y}$ is called the *scalar curl* of the vector field $\mathbf{F}(x,y) = P(x,y)\mathbf{i} + Q(x,y)\mathbf{j}$. It turns out that the scalar curl of a vector field in the plane is one measure of the tendency of the vector field to rotate counterclockwise about a point. To get some intuition for this we consider a matchstick placed in a fluid flow.

Figure 5.3.4, parts (a) and (b), illustrates velocity fields on horizontal matchsticks in a fluid flow. Since we are concerned with rotation induced by the flow, the figures only depict the vertical component $Q(x,y)\mathbf{j}$. The horizontal component will not cause a horizontal matchstick to rotate. The flow of Figure 5.3.4(a) is symmetric, so it would push the stick along without introducing any rotation. The flow of Figure 5.3.4(b) is faster on the right than the left, so as it pushes the stick along it introduces a counterclockwise rotation.

Consider the vector field $\mathbf{F} = \langle P, Q \rangle$ of Figure 5.3.4(b). The fact that the vertical component of $\mathbf{F}$ (a.k.a. $Q(x,y)$) gets longer as you move to the right indicates that Q is an increasing function of x. Thus if $\frac{\partial Q}{\partial x} > 0$, the flow introduces a counterclockwise rotation.

Similarly, a vertical stick rotates counterclockwise if the horizontal component of $\mathbf{F}$ gets more negative (decreases) as y increases (see Figure 5.3.4(c)). The horizontal component of $\mathbf{F}$ is P, so this says that if $-\frac{\partial P}{\partial y} > 0$, then $\mathbf{F}$ induces a local counterclockwise rotation (make sure all the signs are right!).

The scalar curl combines both horizontal and vertical information in one function. When $Q_x - P_y > 0$, the horizontal and vertical components induce local counterclockwise rotations. A negative scalar curl indicates local clockwise rotations are introduced, and zero scalar curl means the stick just moves along without starting to spin.

Let us consider some examples of vector fields with different scalar curls. If $\mathbf{F}(x,y) = \langle x,y \rangle$, then all vectors point away from the origin (see Figure 5.3.5(a)). Intuitively the vector field does not seem to rotate, so a reasonable measure of the rotation of $\mathbf{F}$ should

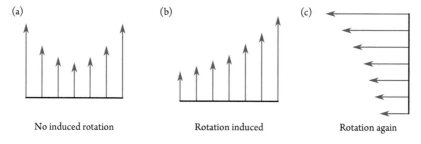

(a) No induced rotation   (b) Rotation induced   (c) Rotation again

Figure 5.3.4 Flows and local rotation

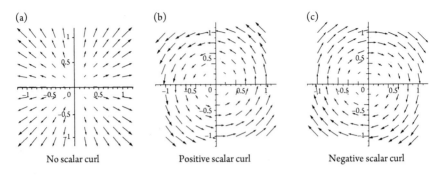

(a) No scalar curl   (b) Positive scalar curl   (c) Negative scalar curl

Figure 5.3.5 Examples of scalar curl

be zero, and direct computation verifies this:

$$\frac{\partial Q}{\partial x} - \frac{\partial P}{\partial y} = \frac{\partial}{\partial x}y - \frac{\partial}{\partial y}x = 0.$$

As a second example, the vector field $\mathbf{F}(x,y) = \langle -y, x\rangle$ seems to rotate in a counter-clockwise fashion, so we would expect a positive scalar curl (see Figure 5.3.5(b)). Indeed, we calculate $\frac{\partial}{\partial x}x - \frac{\partial}{\partial y}(-y) = 2$.

Finally, $\mathbf{F}(x,y) = \langle y, -x\rangle$ rotates in a clockwise fashion, and has scalar curl $\frac{\partial}{\partial x}(-x) - \frac{\partial}{\partial y}y = -2$ (Figure 5.3.5(c)).

These pictures help our intuition, but of course the situation can be more complicated than this. For example, any conservative vector field is of the form $\mathbf{F} = \langle P, Q\rangle = \langle f_x, f_y\rangle$ so the scalar curl is $\frac{\partial Q}{\partial x} - \frac{\partial P}{\partial y} = \frac{\partial f}{\partial x \partial y} - \frac{\partial f}{\partial y \partial x} = 0$! This means that conservative vector fields have no scalar curl.

There is a more precise interpretation of the scalar curl: it is the (infinitesimal) circulation per unit area. In other words, let $\mathbf{a}$ be a point and R_ϵ a disk radius ϵ centered at $\mathbf{a}$. The mean value theorem for integrals, combined with Green's theorem, can be used to show

$$\left(\frac{\partial Q}{\partial x}(\mathbf{a}) - \frac{\partial P}{\partial y}(\mathbf{a})\right) = \lim_{\epsilon \to 0} \frac{1}{\text{Area of } R_\epsilon}\int_C \mathbf{F}\cdot d\mathbf{C}.$$

Rather than an analytic discussion of this fact, we illustrate with a Math App.

Math App 5.3.1. *Interpreting scalar curl*

This Math App allows you to visualize the limit just described, using several different vector fields. Take this opportunity to begin experimenting.

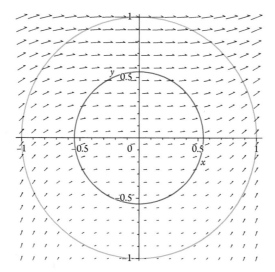

More on Green's Theorem: Now that you've had some experience with the integrand of Green's theorem, we continue with several examples.

First note that Green's theorem is consistent with what we know about integrals of conservative vector fields (whose scalar curl is zero). If **F** is conservative and C is a closed loop in the plane, then Green's theorem gives

$$\int_C \mathbf{F} \cdot d\mathbf{C} = \iint_R \frac{\partial Q}{\partial x} - \frac{\partial P}{\partial y} dA = \iint_R 0 \, dA = 0.$$

This is expected since one of the defining characteristics of a conservative vector field is that integrals around closed loops vanish.

Example 5.3.4. *Using polar coordinates*

Let R be the quarter of the unit circle in the first quadrant, and C its boundary oriented counterclockwise. Evaluate $\int_C \sin x \, dx + \sqrt{x^2 + y^2} \, dy$ (see Figure 5.3.6). The integrand is

$$\frac{\partial}{\partial x}\sqrt{x^2 + y^2} - \frac{\partial}{\partial y}\sin x = \frac{x}{\sqrt{x^2 + y^2}}.$$

The double integral becomes

$$\int_C \sin x \, dx + \sqrt{x^2 + y^2} \, dy = \iint_R \frac{x}{\sqrt{x^2 + y^2}} \, dA,$$

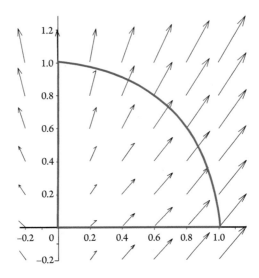

Figure 5.3.6 Using polar coordinates in Green's theorem

which is most easily calculated using polar coordinates. Doing so gives:

$$\int_C \sin x\,dx + \sqrt{x^2+y^2}\,dy = \iint_R \frac{x}{\sqrt{x^2+y^2}}\,dA = \int_0^{\pi/2}\int_0^1 \frac{r\cos\theta}{r}\,r\,dr\,d\theta$$

$$= \int_0^{\pi/2}\int_0^1 r\cos\theta\,dr\,d\theta = \int_0^{\pi/2} \frac{r^2}{2}\cos\theta\,\Big|_{r=0}^{1}\,d\theta$$

$$= \int_0^{\pi/2} \frac{1}{2}\cos\theta\,d\theta = \frac{1}{2}\sin\theta\,\Big|_0^{\pi/2} = \frac{1}{2}. \blacktriangle$$

We now give an interesting application of Green's theorem to calculating area, arising from the fact that the scalar curl of $\mathbf{F}(x,y) = \langle -y,x\rangle$ is 2. Let C be a simple closed loop bounding the region R in the plane. Then Green's theorem gives

$$\int_C -y\,dx + x\,dy = \iint_R 2\,dA = 2 \text{ (the area of } R\text{).}$$

Example 5.3.5. *The area of an ellipse*

Find the area enclosed by the ellipse $\frac{x^2}{9} + \frac{y^2}{4} = 1$.
    Using double integrals, we'd have to calculate

$$\int_{-3}^{3}\int_{-\frac{2}{3}\sqrt{9-x^2}}^{\frac{2}{3}\sqrt{9-x^2}} dy\,dx = \int_{-3}^{3} \frac{4}{3}\sqrt{9-x^2}\,dx,$$

which can be done using several different approaches that may not be immediately apparent. We choose to calculate the area using the line integral.

To calculate the line integral, we first parameterize the ellipse by $\mathbf{C}(t) = (3\cos t, 2\sin t)$, for $0 \leq t \leq 2\pi$. Then $dx = -3\sin t\, dt$ and $dy = 2\cos t\, dt$, so that

$$\int_C -y\,dx + x\,dy = \int_0^{2\pi} -(2\sin t)(-3\sin t)\,dt + 3\cos t\,2\cos t\,dt = \int_0^{2\pi} 6\,dt = 12\pi.$$

The area of the ellipse is half this, so it's 6π. ▲

In the previous examples we have used the double integral of Green's theorem to make evaluating the line integral easier. This is the first example where the line integral told us something about the double integral.

Green's Theorem—Flux Form: We introduce Green's theorem as a method for computing the circulation of a fluid flow around a closed loop, and circulation involved the tangential component of the velocity vector. Section 2.6 also introduced the flux of a flow across a curve by considering the normal component of velocity. Example 2.6.6 calculated the flux across line segments of a uniform flow, while Example 2.6.7 approximated flux of a variable flow across a curve. It might be helpful to review those before continuing. In fact, a quick visit to the *Approximating flux* Math App would be good preparation for the following discussion. We include the hyperlink here for your convenience.

Math App 5.3.2. *Approximating flux—revisited*

This is the same Math App you encountered in Section 2.6, but included here for your convenience.

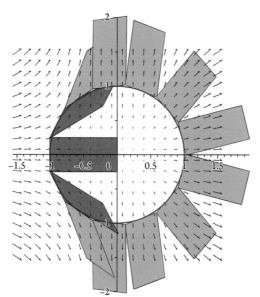

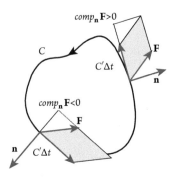

Figure 5.3.7 Flux approximated

Recall that a convention was made to determine which flux was positive. Flow from left to right across C is the positive direction while right to left is negative. When fluid flows out of C in the upper right of Figure 5.3.7, the component of $\mathbf{F}$ in the left-to-right direction is positive, whereas when fluid flows into C the component is negative (see the lower left of Figure 5.3.7). This shows that $comp_{\mathbf{n}}\mathbf{F}$ records the height of the parallelogram with the correct sign to record inward and outward flux. The amount of fluid crossing C per unit time is approximated locally by the area of the parallelogram spanned by $\mathbf{F}$ and $\mathbf{C}'\Delta t$, which is

$$comp_{\mathbf{n}}\mathbf{F}\left(\mathbf{C}(t)\right)\left\|\mathbf{C}'(t)\right\|\Delta t.$$

The total flux is approximated by summing these local contributions and taking a limit as the approximation improves, so

$$Flux = \int_C comp_{\mathbf{n}}\mathbf{F}dC,$$

where we recall that $dC = \left\|\mathbf{C}'(t)\right\| dt$.

We now write the flux integral in differential form in order to apply Green's theorem (the circulation form). Recall that if $\mathbf{C}(t) = (x(t),y(t))$, $a \leq t \leq b$, then $\mathbf{C}' = \langle x',y'\rangle$ and the rightward-pointing normal is $\mathbf{n} = \langle y',-x'\rangle$ (I don't really expect you to remember this, but we discussed it in the fluid flow paragraph of Section 2.6). Note that $\left\|\mathbf{C}'\right\| = \|\mathbf{n}\|$. Letting $\mathbf{F} = \langle P,Q\rangle$ we see that the flux integral in differential notation is

$$\oint_C comp_{\mathbf{n}}\mathbf{F}dC = \int_a^b \mathbf{F}\cdot\frac{\mathbf{n}}{\|\mathbf{n}\|}\left\|\mathbf{C}'\right\| dt$$

$$= \int_a^b \mathbf{F}\cdot\mathbf{n}\, dt \qquad\qquad (\text{since } \left\|\mathbf{C}'\right\| = \|\mathbf{n}\|)$$

$$= \int_a^b \langle P,Q \rangle \cdot \langle y', -x' \rangle \, dt = \int_a^b Py' \, dt - Qx' \, dt$$

$$= \oint_C -Q\,dx + P\,dy. \qquad \text{(since } dx = x'\,dt,\ dy = y'\,dt\text{)}$$

Applying Green's theorem to this integral (i.e. replacing the P, Q of Theorem 5.3.2 with $-Q, P$, respectively) gives the flux form of Green's theorem:

Theorem 5.3.3. *Let $\mathbf{F} = \langle P, Q \rangle$ be a smooth vector field defined on the region R in the plane with boundary $C = C_1 \cup \cdots \cup C_n$, with each component oriented so that R is to its left. Suppose P, Q are continuously differentiable on R. Then the flux of $\mathbf{F}$ across C is given by*

$$\oint_C comp_n \mathbf{F} dC = \oint_C -Q\,dx + P\,dy = \iint_R \frac{\partial P}{\partial x} + \frac{\partial Q}{\partial y} \, dA.$$

The integrand $\frac{\partial P}{\partial x} + \frac{\partial Q}{\partial y}$ of the double integral is called the *divergence* of the vector field $\mathbf{F}$, and will be analyzed in Section 5.5. Also note that for two-dimensional flows, the flow across a curve makes sense because there is one "across" direction—the normal direction to the curve. In three dimensions there are whole planes normal to a curve at a point, and the notion of "across a curve" is ambiguous. Surfaces in space, however, have unique normal directions. Therefore we will consider the flux of three-dimensional flows across surfaces in Section 5.4. We conclude this section with an example.

Example 5.3.6. *Flux from a source*

An ideal fluid flow with source at the origin has velocity field

$$\mathbf{F}(x,y) = \left\langle \frac{x}{x^2 + y^2}, \frac{y}{x^2 + y^2} \right\rangle$$

in Cartesian coordinates (a nice exercise is to verify that $\mathbf{F}$ is the gradient of the velocity potential, whose polar formula is $\varphi = \ln r$). Let C be the circle $(x - 2)^2 + y^2 = 1$ oriented counterclockwise, and compute the flux of $\mathbf{F}$ across C (see Figure 5.3.8(a)).

The only point where $\mathbf{F}$ is not defined is at the origin, which is not in the disk bounded by C, therefore Theorem 5.3.3 can be applied. The divergence of $\mathbf{F}$, the integrand of the double integral, is

$$\frac{\partial P}{\partial x} + \frac{\partial Q}{\partial y} = \frac{y^2 - x^2}{(x^2 + y^2)^2} + \frac{x^2 - y^2}{(x^2 + y^2)^2} = 0.$$

The flux form of Green's theorem implies the flux of $\mathbf{F}$ across C is zero. This means that the same amount of water enters R as leaves it per unit time.

Since the divergence of $\mathbf{F}$ is zero everywhere $\mathbf{F}$ is defined, you might think that the flux around any closed loop is zero. We now directly compute the line integral that gives flux

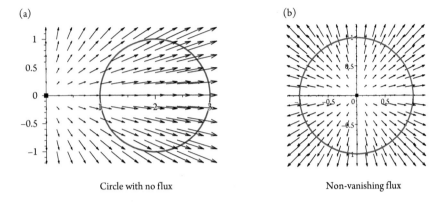

(a)

(b)

Circle with no flux

Non-vanishing flux

Figure 5.3.8 Flux from an ideal source flow

across the unit circle. Figure 5.3.8(b) indicates flows out of the unit circle, which should mean positive flux.

In this case, the curve is parameterized by $C(t) = (\cos t, \sin t)$, $0 \leq t \leq 2\pi$, which implies $dx = -\sin t\, dt$, $dy = \cos t\, dt$. Evaluating the component functions of $\mathbf{F}$ on $C(t)$ gives

$$P(C(t)) = \frac{\cos t}{\cos^2 t + \sin^2 t} = \cos t, \quad \text{and} \quad Q(C(t)) = \frac{\sin t}{\cos^2 t + \sin^2 t} = \sin t.$$

The flux integral, then, is

$$Flux = \oint_C -Q dx + P dy = \int_0^{2\pi} -\sin t(-\sin t) + \cos t \cos t\, dt = \int_0^{2\pi} dt = 2\pi.$$

It makes intuitive sense from the figure that the flux is positive. A question for you: Is this a counter-example to Green's theorem? The divergence of $\mathbf{F}$ vanishes so the double integral should too, but the line integral is non-zero! How can this be? ▲

Exercises

1. Verify that $\int_C P dx = -\iint_R \frac{\partial P}{\partial y} dA$ where C is the boundary of the rectangle $R = [a, b] \times [c, d]$ oriented counterclockwise.

2. Evaluate $\int_C (3x - 2y)dx - xdy$, where C is the boundary of the region under $y = 1 - x^2$ and above the x-axis in two ways:
   (a) By parameterizing C and computing the line integral directly.
   (b) By using Green's theorem.

3. Evaluate $\int_C y\,dx - x^2\,dy$, where C is the boundary of the top half of the unit disk in two ways:

   (a) by parameterizing C and computing the line integral directly,

   (b) by using Green's theorem.

4. Find the work done by the force $\mathbf{F}(x,y) = \langle 2y + \cos x, xy - y^3 \rangle$ on a particle that moves once around the unit circle oriented counterclockwise.

5. Find the circulation of $\mathbf{F}(x,y) = \langle 4y - \ln x, e^y - xy \rangle$ around the unit circle oriented counterclockwise.

6. Find the circulation of $\mathbf{F}(x,y) = \langle x^2 + y^2, x^3 - 3x + 5y^2 \rangle$ around the triangle with vertices $(0,0)$, $(1,0)$, and $(0,2)$ in the xy-plane oriented counterclockwise.

7. Let C be the square with vertices $(0,0)$, $(1,0)$, $(1,1)$, and $(0,1)$ oriented counterclockwise. Find the circulation $\oint_C \mathbf{F} \cdot d\mathbf{C}$ of $\mathbf{F} = \langle 3xy^2, -y^3 \rangle$ around C.

8. Let $\mathbf{F} = \langle 2x, 3y^2 - z, -y \rangle$ and $\mathbf{C}(t) = (\cos t, t, \sin t), 0 \le t \le \pi$. Can you evaluate $\int_C \mathbf{F} \cdot d\mathbf{C}$ using (justify your answer in each case)

   (a) the formula $\int_C \mathbf{F} \cdot d\mathbf{C} = \int_a^b \mathbf{F}(\mathbf{C}(t)) \cdot \mathbf{C}'(t)\,dt$

   (b) the fundamental theorem of line integrals

   (c) Green's theorem.

   (d) Choose one valid method and evaluate $\int_C \mathbf{F} \cdot d\mathbf{C}$.

9. Let C be the boundary of a region R in the plane oriented counterclockwise. Find the work done by the force $\mathbf{F}(x,y) = \langle y, -x \rangle$ on a particle that moves once around $\mathbf{C}$ if the area of R is 7.

10. Evaluate $\int_C \cos y\,dx + x^2\,dy$ where C is the boundary of the rectangle $[0, \pi] \times [0, \pi]$ oriented counterclockwise.

11. Compute the work done by the force $\mathbf{F}(x,y) = \langle x + 2y, xy \rangle$ on a particle traveling once counterclockwise around the triangle with vertices $(0,0)$, $(1,0)$, and $(0,1)$.

12. Let $\mathbf{F}$ be the vector field $\mathbf{F}(x,y) = \langle -y/(x^2 + y^2), x/(x^2 + y^2) \rangle$ and C a simple closed loop not containing the origin. Evaluate $\int_C \mathbf{F} \cdot d\mathbf{C}$ and $\int_{-C} \mathbf{F} \cdot d\mathbf{C}$.

13. Let $\mathbf{F}$ be the vector field $\mathbf{F}(x,y) = \langle (x^2 + y^2 - 2y)/(x^2 + y^2), 2x/(x^2 + y^2) \rangle$.

   (a) Evaluate the circulation of $\mathbf{F}$ around the unit circle directly.

   (b) Show that the circulation of $\mathbf{F}$ around all simple closed curves containing the origin is the same.

14. Find the flux of $\mathbf{F} = \langle x^3 - y^2, 4x^2 + y^3 \rangle$ across the unit circle oriented counterclockwise.

15. Find the flux of $\mathbf{F} = \langle -y, x \rangle$ across any simple closed loop in the plane.

16. Find the flux of $\mathbf{F} = \langle x^3 - 8x, y^2 + y \rangle$ across the triangle with vertices $(0,0), (0,3), (-1,0)$ oriented counterclockwise.

17. Define non-constant component functions for a vector field $\mathbf{F}(x,y)$ that guarantee no flux across any simple closed loop in the plane.

18. Show that the flux of $\mathbf{F}(x,y) = \langle x/(x^2+y^2), y/(x^2+y^2) \rangle$ across any simple closed loop containing the origin is the same.

19. Evaluate $\int_C x^2 y\, dx - y^2 x\, dy$ where C is the unit circle oriented counterclockwise.

20. Let R be the region in the first quadrant between the circles $x^2 + y^2 = 1$ and $x^2 + y^2 = 4$, and let C be its boundary oriented counterclockwise. Evaluate the line integral $\int_C xy^2\, dx + x^2 y\, dy$.

21. Let R be the region in the first quadrant between the circles $x^2 + y^2 = 1$ and $x^2 + y^2 = 4$, and let C be its boundary oriented counterclockwise. Evaluate the line integral $\int_C xy^2\, dx - x^2 y\, dy$.

22. Use the line integral of Green's theorem to find the area enclosed by the ellipse $\frac{x^2}{16} + \frac{y^2}{25} = 1$.

23. Use the line integral of Green's theorem to find the area enclosed by the ellipse $\frac{x^2}{a^2} + \frac{y^2}{b^2} = 1$.

5.4 Surface Integrals of Vector Fields

At this point we have quite a bit of experience with integrating vector fields along curves. To evaluate a line integral we can parameterize the curve and evaluate directly, use the fundamental theorem for line integrals when the vector field is conservative, or use Green's theorem when considering closed loops in the plane. In this section we define the integral of a vector field across a surface and investigate some of its properties.

For line integrals we integrated the tangential component of **F** along a curve C, which made sense because curves have one tangential direction. Since surfaces have tangent *planes*, the notion of "tangential direction" is ambiguous. Surfaces, however, do have unique normal directions (up to a choice of sign), so the integral of a vector field across a surface turns out to be integrating the normal component of **F**.

We consider the velocity field of a fluid flow to motivate the definition of a surface integral. To begin with, suppose a fluid is flowing uniformly with constant velocity vector $2\mathbf{j}$. In other words, each particle in the fluid is moving 2 meters per second in the positive y-direction. Insert a rectangular sieve into the fluid, perpendicular to the flow. Our goal is to compute the volume of water that passes through our sieve in one second—this is the *flux* of the flow across the sieve. Figure 5.4.1(*a*) illustrates the sieve together with a few velocity vectors of the fluid flow.

To find the volume of fluid passing through the sieve in one second (flux), notice that particles on the sieve travel to the ends of the velocity vectors. Particles that were 2 units back will end up on the sieve. Thus the rectangular box in Figure 5.4.1(*b*) represents the fluid that has passed through the sieve in one unit of time. Its volume is the area of the sieve A times the length of the velocity vector $\|\mathbf{v}\|$.

Now consider the case where the sieve is placed at an angle to the flow of the fluid, as in Figure 5.4.2. It is still true that particles starting on the sieve at time zero travel to the tip of the velocity vector; it's just that this vector is not at a right angle to the sieve. Thus the flux of the fluid can be represented by the parallelepiped in Figure 5.4.2(*b*).

The good news is that we learned how to calculate the signed volume of a parallelepiped in Section 2.4. Indeed, the volume of a parallelepiped determined by vectors $\mathbf{v}, \mathbf{u}, \mathbf{w}$ is the

(a)

(b) $V = A \cdot \|\mathbf{V}\|$

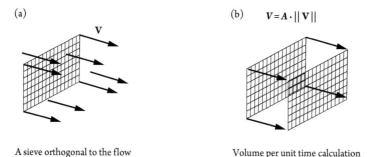

A sieve orthogonal to the flow   Volume per unit time calculation

Figure 5.4.1 Computing volume crossing a surface orthogonal to flow

(a)

(b)

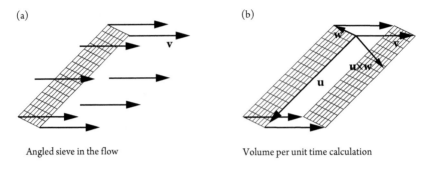

Angled sieve in the flow

Volume per unit time calculation

Figure 5.4.2 Computing volume crossing an angled surface

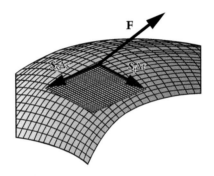

Figure 5.4.3 Local approximation of flux

triple scalar product $\mathbf{v} \cdot (\mathbf{u} \times \mathbf{w})$ (see Theorem 2.4.3). Moreover, this product is positive when $\mathbf{u}, \mathbf{w}, \mathbf{v}$ form a right-handed frame, and negative otherwise.

Let's apply these observations to a variable velocity field (not just a uniform flow), and calculate the flux across an arbitrary surface S (not just a rectangle). The key to this calculation, like most of calculus, is as follows:

Approximate the desired quantity with ones you can calculate, then take a limit as your approximation gets better.

Suppose our surface S is given to us parametrically by $\mathbf{S}(s,t)$, for $a \leq s \leq b$ and $c \leq t \leq d$, and that our velocity field is $\mathbf{F}$. As we've seen when finding surface area, and when defining the integral of a function on a surface, small patches of the surface S can be approximated by the parallelogram determined by the vectors $\mathbf{S}_s \Delta s$ and $\mathbf{S}_t \Delta t$ (see Figure 4.5.2). Thus we approximate the surface $\mathbf{S}$ with parallelogram shingles as in Figure 4.5.3. Now make the simplifying assumption that our velocity field $\mathbf{F}$ is constant on the parallelogram shingles. With these assumptions, we use the triple scalar product to make local calculations, which we add up to approximate the total flux.

The fluid flows across each parallelogram shingle to form a parallelepiped in unit time, and the parallelepiped is defined by the velocity vector $\mathbf{F}$ and the tangential vectors $\mathbf{S}_s \Delta s$ and $\mathbf{S}_t \Delta t$ (see Figure 5.4.3). Thus the volume of fluid flowing across each patch is the triple scalar product

$$\mathbf{F} \cdot (\mathbf{S}_s \Delta s \times \mathbf{S}_t \Delta t) = \mathbf{F} \cdot (\mathbf{S}_s \times \mathbf{S}_t) \Delta s \Delta t.$$

Recalling that $\mathbf{F}$ is evaluated at a point on the surface (i.e. it is assumed locally constant), the approximate flux across the surface is

$$V \approx \sum_{i=1}^{n} \sum_{j=1}^{n} \mathbf{F}(\mathbf{S}(s_i, t_j)) \cdot (\mathbf{S}_s \times \mathbf{S}_t) \Delta s \Delta t.$$

Taking a limit, as our approximation gets better, yields the integral

$$\int_a^b \int_c^d \mathbf{F}(\mathbf{S}(s, t)) \cdot (\mathbf{S}_s \times \mathbf{S}_t) dt ds,$$

which motivates our definition of the integral of a vector field across a surface.

Definition 5.4.1. *Let $\mathbf{F}$ be a vector field and S a surface given parametrically by $\mathbf{S}(s, t)$ over the domain R in the st-plane. The flux of $\mathbf{F}$ across S is the integral*

$$\iint_R \mathbf{F}(\mathbf{S}(s, t)) \cdot (\mathbf{S}_s \times \mathbf{S}_t) dA.$$

By the previous discussion, if $\mathbf{F}$ is the velocity field of a fluid flow then the flux of $\mathbf{F}$ across S can be interpreted as the volume of fluid flowing across S in unit time. A remark about orientation is in order. At each point on $\mathbf{S}$ there are two normal directions given by $\pm \mathbf{S}_s \times \mathbf{S}_t$. By using $\mathbf{S}_s \times \mathbf{S}_t$ in Definition 5.4.1 we've chosen one particular normal direction as the "positive side" of the surface. Depending on the situation, the vector $-\mathbf{S}_s \times \mathbf{S}_t$ may need to be used instead.

Math App 5.4.1. *Flux across a surface*

This Math App gives you a chance to interact with approximations of flux across a surface. Take a look!

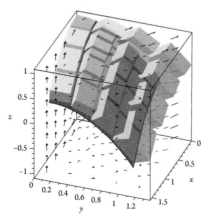

We now outline the steps for computing flux.

Computing Flux Across a Parametric Surface

To compute the flux of the vector field $\mathbf{F}$ across the parametric surface $\mathbf{S}(s,t)$, for a region R in the st-plane:

1. Calculate the normal vector $\mathbf{S}_s \times \mathbf{S}_t$.

2. Evaluate $\mathbf{F}$ on $\mathbf{S}$ and determine the integrand $\mathbf{F}(\mathbf{S}(s,t)) \cdot (\mathbf{S}_s \times \mathbf{S}_t)$.

3. Evaluate the integral $\iint_R \mathbf{F}(\mathbf{S}(s,t)) \cdot (\mathbf{S}_s \times \mathbf{S}_t)dA$.

One should notice the similarity between the procedure for calculating a flux integral and a line integral. Because of the similarities with line integrals, and for simplicity, we introduce the notation

$$\iint_S \mathbf{F} \cdot d\mathbf{S} = \iint_R \mathbf{F}(\mathbf{S}(s,t)) \cdot (\mathbf{S}_s \times \mathbf{S}_t)dA,$$

for the surface integral of $\mathbf{F}$ across the surface S. For parametric surfaces, then, the area form $d\mathbf{S}$ is shorthand notation for $(\mathbf{S}_s \times \mathbf{S}_t)dA$.

Example 5.4.1. *Flux across a parametric surface*

Evaluate the flux of $\mathbf{F}(x,y,z) = \langle zx, zy, z \rangle$ across the parametric surface $\mathbf{S}(s,t) = \langle s\cos t, s\sin t, 1 - s^2 \rangle$, for $0 \leq s \leq 1$ and $0 \leq t \leq \pi/2$ (see Figure 5.4.4).

First we find the normal vector to the surface:

$$\mathbf{S}_s \times \mathbf{S}_t = \begin{vmatrix} \mathbf{i} & \mathbf{j} & \mathbf{k} \\ \cos t & \sin t & -2s \\ -s\sin t & s\cos t & 0 \end{vmatrix} = \langle 2s^2 \cos t, 2s^2 \sin t, s \rangle.$$

Evaluating $\mathbf{F}(s\cos t, s\sin t, 1 - s^2) = \langle (1 - s^2)s\cos t, (1 - s^2)s\sin t, 1 - s^2 \rangle$, so the integrand of the flux integral is

$$\begin{aligned} \mathbf{F} \cdot \mathbf{S}_s \times \mathbf{S}_t &= \langle (1 - s^2)s\cos t, (1 - s^2)s\sin t, 1 - s^2 \rangle \cdot \langle 2s^2 \cos t, 2s^2 \sin t, s \rangle \\ &= (1 - s^2)\left(2s^3 \cos^2 t + 2s^3 \sin^2 t + s - s^3\right) \\ &= (1 - s^2)(s + s^3) = s - s^5. \end{aligned}$$

The flux, then, is given by the integral

$$\int_0^{pi/2} \int_0^1 s - s^5 \, ds\, dt = \int_0^{\pi/2} \frac{s^2}{2} - \frac{s^6}{6}\bigg|_0^1 \, dt = \int_0^{\pi/2} \frac{1}{3} \, dt = \frac{\pi}{6}. \; \blacktriangle$$

Since the graph of the function $z = f(x,y)$ can be thought of as the parametric surface $\mathbf{S}(x,y) = (x,y,f(x,y))$, the formula for flux across a parametric surface translates into one

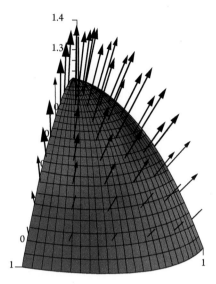

Figure 5.4.4 Flux across a parametric surface

for flux across the graph of a function. Straightforward substitution shows that the flux of $\mathbf{F}(x,y,z)$ across the surface S given as the graph of $z = f(x,y)$ above the region R in the plane is:

$$\iint_S \mathbf{F} \cdot d\mathbf{S} = \iint_R \mathbf{F}(x,y,f(x,y)) \cdot \left(-f_x(x,y), -f_y(x,y), 1\right) dA.$$

This formula leads to the following process for evaluating flux across a graph.

Computing Flux Across the Graph of a Function

To compute the flux of the vector field $\mathbf{F}$ across the surface $z = f(x,y)$ over a region R in the xy-plane:

1. Calculate the normal vector $\left(-f_x, -f_y, 1\right)$.
2. Evaluate $\mathbf{F}$ on the surface $z = f(x,y)$ and determine the integrand $\mathbf{F}(x,y,f(x,y)) \cdot \left(-f_x, -f_y, 1\right)$.
3. Evaluate the integral $\iint_R \mathbf{F}(x,y,f(x,y)) \cdot \left(-f_x(x,y), -f_y(x,y), 1\right) dA$.

Example 5.4.2. *Flux across the graph of a function*

Compute the flux of $\mathbf{F} = \left(z, yz, x\right)$ across the portion of the plane $3x + 2y + z = 6$ above the rectangle $[0,1] \times [0,1]$ (see Figure 5.4.5).

Solving for z, we get the plane is the graph of $f(x,y) = 6 - 3x - 2y$. Thus the normal vector is $\left(-f_x, -f_y, 1\right) = \langle 3, 2, 1 \rangle$. Further, evaluating the vector field on the surface gives

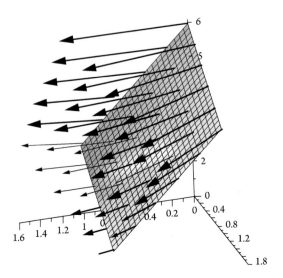

Figure 5.4.5 Flux across a plane

$\mathbf{F}(x,y,6-3x-2y) = \langle 6-3x-2y, 6y-3xy-2y^2, x \rangle$, so the flux is given by

$$\iint_S \mathbf{F} \cdot d\mathbf{S} = \int_0^1 \int_0^1 \langle 6-3x-2y, 6y-3xy-2y^2, x \rangle \cdot \langle 3,2,1 \rangle \, dy dx$$

$$= \int_0^1 \int_0^1 18 - 8x + 6y - 6xy - 4y^2 \, dy dx$$

$$= \int_0^1 (18-8x)y + 3y^2 - 3xy^2 - \frac{4y^3}{3} \Big|_{y=0}^{1} dx$$

$$= \int_0^1 \frac{59}{3} - 11x \, dx = \frac{59}{3}x - \frac{11}{2}x^2 \Big|_0^1 = \frac{85}{6}. \ \blacktriangle$$

Remark: The flux of a vector field across a surface measures how much of the vector field points across the surface. More precisely, it's the integral of the component $comp_{\mathbf{n}}\mathbf{F}$ of $\mathbf{F}$ in the direction $\mathbf{n}$ normal to the surface. If the vector field $\mathbf{F}$ is always tangent to the surface, then the component $comp_{\mathbf{n}}\mathbf{F}$ is zero everywhere, and the flux is as well. We formalize these observations in the following results.

Theorem 5.4.1. *Let $\mathbf{F}$ be a smooth vector field and S a smooth oriented surface with normal direction $\mathbf{n}$ on which $\mathbf{F}$ is defined. The flux of $\mathbf{F}$ across S is the integral on S of the scalar component of $\mathbf{F}$ in the $\mathbf{n}$ direction. In symbols*

$$\iint_S \mathbf{F} \cdot d\mathbf{S} = \iint_S comp_{\mathbf{n}}\mathbf{F} dS.$$

Proof. We prove it for the parametric surface $\mathbf{S}(s,t)$ defined over the region R in the st-plane with normal vector $\mathbf{S}_s \times \mathbf{S}_t$. By definition 5.4.1, some algebraic trickery, and the fact that $dS = \|\mathbf{S}_s \times \mathbf{S}_t\| \, dA$, we have

$$\iint_S \mathbf{F} \cdot d\mathbf{S} = \iint_R \mathbf{F}(\mathbf{S}(s,t)) \cdot (\mathbf{S}_s \times \mathbf{S}_t) dA$$

$$= \iint_R \mathbf{F}(\mathbf{S}(s,t)) \cdot (\mathbf{S}_s \times \mathbf{S}_t) \frac{\|\mathbf{S}_s \times \mathbf{S}_t\|}{\|\mathbf{S}_s \times \mathbf{S}_t\|} dA$$

$$= \iint_R \mathbf{F}(\mathbf{S}(s,t)) \cdot \frac{(\mathbf{S}_s \times \mathbf{S}_t)}{\|\mathbf{S}_s \times \mathbf{S}_t\|} \|\mathbf{S}_s \times \mathbf{S}_t\| \, dA$$

$$= \iint_S \mathbf{F}(\mathbf{S}(s,t)) \cdot \frac{(\mathbf{S}_s \times \mathbf{S}_t)}{\|\mathbf{S}_s \times \mathbf{S}_t\|} dS$$

$$= \iint_S comp_{\mathbf{n}} \mathbf{F} dS. \blacksquare$$

Note that Theorem 5.4.1 interprets the flux of a vector field across a surface as the integral of a scalar function over the surface as defined in Section 4.6. Thus we see that the integral of a function over a surface can provide interesting information, depending on what the integrand is.

Remark: We take the time to point out two consequences of Theorem 5.4.1.

- If $\mathbf{F}$ is always tangent to the surface S, then the component of $\mathbf{F}$ in the normal direction is zero, so $\iint_S \mathbf{F} \cdot d\mathbf{S} = 0$.
- If $comp_{\mathbf{n}}\mathbf{F} = c$ and S has area A, then $\iint_S \mathbf{F} \cdot d\mathbf{S} = \iint_S c \, dS = cA$ since $\iint_S dS$ is the area of S.

We now consider several applications of these remarks.

Example 5.4.3. *Tangential* $\mathbf{F}$ *means* $\iint_S \mathbf{F} \cdot d\mathbf{S} = 0$

Show that the flux of the vector field $\mathbf{F}(x,y,z) = \langle x, y, -2z \rangle$ across any portion of the surface $(x^2 + y^2)z = 1$ is zero.

Our approach will be to show that $comp_{\mathbf{n}}\mathbf{F} = 0$. The surface is the level surface of the function $g(x,y,z) = (x^2 + y^2)z$ at level one. Since gradients are perpendicular to level surfaces, a normal vector to the surface is

$$\mathbf{n} = \nabla g = \langle 2xz, 2yz, x^2 + y^2 \rangle.$$

But then

$$\mathbf{F} \cdot \mathbf{n} = \langle x, y, -2z \rangle \cdot \langle 2xz, 2yz, x^2 + y^2 \rangle = 2x^2 z + 2y^2 z - 2z \left(x^2 + y^2 \right) = 0,$$

which implies $\mathbf{F}$ is tangent to the surface everywhere, and the flux is zero (see Figure 5.4.6. ▲

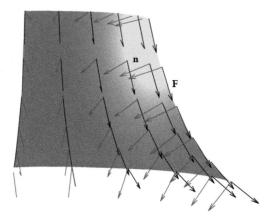

Figure 5.4.6 The flux of tangent field is zero

In this last example the gradient ∇g was perpendicular to $\mathbf{F}$ for *all* points (x, y, z). We never had to use the fact that $(x^2 + y^2)z = 1$ to make the dot product $\mathbf{F} \cdot \mathbf{n}$ equal zero. This means that the flux across *every* level surface of g will be zero, not just $g(x, y, z) = 1$. We include one more example which requires one to evaluate $\mathbf{F}$ on the given surface.

Example 5.4.4. *Another tangent field with zero flux.*

Show that the vector field $\mathbf{F} = \langle x - yz, y + xz, z \rangle$ has zero flux across the cone $z = \sqrt{x^2 + y^2}$.

To do this we'll show that $\mathbf{F}$ is tangent to the cone by showing it is perpendicular to the normal vector. Since the cone is the graph of a function $f(x, y) = \sqrt{x^2 + y^2}$, normal vectors are scalar multiples of

$$\langle -f_x, -f_y, 1 \rangle = \left\langle -\frac{x}{\sqrt{x^2 + y^2}}, -\frac{y}{\sqrt{x^2 + y^2}}, 1 \right\rangle.$$

We multiply by the scalar $\sqrt{x^2 + y^2}$ and use the normal vector

$$\mathbf{n} = \left\langle -x, -y, \sqrt{x^2 + y^2} \right\rangle.$$

Calculating the dot product gives

$$\mathbf{F} \cdot \mathbf{n} = \langle x - yz, y + xz, z \rangle \cdot \left\langle -x, -y, \sqrt{x^2 + y^2} \right\rangle = -x^2 - y^2 + z\sqrt{x^2 + y^2}.$$

This is not zero everywhere, but we are only concerned with the field on the cone. Therefore we substitute $\sqrt{x^2 + y^2}$ for z in the dot product, obtaining

$$\mathbf{F} \cdot \mathbf{n} = -x^2 - y^2 + z\sqrt{x^2 + y^2} = -x^2 - y^2 + \left(\sqrt{x^2 + y^2}\right)^2 = 0. \ \blacktriangle$$

Example 5.4.5. *Constant normal component*

Compute the flux of

$$\mathbf{F}(x, y, z) = \left\langle \frac{x}{\sqrt{x^2 + y^2 + z^2}}, \frac{y}{\sqrt{x^2 + y^2 + z^2}}, 0 \right\rangle$$

across the portion of the cone $z = \sqrt{x^2 + y^2}$ beneath the plane $z = 1$.

To do so we will show that $\mathbf{F}$ has a constant component in the normal direction to the cone, then multiply by the area.

A normal vector to the surface $z = \sqrt{x^2 + y^2}$ is

$$\mathbf{n} = \left\langle -\frac{x}{\sqrt{x^2 + y^2}}, -\frac{y}{\sqrt{x^2 + y^2}}, 1 \right\rangle,$$

which has length $\|\mathbf{n}\| = \sqrt{2}$ (verify this!). Evaluating $\mathbf{F}$ on the cone $z = \sqrt{x^2 + y^2}$ gives

$$\mathbf{F}(x, y, \sqrt{x^2 + y^2}) = \left\langle \frac{x}{\sqrt{2(x^2 + y^2)}}, \frac{y}{\sqrt{2(x^2 + y^2)}}, 0 \right\rangle.$$

Therefore the component of $\mathbf{F}$ normal to the surface is

$$comp_n\mathbf{F} = \mathbf{F}(x, y, \sqrt{x^2 + y^2}) \cdot \frac{\mathbf{n}}{\|\mathbf{n}\|}$$

$$= \left\langle \frac{x}{\sqrt{2(x^2 + y^2)}}, \frac{y}{\sqrt{2(x^2 + y^2)}}, 0 \right\rangle \cdot \frac{1}{\sqrt{2}} \left\langle -\frac{x}{\sqrt{x^2 + y^2}}, -\frac{y}{\sqrt{x^2 + y^2}}, 1 \right\rangle$$

$$= -\frac{1}{\sqrt{2}} \frac{x^2 + y^2}{\sqrt{2(x^2 + y^2)}} = -\frac{1}{2}$$

Therefore the component of $\mathbf{F}$ in the normal direction is constant. Since the lateral surface area of a cone radius r and height h is $A_L = 2\pi r\sqrt{r^2 + h^2}$, and in our case $r = h = 1$, we calculate the flux to be

$$\iint_S \mathbf{F} \cdot d\mathbf{S} = -\frac{1}{2} 2\pi \sqrt{2} = -\pi\sqrt{2}. \ \blacktriangle$$

We conclude with an important example, the flux of an electric field due to a point charge across a sphere.

Example 5.4.6. *Electric flux across a sphere*

Find the flux of

$$\mathbf{E}(x,y,z) = \left\langle \frac{x}{\left(x^2+y^2+z^2\right)^{3/2}}, \frac{y}{\left(x^2+y^2+z^2\right)^{3/2}}, \frac{z}{\left(x^2+y^2+z^2\right)^{3/2}} \right\rangle$$

across the sphere S radius 2, centered at the origin, with outward-pointing normal.

To do so, we calculate the component of $\mathbf{E}$ in the normal direction of S. On the sphere we have $x^2+y^2+z^2 = 4$, so evaluating $\mathbf{E}$ on S gives

$$\mathbf{E}(x,y,z) = \left\langle \frac{x}{(4)^{3/2}}, \frac{y}{(4)^{3/2}}, \frac{z}{(4)^{3/2}} \right\rangle = \frac{1}{8}\langle x,y,z \rangle.$$

Moreover, an outward-pointing normal vector is $\mathbf{n} = \langle x,y,z \rangle$, with length $\|\mathbf{n}\| = 2$ since (x,y,z) is on S. Using these observations we calculate

$$comp_{\mathbf{n}}\mathbf{E} = \mathbf{E} \cdot \frac{\mathbf{n}}{\|\mathbf{n}\|} = \frac{1}{8}\langle x,y,z \rangle \cdot \frac{\langle x,y,z \rangle}{2} = \frac{1}{16}\left(x^2+y^2+z^2\right) = \frac{1}{4}.$$

The last equality follows from evaluating the component on the surface (since the surface, in this case, is $x^2+y^2+z^2 = 4$). Since $comp_{\mathbf{n}}\mathbf{E}$ is constant we can multiply it by surface area to evaluate the flux. Further, the area of S is 16π, so

$$\iint_S \mathbf{E} \cdot d\mathbf{S} = cA = \frac{1}{4}16\pi = 4\pi. \ \blacktriangle$$

Exercises

1. Evaluate the flux of $\mathbf{F} = \langle 1,1,1 \rangle$ across the portion of the plane $z = xy$ above the unit disk.

2. Evaluate the flux of $\mathbf{F} = \langle 2x - z, x + 3z, y - z \rangle$ across the paraboloid $z = 4 - x^2 - y^2$ above the rectangle $[-1,1] \times [-1,1]$.

3. Evaluate the flux of $\mathbf{F} = \langle x, z^2, 2y - x \rangle$ across the portion of the cone $z = 1 - \sqrt{x^2+y^2}$ above the unit disk.

4. Evaluate the flux of $\mathbf{F} = \langle xy, z^2 - y, x^2 - 2zy \rangle$ across the portion of the plane $2x + 3y + z = 6$ in the first octant.

5. Evaluate the flux of $\mathbf{F} = \langle x, y, z \rangle$ across the portion of the paraboloid $z = 3 + x^2 + y^2$ above the unit disk.

6. Evaluate the flux of $\mathbf{F} = \langle 3x + 2y, z, y \rangle$ across the portion of the plane $2x - y + z = 6$ above the rectangle $[0, 1] \times [0, 1]$.

7. Evaluate the flux of $\mathbf{F} = \langle x - 2y, x + y, z \rangle$ across the portion of the plane $y + z = 1$ inside the cylinder $x^2 + y^2 = 1$.

8. Evaluate the flux of $\mathbf{F} = \langle y + z, x + z, y \rangle$ across the portion of the saddle $z = x^2 - y^2$ above the rectangle $[0, 1] \times [0, 1]$.

9. Find the flux of $\mathbf{F} = \langle x^2, y + z, x - z \rangle$ across the parallelogram with vertices $(0, 0, 0)$, $(1, 2, -1)$, $(3, 2, 1)$ and $(4, 4, 0)$.

10. Find the flux of $\mathbf{F} = \langle 2x - y, 3y + 2z, z - 2x \rangle$ across the parallelogram with vertices $(1, 0, -3)$, $(2, 4, 0)$, $(-2, 1, 3)$, and $(-1, 5, 0)$.

11. Parameterize the portion of the cone $z = \sqrt{x^2 + y^2}$ above the unit disk, and find the flux of $\mathbf{F} = \langle y, -z, x \rangle$ across it.

12. Evaluate the flux of $\mathbf{F} = \langle 0, 0, 3 \rangle$ across the parametric surface $S(s, t) = (2s - 3t + 4, s + 2t, 3s - t)$ for $0 \le s \le 1$ and $0 \le t \le 1$.

13. Evaluate the flux of $\mathbf{F} = \langle 1, 1, 2 \rangle$ across the parametric surface $S(s, t) = (t + s, t - s, s)$ for $0 \le s \le 1$ and $0 \le t \le 1$.

14. Evaluate the flux of $\mathbf{F} = \langle xy, 2z, z + 2x \rangle$ across the parametric surface $S(s, t) = (2 - t + 3s, s, t - s)$ for $0 \le s \le 1$ and $0 \le t \le 1$.

15. Evaluate the flux of $\mathbf{F} = \langle y^2 - x, z, y \rangle$ across the parametric surface $S(s, t) = (t^2 - s, st, 2 + s^2)$ for $0 \le s, t \le 1$.

16. Evaluate the flux of $\mathbf{F} = \frac{1}{(x^2 + y^2 + z^2)^{3/2}} \langle -x, -y, -x \rangle$ across the parametric surface $S(s, t) = (\cos t, \sin t, s)$ for $-1 \le s \le 1$ and $0 \le t \le \pi/2$.

17. Evaluate the flux of $\mathbf{F} = \langle y, -x, z^2 \rangle$ across the parametric surface $S(s, t) = (s \cos t, s \sin t, t)$ for $0 \le s \le 1$ and $0 \le t \le \pi$.

18. Evaluate the flux of $\mathbf{F} = \langle xz^2, yz^2, z^3 \rangle$ across the unit sphere $S(s, t) = (\sin s \cos t, \sin s \sin t, \cos s)$, for $0 \le s \le \pi$ and $0 \le t \le 2\pi$.

19. Find the flux of $\mathbf{F} = \langle z, xy, x + y + 2 \rangle$ across the portion of the paraboloid $z = x^2 + y^2$ that lies above the unit disk.

20. A current moving positively along the y-axis induces the magnetic field

$$\mathbf{B}(x, y, z) = \left\langle \frac{z}{x^2 + z^2}, 0, \frac{-x}{x^2 + z^2} \right\rangle.$$

Find the magnetic flux of $\mathbf{B}$ across the surface $S(s, t) = (t + 1, s, t - 1)$ for $0 \le s, t \le 1$.

21. A point charge placed at the origin induces (a scalar multiple of) the electric field

$$\mathbf{E}(x,y,z) = \left(\frac{x}{\left(x^2 + y^2 + z^2\right)^{3/2}}, \frac{y}{\left(x^2 + y^2 + z^2\right)^{3/2}}, \frac{z}{\left(x^2 + y^2 + z^2\right)^{3/2}} \right).$$

Find the flux of $\mathbf{E}$ across a sphere radius R centered at the origin.

22. Show that the flux of $\mathbf{F}(x,y,z) = \langle zy, xz, 2xy \rangle$ across any portion of the one-sheeted hyperboloid $x^2 + y^2 - z^2 = 1$ is zero.

23. Show that the flux of $\mathbf{F}(x,y,z) = \langle x,y,z \rangle$ across any portion of the cone $S(s,t) = (s\cos t, s\sin t, s)$, $0 \le s$, $0 \le t \le 2\pi$ is zero.

24. Show that the flux of $\mathbf{F}(x,y,z) = \langle x,y,2z \rangle$ across any portion of the paraboloid $z = x^2 + y^2$ is zero.

25. Show that the flux of $\mathbf{F}(x,y,z) = \langle 1 - z - x^2, -xy, x - xz \rangle$ across any portion of the unit sphere is zero.

26. Show that the flux of $\mathbf{F}(x,y,z) = \langle -xy, 1 - z - y^2, y - yz \rangle$ across any portion of the unit sphere is zero. (The vector field restricted to the sphere is the image of $\mathbf{j}$ under stereographic projection.)

5.5 Differentiating Vector Fields

The physical notions of work and flux have motivated integration of vector fields along curves and across surfaces. The gradient of a function is its derivative, and The fundamental theorem for line integrals shows that integrating a conservative vector field amounts to finding an antiderivative. There is a similar interpretation of Green's theorem. The line integral of Green's theorem can be thought of as resulting from performing one integration of the double integral. In this sense, integrating the scalar curl just gives the vector field back (take a look at Theorem 5.3.2 while reading that sentence). Since integration is differentiation backwards, this means that the scalar curl is the derivative of a vector field. Analogously, the flux form of Green's theorem suggests that divergence is another way to differentiate vector fields (see Theorem 5.3.3). In this section we make this more formal, and take a look at vector differentiation.

One notational convention that is surprisingly useful in defining vector derivatives is the "del" operator:

$$\nabla = \frac{\partial}{\partial x}\mathbf{i} + \frac{\partial}{\partial y}\mathbf{j} + \frac{\partial}{\partial z}\mathbf{k}$$

It is a vector whose components are differential operators, so ∇ by itself really has no geometric meaning. Algebraically, however, we will treat it like a vector. The operations of scalar multiplication, dot product, and cross product all have nice interpretations when generalized to the del operator. Each of these products—scalar multiplication, dot product, cross product—with the del operator define different vector derivatives. While the definitions can be reduced to algebraic manipulation, it is worth the effort to develop some intuition for the meaning of each.

Scalar Multiplication—the Gradient Revisited: In order to define a scalar times ∇ we remember that ∇ "multiplies" functions. Given a scalar-valued function $f(x,y,z)$, we define the product ∇f by

$$\nabla f = \left(\frac{\partial}{\partial x}\mathbf{i} + \frac{\partial}{\partial y}\mathbf{j} + \frac{\partial}{\partial z}\mathbf{k}\right)f = \frac{\partial f}{\partial x}\mathbf{i} + \frac{\partial f}{\partial y}\mathbf{j} + \frac{\partial f}{\partial z}\mathbf{k}.$$

This is, of course, our old friend the gradient of f. That's good, because we don't want the notation ∇f to mean something different in this context.

Multiplying the del operator by the scalar function f provides a familiar setting from which to see how ∇ behaves algebraically. The del operator behaves like a vector, but multiplication becomes differentiation. Instead of multiplying each component of ∇ by the scalar f, each component of ∇f is the appropriate derivative of f—multiplication becomes differentiation.

Even though Chapter 3 introduces the gradient from a computational point of view, we saw that it encodes a lot of geometric information. In particular, it facilitates the chain rule for differentiation and determines the rate of change of f in any given direction. Thus an algebraic convenience has geometric and physical significance.

Dot Products—Divergence: We now interpret the dot product of ∇ with a vector. While interpreting scalar multiplication with the del operator we saw that "scalars" are really functions of several variables. Analogously, the "vector" to be dotted with ∇ should be a *vector field*. Given a vector field $\mathbf{F}$, the dot product $\nabla \cdot \mathbf{F}$ will be called the *divergence* of $\mathbf{F}$. We begin with an algebraic example and definition of divergence, then provide some intuition for what the divergence of a vector field calculates.

Example 5.5.1. *A first look at divergence*

Let $\mathbf{F} = \left\langle x^2 y, \sin z e^{y^2}, 3z^2 \cos x \right\rangle$, and evaluate $\nabla \cdot \mathbf{F}$. Remembering to differentiate rather than multiply, we see

$$
\begin{aligned}
\nabla \cdot \mathbf{F} &= \left(\frac{\partial}{\partial x}, \frac{\partial}{\partial y}, \frac{\partial}{\partial z} \right) \cdot \left\langle x^2 y, e^{y^2} \sin z, 3z^2 \cos x \right\rangle \\
&= \frac{\partial}{\partial x} \left(x^2 y \right) + \frac{\partial}{\partial y} \left(e^{y^2} \sin z \right) + \frac{\partial}{\partial z} \left(3z^2 \cos x \right) \\
&= 2xy + 2y e^{y^2} \sin z + 6z \cos x. \ \blacktriangle
\end{aligned}
$$

Thus the "del operator" performs specified differentiation operations on vector fields. We make the following definition:

Definition 5.5.1. *Let* $\mathbf{F} = \langle P, Q, R \rangle$ *be a vector field. The divergence* $\mathrm{div}\mathbf{F}$ *of* $\mathbf{F}$ *is*

$$
\mathrm{div}\mathbf{F} = \nabla \cdot \mathbf{F} = \left(\frac{\partial}{\partial x}, \frac{\partial}{\partial y}, \frac{\partial}{\partial z} \right) \cdot \langle P, Q, R \rangle = \frac{\partial P}{\partial x} + \frac{\partial Q}{\partial y} + \frac{\partial R}{\partial z}.
$$

For vector fields in the plane, just ignore the third coordinate in the above definition. If our vector field is the velocity field of a fluid, one can think of divergence as a measure of the tendency of the fluid to compress or expand at a given point. If $\nabla \cdot \mathbf{F} = 0$, the field $\mathbf{F}$ is sometimes called *incompressible* for this reason. Such a field is also called divergence free.

Example 5.5.2. *Some vector fields in the plane*

When discussing the scalar curl of a vector field we considered the three simple cases of Figure 5.3.5. The divergence of $\mathbf{F} = \langle x, y \rangle$ is

$$
\mathrm{div}\mathbf{F} = \nabla \cdot \mathbf{F} = \frac{\partial}{\partial x}(x) + \frac{\partial}{\partial y}(y) = 2,
$$

indicating that the field tends to expand everywhere. This makes intuitive sense by considering Figure 5.3.5(a).

(a)   (b)

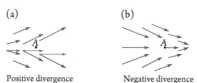

Positive divergence   Negative divergence

Figure 5.5.1 Divergence intuition

One also immediately calculates that $\mathbf{F} = \langle -y, x \rangle$ is divergence free since

$$\mathrm{div}\mathbf{F} = \nabla \cdot \mathbf{F} = \frac{\partial}{\partial x}(-y) + \frac{\partial}{\partial y}(x) = 0,$$

so that Figure 5.3.5(b) depicts an incompressible field. ▲

It is worthwhile to see how the partial derivatives in the divergence formula might measure changes in fluid density (i.e. fluid compression or expansion). To do so, consider the flow whose velocity field $\mathbf{F} = \langle P, Q \rangle$ is depicted in 5.5.1(a). Evidently the fluid is accelerating as you move from left to right, causing expansion near the point A. We wish to consider the partial derivative P_x at A. Now P is the horizontal component of $\mathbf{F}$, which is the signed lengths of the projections in 5.5.1(a). These lengths become more and more positive as x increases, therefore P is an increasing function of x and $\partial P / \partial x > 0$.

A similar analysis on the vertical component shows that if $\partial Q / \partial y > 0$ then the flow is expanding. (Just turn your head sideways and look at Figure 5.5.1 –or keep the figure as is, sketch in the vertical components and make sure their *signed* lengths are increasing as you move up past A.) This reasoning makes it plausible to conclude that if

$$\mathrm{div}\mathbf{F} = \frac{\partial P}{\partial x} + \frac{\partial Q}{\partial y} > 0,$$

the fluid is expanding.

On the other hand, Figure 5.5.1(b) shows a vector field that is compressing near the point A. The horizontal components (not pictured) are seen to be decreasing as you travel from left to right. This provides intuitive evidence that the flow is compressing when $P_x < 0$. Applying similar reasoning to the vertical component indicates that if div $\mathbf{F} < 0$ at a point, the fluid is compressing there.

Math App 5.5.1. *Area and incompressibility*

If a fluid is incompressible, then the area of a region remains constant as it flows, even though its shape can change dramatically. The following Math App takes a brief look at this phenomenon.

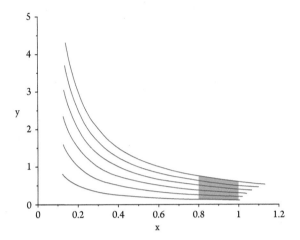

Cross products—curl: In Section 5.3 we spent some time showing that the scalar curl $Q_x - P_y$ measured the tendency of a fluid flow to induce local rotations. We now generalize the notion of two-dimensional scalar curl to the curl of a vector field in $\mathbb{R}^3$. It turns out that this corresponds to the cross product $\nabla \times \mathbf{F}$ of the del operator with a vector field. We motivate the curl using rotations, then give the algebraic definition.

Rotation in space can be considered a vector quantity, call it $\boldsymbol{\omega}$. The vector $\boldsymbol{\omega}$ points along the axis of rotation satisfying the right-hand rule (so that as your right thumb points along $\boldsymbol{\omega}$, your fingers curl around the axis in the direction of the rotating fluid). The length $\|\boldsymbol{\omega}\|$ is the angular velocity of the rotation.

When the vector field $\mathbf{F} = \langle P, Q, R \rangle$ is the velocity of a flow, for example water in a pool or air in a room, it can induce local rotations. Indeed, imagine a tiny spherical balloon with enough helium to hover at exactly one point. Imagine it stays in the same place, but air flow could cause it to rotate while hovering there. Our goal is to find a vector that measures that rotation, which we will call the curl of $\mathbf{F}$ and denote by curl$\mathbf{F}$.

Since curl$\mathbf{F}$ is a vector, we can write it as a linear combination of the standard basis vectors $\mathbf{i}, \mathbf{j}, \mathbf{k}$:

$$\text{curl}\mathbf{F} = A\mathbf{i} + B\mathbf{j} + C\mathbf{k}.$$

Now curl$\mathbf{F}$ is measuring rotation, so each component in this sum measures a part of that rotation. Let's describe the rotation that $C\mathbf{k}$ measures first.

Pointing your right thumb along the vector $\mathbf{k}$, the fingers of your right hand curl counterclockwise in the xy-plane (from the positive x-axis to the positive y-axis). Thus the component $C\mathbf{k}$ should measure the rotation of curl$\mathbf{F}$ in the xy-plane. This is exactly what the scalar curl of Section 5.3 does! Therefore

$$C = \frac{\partial Q}{\partial x} - \frac{\partial P}{\partial y}.$$

Similar reasoning can be used to find the **i** and **j** components of curl**F**. Placing your right thumb along **j** your fingers curl from the positive z-axis to the positive x-axis (don't be lazy, try this for yourself). Now reason as above replacing x, y with z, x in that order because of the direction your fingers are curling in each case. Making this substitution, and arguing analogously for **i**, we get

$$B = \frac{\partial P}{\partial z} - \frac{\partial R}{\partial x} \quad \text{and} \quad A = \frac{\partial R}{\partial y} - \frac{\partial Q}{\partial z}.$$

You should point your right thumb in the **i** direction and confirm that your fingers curl in such a way as to give the A above.

Our goal was to quantify local rotations induced by a fluid flow. The idea was to decompose curl**F** into a sum of three components, each component measuring a two-dimensional rotation. Then we used scalar curl, which measures two-dimensional rotation, to quantify each component. The result is

$$\text{curl}\mathbf{F} = \left(\frac{\partial R}{\partial y} - \frac{\partial Q}{\partial z} \right) \mathbf{i} + \left(\frac{\partial P}{\partial z} - \frac{\partial R}{\partial x} \right) \mathbf{j} + \left(\frac{\partial Q}{\partial x} - \frac{\partial P}{\partial y} \right) \mathbf{k}. \tag{5.5.1}$$

It turns out that the del operator gives a particularly handy way to remember this formula, which we use as the definition of curl.

Definition 5.5.2. *The curl* curl**F** *of the vector field* **F** *is defined to be*

$$\text{curl}\mathbf{F} = \nabla \times \mathbf{F}.$$

The above definition is the easy way to remember how to calculate Equation 5.5.1. I don't memorize the formula; I just remember that curl is a determinant. For completeness we illustrate the calculation in general. If $\mathbf{F} = \langle P, Q, R \rangle$, then we compute

$$\text{curl}\mathbf{F} = \nabla \times \mathbf{F} = \begin{vmatrix} \mathbf{i} & \mathbf{j} & \mathbf{k} \\ \frac{\partial}{\partial x} & \frac{\partial}{\partial y} & \frac{\partial}{\partial z} \\ P & Q & R \end{vmatrix}$$

$$= \left(\frac{\partial R}{\partial y} - \frac{\partial Q}{\partial z} \right) \mathbf{i} + \left(\frac{\partial P}{\partial z} - \frac{\partial R}{\partial x} \right) \mathbf{j} + \left(\frac{\partial Q}{\partial x} - \frac{\partial P}{\partial y} \right) \mathbf{k}. \blacktriangle$$

Again, remember that instead of multiplying we differentiate when the del operator is involved.

Thus the curl of a vector field is another vector field. It is not surprising that if $\mathbf{F} = \langle P, Q, 0 \rangle$ is a vector field in the plane (so P and Q are functions of only x and y), then

$$\text{curl}\mathbf{F} = \left(\frac{\partial Q}{\partial x} - \frac{\partial P}{\partial y} \right) \mathbf{k}.$$

Analytically, this follows from the fact that all the partial derivatives in the first two components of curlF are zero in this instance. Therefore the scalar curl of a planar vector field **F** is the coefficient of **k** in curlF. Conceptually this follows from our derivation of curlF as a sum of two-dimensional curls.

Example 5.5.3. *Calculating curl*

Let $\mathbf{F} = \langle xy^2, x^2z, 3yz\rangle$ and calculate curlF.

$$\text{curl}\mathbf{F} = \nabla \times \mathbf{F} = \begin{vmatrix} \mathbf{i} & \mathbf{j} & \mathbf{k} \\ \frac{\partial}{\partial x} & \frac{\partial}{\partial y} & \frac{\partial}{\partial z} \\ xy^2 & x^2z & 3yz \end{vmatrix}$$

$$= \left(\frac{\partial}{\partial y}3yz - \frac{\partial}{\partial z}x^2z\right)\mathbf{i} + \left(\frac{\partial}{\partial z}xy^2 - \frac{\partial}{\partial x}3yz\right)\mathbf{j} + \left(\frac{\partial}{\partial x}x^2z - \frac{\partial}{\partial y}xy^2\right)\mathbf{k}$$

$$= (3z - x^2)\mathbf{i} + (2xz - 2xy)\mathbf{k}. \ \blacktriangle$$

The curl of a vector field measures the direction and speed of its tendency to rotate. Intuitively, if **F** is the velocity field of a fluid and we place a top in it, the top would tilt to a specific direction and start spinning. The vector curlF points in the direction that the top's axis would point, and the length of curlF measures the speed of rotation. For this reason, if curlF $= \mathbf{0}$ the vector field **F** is called *irrotational*.

Second Derivatives and the Laplace Operator: The operations of scalar multiplication, dot product, and cross product with the del operator all led to first derivatives. The objects being differentiated differ (scalar multiplication differentiates functions, while dot and cross products differentiate vector fields), but each operation involves a single level of differentiation. We briefly comment on second derivatives and the del operator.

Not every symbolic second derivative makes sense. For example, the derivative $\nabla \times (\nabla \cdot \mathbf{F})$ can't be calculated because the divergence $\nabla \cdot \mathbf{F}$ is a scalar function, which doesn't have a curl. Letting f be a scalar-valued function and **F** a vector field on $\mathbb{R}^3$, the five second derivatives that make sense are

- Divergence of the gradient: $\nabla \cdot (\nabla f)$.
- Curl of the gradient: $\nabla \times (\nabla f)$.
- Gradient of the divergence: $\nabla (\nabla \cdot \mathbf{F})$.
- Divergence of the curl: $\nabla \cdot (\nabla \times \mathbf{F})$.
- Curl of the curl: $\nabla \times (\nabla \times \mathbf{F})$.

You should make sure that the above operations are all defined. Make sure, for example, that you never take the curl of a function or the gradient of a vector field in the process.

Two of these second derivatives are always zero, namely

$$\nabla \times (\nabla f) = \mathbf{0} \quad \text{and} \quad \nabla \cdot (\nabla \times \mathbf{F}) = 0.$$

The proofs are left as exercises. Note that one of the derivatives is a vector while the other is a scalar.

Stokes' theorem, to be discussed in Section 5.6, proves a converse of the fact that $\nabla \times (\nabla f) = \mathbf{0}$. More precisely, if $\mathbf{F}$ is a vector field with $\operatorname{curl}\mathbf{F} = \nabla \times \mathbf{F} = \mathbf{0}$, then $\mathbf{F}$ is conservative and has a potential function (i.e. $\mathbf{F} = \nabla f$). This gives another way to test if $\mathbf{F}$ is conservative or not. If the curl is the zero vector, then $\mathbf{F}$ is conservative; otherwise, it's not.

The divergence of the gradient, $\nabla \cdot (\nabla f)$, is so important that it gets its own name and notation. It's called the *Laplacian* of f, and denoted $\nabla^2 f$. Thus the Laplacian of a function is

$$\nabla^2 f = \nabla \cdot (\nabla f) = \nabla \cdot \left(\frac{\partial f}{\partial x}, \frac{\partial f}{\partial y}, \frac{\partial f}{\partial z} \right) = \frac{\partial^2 f}{\partial x^2} + \frac{\partial^2 f}{\partial y^2} + \frac{\partial^2 f}{\partial x^2}.$$

The operator ∇^2 is called the Laplace operator when no reference is made to a particular function.

Example 5.5.4. *Electric potential*

The electric potential V a distance ρ units from a point charge Q is

$$V = \frac{kQ}{\rho},$$

where k is Coulomb's constant. Find the Laplacian of $\nabla^2 V$.

If Q is at the origin, then in Cartesian coordinates we have

$$V = \frac{kQ}{\sqrt{x^2 + y^2 + z^2}}.$$

Calculating the first partial derivative with respect to x gives

$$\frac{\partial V}{\partial x} = -2kQ \frac{x}{(x^2 + y^2 + z^2)^{3/2}},$$

and the second partial derivative with respect to x is

$$\frac{\partial^2 V}{\partial x^2} = 2kQ \frac{2x^2 - y^2 - z^2}{(x^2 + y^2 + z^2)^{5/2}}.$$

Similar calculations show that the repeated partials with respect to y and z are

$$\frac{\partial^2 V}{\partial y^2} = 2kQ \frac{2y^2 - x^2 - z^2}{(x^2 + y^2 + z^2)^{5/2}} \quad \text{and} \quad \frac{\partial^2 V}{\partial z^2} = 2kQ \frac{2z^2 - x^2 - y^2}{(x^2 + y^2 + z^2)^{5/2}}.$$

The numerators of these second partials sum to zero, so the Laplacian of V is

$$\nabla^2 V = \nabla \cdot (\nabla f) = \frac{\partial^2 V}{\partial x^2} + \frac{\partial^2 V}{\partial y^2} + \frac{\partial^2 V}{\partial x^2} = 0. \ \blacktriangle$$

The electric potential is a special kind of function, its Laplacian is zero. In general, the equation $\nabla^2 f = 0$ is called Laplace's equation, and any solution to it is called a *harmonic function*. The above example illustrates that the electric potential due to a point charge is a harmonic function.

It should also be mentioned that there is a Laplace operator in each dimension. If $f(x)$ is a function of one variable, then $\nabla^2 f$ is just the second derivative of f. The Laplacian of a function $f(x,y)$ of two variables is

$$\nabla^2 f = \frac{\partial^2 f}{\partial x^2} + \frac{\partial^2 f}{\partial y^2}.$$

Example 5.5.5. *Two-variable Laplacian*

Calculate the Laplacian of $f(x,y) = \frac{1}{\sqrt{x^2+y^2}}$.
The first partial derivative with respect to x is

$$\frac{\partial f}{\partial x} = -\frac{x}{(x^2+y^2)^{3/2}},$$

and the second is

$$\frac{\partial^2 f}{\partial x^2} = \frac{2x^2 - y^2}{(x^2+y^2)^{5/2}}.$$

Analogously differentiating with respect to y gives

$$\begin{aligned}
\nabla^2 f &= \frac{2x^2 - y^2}{(x^2+y^2)^{5/2}} + \frac{-x^2 + 2y^2}{(x^2+y^2)^{5/2}} \\
&= \frac{x^2 + y^2}{(x^2+y^2)^{5/2}} = \frac{1}{(x^2+y^2)^{3/2}}.
\end{aligned}$$

Thus a two-dimensional analog of electric potential does not satisfy Laplace's equation. In other words, this function is not harmonic. $\blacktriangle$

The Laplace operator makes an appearance in many physical settings. It plays a role in the heat equation, the wave equation, fluid flow, and many other settings. Harmonic functions, which are solutions to $\nabla^2 f = 0$, have the interesting property that any extreme values they attain are realized on the boundary of their domains.

Another way to say this is harmonic functions cannot achieve extreme values in the interior of their domain. This follows from the fact that if f is harmonic, then its value at each point is the average of the values around that point. This, in turn, is a nice application of the flux form of Green's theorem, so we present it here. This is really for your enjoyment, so don't stress about the details and try to get the gist of the argument.

Suppose $f(x,y)$ is harmonic near the origin, and let C_r be the circle radius r centered at the origin. The average value of f on C_r is

$$\frac{1}{2\pi r} \oint_{C_r} f \, dC,$$

where this is the integral of a function on a curve defined in Section 4.6.

Using the standard parameterization $\mathbf{C}(t) = (r\cos t, r\sin t)$, so that $\mathbf{C}'(t) = \langle -r\sin t, r\cos t \rangle$, we get that $dC = \|\mathbf{C}'\| = rdt$. The average value of f on C_r becomes

$$\frac{1}{2\pi r} \oint_{C_r} f \, dC = \frac{1}{2\pi r} \int_0^{2\pi} f(r\cos t, r\sin t) r \, dt = \frac{1}{2\pi} \int_0^{2\pi} f(r\cos t, r\sin t) dt,$$

where the last equality comes from factoring the constant r out of the integral. The trick is to show that the average value doesn't depend on r, by showing its derivative with respect to r is zero. Combining the constant nature of the average value with the fact that $f(0,0) = \lim_{r\to 0} \frac{1}{2\pi r} \oint_{C_r} f \, dC$ shows that $f(0,0)$ is the average value of f around small circles near the origin. We proceed with the argument.

To show the average value is constant with respect to r, we differentiate the integral with respect to r

$$\frac{d}{dr} \frac{1}{2\pi r} \oint_{C_r} f \, dC = \frac{d}{dr} \frac{1}{2\pi} \int_0^{2\pi} f(r\cos t, r\sin t) dt$$

$$= \frac{1}{2\pi} \int_0^{2\pi} \nabla f(r\cos t, r\sin t) \cdot \langle \cos t, \sin t \rangle \, dt. \qquad (5.5.2)$$

The final equality is an application of the chain rule which says that

$$\frac{d}{dr} f(r\cos t, r\sin t) = \nabla f(r\cos t, r\sin t) \cdot \langle x_r, y_r \rangle = \nabla f(r\cos t, r\sin t) \cdot \langle \cos t, \sin t \rangle.$$

Returning to Equation 5.5.2, note that $\langle \cos t, \sin t \rangle$ is a unit normal vector to the curve C_r because it points radially outward from the circle. Thus the integrand is simply $comp_\mathbf{n} \nabla f$, which is exactly the integrand for the line integral in Theorem 5.3.3. The component functions for ∇f are, of course, $P = f_x$ and $Q = f_y$. Applying the flux form of Green's theorem we find

$$\frac{d}{dr}\frac{1}{2\pi r}\oint_{C_r} f\,dC = \frac{1}{2\pi}\int_0^{2\pi} \nabla f(r\cos t, r\sin t)\cdot\langle\cos t,\sin t\rangle\,dt$$

$$= \frac{1}{2\pi}\oint_{C_r} comp_{\mathbf{n}}\nabla f\,dC$$

$$= \frac{1}{2\pi}\iint_R \frac{\partial}{\partial x}f_x + \frac{\partial}{\partial y}f_y\,dA \qquad \text{(applying Green's)}$$

$$= \frac{1}{2\pi}\iint_R \frac{\partial^2 f}{\partial x^2} + \frac{\partial^2 f}{\partial y^2}\,dA = 0. \qquad \text{(since } \nabla^2 f = 0\text{)}$$

Whew! Now since the average value of f is the same on all C_r, and since $f(0,0)$ is the limiting average value, we have

$$f(0,0) = \frac{1}{2\pi r}\oint_{C_r} f\,dC.$$

The upshot of this argument is this:

The value of a harmonic function is the average of its nearby values.

An average is never a maximum or minimum (for non-constant functions), so this implies that harmonic functions do not have local extreme values at interior points of their domains.

This makes it seem like harmonic functions are special, and they are. There is one quick method for creating harmonic functions using complex functions. Take any differentiable complex function you like and separate it into its real and imaginary parts. Each of these will be harmonic functions, and in fact they are what's called *harmonic conjugates* because their gradients are perpendicular.

Example 5.5.6. *Harmonic conjugates*

Find the harmonic conjugates corresponding to $f(z) = z^3$.
First we think of $z = x + iy$, and expand

$$z^3 = (x+iy)^3 = x^3 + 3ix^2y - 3xy^2 - iy^3 = x^3 - 3xy^2 + i(3x^2y - y^3).$$

The harmonic conjugates are $\varphi = x^3 - 3xy^2$ and $\psi = 3x^2y - y^3$. We leave it to you to check that they satisfy Laplace's equation. Note that these are the velocity potential and stream functions for an ideal flow around a $\pi/3$-corner. ▲

Fluid flow: The notions of divergence and curl finally give mathematical precision to ideal fluid flows. An ideal fluid flow is one in which the net flux in every volume is zero. Think of hot water flowing through a tea ball. Any water flowing into the ball forces the same amount to flow out somewhere else. Thus the net flux into or out of the ball is zero. If the vector field $\mathbf{F}$ represents the velocity field of an ideal fluid flow, then $\text{div}\mathbf{F} = 0$

everywhere (we prove this in Section 5.7). An ideal flow, by definition, also introduces no local rotation. Thus curl$\mathbf{F} = \mathbf{0}$ everywhere when $\mathbf{F}$ is an ideal flow.

This is a concise description of an ideal flow in terms of its velocity field. The field $\mathbf{F}$ represents an ideal flow when both its curl and divergence vanish.

In Section 1.2 an ideal flow was described in terms of its velocity potential and stream functions, while in Section 2.6 we used velocity fields to describe them. Since ideal flows have vanishing curl, they also have potential functions, thus justifying both perspectives on ideal flows. Moreover, this shows that the velocity field of an ideal flow is the gradient of the velocity potential (actually some consider $-\nabla\varphi$ instead, similar to the electric field being the negative gradient of potential).

We remark that there are many important flows that are not ideal. The flow of air across an airplane's wing compresses, for example, and will have non-zero divergence. Turbulent flows are common in engineering applications, and are not ideal either. In any case, we now have a quick way to determine if a flow is ideal or not from its velocity field.

Example 5.5.7. *An ideal flow*

Which of the following vector fields could model ideal fluid flow where they are defined?

(a) $\mathbf{F}(x,y) = \langle 6xy, 3x^2 - 3y^2 \rangle$
   Direct calculation shows

$$\text{div}\mathbf{F} = \frac{\partial}{\partial x}6xy + \frac{\partial}{\partial y}\left(3x^2 - 3y^2\right) = 6y - 6y = 0,$$

and, since $\mathbf{F}$ is a vector field in the plane we need only compute the scalar curl, or vertical component of curl$\mathbf{F}$:

$$\frac{\partial Q}{\partial x} - \frac{\partial P}{\partial y} = 6x - 6x = 0.$$

Since both curl and divergence vanish, $\mathbf{F}$ models an ideal fluid flow. In fact it is a multiple of the flow around a $\pi/2$-corner that we encountered in Example 1.2.10.

(b) $\mathbf{F}(x,y) = \langle -y, x \rangle$
   The divergence of $\mathbf{F}$ is indeed zero, but the scalar curl is not (what is it?). Therefore this doesn't represent an ideal fluid flow.

(c) $\mathbf{F}(x,y) = \left\langle \frac{x}{\sqrt{x^2+y^2}}, \frac{y}{\sqrt{x^2+y^2}} \right\rangle$
   A tedious calculation shows that the divergence of $\mathbf{F}$ is

$$\text{div}\mathbf{F} = P_x + P_y = \frac{y^2}{(x^2+y^2)^{3/2}} + \frac{x^2}{(x^2+y^2)^{3/2}} = \frac{1}{\sqrt{x^2+y^2}},$$

so it cannot be the velocity field of an ideal flow. ▲

Exercises

1. Determine the curl and divergence of the following vector fields.
   (a) $\mathbf{F} = \langle x^2 - y, 2xz + y^2, z - 2xy \rangle$
   (b) $\mathbf{F} = \langle \sin x, e^y, z^2 \rangle$
   (c) $\mathbf{F} = \langle y^3 z, xz^2, x + 2y \rangle$
   (d) $\mathbf{F} = \langle x - y + 2z, 3x + y - 2z, 3x - 3y + 2z \rangle$

2. Determine whether the following vector fields are conservative. If so, find their potential functions.
   (a) $\mathbf{F} = \langle x^3, y^2, z \rangle$
   (b) $\mathbf{F} = \langle x + y, x - y + z, x + y - z \rangle$

3. Let $f(x), g(y)$, and $h(z)$ be single-variable functions. Show that $\mathbf{F} = \langle f(x), g(y), h(z) \rangle$ is irrotational.

4. Let $f(y,z)$, $g(x,z)$, and $h(x,y)$ be two-variable functions. Show that $\mathbf{F} = \langle f(y,z), g(x,z), h(x,y) \rangle$ is incompressible.

5. Let $\mathbf{F} = \langle P, Q, R \rangle$, where P, Q, and R are all linear functions of x, y, and z (i.e. $P(x,y,z) = Ax + By + Cz$ for some constants A, B, and C). Find linear functions for P, Q, and R so that
   (a) $\mathbf{F}$ is irrotational but not incompressible.
   (b) $\mathbf{F}$ is incompressible but not irrotational.
   (c) $\mathbf{F}$ is both incompressible and irrotational.
   (d) $\mathbf{F}$ is neither incompressible nor irrotational.

6. Show that the divergence of the curl vanishes: $\nabla \cdot (\nabla \times \mathbf{F}) = 0$.

7. Show that the curl of the gradient vanishes: $\nabla \times (\nabla f) = \mathbf{0}$.

8. Which of the following satisfy Laplace's equation?
   (a) $f(x,y) = x^2 y - y^3$.
   (b) $f(x,y) = e^x \cos y$.
   (c) $f(x,y) = \ln(x^2 + y^2)$.
   (d) $f(x,y) = \sin x \cos y$.

9. Show that if f and g are harmonic, so is $f + g$. Argue that the electric potential

$$V = \sum \frac{kQ_i}{\rho_i}$$

for a system of fixed point charges is harmonic.

10. Find harmonic conjugates corresponding to the real and imaginary parts of $f(z) = z^2$.

11. Find harmonic conjugates corresponding to the real and imaginary parts of $f(z) = 1/z$.

12. Find harmonic conjugates corresponding to the real and imaginary parts of $f(z) = e^z$.

13. Find the velocity field for a point source at the origin with velocity potential $\varphi = \ln r$. Show that it is both irrotational and incompressible.

14. Find the velocity field for a point vortex at the origin with velocity potential $\varphi = \theta$. Show that it is both irrotational and incompressible.

15. Find the velocity field, and show it's both irrotational and incompressible, for flow around a π/n-corner with velocity potential $\varphi = r^n \cos n\theta$.

16. Suppose the single-variable function $f(x)$ satisfies Laplace's equation on the interval $[a, b]$ (so $f''(x) = 0$ on $[a, b]$). Show that f does not achieve extreme values in the interior of the interval.

5.6 Stokes' Theorem

We are now in position to discuss Stokes' theorem, which generalizes Green's theorem to surfaces in three dimensions. Recall that Green's theorem states that under the right circumstances

$$\int_C \mathbf{F} \cdot d\mathbf{S} = \iint_R \frac{\partial Q}{\partial x} - \frac{\partial P}{\partial y} \, dA.$$

Green's theorem equates the circulation of $\mathbf{F}$ around a closed loop C to the integral of the scalar curl over the region in the plane that C bounds. Thus it can be thought of as equating a line integral along a closed loop to a surface integral over a planar surface that the loop bounds. In three dimensions the surface need not be planar, and the scalar curl is replaced by curl$\mathbf{F}$.

In Green's theorem it was assumed that C was oriented so that the region R lies to your left. Similarly, there are orientation issues involved with Stokes' theorem. Given a surface S in space, there are two options for choosing a normal direction. For simplicity, we'll say up and down. Let $\mathbf{n}$ be a choice of normal direction for S, then $\mathbf{n}$ induces an orientation on the boundary $C = \partial S$ as follows: choose a direction on C so that if you walk that direction along C, standing in the $\mathbf{n}$ direction, the surface S lies to your left. Given a surface S and choice of normal direction $\mathbf{n}$, we call this the *induced orientation* on C.

Theorem 5.6.1. *Let $\mathbf{F}$ be a smooth vector field and S an oriented surface with boundary curve C having the orientation induced by S. Given a one-to-one parameterization $\mathbf{S}$ of S, we have*

$$\oint_C \mathbf{F} \cdot d\mathbf{C} = \iint_S (curl\mathbf{F}) \cdot d\mathbf{S}.$$

To simplify the proof, we consider the case where S is the graph of a function $z = g(x,y)$ over a region R in the xy-plane. We also assume that S is oriented with the upward-pointing normal $\mathbf{n} = \langle -g_x, -g_y, 1 \rangle$ so that the boundary ∂R of the region R oriented clockwise gets mapped to C, the boundary of S, with its orientation. With these conventions we proceed with the proof.

Proof. The idea of the proof is to translate both integrals into integrals in the plane and apply Green's theorem. There's some serious book-keeping involved, particularly with applications of the chain rule.

Let $\mathbf{F} = \langle P, Q, R \rangle$ be the component functions of $\mathbf{F}$. We calculate the circulation integral of Stokes' theorem first, which is definitely the more challenging calculation.

To evaluate $\oint_C \mathbf{F} \cdot d\mathbf{C}$ note that if $(x(t), y(t))$, $a \le t \le b$, parameterizes ∂R in the xy-plane, then $(x(t), y(t), g(x(t), y(t)))$ parameterizes C (apply the technique of Example 1.5.2). The line integral becomes

$$\oint_C \mathbf{F} \cdot d\mathbf{C} = \int_a^b Px' + Qy' + Rz' \, dt$$

$$= \int_a^b Px' + Qy' + R(g_x x' + g_y y') \, dt \quad \text{(Chain rule on } z' = \frac{d}{dt} g(x(t), y(t)))$$

$$= \oint_{\partial R} (P + Rg_x) \, dx + (Q + Rg_y) \, dy.$$

This last integral is an integral over the curve ∂R in the plane oriented counterclockwise, so we can apply Theorem 5.3.2 to it:

$$\oint_C \mathbf{F} \cdot d\mathbf{C} = \oint_{\partial R} (P + Rg_x) \, dx + (Q + Rg_y) \, dy$$

$$= \iint_R \frac{\partial}{\partial x}(Q + Rg_y) - \frac{\partial}{\partial y}(P + Rg_x) \, dA. \tag{5.6.1}$$

Now when taking partial derivatives that P, Q, R are functions of x, y, and $z = g(x,y)$, so the chain rule must be used. Direct calculation gives

$$\frac{\partial}{\partial x}Q = Q_x + Q_z g_x \quad \text{and} \quad \frac{\partial}{\partial x}Rg_y = (R_x + R_z g_x)g_y + Rg_{yx}.$$

$$\frac{\partial}{\partial y}P = P_y + P_z g_y \quad \text{and} \quad \frac{\partial}{\partial y}Rg_x = (R_y + R_z g_y)g_x + Rg_{xy}.$$

The integrand in Equation 5.6.1 is the top line minus the bottom, and you notice lots of stuff cancels on the right (recall that mixed partials are equal). What remains gives

$$\oint_C \mathbf{F} \cdot d\mathbf{C} = \iint_R Q_x - P_y + Q_z g_x - P_z g_y + R_x g_y - R_y g_x \, dA. \tag{5.6.2}$$

With the difficult calculation in hand we turn our attention to the flux integral of Stokes' theorem and recall that

$$\text{curl}\mathbf{F} = \langle R_y - Q_z, P_z - R_x, Q_x - P_y \rangle.$$

The double integral becomes

$$\iint_S (\text{curl}\mathbf{F}) \cdot d\mathbf{S} = \iint_R \langle R_y - Q_z, P_z - R_x, Q_x - P_y \rangle \cdot \langle -g_x, -g_y, 1 \rangle \, dA$$

$$= \iint_R -(R_y - Q_z)g_x - (P_z - R_x)g_y + (Q_x - P_y) \, dA. \tag{5.6.3}$$

which is seen to equal the circulation integral of Equation 5.6.2. □

So that was the easy version of the proof. The proof for parametric surfaces is notationally more cumbersome, but is basically the same. We content ourselves with the case just described.

Remarks: Here are some conceptual observations before continuing with examples.

- Note that Green's theorem really is a special case of Stokes' theorem. If S is a region in the plane and $\mathbf{F} = \langle P, Q, 0 \rangle$ (with P, Q independent of z), then the normal vector to S is $\mathbf{n} = \langle 0, 0, 1 \rangle$ and the integrand of the surface integral is

$$\text{curl}\mathbf{F} \cdot \mathbf{n} = \langle 0, 0, Q_x - P_y \rangle \cdot \langle 0, 0, 1 \rangle = Q_x - P_y.$$

Thus Stokes' theorem reduces to Green's when the surface is planar.

- In Section 5.3 we mentioned that Green's theorem allows us to interpret the scalar curl as circulation per unit area. In particular we saw

$$\left(\frac{\partial Q}{\partial x}(\mathbf{a}) - \frac{\partial P}{\partial y}(\mathbf{a}) \right) = \lim_{\epsilon \to 0} \frac{1}{\text{Area of } R_\epsilon} \int_C \mathbf{F} \cdot d\mathbf{C},$$

where $\mathbf{a}$ is the center of the disk R_ϵ of radius epsilon.

Stokes' theorem yields a similar interpretation for $\text{curl}\mathbf{F}$. In the planar case, the scalar curl is the component of $\text{curl}\mathbf{F}$ in the normal direction to the surface. In the general case, if $\mathbf{a}$ is a point on the surface S_ϵ and $\mathbf{n}$ the unit normal to the surface at $\mathbf{a}$, analogous reasoning shows

$$\text{curl}\mathbf{F}(\mathbf{a}) \cdot \mathbf{n} = \lim_{\epsilon \to 0} \frac{1}{\text{Area of } S_\epsilon} \int_C \mathbf{F} \cdot d\mathbf{C}.$$

If $\mathbf{F}$ is the velocity of a flow and S a sieve in the flow, attach a pinwheel to S at $\mathbf{a}$. This observation shows the component of $\text{curl}\mathbf{F}$ normal to the surface measures the rate at which the pinwheel spins.

Math App 5.6.1. *Vector curl*

This Math App allows you to visualize how a fluid flow rotates a pinwheel on a fixed axis, and compare how changing the axis (but leaving the flow unchanged) affects how fast the pinwheel spins. Enjoy.

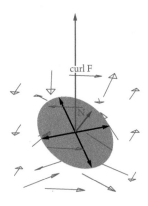

- We make a final remark regarding Stokes' theorem. Notice that the same curve C is the boundary of many surfaces, and that Stokes' theorem applies to all of them that are piecewise smooth and oriented correctly. Suppose that S and $\tilde{S}$ are oriented surfaces that share the same boundary curve C and induce the same orientation on it. Stokes' theorem implies

$$\iint_S (\text{curl}\mathbf{F}) \cdot d\mathbf{S} = \oint_C \mathbf{F} \cdot d\mathbf{C} = \iint_{\tilde{S}} (\text{curl}\mathbf{F}) \cdot d\mathbf{S},$$

or that the flux integral of curl$\mathbf{F}$ across them is the same value—that of the line integral $\oint_C \mathbf{F} \cdot d\mathbf{C}$.

Example 5.6.1. *Verifying Stokes' theorem*

Let $\mathbf{F} = \langle x+y, y-z, z-x \rangle$ and S be the portion of the plane $x+y-z=1$ inside the cylinder $x^2 + y^2 = 1$ oriented by the upward-pointing normal. The boundary curve C is the intersection of the plane and cylinder, and the induced orientation on C is counterclockwise when viewed from above (see Figure 5.6.1). Verify Stokes' theorem in this case by computing both the line and surface integrals.

Computing $\oint_C \mathbf{F} \cdot d\mathbf{C}$

To compute the line integral directly, we first parameterize the boundary curve. The x and y coordinates satisfy $x^2 + y^2 = 1$, so they can be parameterized by $x = \cos t$ and $y = \sin t$, for $0 \le t \le 2\pi$. Now substitute these into the equation $z = x+y-1$ of the plane to find parametric equations for the ellipse. Thus our curve, and its tangent vector, are

$$\mathbf{C}(t) = (\cos t, \sin t, \cos t + \sin t - 1) \text{ for } 0 \le t \le 2\pi$$
$$\mathbf{C}'(t) = \langle -\sin t, \cos t, \cos t - \sin t \rangle.$$

(a)

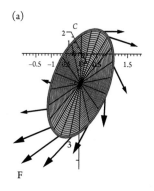

Circulation of **F** around C

(b)

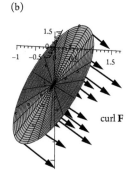

Flux of curl **F** across S

Figure 5.6.1 Verifying Stokes' theorem

The integrand for the line integral is

$$\mathbf{F}(\mathbf{C}(t)) \cdot \mathbf{C}'(t) = \langle \cos t + \sin t, 1 - \cos t, \sin t - 1 \rangle \cdot \langle -\sin t, \cos t, \cos t - \sin t \rangle$$

$$= \sin t - \sin^2 t - 1 = \sin t + \frac{1}{2}\cos 2t - \frac{3}{2},$$

where the last equality follows from the half-angle formula $\sin^2 t = \frac{1}{2}(1 - \cos 2t)$. Integrating gives

$$\int_0^{2\pi} \sin t + \frac{1}{2}\cos 2t - \frac{3}{2}\, dt = -\cos t + \frac{1}{4}\sin 2t - \frac{3}{2}t \Big|_0^{2\pi} = -3\pi.$$

Computing $\iint_S (\mathrm{curl}\mathbf{F}) \cdot d\mathbf{S}$

To compute the surface integral we need to describe S parametrically or as the graph of a function of two variables. In this case, S is the portion of the plane $z = x + y - 1$ lying over the unit disk with upward-pointing normal. We want to evaluate

$$\iint_{\text{unit disk}} \mathrm{curl}\mathbf{F} \cdot \langle -f_x, -f_y, 1 \rangle\, dA.$$

Since in our case $z = f(x,y) = x + y - 1$, we observe that $\langle -f_x, -f_y, 1 \rangle = \langle -1, -1, 1 \rangle$. Calculating the curl gives

$$\mathrm{curl}\mathbf{F} = \begin{vmatrix} \mathbf{i} & \mathbf{j} & \mathbf{k} \\ \frac{\partial}{\partial x} & \frac{\partial}{\partial y} & \frac{\partial}{\partial z} \\ x+y & y-z & z-x \end{vmatrix} = \langle 1, 1, -1 \rangle.$$

Thus the surface integral in Stokes' theorem becomes

$$\iint_{\text{unit disk}} \langle 1, 1, -1 \rangle \cdot \langle -1, -1, 1 \rangle\, dA = \iint_{\text{unit disk}} -3\, dA = -3\pi,$$

since the integral is -3 times the area of the unit disk. Thus, in this case, Stokes' theorem has been verified. ▲

Take care to remember which vector field you're integrating. When using the surface integral of Stokes' theorem you integrate $\mathrm{curl}\mathbf{F}$, but for the line integral you integrate $\mathbf{F}$. Recall that $\mathrm{curl}\mathbf{F}$ measures rotation near points, so placing your right thumb along $\mathrm{curl}\mathbf{F}$ in Figure 5.6.1(b) has your fingers curling clockwise from our vantage point. Points on the surface are spinning in clockwise fashion when viewed from above. Thus we expect $\mathbf{F}$ to travel around the boundary curve C in clockwise fashion, an expectation which is verified in Figure 5.6.1(a). Intuitively, local spinning on the surface induces circulation around the boundary.

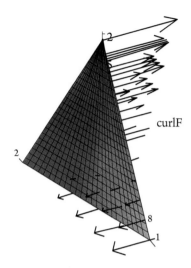

curlF

Figure 5.6.2 Using the surface integral to find circulation

Also note that the surface integral in the previous example was considerably easier than the line integral. The reason is that the curl was particularly simple. If not directed to choose a specific integral to evaluate, investigate which might be easier before embarking on integration.

Example 5.6.2. *Using the surface integral of Stokes' Theorem*

Use the surface integral of Stokes' theorem to evaluate $\int_C \mathbf{F} \cdot d\mathbf{C}$ in the following cases.

(a) Let C be the triangle with vertices $(-1,0,2)$, $(1,1,-3)$, and $(0,3,5)$ oriented counterclockwise when viewed from above and let $\mathbf{F} = \langle yz, xz, xy \rangle$.
   In this case

$$\text{curl}\mathbf{F} = \begin{vmatrix} \mathbf{i} & \mathbf{j} & \mathbf{k} \\ \frac{\partial}{\partial x} & \frac{\partial}{\partial y} & \frac{\partial}{\partial z} \\ yz & xz & xy \end{vmatrix} = \langle 0,0,0 \rangle,$$

so the surface integral of Stokes' theorem is always zero. Thus $\int_C \mathbf{F} \cdot d\mathbf{C} = 0$. Of course, we've seen that curl$\mathbf{F} = \mathbf{0}$ means that $\mathbf{F}$ is conservative, so the integral around any closed loop will be zero.

(b) Let C be the triangle with vertices $(2,0,0)$, $(0,1,0)$, and $(0,0,2)$ oriented counterclockwise when viewed from above and let $\mathbf{F} = \langle x^2, z^2, y^2 \rangle$.
   First we calculate curl, in case that makes our lives easier. This time we get

$$\text{curl}\mathbf{F} = \begin{vmatrix} \mathbf{i} & \mathbf{j} & \mathbf{k} \\ \frac{\partial}{\partial x} & \frac{\partial}{\partial y} & \frac{\partial}{\partial z} \\ x^2 & z^2 & y^2 \end{vmatrix} = \langle 2y - 2z, 0, 0 \rangle.$$

Now we must describe our surface, given by the vertices of a triangle, parametrically or as the graph of a function of two variables. The triangle is part of a plane with the given intercepts. It isn't too hard to verify an equation for the plane is $x + 2y + z = 2$, so solving for z we get that the surface S is a portion of the graph of $z = 2 - x - 2y$. In particular, it's the portion above the triangle in the first quadrant of the xy-plane and bounded by $x + 2y = 2$. The upward-pointing normal vector is $(-f_x, -f_y, 1) = (1, 2, 1)$.

We now evaluate curl $\mathbf{F}$ on the surface, which means analytically that we use the equation of the surface to eliminate z from the curl, giving

$$\text{curl}\mathbf{F} = (2y - 2(2 - x - 2y), 0, 0) = (-4 + 6y + 2x, 0, 0).$$

The surface integral is

$$\iint_S (-4 + 6y + 2x, 0, 0) \cdot (1, 2, 1)\, dA = \int_0^1 \int_0^{2-2y} -4 + 6y + 2x\, dxdy$$

$$= \int_0^1 (-4 + 6y)x + x^2\big|_{x=0}^{2-2y}\, dy$$

$$= \int_0^1 (-4 + 6y)(2 - 2y) + (2 - 2y)^2\, dy$$

$$= \int_0^1 -2y^2 + 3y - 1\, dy = -\frac{2}{3}. \blacktriangle$$

Example 5.6.3. *Using the line integral of Stokes' theorem*

Use Stokes' Theorem to evaluate $\iint_S \text{curl}\mathbf{F} \cdot d\mathbf{S}$ where $\mathbf{F} = (xz, xy, yz)$ and S is the portion of the graph of $z = 1 - \frac{x^2}{4} - y^2$ lying above the xy-plane with upward-pointing normal.

To evaluate the line integral of Stokes' theorem, we need to parameterize the boundary of S, which is the intersection of the surface $z = 1 - \frac{x^2}{4} - y^2$ with the xy-plane. This is the ellipse $\frac{x^2}{4} + y^2 = 1$ which is parameterized by $\mathbf{C}(t) = (2\cos t, \sin t, 0)$ (recall that the curve is in $\mathbb{R}^3$, so we need the third coordinate even though it's always zero). The integrand for the line integral is

$$\mathbf{F}(\mathbf{C}(t)) \cdot \mathbf{C}'(t) = (0, 2\cos t \sin t, 0) \cdot (-2\sin t, \cos t, 0) = 2\cos^2 t \sin t.$$

Evaluating we have

$$\int_C \mathbf{F} \cdot d\mathbf{C} = \int_0^{2\pi} 2\cos^2 t \sin t\, dt = -\frac{2}{3}\cos^3 t\Big|_0^{2\pi} = 0.$$

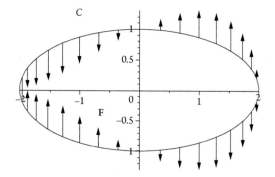

Figure 5.6.3 Using the line integral to find flux of curl**F**

Notice that the circulation is zero because the positive and negative circulation cancel each other out. Just because the circulation is zero that doesn't mean **F** is orthogonal to C everywhere (see Figure 5.6.3). ▲

This section concludes with an example where switching surfaces is advantageous.

Example 5.6.4. *Switching surfaces*

Let S be the sides of the pyramid that has base $[0,2] \times [0,2]$ in the xy-plane, and top vertex at the point $(1,1,1)$. Thus S consists of four triangles meeting at $(1,1,1)$. Suppose S is oriented with an upward-pointing normal, then C is the square in the xy-plane bounding the base of the pyramid. Let $\mathbf{F} = \langle xz^2, x^2 + y^2z^2, x^3 \rangle$, and evaluate $\iint_S (\text{curl}\mathbf{F}) \cdot d\mathbf{S}$.

By the remarks preceeding the example, we can exchange S for a more convenient surface S' with the same boundary to do the calculation. We choose the base of the pyramid, with upward pointing normal $\langle 0,0,1 \rangle$. We compute the curl, evaluate it on the surface S', and compute the surface integral.

$$\text{curl}\mathbf{F} = \begin{vmatrix} \mathbf{i} & \mathbf{j} & \mathbf{k} \\ \frac{\partial}{\partial x} & \frac{\partial}{\partial y} & \frac{\partial}{\partial z} \\ xz^2 & x^2 + y^2z^2 & x^3 \end{vmatrix} = \langle -2y^2z, 2xz - 3x^2, 2x \rangle.$$

Evaluating the curl on the surface $z = 0$ gives $\text{curl}\mathbf{F} = \langle 0, -3x^2, 2x \rangle$, and the integrand is $\text{curl}\mathbf{F} \cdot \langle 0,0,1 \rangle = \langle 0, -3x^2, 2x \rangle \cdot \langle 0,0,1 \rangle = 2x$. We have

$$\iint_{S'} (\text{curl}\mathbf{F}) \cdot d\mathbf{S} = \int_0^2 \int_0^2 2x\, dx\, dy = \int_0^2 4\, dy = 8.$$

This integral was considerably easier than finding equations for the triangular faces of the pyramid, or even parameterizing the curve and computing the line integral directly. ▲

Exercises

1. Verify Stokes' theorem for the constant vector field $\mathbf{F}(x,y,z) = \langle 2,1,3 \rangle$ and the portion of the paraboloid $z = 4 - x^2 - y^2$ lying on or above the xy-plane.

2. Verify Stokes' theorem for the constant vector field $\mathbf{F}(x,y,z) = \langle 1,4,7 \rangle$ and the portion of the cone $z = 1 - \sqrt{x^2 + y^2}$ lying on or above the xy-plane.

3. Verify Stokes' theorem for the vector field $\mathbf{F}(x,y,z) = \langle y, 2z, 3x \rangle$ and the portion of the saddle $z = x^2 - y^2$ lying "above" the unit disk. Note the boundary curve is $C(t) = (\cos t, \sin t, \cos^2 t - \sin^2 t)$.

4. Verify Stokes' theorem for the vector field $\mathbf{F}(x,y,z) = \langle 2z, 3x, y \rangle$ and the portion of the plane $z = 2x + 3y$ lying "above" the unit disk. Note the boundary curve is $C(t) = (\cos t, \sin t, 2\cos t + 3\sin t)$.

   In exercises 5–7 use the surface integral of Stokes' theorem to evaluate $\int_C \mathbf{F} \cdot d\mathbf{C}$, and let C be the surface boundary with induced orientation.

5. $\mathbf{F} = \left\langle \cos x, y^3, \frac{1}{1+z^2} \right\rangle$, S is the portion of the plane $x - 2y + z = 4$ inside the cylinder $x^2 + y^2 = 1$ with upward-pointing normal.

6. $\mathbf{F} = \langle x + 2y, 3x - 4y, 2z + 4x \rangle$, S is the portion of the surface $z = 1 - x^2 - y^2$ above the xy-plane with upward-pointing normal.

7. $\mathbf{F} = \langle xz, yz, xy \rangle$, S is the surface $\mathbf{S}(s,t) = (s\cos t, s\sin t, t)$, for $0 \le s \le 1, 0 \le t \le \pi/2$ with normal vector $\mathbf{S}_s \times \mathbf{S}_t$.

   In exercises 8–10 use the line integral of Stokes' theorem to evaluate $\iint_S \operatorname{curl}\mathbf{F} \cdot d\mathbf{S}$.

8. $\mathbf{F} = \langle x, z + 2x - 1, x^2 + y^2 \rangle$, S is the portion of the plane $2x + y + z = 1$ inside the cylinder $x^2 + y^2 = 1$ with upward-pointing normal.

9. $\mathbf{F} = \langle -y, x, z \rangle$, and S is the portion of the sphere $x^2 + y^2 + z^2 = 4$ lying above the plane $z = 1$ with upward-pointing normal.

10. $\mathbf{F} = \langle xy + z, 2z, y^2 + 2z \rangle$, and S is the portion of the ellipsoid $\frac{x^2}{9} + y^2 + \frac{z^2}{16} = 1$ lying above the xy-plane with upward-pointing normal.

11. Evaluate the integral $\iint_S \operatorname{curl}\mathbf{F} \cdot d\mathbf{S}$ where $\mathbf{F} = \langle 3x - 2y, y + z, 2z - x \rangle$, and S is the sides (without the base) of the tetrahedron with vertices $(0,0,0)$, $(1,0,0)$, $(0,1,0)$, $(0,0,1)$ with upward-pointing normal. (Hint: switch surfaces.)

12. Evaluate the integral $\iint_S \operatorname{curl}\mathbf{F} \cdot d\mathbf{S}$ where $\mathbf{F} = \langle x + y - 2, x^2 + z, y - z \rangle$, and S is the portion of the paraboloid $z = 1 - x^2 - y^2$ lying on or above the xy-plane. (Hint: switch surfaces.)

13. Evaluate the integral $\iint_S \operatorname{curl}\mathbf{F} \cdot d\mathbf{S}$ where $\mathbf{F} = \langle yz^2 + xz, xy, xy^2z^3 \rangle$, and S is the portion of the cone $z = 1 - \sqrt{x^2 + y^2}$ lying on or above the xy-plane. (Hint: switch surfaces.)

14. Evaluate the integral $\iint_S \operatorname{curl}\mathbf{F} \cdot d\mathbf{S}$ where $\mathbf{F} = \langle 3x - y, x + z^2, x - y \rangle$, and S is the hemisphere $z = \sqrt{1 - x^2 - y^2}$. (Hint: switch surfaces.)

15. Let S be the parametric surface $S(s,t) = (3+s, 2-t+3s, 1-2s+t)$, $0 \leq s, t \leq 1$, and $\mathbf{F} = (x^3 - 2x, yx^2, z^3 - 2z)$. Compare and contrast evaluating $\oint_C \mathbf{F} \cdot d\mathbf{C}$ directly with evaluating the surface integral $\iint_S \text{curl}\mathbf{F} \cdot d\mathbf{S}$ or switching surfaces. Now choose one method to evaluate the integral.

16. Let S be the portion of surface $z = x^2 + y^2 - 1$ below the xy-plane, and $\mathbf{F} = (x^2 y, xy^2, z - \cos x \sin y)$. Compare and contrast evaluating $\oint_C \mathbf{F} \cdot d\mathbf{C}$ directly with evaluating the surface integral $\iint_S \text{curl}\mathbf{F} \cdot d\mathbf{S}$ or switching surfaces. Now choose one method to evaluate the integral.

17. Let S be the hemisphere $z = \sqrt{1 - x^2 - y^2}$, and $\mathbf{F} = (-y, x, x^2 - 3yz)$. Compare and contrast evaluating $\oint_C \mathbf{F} \cdot d\mathbf{C}$ directly with evaluating the surface integral $\iint_S \text{curl}\mathbf{F} \cdot d\mathbf{S}$ or switching surfaces. Now choose one method to evaluate the integral.

18. This exercise involves vector fields with constant curl.
    (a) Show that if $\mathbf{F} = (Bz - Cy, Cx - Az, Ay - Bx)$, then $\text{curl}\mathbf{F} = (2A, 2B, 2C)$.
    (b) Find a vector field $\mathbf{F}$ such that $\text{curl}\mathbf{F} = (-2, 4, 6)$.
    (c) Let S be the upper unit hemisphere $z = \sqrt{1 - x^2 - y^2}$, and $\mathbf{F}$ your solution to part (b). Compare and contrast evaluating $\oint_C \mathbf{F} \cdot d\mathbf{C}$ directly with evaluating the surface integral $\iint_S \text{curl}\mathbf{F} \cdot d\mathbf{S}$.

19. Let S be a closed surface, say a sphere. Argue that $\iint_S \text{curl}\mathbf{F} d\mathbf{S} = 0$ for any vector field $\mathbf{F}$. (Hint: let C be an equator of the sphere and use Stokes' theorem on the hemispheres.)

5.7 The Divergence Theorem

The divergence of a vector field was first encountered as the integrand in the flux formulation of Green's theorem (see Section 5.3). In this section we introduce Gauss's divergence theorem, which generalizes Theorem 5.3.3 to three dimensions. The divergence theorem relates the flux of a vector field across a closed surface to the integral of its divergence over the solid that the surface bounds. More precisely we have:

Theorem 5.7.1. *Let $S = S_1 \cup \cdots \cup S_n$ be a union of piecewise smooth closed surfaces bounding the solid W, each with outward-pointing normal, and let $\mathbf{F}$ be a smooth vector field defined on a neighborhood containing W. Then*

$$\iint_S \mathbf{F} \cdot d\mathbf{S} = \iiint_W \operatorname{div}\mathbf{F} \, dV.$$

The proof of the divergence theorem, or Gauss's theorem as it is also called, is pretty much brute force. One just integrates both sides of the equation to a certain point, and observes that the results are the same. This is similar in flavor to the proof of Green's theorem. We prove it in the special case where W is a rectangular box.

Proof. Let W be the box $[a,b] \times [c,d] \times [e,f]$ and S its boundary with outward pointing normal, let $\mathbf{F}(x,y,z) = \langle P(x,y,z), Q(x,y,z), R(x,y,z) \rangle$, and consider the triple integral of the divergence theorem first. With these conventions we have

$$\iiint_W \operatorname{div}\mathbf{W} \, dV = \int_e^f \int_c^d \int_a^b \frac{\partial P}{\partial x} + \frac{\partial Q}{\partial y} + \frac{\partial R}{\partial z} dxdydz.$$

To prove the theorem, we calculate the contributions of each component function of $\mathbf{F}$ separately. Calculating the contribution from $\frac{\partial P}{\partial x}$ to the triple integral we see

$$\int_e^f \int_c^d \int_a^b \frac{\partial P}{\partial x} dxdydz = \int_e^f \int_c^d P(x,y,z)\big|_a^b \, dydz$$

$$= \int_e^f \int_c^d P(b,y,z) - P(a,x,y)dydz. \qquad (5.7.1)$$

We now show that the contribution of P to the flux integral of the divergence theorem is the same as Equation 5.7.1. Since W is a rectangular box, the boundary surface S is the union of six rectangles: the back and front, $S_1 = \{a\} \times [c,d] \times [e,f]$ and $S_2 = \{b\} \times [c,d] \times [e,f]$ respectively, the left S_3, right S_4, bottom S_5, and top S_6. The outward-pointing normal to S_1 is $-\mathbf{i}$ and to S_2 is $\mathbf{i}$. Similarly, the normals to the other faces are parallel to coordinate axes. This helps in calculating the surface integral in the divergence theorem.

To calculate the contribution of $P(x,y,z)$ to the flux, we use the observations regarding normal vectors to the faces of the cube to see

$$\iint_S \mathbf{F} \cdot d\mathbf{S} = \iint_{S_1} \mathbf{F} \cdot (-\mathbf{i}) \, dA + \iint_{S_2} \mathbf{F} \cdot \mathbf{i} \, dA + \iint_{S_3} \mathbf{F} \cdot (-\mathbf{j}) \, dA$$

$$+ \iint_{S_4} \mathbf{F} \cdot \mathbf{j} \, dA + \iint_{S_5} \mathbf{F} \cdot (-\mathbf{k}) \, dA + \iint_{S_6} \mathbf{F} \cdot \mathbf{k} \, dA.$$

The component function $P(x,y,z)$ only contributes to the first two integrals, since it vanishes in the remaining dot products. The integrand in each will be $P(x,y,z)$ evaluated on the surface. Since S_1 is the surface $x = a$, we integrate $-P(a,y,z)$ for the first integral. Similarly the second integrand is $P(b,y,z)$. Thus the contribution from $P(x,y,z)$ to the surface integral is

$$\iint_{S_1} \mathbf{F} \cdot (-\mathbf{i}) \, dA + \iint_{S_2} \mathbf{F} \cdot (\mathbf{i}) \, dA$$

$$= \int_e^f \int_c^d -P(a,y,z) \, dy \, dz + \int_e^f \int_c^d P(b,y,z) \, dy \, dz. \qquad (5.7.2)$$

Comparing the contributions of P to the triple integral in Equation 5.7.1 with the contribution in Equation 5.7.2, we see that they are the same. Similar calculations show that the component functions Q and R both contribute the same to the triple and surface integrals. This proves the divergence theorem for the special case of a rectangular box. The arguments generalize to dx-, dy-, or dz-solids using similar techniques, and to even more complicated domains by decomposing them into simpler pieces and adding them. We do not treat the more general settings here. $\square$

Math App 5.7.1. *Divergence theorem*

Before embarking on calculations, a little intuition might be in order. The following Math App endeavors to provide just that. Take a look.

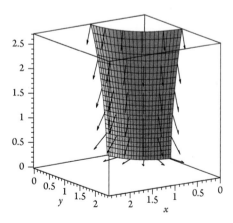

Remarks: We make some initial observations.

- First we compare the important theorems of vector analysis. Recall that both curl and divergence are ways to differentiate vector fields, and that integrals involve antidifferentiation. Notice that Green's, Stokes', and Gauss's theorems all relate the integral of a derivative over a region to the integral of its antiderivative over the boundary of the region. For example, to integrate divF (a derivative of **F**) over a solid W, you integrate its antiderivative **F** over the boundary of W. This is analogous to the one-dimensional case, where the integral $\int_a^b f(x)\,dx$ of the derivate $f(x) = F'(x)$ over the interval $[a, b]$ is related to the antiderivative evaluated on its boundary. In this case, the boundary of an interval is its endpoints. Thus we get $\int_a^b f(x)\,dx = F(b) - F(a)$.

  All of the important integrals we've discussed, then, follow the same general pattern: integrating a derivative over a region equals the integral of the antiderivative over the boundary of the region.

- The divergence theorem gives us a local interpretation of divF in much that same way that Stokes' theorem interprets curlF. In particular, if W_ϵ is a solid ball centered at **a** radius ϵ, with boundary surface S_ϵ, then the mean value theorem for integrals implies

$$\text{divF}(\mathbf{a}) = \lim_{\epsilon \to 0} \frac{1}{\text{Volume of } W_\epsilon} \iint_{S_\epsilon} \mathbf{F} \cdot d\mathbf{S}.$$

  One interprets divF, then, as flux per unit volume. Sometimes a point of positive divergence is called a source, while a sink is a point of negative divergence. It might be more intuitive to think of a gas expanding near sources and compressing near sinks. This formalizes the intuition we gave in Section 5.5, making the notion of an incompressible flow more precise.

- Theorem 5.7.1 allows for solids with several boundary surfaces, as long as all normal vectors are chosen to be outward-pointing. For example, let W be the solid between concentric spheres S_1, S_2, of radius 1 and 2 centered at the origin. The solid W defines the outward-pointing normal on its boundary surfaces. In this case, the outward normal on S_2 points away from the origin, while on S_1 it points toward the origin. If we let $-S_1$ denote S_1 with this inward orientation, then the boundary of W is $\partial W = -S_1 \cup S_2$. The divergence theorem then implies

$$\iiint_W \text{divF}\,dV = \iint_{-S_1 \cup S_2} \mathbf{F} \cdot d\mathbf{S} = \iint_{S_2} \mathbf{F} \cdot d\mathbf{S} - \iint_{S_1} \mathbf{F} \cdot d\mathbf{S},$$

since changing the direction of the normal vector on S_1 changes the sign of the flux integral. Note that if $\text{divF} = 0$ on the solid W, this implies

$$\iint_{S_2} \mathbf{F} \cdot d\mathbf{S} = \iint_{S_1} \mathbf{F} \cdot d\mathbf{S}.$$

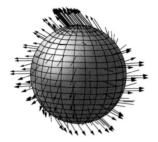

Figure 5.7.1 Flux across a sphere

The surfaces needn't be spherical, of course, and the remarks apply to solids with several boundary components as well.

Applications of Gauss's theorem usually involve using a triple integral to evaluate a related surface integral. We illustrate with a couple examples.

Example 5.7.1. *Calculating flux across a sphere*

Let $\mathbf{F}(x,y,z) = (x^3 - y^3, y^3 - z^3, z^3 - x^3)$. Calculate the flux across the unit sphere.

By the divergence theorem, we will calculate the flux over the sphere by integrating the divergence over the entire ball. Moreover, the domain of integration and the divergence both indicate that spherical coordinates will be the easiest way to evaluate the integral. The divergence is seen to be

$$\text{div}\mathbf{F} = \frac{\partial(x^3 - y^3)}{\partial x} + \frac{\partial(y^3 - z^3)}{\partial y} + \frac{\partial(z^3 - x^3)}{\partial z} = 3(x^2 + y^2 + z^2).$$

Applying the theorem we have

$$\iint_S \mathbf{F} \cdot d\mathbf{S} = \iiint_W \text{div } \mathbf{F} dV = \iiint_W 3(x^2 + y^2 + z^2)\, dV$$

$$= \int_0^{2\pi} \int_0^{\pi} \int_0^1 3\rho^4 \sin\phi\, d\rho\, d\phi\, d\theta = \int_0^{2\pi} \int_0^{\pi} \frac{3\rho^5}{5} \sin\phi \Big|_0^1 d\phi\, d\theta$$

$$= \frac{3}{5} \int_0^{2\pi} \int_0^{\pi} \sin\phi\, d\phi\, d\theta = \frac{3}{5} \int_0^{2\pi} -\cos\phi|_0^{\pi}\, d\theta$$

$$= \frac{6}{5} \int_0^{2\pi} d\theta = \frac{12\pi}{5}. \ \blacktriangle$$

Example 5.7.2. *Calculating flux across a cone*

Calculate the flux of $\mathbf{F}(x,y,z) = (xy^2, yz^2, zx^2)$ across the right circular cone of height 1 with unit disk as base. Thus the solid is below the cone $z = 1 - \sqrt{x^2 + y^2}$ and above the xy-plane.

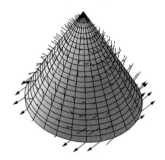

Figure 5.7.2 Flux across a cone

The triple integral of the divergence theorem will be easier to do using cylindrical coordinates. In cylindrical coordinates the solid is described by the system of inequalities

$$0 \leq z \leq 1 - r$$
$$0 \leq r \leq 1$$
$$0 \leq \theta \leq 2\pi$$

A quick calculation shows $\text{div}\mathbf{F} = y^2 + z^2 + x^2$, so the flux is given by

$$\iint_S \mathbf{F} \cdot d\mathbf{S} = \int_0^{2\pi} \int_0^1 \int_0^{1-r} (r^2 + z^2) r \, dz \, dr \, d\theta = \int_0^{2\pi} \int_0^1 r^3 z + r\frac{z^3}{3}\Big|_0^{1-r} dr \, d\theta$$

$$= \int_0^{2\pi} \int_0^1 r(1-r)\left(r^2 + \frac{(1-r)^2}{3}\right) dr \, d\theta$$

$$= \frac{1}{3}\int_0^{2\pi} \int_0^1 -4r^4 + 6r^3 - 3r^2 + r \, dr \, d\theta$$

$$= \frac{1}{3}\int_0^{2\pi} -\frac{4r^5}{5} + \frac{3r^4}{2} - r^3 + \frac{r^2}{2}\Big|_0^1 d\theta = \frac{1}{3}\int_0^{2\pi} \frac{1}{5} d\theta = \frac{2\pi}{15}. \blacktriangle$$

We now consider the flux of an electric field due to a point charge, and show that it induces the same flux across any surface containing it.

Example 5.7.3. *Electric flux*

Suppose a point charge Q is placed at the origin generating the electric field $\mathbf{E} = \frac{kQ}{\rho^2}\widehat{\rho}$. Show that the flux across any sphere centered at the origin with outward-pointing normal is the same.

If V is the electric potential, we know $\mathbf{E} = -\nabla V$. Thus, away from the origin where $\mathbf{E}$ is not defined, the divergence of $\mathbf{E}$ is given by

$$\text{div}\mathbf{E} = \nabla(-\nabla V) = -\nabla^2 V = 0,$$

where the last equality follows from Example 5.5.4 where we showed that V satisfies Laplace's equation. Since $\text{div}\mathbf{E} = 0$ on W, the remark following Theorem 5.7.1 shows the flux across any two spheres centered at the origin to be the same.

The same argument generalizes to any closed surface bounding a solid containing the origin (see Example 5.3.6 for a two-dimensional example). ▲

Exercises

1. Find the flux of $\mathbf{F}(x,y,z) = \langle 3,1,-4 \rangle$ across the unit sphere with outward-pointing normal direction.

2. Find the flux of $\mathbf{F}(x,y,z) = \langle 1,4,-5 \rangle$ across the surface of the cube $W = [0,1] \times [0,1] \times [0,1]$ with outward-pointing normal.

3. Find the flux of $\mathbf{F}(x,y,z) = \langle x^2,1,z \rangle$ across the boundary of the solid inside $y^2 + z^2 = 1$ and between $x = 0$ and $x + z = 2$ with outward-pointing normal.

4. Find the flux of $\mathbf{F}(x,y,z) = \langle 3x + 2y, z - y, x + y - z \rangle$ across the boundary of the solid inside $x^2 + y^2 = 1$ and between $0 \le z \le 1$ with outward-pointing normal.

5. Find the flux of $\mathbf{F}(x,y,z) = \left\langle xz, x^2 yz, \frac{y^2 z^2}{2} \right\rangle$ across the cone $\sqrt{x^2 + y^2} \le z \le 1$ and between $x = 0$ and $x + z = 2$ with outward-pointing normal.

6. Find the flux of $\mathbf{F}(x,y,z) = \langle 7x - y + 2z, 3 + 4z, x - 2y + z \rangle$ across the boundary of the solid in the first octant under both $x + z = 1$ and $x + y + z = 2$ with outward-pointing normal.

7. Find the flux of $\mathbf{F}(x,y,z) = \langle xy^2 - yz^2, yz^2 - zx^2, zx^2 - xy^2 \rangle$ across the surfaces (with outward-pointing normal vectors):
   (a) The unit sphere.
   (b) The cylindrical can with lateral surface $x^2 + y^2 = 1$, top at $z = 1$ and bottom at $z = 0$.

8. Find the flux of $\mathbf{F}(x,y,z) = \langle xy - yz, x^2 z + 2y, 3y - 2z \rangle$ across the cube $[0,1] \times [0,1] \times [0,1]$ with outward-pointing normal.

9. Find the flux of $\mathbf{F}(x,y,z) = \langle 3xz^2, y^3, 3x - 2y \rangle$ across the surface between $x = 0$ and $x = 1$, above the xy-plane and below the cylinder $y^2 + z^2 = 1$.

10. Show that the flux of a constant vector field across a closed surface is zero.

11. Let W be a solid with boundary S. Show that the flux of $\mathbf{F}(x,y,z) = \langle 3x, x^2 - y, y - z + 3x \rangle$ across S is the volume of W.

12. Let W be the cylinder $x^2 + y^2 \le R^2$ and $0 \le z \le H$, and let S be the boundary of W with outward-pointing normal. Evaluate the flux of

$$\mathbf{F}(x,y,z) = \langle \sin(xy), x^2 + 3y, -zy\cos(xy) \rangle$$

across S without integrating.

13. Show that the flux of $\mathbf{F}(x,y,z) = \langle f(y,z), g(x,z), h(x,y) \rangle$ across any closed surface is zero.

14. Let W be a solid with boundary S and volume $vol(W)$. Show that if the component functions of $\mathbf{F}$ are linear, then the flux of $\mathbf{F}$ across S can be calculated without integrating, just using $vol(W)$.

15. Calculate the flux of $\mathbf{E} = \frac{kQ}{\rho^2}\widehat{\rho}$ across the unit sphere.

16. Let S be the boundary of a solid W that does not contain the origin, and let $\mathbf{E} = \frac{kQ}{\rho^2}\widehat{\rho}$ be the electric field due to a point charge Q at the origin. Argue that $\iint_S \mathbf{E} \cdot d\mathbf{S} = 0$.

5.8 Curvilinear Coordinates

In this section we derive cylindrical formulas for the gradient, divergence, Laplacian, and curl. These formulas are quite useful in applications, and therefore warrant discussion. We begin by introducing *coordinate vector fields* for each coordinate, which are unit vectors that point in the direction of increase for their corresponding coordinate. Coordinate vector fields will be used in all subsequent derivations. We then derive cylindrical formulas for the aforementioned derivatives. The important point here is that the curvilinear formulas are consequences of Stokes' and the divergence theorems. The arguments presented are intuitive, and not meant to be precise. The section ends with a list of spherical formulas as well as some applications.

Coordinate Vector Fields: In this paragraph we introduce coordinate vector fields in cylindrical and spherical coordinates as an alternative to the standard basis vectors. In this context we think of the vector **i** as the vector field that assigns to each point the unit vector in the direction of increasing x-values. To emphasize that we're thinking of it as a vector *field* pointing one unit in the positive x-direction at every point, we might use the notation $\widehat{x}$. Analogously, we let $\widehat{r}$ denote the vector field that assigns to each point of space the unit vector at that point in the direction of increasing r. More generally the *coordinate vector fields* $\widehat{z}, \widehat{\theta}, \widehat{\rho}, \widehat{\phi}$ are unit vector fields that point in the direction of coordinate increase (see Figure 5.8.1). Thus the fields **i, j, k** are the coordinate vector fields arising from the Cartesian coordinate system on $\mathbb{R}^3$.

Math App 5.8.1. *Coordinate vector fields*

This Math App provides the opportunity to interact with coordinate vector fields, and see how they change as you move through space.

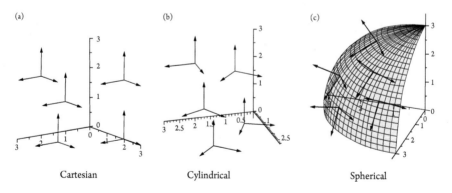

(a) Cartesian (b) Cylindrical (c) Spherical

Figure 5.8.1 Coordinate vector fields

Typically we expressed a general vector field $\mathbf{F}$ as a linear combination of standard basis vector fields using component functions, as in

$$\mathbf{F}(x,y,z) = \langle P, Q, R \rangle = P\mathbf{i} + Q\mathbf{j} + R\mathbf{k},$$

where P, Q, R are functions of x, y, z. Formulas for dot and cross products, gradients, divergence, and curl all depended on decomposing $\mathbf{F}$ this way. For example, before calculating the divergence of $\mathbf{F}$ we have to know the component functions P, Q, R. The properties of the standard basis vector *fields* (as we now think of them) that make them convenient is that they are unit vectors, they are mutually orthogonal, and they are constant. The coordinate vector fields corresponding to cylindrical and spherical coordinates share two of these three properties. At each point in space, the cylindrical coordinate vector fields $\widehat{r}, \widehat{\theta}, \widehat{z}$ are mutually orthogonal, as are the spherical coordinate fields $\widehat{\rho}, \widehat{\theta}, \widehat{\phi}$. By definition, they are also unit vectors. They are *not* constant vector fields, however, as we now demonstrate.

Example 5.8.1. *Relating coordinate vector fields*

Express the following coordinate vector fields as linear combinations of the given system.

(a) Express $\widehat{r}$ as a linear combination of Cartesian coordinate vector fields.
    Figure 5.8.2(a) demonstrates how to construct right triangles to find the Cartesian components of $\widehat{r}$. Since the hypotenuse of these triangles is a unit vector, namely $\widehat{r}$, the components are seen to be $\cos\theta$ and $\sin\theta$, so

$$\widehat{r} = \cos\theta\,\mathbf{i} + \sin\theta\,\mathbf{j}.$$

This is certainly true, but it might be unsatisfying because the coefficient functions P and Q are given in polar coordinates. This turns out to be useful sometimes,

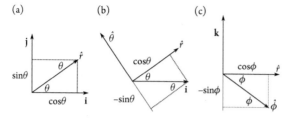

Figure 5.8.2 Relating coordinate vector fields

but we can find a completely Cartesian description noting $\cos\theta = r\cos\theta/r = x/\sqrt{x^2+y^2}$. A similar analysis for $\sin\theta$ gives

$$\widehat{r} = \cos\theta\, \mathbf{i} + \sin\theta\, \mathbf{j} = \frac{x}{\sqrt{x^2+y^2}}\mathbf{i} + \frac{y}{\sqrt{x^2+y^2}}\mathbf{j}.$$

(b) Express $\mathbf{i}$ as a linear combination of cylindrical coordinate vector fields.

Now our goal is to decompose a Cartesian coordinate field into its $\widehat{r}, \widehat{\theta}, \widehat{z}$ components. Since $\mathbf{i}$ has no vertical component it will be a linear combination of $\widehat{r}$ and $\widehat{\theta}$. A glance at Figure 5.8.2(b) shows that

$$\mathbf{i} = \cos\theta\, \widehat{r} - \sin\theta\, \widehat{\theta}.$$

(c) Express $\widehat{\phi}$ as a linear combination of Cartesian coordinate vector fields.

The vector $\widehat{\phi}$ lies in a half-plane (the constant coordinate surface obtained by fixing θ). The right triangles of Figure 5.8.2(c) decompose $\widehat{\phi}$ into its $\widehat{r}$ and $\mathbf{k}$ components. Since $\widehat{\phi}$ is perpendicular to the ray from the origin, the angles are as labeled (convince yourself of this!). Trigonometry, together with the fact that $\widehat{r} = \cos\theta\, \mathbf{i} + \sin\theta\, \mathbf{j}$, yields

$$\widehat{\phi} = \cos\phi\, \widehat{r} - \sin\phi\, \mathbf{k} = \cos\phi\cos\theta\, \mathbf{i} + \cos\phi\sin\theta\, \mathbf{j} - \sin\phi\, \mathbf{k}.$$

While we could find Cartesian versions of the component functions, they are cumbersome and not very enlightening, so we refrain from doing so. ▲

Dot products and cross products of coordinate vector fields are quite straightforward, so we briefly discuss them here. Analysis similar to that of the previous example shows that $\widehat{\theta} = -\sin\theta\, \mathbf{i} + \cos\theta\, \mathbf{j}$, and direct computation of the dot product gives

$$\widehat{r}\cdot\widehat{\theta} = \big(\cos\theta\, \mathbf{i} + \sin\theta\, \mathbf{j}\big)\cdot\big(-\sin\theta\, \mathbf{i} + \cos\theta\, \mathbf{j}\big) = \cos\theta(-\sin\theta) + \sin\theta\cos\theta = 0.$$

This confirms our earlier statement that these vectors are orthogonal. Similar computations show that, within a given coordinate system, coordinate vector fields are mutually orthogonal everywhere.

Knowing that coordinate vector fields are mutually orthogonal, and unit vectors, allows us to compute cross products just using the right-hand rule. Recall that to compute $\mathbf{i} \times \mathbf{j}$, simultaneously place your right index finger along $\mathbf{i}$ and your right middle finger along $\mathbf{j}$. Your right thumb then points in the direction of $\mathbf{i} \times \mathbf{j}$. In this case we find $\mathbf{i} \times \mathbf{j} = \mathbf{k}$. In the same way one computes $\widehat{\boldsymbol{\theta}} \times \widehat{\boldsymbol{r}} = -\widehat{\boldsymbol{z}}, \widehat{\boldsymbol{\theta}} \times \widehat{\boldsymbol{\rho}} = \widehat{\boldsymbol{\phi}}$, as well as cross products of any two coordinate vectors from the same coordinate system.

More generally, given any vector field it can be described as linear combinations of coordinate fields in any coordinate system. Choosing appropriate coordinate systems can sometimes be illuminating.

Example 5.8.2. *Polar vortex velocity field*

A vortex in an ideal fluid flow is most easily described using polar functions for its velocity potential and stream function. The equations are

$$\varphi = \theta, \ \psi = -\ln r.$$

In Example 2.6.4 we saw the Cartesian form of the velocity field for this flow is

$$\mathbf{F}(x,y) = \left(\frac{-y}{x^2+y^2}, \frac{x}{x^2+y^2} \right) = \frac{-y}{x^2+y^2}\mathbf{i} + \frac{x}{x^2+y^2}\mathbf{j}.$$

Rewrite $\mathbf{F}$ as a linear combination of polar coordinate vector fields $\widehat{\boldsymbol{r}}$ and $\widehat{\boldsymbol{\theta}}$.

To do this we must rewrite the component functions *and* the vectors $\mathbf{i}, \mathbf{j}$ in polar form. Since

$$\mathbf{i} = \cos\theta\widehat{\boldsymbol{r}} - \sin\theta\widehat{\boldsymbol{\theta}} \quad \text{and} \quad \mathbf{j} = \sin\theta\widehat{\boldsymbol{r}} + \cos\theta\widehat{\boldsymbol{\theta}},$$

and since

$$\frac{-y}{x^2+y^2} = \frac{-\sin\theta}{r} \quad \text{and} \quad \frac{x}{x^2+y^2} = \frac{\cos\theta}{r},$$

we substitute and find

$$\begin{aligned} \mathbf{F}(r,\theta) &= \frac{-y}{x^2+y^2}\mathbf{i} + \frac{x}{x^2+y^2}\mathbf{j} \\ &= \frac{-\sin\theta}{r}\left(\cos\theta\widehat{\boldsymbol{r}} - \sin\theta\widehat{\boldsymbol{\theta}}\right) + \frac{\cos\theta}{r}\left(\sin\theta\widehat{\boldsymbol{r}} + \cos\theta\widehat{\boldsymbol{\theta}}\right) \\ &= \left(\frac{-\sin\theta\cos\theta}{r} + \frac{\cos\theta\sin\theta}{r}\right)\widehat{\boldsymbol{r}} + \left(\frac{\sin^2\theta}{r} + \frac{\cos^2\theta}{r}\right)\widehat{\boldsymbol{\theta}} \\ &= \frac{1}{r}\widehat{\boldsymbol{\theta}}. \end{aligned}$$

The polar form of $\mathbf{F}$ is much more natural than the original Cartesian form. ▲

The velocity field in the previous example is just the gradient of the velocity potential function. In Cartesian coordinates the velocity potential is $\varphi = \tan^{-1}(y/x)$, from which you can calculate that $\mathbf{F} = \nabla\varphi$. Introducing Cartesian coordinates allowed us to calculate the gradient. The example shows, however, that in polar coordinates $\nabla\varphi = \frac{1}{r}\widehat{\boldsymbol{\theta}}$. There should be a more efficient way of calculating this directly, without transitioning to Cartesian coordinates—and there is!

Cylindrical Gradients: In this paragraph we describe how to calculate gradients of functions given in cylindrical coordinates. If the original function is cylindrical, the resulting gradient is a linear combination of $\widehat{\boldsymbol{r}},\widehat{\boldsymbol{\theta}},\widehat{\boldsymbol{z}}$.

Let $f(r,\theta,z)$ be a polar function, we wish to find component functions U,V,W so that

$$\nabla f(r,\theta,z) = U\widehat{\boldsymbol{r}} + V\widehat{\boldsymbol{\theta}} + W\widehat{\boldsymbol{z}}.$$

The vector $\nabla f = \langle f_x, f_y, f_z \rangle$ is already defined as a linear combination of Cartesian coordinate vector fields, so the gradient exists without another formula. The point of this discussion is not to define the gradient, but to find it directly as a linear combination of cylindrical coordinate vectors.

To derive a cylindrical gradient formula we will approximate changes in function values in two ways. Consider Δf, the change in function value as one moves from the point (r,θ,z) to $(r+\Delta r, \theta+\Delta\theta, z+\Delta z)$. We will approximate Δf using "tangent plane approximations" and using directional derivatives—we begin with the tangent "plane" approach.

When discussing functions of two variables, Equation 3.4.2 showed that changes in function values were approximated by

$$\Delta f \approx f_x \Delta x + f_y \Delta y.$$

The same is true for functions of three variables (in any coordinate system), so we see

$$\Delta f \approx f_r \Delta r + f_\theta \Delta\theta + f_z \Delta z. \qquad (5.8.1)$$

To derive the directional derivative approximation, recall that the directional derivative of f is how fast f is changing in that direction, per unit distance. Multiplying the directional derivative times the distance traveled, then, gives us another way to approximate Δf.

We must find the vector $\mathbf{v}$ from (r,θ,z) to $(r+\Delta r, \theta+\Delta\theta, z+\Delta z)$, then use the directional derivative of f in the $\mathbf{v}$ direction to estimate Δf. The key will be to find $\mathbf{v}$ in terms of $\widehat{\boldsymbol{r}},\widehat{\boldsymbol{\theta}},\widehat{\boldsymbol{z}}$.

When discussing integration in cylindrical coordinates we considered the solid generated by changing each variable a little (see Figure 5.8.3). In the present context, we consider the changes in coordinates so small that the solid is essentially a rectangular box. The point (r,θ,z) is the lower left of the box, while $(r+\Delta r, \theta+\Delta\theta, z+\Delta z)$ is the upper

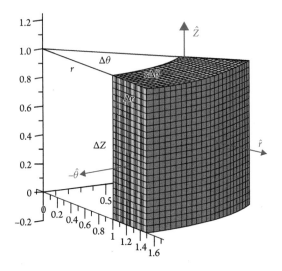

Figure 5.8.3 An infinitesimal cylindrical solid

right corner diagonally opposite. The vector **v** is the diagonal of this "box". As a vector, then, it is the sum of the edges of the box (thought of as vectors).

The edges of the box are parallel to coordinate vectors $\widehat{r}, \widehat{\theta}, \widehat{z}$, and have lengths $\Delta r, r\Delta\theta, \Delta z$ respectively. Thus the vector **v** in cylindrical coordinates is

$$\mathbf{v} = \Delta r \widehat{r} + r\Delta\theta \widehat{\theta} + \Delta z \widehat{z}.$$

The directional derivative, or rate of change of f, in this direction is

$$comp_{\mathbf{v}}\nabla f = \nabla f \cdot \frac{\mathbf{v}}{\|\mathbf{v}\|}.$$

This is how fast f changes, per unit distance, in the **v** direction. To estimate the change in function value, multiply the rate times the distance traveled, which is $\|\mathbf{v}\|$. Using this estimate we get

$$\Delta f \approx \left(comp_{\mathbf{v}}\nabla f\right)\|\mathbf{v}\| = \nabla f \cdot \frac{\mathbf{v}}{\|\mathbf{v}\|}\|\mathbf{v}\| = \nabla f \cdot \mathbf{v}.$$

Thinking of $\nabla f = U\widehat{r} + V\widehat{\theta} + W\widehat{z}$, this approximation becomes

$$\begin{aligned}\Delta f &\approx \nabla f \cdot \mathbf{v}\\ &= \left(U\widehat{r} + V\widehat{\theta} + W\widehat{z}\right) \cdot \left(\Delta r\widehat{r} + r\Delta\theta\widehat{\theta} + \Delta z\widehat{z}\right)\\ &= U\Delta r + rV\Delta\theta + W\Delta z.\end{aligned}\qquad(5.8.2)$$

Equating the tangent plane approximation of Equation 5.8.1 with the directional derivative approximation of Equation 5.8.2, we see

$$f_r \Delta r + f_\theta \Delta \theta + f_z \Delta z = U \Delta r + rV \Delta \theta + W \Delta z.$$

This is satisfied when $U = f_r$, $V = \frac{1}{r} f_\theta$, and $W = f_z$. Thus the cylindrical gradient formula is

$$\nabla f = f_r \widehat{r} + \frac{1}{r} f_\theta \widehat{\theta} + f_z \widehat{z}. \qquad (5.8.3)$$

Example 5.8.3. *Cylindrical gradients*

(a) Let $\varphi(r,\theta) = \theta$ be the velocity potential of a vortex flow. Compute $\nabla \varphi$ directly.
Since $\varphi_r = 0$ and $\varphi_\theta = 1$, the cylindrical formula for the gradient gives

$$\nabla \varphi = \varphi_r \widehat{r} + \frac{1}{r} \varphi_\theta \widehat{\theta} = \frac{1}{r} \widehat{\theta}.$$

This is a clear improvement over the method of Example 5.8.2.

(b) Let $f(r,\theta,z) = r^2 + z^2$, find ∇f.
In this case direct calculation shows

$$\nabla f = 2r\widehat{r} + 2z\widehat{z}. \; \blacktriangle$$

Cylindrical divergence: One of the remarks following Theorem 5.7.1, the divergence theorem, pointed out that the theorem provides an interpretation of divergence at a point. It is the infinitesimal flux per unit volume. Given a vector field $\mathbf{F}$ and a small solid X with boundary surface S, this means

$$\text{Flux of } \mathbf{F} \text{ across } S = \text{div}\mathbf{F} \cdot Vol(X).$$

To find a cylindrical form for div$\mathbf{F}$ we'll use this (infinitesimal) equation and a judicious choice of X for which we can calculate its volume and the flux across its boundary. The coefficient of $Vol(X)$ will be the divergence of $\mathbf{F}$.

Let X be the solid obtained by starting with a point (r,θ,z) and increasing each coordinate a small amount, and let S be its boundary with outward-pointing normal. As the changes in coordinates decrease, the solid X approaches a rectangular box with dimensions Δr, $r\Delta\theta$, and Δz. Thus $Vol(X) = r\Delta r\Delta\theta \Delta z$.

It remains to compute the flux of $\mathbf{F}$ across the boundary S of X. Note that S is made up of six faces, and we'll compute the contribution of flux from the inner and outer faces (closest to and farthest from the z-axis). The introduction of Section 5.4 showed that flux across a surface is the normal component of $\mathbf{F}$ times the area of S. Let us compute the flux of $\mathbf{F} = U\widehat{r} + V\widehat{\theta} + W\widehat{z}$ across the inner face of S. As in Figure 5.8.3, the unit normal vector to the inner face is $-\widehat{r}$, and its area is $r\Delta\theta \Delta z$. The flux of $\mathbf{F}$ across the inner face is

$$comp_{-\widehat{r}}\mathbf{F}(\text{Area}) = \left(U\widehat{r} + V\widehat{\theta} + W\widehat{z} \right) \cdot (-\widehat{r})r\Delta\theta\,\Delta z = -rU\Delta\theta\,\Delta z.$$

We've used the fact that $\widehat{r}, \widehat{\theta}, \widehat{z}$ are unit vectors, so $\widehat{r} \cdot \widehat{r} = 1$, and mutually orthogonal so $\widehat{r} \cdot \widehat{\theta} = 0 = \widehat{r} \cdot \widehat{z}$.

Two things change when computing the flux across the outer face. The normal vector is $\widehat{r}$ rather than $-\widehat{r}$, and the functions are evaluated at $(r + \Delta r, \theta, z)$ rather than (r, θ, z). The sum of the flux contributions across inner and outer faces is

$$(r + \Delta r)U(r + \Delta r, \theta, z)\Delta\theta\,\Delta z - rU(r, \theta, z)\Delta\theta\,\Delta z = \frac{\partial}{\partial r}(rU)\,\Delta r\Delta\theta\,\Delta z, \qquad (5.8.4)$$

where the mean value theorem has been applied to get

$$(r + \Delta r)U(r + \Delta r, \theta, z) - rU(r, \theta, z) = \frac{\partial}{\partial r}(rU)\,\Delta r.$$

The right-hand side of Equation 5.8.4 is the contribution of the inner and outer faces to the flux of $\mathbf{F}$ across S, so we want it to look like $\text{div}\mathbf{F} \cdot Vol(X)$. The last bit is almost $Vol(X)$, but is off by a factor of r. Multiplying and dividing by r shows that the inner and outer faces contribute

$$\frac{\partial}{\partial r}(rU)\,\Delta r\Delta\theta\,\Delta z = \frac{1}{r}\frac{\partial}{\partial r}(rU)\,r\Delta r\Delta\theta\,\Delta z = \frac{1}{r}\frac{\partial}{\partial r}(rU)\,Vol(X). \qquad (5.8.5)$$

The top and bottom faces have normal vectors $\pm\widehat{z}$ and both have area $r\Delta\theta\,\Delta r$. Functions on the top face are evaluated at $(r, \theta, z + \nabla z)$, while on the bottom face evaluate functions at (r, θ, z). Taking these considerations into account, the flux contribution is

$$rW(r, \theta, z + \Delta z)\Delta\theta\,\Delta r - rW(r, \theta, z)\Delta\theta\,\Delta r = \frac{\partial}{\partial z}(rW)\,\Delta r\Delta\theta\,\Delta z$$

$$= \frac{\partial}{\partial z}(W)\,r\Delta r\Delta\theta\,\Delta z = \frac{\partial}{\partial z}(W)\,Vol(X). \qquad (5.8.6)$$

Since r is a constant with respect to z, we were able to slide it past the derivative and have it contribute to the volume portion of the expression.

The sides have normal vectors $\pm\widehat{\theta}$ with area $\Delta r\Delta z$. The flux contributions are

$$\mathbf{F} \cdot \widehat{\theta}\Delta r\Delta z - \mathbf{F} \cdot \widehat{\theta}\Delta r\Delta z,$$

where the first term is evaluated at $(r, \theta + \Delta\theta, z)$ and the second at (r, θ, z). Another application of the mean value theorem, together with multiplying and dividing by r, shows that this flux contribution is

$$\frac{\partial V}{\partial\theta}\Delta r\Delta\theta\,\Delta z = \frac{1}{r}\frac{\partial V}{\partial\theta}r\Delta r\Delta\theta\,\Delta z = \frac{1}{r}\frac{\partial V}{\partial\theta}Vol(X). \qquad (5.8.7)$$

Summing the flux contributions of Equations 5.8.5, 5.8.6, and 5.8.7 we get

$$\text{Flux of }\mathbf{F}\text{ across }S = \left(\frac{1}{r}\frac{\partial}{\partial r}(rU) + \frac{\partial W}{\partial z} + \frac{1}{r}\frac{\partial V}{\partial \theta}\right)Vol(X).$$

The coefficient of $Vol(X)$ is the divergence, so in cylindrical coordinates

$$\text{div}\mathbf{F} = \frac{1}{r}\frac{\partial}{\partial r}(rU) + \frac{1}{r}\frac{\partial V}{\partial \theta} + \frac{\partial W}{\partial z}. \tag{5.8.8}$$

Example 5.8.4. *Cylindrical divergence*

Evaluate the divergence of $\mathbf{F} = r\widehat{r} + z\widehat{z}$.
Since $U = r$, $V = 0$ and $W = z$, we have

$$\text{div}\mathbf{F} = \frac{1}{r}\frac{\partial}{\partial r}r^2 + \frac{\partial}{\partial z}z = \frac{2r}{r} + 1 = 3.$$

The vector $r\widehat{r} + z\widehat{z}$ at the point (r,θ,z) does not move in the $\widehat{\theta}$ direction, so it is in the same half-plane as the position vector to (r,θ,z). In fact, $\mathbf{F}$ goes the same distance away from the z-axis (distance r) and the same vertical distance (distance z) as the position vector. Thus $\mathbf{F}$ is the vector field that assigns the position vector to each point. This has Cartesian description $\mathbf{F} = \langle x,y,z\rangle$ from which we find $\text{div}\mathbf{F} = 3$. This confirms our cylindrical calculation. ▲

Cylindrical Laplacian: Recall the Laplacian operator ∇^2 is a particular second derivative of a function of several variables. Given a function $f : \mathbb{R}^n \to \mathbb{R}$, the Laplacian $\nabla^2 f$ is the divergence of the gradient of f. At this point we have derived a cylindrical formula for the gradient of a function and for the divergence of a vector field. Composing these operations, together with product rules, gives a formula for the Laplacian of cylindrical functions.

Combining Equations 5.8.3 and 5.8.8 yields the cylindrical form of the Laplacian:

$$\nabla^2 f = \frac{1}{r}\frac{\partial}{\partial r}(rf_r) + \frac{1}{r}\frac{\partial}{\partial \theta}\left(\frac{1}{r}f_\theta\right) + \frac{\partial}{\partial z}(f_z)$$

$$= \frac{1}{r}f_r + f_{rr} + \frac{1}{r^2}f_{\theta\theta} + f_{zz}. \tag{5.8.9}$$

Recall that functions satisfying Laplace's equation $\nabla^2 f = 0$ are called harmonic. We showed that harmonic functions are special. In particular, they have no local extreme values. Let's use the curvilinear formulas just derived for some calculations.

Example 5.8.5. *Vortex flow*

The velocity potential of a vortex flow in the plane is $\varphi = \theta$. Compute the Laplacian $\nabla^2 \varphi$.

In this case we use the two-dimensional polar formula for the Laplacian, obtained by ignoring the z-portion of Equation 5.8.9. Since the partial derivatives φ_r, φ_{rr}, and $\varphi_{\theta\theta}$ all vanish, the velocity potential satisfies $\nabla^2\varphi = 0$ and is harmonic. ▲

Example 5.8.6. *Another harmonic function*

Show that the function $f(r,\theta) = \frac{1}{r}\cos\theta$ is harmonic.

This example should be a little more interesting, since the partial derivatives won't all vanish. Direct calculation shows

$$f_r = -\frac{1}{r^2}\cos\theta; \quad f_{rr} = \frac{2}{r^3}\cos\theta; \quad \text{and } f_{\theta\theta} = -\frac{1}{r}\cos\theta.$$

Substituting into the cylindrical Laplacian gives

$$\nabla f = -\frac{1}{r^3}\cos\theta + \frac{2}{r^3}\cos\theta - \frac{1}{r^3}\cos\theta = 0. \; ▲$$

Cylindrical Curl: One consequence of Stokes' theorem is an interpretation of the curl of a vector field as circulation per unit area. Thus the circulation is the product of curl and area. The remarks following Theorem 5.6.1 make this more precise. If the (infinitesimal) surface S has unit normal $\mathbf{n}$, then the circulation of $\mathbf{F}$ around the boundary of S is the area of S times $comp_\mathbf{n}\mathbf{F}$. More symbolically,

$$\oint_C \mathbf{F} \cdot d\mathbf{C} = (comp_\mathbf{n}\text{curl}\mathbf{F})(\text{Area of } S).$$

This equation will be used to find a cylindrical form for curl just as interpreting divergence as flux per volume led to a cylindrical divergence formula. A well-chosen infinitesimal surface S allows us to find curl$\mathbf{F}$ one component at a time.

Let $\mathbf{F} = U\hat{\mathbf{r}} + V\hat{\boldsymbol{\theta}} + W\hat{\mathbf{z}}$ and curl$\mathbf{F} = P\hat{\mathbf{r}} + Q\hat{\boldsymbol{\theta}} + R\hat{\mathbf{z}}$. To find R in terms of U, V, W, consider the patch of surface S obtained by holding z fixed and allowing small changes in r and θ. The area of S is $r\Delta\theta\,\Delta r$ and $\hat{\mathbf{z}}$ a unit normal vector, so

$$\oint_C \mathbf{F} \cdot d\mathbf{C} = \left(\text{curl}\mathbf{F} \cdot \hat{\mathbf{z}}\right) r\Delta\theta\,\Delta r = rR\Delta\theta\,\Delta r = R(\text{Area of } S).$$

To find R we must calculate the circulation of $\mathbf{F}$ around the boundary of S. This consists of four curves around the patch in Figure 5.8.4. In the infinitesimal limit, the segment C_1 from (r,θ,z) to $(r+\Delta r,\theta,z)$ is the linear displacement $\Delta r\hat{\mathbf{r}}$. Since $\mathbf{F}$ is essentially constant there, the line integral along C_1 is just its dot product with the displacement vector. Thus

$$\int_{C_1} \mathbf{F} \cdot d\mathbf{C} = \mathbf{F} \cdot \left(\Delta r\hat{\mathbf{r}}\right) = U\Delta r,$$

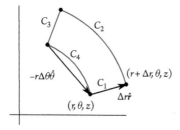

Figure 5.8.4 Computing curl for fixed z

where U is evaluated at (r,θ,z). The segment C_3 is the displacement vector $-\Delta r\hat{r}$, so

$$\int_{C_3} \mathbf{F}\cdot d\mathbf{C} = -U\Delta r,$$

where U is evaluated at $(r,\theta+\Delta\theta,z)$. Summing the two, and applying the mean value theorem gives their contribution to the circulation is

$$U(r,\theta,z)\Delta r - U(r,\theta+\Delta\theta,z)\Delta r = -\frac{\partial U}{\partial\theta}\Delta\theta\Delta r = -\frac{1}{r}\frac{\partial U}{\partial\theta}(\text{Area of } S).$$

Now consider the segments C_2 and C_4 given by vectors $r\Delta\theta\hat{\boldsymbol\theta}$ and $-r\Delta\theta\hat{\boldsymbol\theta}$. Dotting with $\mathbf{F}$ to compute the line integrals shows that their contribution to the circulation is

$$\mathbf{F}\cdot(r\Delta\theta\hat{\boldsymbol\theta}) - \mathbf{F}\cdot(r\Delta\theta\hat{\boldsymbol\theta}) = ((r+\Delta r)V(r+\Delta r,\theta,z)-rV(r,\theta,z))\,\Delta\theta$$

$$= \frac{\partial}{\partial r}(rV)\Delta r\Delta\theta = \frac{1}{r}\frac{\partial}{\partial r}(rV)(\text{Area of } S).$$

These computations add to give the circulation

$$\oint_C \mathbf{F}\cdot d\mathbf{C} = \left(-\frac{1}{r}\frac{\partial U}{\partial\theta}+\frac{1}{r}\frac{\partial}{\partial r}(rV)\right)(\text{Area of } S).$$

Since R is the coefficient of the area of S, this calculation shows

$$R = \frac{1}{r}\left(\frac{\partial}{\partial r}(rV)-\frac{\partial U}{\partial\theta}\right).$$

If you now consider surfaces obtained by holding r fixed and then θ fixed, similar calculations derive the cylindrical curl formula

$$\nabla\times\mathbf{F} = \left(\frac{1}{r}W_\theta - V_z\right)\hat{r} + (U_z - W_r)\hat{\boldsymbol\theta} + \frac{1}{r}\left(\frac{\partial}{\partial r}(rV)-U_\theta\right)\hat{z}. \qquad (5.8.10)$$

Example 5.8.7. *An irrotational vector field*

Let $\mathbf{F} = r^3\widehat{\mathbf{r}} - \frac{1}{r}\widehat{\boldsymbol{\theta}} + 4z^2\widehat{\mathbf{z}}$, and compute the curl of $\mathbf{F}$.
A look at Formula 5.8.10, shows that all the derivatives vanish, even

$$\frac{\partial}{\partial r}(rV) = \frac{\partial}{\partial r}(-1) = 0.$$

Therefore the vector field is irrotational. ▲

Spherical Formulas: The gradient, divergence, Laplacian, and curl in spherical coordinates can be derived similarly to those in cylindrical. The derivation gets messier without adding much insight, so we choose simply to list the formulas and apply them. There is a nice general derivation in Appendix A of Griffiths' *Introduction to Electrodynamics* for the interested reader ([1]). Now let f be a spherical function and $\mathbf{F} = U\widehat{\boldsymbol{\rho}} + V\widehat{\boldsymbol{\theta}} + W\widehat{\boldsymbol{\phi}}$ a spherical vector field. The formulas for gradient, divergence, Laplacian, and curl are:

$$\nabla f = f_\rho \widehat{\boldsymbol{\rho}} + \frac{1}{\rho\sin\phi}f_\theta\widehat{\boldsymbol{\theta}} + \frac{1}{\rho}f_\phi\widehat{\boldsymbol{\phi}}. \tag{5.8.11}$$

$$\text{div}\mathbf{F} = \frac{1}{\rho^2}\frac{\partial}{\partial\rho}\left(\rho^2 U\right) + \frac{1}{\rho\sin\phi}\frac{\partial V}{\partial\theta} + \frac{1}{\rho\sin\phi}\frac{\partial}{\partial\phi}\left(\sin\phi W\right). \tag{5.8.12}$$

$$\nabla^2 f = \frac{2}{\rho}f_\rho + f_{\rho\rho} + \frac{1}{\rho^2\sin^2\phi}f_{\theta\theta} + \frac{\cos\phi}{\rho^2\sin\phi}f_\phi + \frac{1}{\rho^2}f_{\phi\phi}. \tag{5.8.13}$$

$$\nabla\times\mathbf{F} = \frac{1}{\rho\sin\phi}\left(\frac{\partial}{\partial\phi}(\sin\phi V) - W_\theta\right)\widehat{\boldsymbol{\rho}}$$

$$+\frac{1}{\rho}\left(\frac{\partial}{\partial\rho}(\rho W) - U_\phi\right)\widehat{\boldsymbol{\theta}} + \frac{1}{\rho}\left(\frac{1}{\sin\phi}U_\theta - \frac{\partial}{\partial\rho}(\rho V)\right)\widehat{\boldsymbol{\phi}} \tag{5.8.14}$$

This section concludes with several applications of these formulas.

Example 5.8.8. *A spherical gradient*

Find the gradient of $f(\rho,\theta,\phi) = \rho\sin\phi\cos\theta$.
Calculating partial derivatives gives

$$f_\rho = \sin\phi\cos\theta; \qquad f_\theta = -\rho\sin\phi\sin\theta; \qquad \text{and } f_\phi = \rho\cos\phi\cos\theta.$$

Substituting into the spherical gradient formula gives

$$\nabla f = \sin\phi\cos\theta\,\widehat{\boldsymbol{\rho}} - \sin\theta\widehat{\boldsymbol{\theta}} + \cos\phi\cos\theta\widehat{\boldsymbol{\phi}}.$$

It is interesting to note that in Cartesian coordinates $f(x,y,z) = x$, so the gradient is $\nabla f = \mathbf{i}$. Since these are the same vector we now have $\mathbf{i}$ written as a linear combination of spherical coordinate vectors $\widehat{\boldsymbol{\rho}}, \widehat{\boldsymbol{\theta}}$, and $\widehat{\boldsymbol{\phi}}$. ▲

Example 5.8.9. *Spherical divergence*

Find the divergence of $\mathbf{F} = \widehat{\phi}$.
In this case $U = V = 0$ and $W = 1$, so applying the spherical divergence formula gives

$$\nabla \widehat{\phi} = \frac{1}{\rho \sin \phi} \frac{\partial}{\partial \phi} \sin \phi = \frac{\cos \phi}{\rho \sin \phi}.$$

The fact that $\widehat{\phi}$ has non-zero divergence reiterates that it is *not* a constant vector field. Spherical coordinate vector fields change as you move from point to point, in contrast to Cartesian coordinate vector fields which are constant everywhere. ▲

Example 5.8.10. *Electric potential*

In Example 5.5.4 we showed, by converting to Cartesian coordinates, that the electric potential $V = kQ/\rho$ is harmonic. Let us do this directly now that we have a spherical formula for the Laplace operator.

A quick calculation shows $V_\rho = -\frac{kQ}{\rho^2}$ and $V_{\rho\rho} = \frac{2kQ}{\rho^3}$, while all other partials vanish. Substituting these into Equation 5.8.13 yields

$$\nabla^2 V = \frac{2}{\rho} V_\rho + V_{\rho\rho} = -\frac{2}{\rho} \frac{kQ}{\rho^2} + \frac{2kQ}{\rho^3} = 0.$$

The simplicity of this calculation shows that spherical coordinates are the best way to describe electric potential for point charges. ▲

Example 5.8.11. $\nabla \times \widehat{\phi}$

As an application of the spherical curl formula we compute $\nabla \times \widehat{\phi}$. In this case $U = V = 0$ and $W = 1$. The only non-zero contribution in the spherical curl formula is

$$\frac{\partial}{\partial \rho}(\rho W) = \frac{\partial}{\partial \rho} \rho = 1.$$

Thus Equation 5.8.14 gives

$$\nabla \times \widehat{\phi} = \frac{1}{\rho} \widehat{\theta}. \ ▲$$

Exercises

1. Sketch a diagram that expresses $\widehat{\theta}$ as a linear combination of Cartesian coordinate vector fields.

2. Sketch a diagram that expresses $\widehat{\rho}$ as a linear combination of Cartesian coordinate vector fields.

3. Express $\hat{\rho}, \hat{\theta}$ and $\hat{\phi}$ as linear combinations of the standard basis vectors. Use this to show the spherical coordinate vector fields are mutually orthogonal everywhere.

4. Express $\mathbf{j}$ as a linear combination of cylindrical and spherical coordinate vector fields. (Hint: apply the gradient formulas to the appropriate function.)

5. Express $\mathbf{k}$ as a linear combination of cylindrical and spherical coordinate vector fields. (Hint: apply the gradient formulas to the appropriate function.)

6. Using the right-hand rule, compute $\hat{r} \times \hat{\theta}, \hat{r} \times \hat{z}$ and $\hat{\theta} \times \hat{z}$.

7. Using the right-hand rule, compute $\hat{\rho} \times \hat{\theta}, \hat{\rho} \times \hat{\phi}$ and $\hat{\theta} \times \hat{\phi}$.

8. Find the gradient of $f(r,\theta,z) = z - r$. Describe some level surfaces of f. Is ∇f orthogonal to them?

9. Show the gradient of $f(\rho,\theta,\phi) = \theta$ is orthogonal to its level surfaces.

10. Find the gradient of $f(\rho,\theta,\phi) = \rho^2$.

11. Compute the divergence of $\hat{r}$. Your answer should be a cylindrical function.

12. Compute the divergence of $\hat{\rho}$. Your answer should be a spherical function.

13. Compute the divergence of $\hat{\theta}$. What type of function should this be?

14. The velocity potential of a source is $\varphi = \ln r$, find its velocity field. Is φ a harmonic function?

15. Find the velocity field of an ideal flow around a π/n-corner with velocity potential $\varphi = r^n \cos n\theta$. Is this potential harmonic?

16. Compute the gradient and Laplacian of $f(r,\theta) = \frac{1}{r}\sin\theta$.

17. Compute the Laplacian of $f(\rho,\theta,\phi) = \rho \cos\phi$. Now rewrite f in Cartesian coordinates. Does your answer make sense?

18. Compute the Laplacian of $f(\rho,\theta,\phi) = \rho \sin\phi$. Now rewrite f in cylindrical coordinates and find its cylindrical Laplacian.

19. Compute the curl $\nabla \times \hat{\theta}$ in cylindrical coordinates.

20. Compute the curl $\nabla \times \hat{r}$ in cylindrical coordinates.

21. Suppose $\mathbf{F} = U\hat{r} + V\hat{\theta} + W\hat{z}$ is a cylindrical vector field where U, V, W are functions of only r, θ, z respectively. Find $\nabla \times \mathbf{F}$ in terms of its component functions.

22. Compute the curl $\nabla \times \hat{\theta}$ in spherical coordinates. Does it coincide with your cylindrical coordinate calculation?

23. Compute the curl $\nabla \times \hat{\rho}$ in spherical coordinates.

References

[1] P. Griffiths, *Introduction to Electrodynamics*, 4th ed., Cambridge University Press, 2017, Cambridge, UK.

[2] S. Lang, *Calculus of Several Variables*, 3rd ed., Springer Verlag Undergraduate Texts in Mathematics, 1987, New York, NY.

[3] J. Marsden, A. Tromba, *Vector Calculus*, 6th ed., W.H. Freeman and Company, 2012, New York, NY.

[4] J. Marsden, A. Weinstein, *Calculus II*, Springer-Verlag, 1985, New York, NY.

[5] T. Needham, *Visual Complex Analysis*, Oxford University Press, 2012, Oxford, UK.

[6] G. Polya, *Mathematics and plausible reasoning, volume 1: Induction and analogy in mathematics*, Princeton University Press, 1954, Princeton, NJ.

[7] A. Sauer, *Field lines of two point charges: exact formulas*, Eur. J. Phys., **39** (2018), 11pp. DOI 10.1088/1361-6404/aad21e

[8] T. Shifrin, *Multivariable Mathematics: Linear Algebra, Multivariable Calculus, and Manifolds*, John Wiley and Sons, Inc., 2005, Hoboken, NJ.

[9] V. Streeter, E.B. Wylie, *Fluid Mechanics*, 8th ed., McGraw Hill, 1985, Hoboken, New York, NY.

Index